Human Missions to Mars
Enabling Technologies for Exploring the Red Planet

Donald Rapp

Human Missions to Mars

Enabling Technologies for Exploring the Red Planet

 Springer

Published in association with
Praxis Publishing
Chichester, UK

Dr Donald Rapp
Independent Contractor
South Pasadena
California
USA

SPRINGER–PRAXIS BOOKS IN ASTRONAUTICAL ENGINEERING
SUBJECT *ADVISORY EDITOR*: John Mason, M.Sc., B.Sc., Ph.D.

ISBN 978-3-540-72938-9 Springer Berlin Heidelberg New York

Springer is part of Springer-Science + Business Media (springer.com)

Library of Congress Control Number: 2007932001

Cover design: Jim Wilkie
Project management: Originator Publishing Services Ltd, Gt Yarmouth, Norfolk, UK

Printed on acid-free paper

Contents

Preface

The basis of science is observation. By making repeated observations of relevant phenomena, we discover patterns in the behavior of natural systems, and from this we formulate theories that allow us to predict the outcome of future observations.

While this description of science works well for topics that have been heavily explored and are fairly well understood, there are many occasions in the real world at the cutting edge of knowledge where we have not yet made enough observations to draw conclusions, or where the phenomena of interest occurred so long ago that it is extremely difficult to fathom what might have actually taken place in ancient epochs.

One important case in point is the question of how, where, and when life originated from inanimate matter. The one basic piece of data that we have is that we know that life existed in a primitive form on Earth over 3 billion years ago (BYA). This was determined from fossil remains of early forms of life in dated deposits. Some of the questions that arise include:

- Where and when did life on Earth begin (on Earth, or somewhere else)?
- What was the process by which inanimate matter formed life in its primitive form?
- Is there life elsewhere in the solar system or external to the solar system?

All of these questions are in some sense subservient to the big question:

- Is the formation of life from inanimate matter a likely (or even a deterministic) process, given a reasonable length of time and a warm climate, liquid water, and a smattering of chemical elements in the lower periodic table?

In addressing these questions, some scientists have applied analysis and imagination to conceive of a wide variety of hypothetical scenarios for the formation of life, none of which is supported by evidence, and all of which seem highly dubious in retrospect. The world of science does not seem to be able to flatly say: "We simply do not understand much about the formation of life," and leave it at that. Just as nature abhors a vacuum, so does science abhor the absence of answers to vital questions. As a result, the science community has jumped on a series of bandwagons over the past several decades purporting to "explain" how life began, although many of these have now fallen by the wayside. Nevertheless, it is widely believed that formation of life from inanimate matter is a likely (or even a deterministic) process, and, given that belief, the most natural place to seek extraterrestrial life is Mars. Thus, the NASA exploration programs are heavily centered on a search for life on Mars.

Human missions to Mars represent the pinnacle of solar system exploration for the next half-century. In addition to providing a means of searching for life on Mars, such missions would represent an inspiring engineering achievement, and create a new era of expansion of humanity into space. Because such missions would require a major technological effort as well as very large expenditures, they remain for the moment as futuristic concepts embodied in paper studies by advocates and enthusiasts.

In the world of science and engineering, there is room for advocates and skeptics. Advocates play an important role in imagining what might be, and stubbornly pursue a dream that may be difficult to realize, but which in the end may be achievable. Skeptics identify the barriers, difficulties, pitfalls, and unknowns that impede the path, and point out the technical developments needed to enable fulfillment of the dream.

In the realm of human missions to Mars, we have a number of studies by advocates and enthusiasts, but there seems to be a total absence of skepticism in this arena. This book represents the first skeptical analysis of human missions to Mars, and it is offered as a counter-balance to the optimism so widely promulgated by the NASA Johnson Space Center, the Mars Society, and others.

Donald Rapp
April 2007

Figures

Tables

Abbreviations and acronyms

ALS	Advanced Life Support
AEDL	Aero Entry, Descent, and Landing (the process of using a aerodynamic forces to decelerate spacecraft at Mars or Earth for orbit insertion or descent)
AIAA	American Institute of Aeronautics and Astronautics
ALARA	As Low As Reasonably Achievable (refers to a philosophy for controlling radiation to astronauts)
ALS	Advanced Life Support (NASA program to develop advanced systems for life support)
Am	Amorphous
AM0	Area Mass
APL	Applied Physics Laboratory
ARC	NASA-Ames Research Center
AS	Ascent Stage (propulsion stage to ascend from Moon or Mars)
ASEE	American Society of Electrical Engineers
ASME	American Society of Mechanical Engineers
AU	Astronomical Unit
BAA	Broad Agency Announcement
BFO	Blood-Forming Organ (critical human element subjected to radiation).
BYA	Billion Years Ago
CaLV	Cargo Launch Vehicle (used to deliver ~125 mT of cargo to LEO)
CEV	Crew Exploration Vehicle (holds crew for transfer to lunar orbit—includes both CM and SM)
CI	Confidence Interval
CL	Cargo Lander

CLV	Crew Launch Vehicle (used to deliver crew in CEV to LEO)
CM	Crew Module (part of CEV that holds crew); Crew-Member
CRV	Crew Return Vehicle (a small crew capsule that tags along with the Habitat on the outbound trip to Mars to provide the crew with a back-up spacecraft)
CSI	Control–Structure Interaction (technical field dealing with control systems and structures)
CTV	Crew Transfer Vehicle (used to transfer crew from LEO to Mars and back)
DAV	Descent/Ascent Vehicle (combination of DS and AS)
DDT&E	Design, Development, Test, & Evaluation
DGB	Disk Gap Band
Di	Diverse
DIPS	Dynamic Isotope Power System
DLE	Double-Layer Ejecta
DOD	Department Of Defense
DOE	Department Of Energy
DPT	Decadal Planning Team (NASA committee)
DRM	Design Reference Mission (a paper analysis of requirements and potential systems for carrying out an end-to-end human mission to the Moon or Mars)
DS	Descent Stage (the propulsion stage used to descend from orbit to the surface)
ECLSS	Environmental Control & Life Support System (the system that controls the human environment in a Habitat and recycles resources)
EDL	Entry, Descent, and Landing (the process for orbit insertion and descent and landing at the Moon or Mars)
EDS	Earth Departure System (propulsion system for departure from LEO to go toward the Moon or Mars)
EHDP	bisphosphonate etidronate
EIRA	ESAS Initial Reference Architecture
ELV	Expendable Launch Vehicle
EOR	Earth Orbit Rendezvous (assembly point prior to departure for the Moon or Mars)
ERV	Earth Return Vehicle
ES	Exploration Systems (branch of NASA for human exploration)
ESA	European Space Agency
ESAS	Exploration Systems Architecture Study (2005 study of architecture for human return to Moon)
ESM	Equivalent System Mass
ESMD	Exploration Systems Mission Directorate (same as ES)
ESRT	Exploration Systems Research and Technology
ET	External Tank

ETSI	ExtraTerrestrial Solar Intensity
EVA	Extra-Vehicular Activity (operations by astronauts in suits outside of Habitats)
FOM	Figure Of Merit
GCR	Galactic Cosmic Ray (source of high-energy, low-level radiation in space)
GEO	Geostationary Earth Orbit
GRC	Glenn Research Center (NASA)
GSFC	Goddard Space Flight Center (NASA)
H	Habitat
HEO	High Earth Orbit (typically >10,000 km altitude)
HIMS	High Resolution Image Management System
HLLV	Heavy Lift Launch Vehicle
HLR	Human Lunar Return (a paper study by JSC in the mid-1990s)
HLV	Heavy Lift Vehicle
HQ	HeadQuarters
HZE	High charge and Energy for $Z > 3$
ICP	Intramural Call for Proposals
IMLEO	Initial Mass in Low Earth Orbit (total mass that must be transported to LEO from Earth to implement a space mission)
ISPP	*In Situ* Propellant Production (production of ascent propellants on the Moon or Mars from indigenous resources)
ISRU	*In Situ* Resource Utilization (production of useful products—e.g., ascent propellants—on the Moon or Mars from indigenous resources)
ISS	International Space Station
ITV	Interplanetary Transfer Vehicle (used to transfer crew from LEO to Mars and back)
JANNAF	Joint Army, Navy, NASA, Air Force Interagency Propulsion Committee
JIMO	Jupiter Icy Moon Orbiter (space mission to explore the icy moons of Jupiter using NEP; mission has not been funded)
JLMS	JPL/Lockheed-Martin Study in 2004–2005 (estimated masses and power needs of ISRU systems for Mars).
JPL	Jet Propulsion Laboratory (NASA)
JSC	Johnson Space Center (NASA)
L/D	Length/Diameter
L2	Libration Point
LaRC	Langley Research Center (NASA)
LCC	Life Cycle Cost (cost of a system over its life).
LCH_4	Liquid methane

LEO	Low Earth Orbit (typically, a circular orbit with altitude in the range 200 to 400 km)
LET	Linear Energy Transfer (process to extrapolate radiation impact data from one energy regime to another)
LH_2	Liquid hydrogen
LANL	Los Alamos National Laboratory
LLO	Low Lunar Orbit (typically, a circular orbit of altitude 100 km)
LLT	LL1-to-LEO Tanker
LMO	Low Mars Orbit
LOC	Loss Of Crew (mission failure where crew dies)
LOI	Lunar Orbit Insertion (process to decelerate a spacecraft so as to inject it into lunar orbit)
LOM	Loss Of Mission (mission failure where crew is saved)
LOR	Lunar Orbit Rendezvous (process of transfer of crew from Ascent Vehicle to Earth Return Vehicle in lunar orbit)
LOX	Liquid OXygen
LPS	Lunar Power System
LRO	Lunar Reconnaissance Orbiter (space mission to observe the Moon from orbit)
LS	Lunar Surface
LSAM	Lunar Surface Access Module (transports crew from lunar orbit to lunar surface and return)
LSH	Landing and Surface Habitat
LSR	Lunar Surface Rendezvous
LSS	Life Support System (provides food, water, air, and waste disposal to Habitats)
LST	Local Solar Time
LV	Launch Vehicle
LWT	Lunar Water Tanker
MAE	Materials Adherence Experiment
MARIE	Mars Radiation Environment Experiment
MARSIS	Mars Advanced Radar for Subsurface and Ionosphere Sounding
MAV	Mars Ascent Vehicle (vehicle used to transport crew from the surface to Mars orbit)
MEP	Mars Exploration Program
MEPAG	Mars Exploration Program Advisory Group
MER	Mars Exploration Rovers (robotic mission to Mars in 2004–2006)
MEx	Mars Express
MGS	Mars Global Surveyor (Mars orbiter)
MIT	Massachusetts Institute of Technology
MLE	Multiple-Layer Ejecta
MLI	Multi-Layer Insulation

MMH	Mono-Methyl Hydrazine (space-storable propellant)
MMOD	MicroMeteoroid/Orbital Debris
MMSSG	Moon–Mars Science Linkage Science Steering Group
MOB	Mars Or Bust (student project)
MOC	Mars Orbiter Camera
MOI	Mars Orbit Insertion (process to decelerate a spacecraft so as to inject it into Mars orbit)
MOLA	Mars Orbiter Laser Altimeter (measures altitude of surface on Mars)
MOR	Mars Orbit Rendezvous (process of transfer of crew from Ascent Vehicle to Earth Return Vehicle in Mars orbit)
MPLO	Mars PayLoad Orbit
MSFC	NASA-Marshall Space Flight Center
MSM	Mars Society Mission (DRM)
mT	metric Tons
MTV	Mars Transit Vehicle (serves as the interplanetary support vehicle for the crew for a round-trip mission to Mars orbit and back to Earth)
NASA	National Aeronautics and Space Administration
NASP	National AeroSpace Place
NCRP	National Council on Radiation Protection and Measurement
NEP	Nuclear Electric Propulsion (propulsion system using a nuclear reactor to generate electric power for electric propulsion).
NERVA	Nuclear Engine for Rocket Vehicle Application
NExT	NASA Exploration Team (NASA analysis team c. 2001)
NRC	National Research Council
NRPS	Nuclear Reactor Power System (used to supply power to outposts)
NS	Neutron Spectrometer (used to detect hydrogen in planetary surfaces)
NTO	Nitrogen TetrOxide (space-storable oxidant for rockets)
NTP	Nuclear Thermal Propulsion (same as NTR)
NTR	Nuclear Thermal Rocket (rocket that employs a nuclear reactor to heat hydrogen to a high temperature prior to efflux from a rocket nozzle)
OD	Optical Depth
OExP	Office of ExPloration
OOS	On-Orbit Staging
ORU	Orbital Replacement Unit
OSHA	Occupational Safety and Health Administration
P(LOC)	Probability of Loss Of Crew
P(LOM)	Probability of Loss Of Mission
P/C	PhysicoChemical

PEM	Proton Exchange Membrane
PF	Mars PathFinder Mission
PMAD	Power Management And Distribution
Pn	Pancake
POD	Point Of Departure
pp	partial pressure
pr μm	precipitable microns
PSR	Precision Segmented Reflector (a now-defunct NASA program to develop a sub-millimeter telescope)
PVA	PhotoVoltaic Array
R&T	Research & Technology
RCS	Reaction Control System (propulsion system used for minor orbit corrections)
Rd	Radial delineated
REID	Risk of Exposure-Induced Death (effect of exposure of humans to radiation)
RFC	Regenerative proton exchange membrane Fuel Cell
RLEP	Robotic Lunar Exploration Program (NASA program to utilize robotic precursors to gain information prior to human landings on the Moon)
RQ	ReQuirements Division
RTG	Radioisotope Thermal Generator (device to convert heat from radioisotopes into electric power on spacecraft)
RTOP	Research/Technology Operation Plan
RWGS	Reverse Water–Gas Shift (chemical process to convert CO_2 to O_2)
S/E	Sabatier/Electrolysis
SAE	Society of Automotive Engineers
sccm	standard cubic centimeters per minute
SDHLV	Shuttle-Derived Heavy Lift Vehicle
SEI	Space Exploration Initiative (an abortive attempt by NASA to establish a human exploration initiative in the early 1990s)
SEOTV	Solar Electric Orbital Transfer Vehicle
SEP	Solar Electric Propulsion (propulsion system using solar arrays to generate electric power for electric propulsion)
SH	Surface Habitat
SHAB	Surface HABitat
SHARAD	SHAllow RADar on 2005 Mars Reconnaissance Orbiter
SI	*Système International* (international system of units of measurement)
SICSA	Sasakawa International Center for Space Architecture
SLE	Single-Layer Ejecta
SLH	SLush Hydrogen

SM	Service Module (part of CEV that supplies resources to CM and propels CM back to Earth)
SOE	Solid-Oxide Electrolysis
SPE	Solar Particle Event (sudden process in which large flux of protons is emitted by the Sun that travel through the solar system)
SPS	Solar Power Satellite
SRM	Solid Rocket Motor (used to boost capability of Launch Vehicles)
SS	Space Shuttle
SSE	Space Science Enterprise
SSME	Space Shuttle Main Engine
STS	Space Transportation System
TAO	Take-off and Ascent to Orbit (TAO)
TEI	Trans-Earth Injection (the process of using propulsion to depart from Mars orbit or lunar orbit and head toward Earth)
TES	Thermal Emission Spectrometer
THEMIS	Thermal Emission Imaging System
TLI	Trans-Lunar Injection (the process of using propulsion to depart from LEO and head toward the Moon)
TMI	Trans-Mars Injection (the process of using propulsion to depart from LEO and head toward Mars)
TPH	Triple-Point hydrogen
TPS	Thermal Protection System (protects spacecraft from extreme heat generated in aero-entry)
TRL	Technology Readiness Level
TSH	Transfer and Surface Habitat
UVVIS	UltraViolet and VISible
W&R	Whedon & Rambaut (authors of paper on effects of exposure to zero-g)
WEH	Water-Equivalent Hydrogen
WEU	Water Extraction Unit (conceptual system to remove water from putative ice-containing regolith on Moon)
YBP	Years Before Present
YSZ	Yttria-Stabilized Zirconia (used for solid-state electrolysis of CO_2)
ZBO	Zero Boil-Off

1

Why explore Mars?

1.1 ROBOTIC EXPLORATION—THE ESTABLISHMENT'S VIEW

JPL manages the NASA Mars Exploration Program (MEP) for NASA. This
Program has been carrying out a series of robotic missions for a number of years
to explore Mars. The great success of the Mars Pathfinder and Mars Exploration
Rover missions has shown that autonomous rovers can traverse the surface of Mars
and make important scientific observations within limited areas. As a result, an
ambitious long-range plan for *in situ* exploration has been developed by the MEP
based on a consensus of views of leading Mars scientists. The highest priority goal is
the search for life (past or present) on Mars. For example, a JPL website[1] deals with:
"Why Explore Mars?" This website argues:

> "After Earth, Mars is the planet with the most hospitable climate in the solar
> system. So hospitable that it may once have harbored primitive, bacteria-like life.
> Outflow channels and other geologic features provide ample evidence that billions
> of years ago liquid water flowed on the surface of Mars. Although liquid water
> may still exist deep below the surface of Mars, currently the temperature is too low
> and the atmosphere too thin for liquid water to exist at the surface.
>
> What caused the change in Mars' climate? Were the conditions necessary for
> life to originate ever present on Mars? Could there be bacteria in the subsurface
> alive today? These are the questions that lead us to explore Mars. The climate of
> Mars has obviously cooled dramatically ... As we begin to explore the universe
> and search for planets in other solar systems, we must first ask the questions:
>
> - Did life occur on another planet in our own solar system?
> - What are the minimal conditions necessary for the formation of life?"

[1] *http://mars.jpl.nasa.gov/msp98/why.html*

The NASA Mars Exploration Program[2] places the greatest emphasis on four themes:

- Search for evidence of past life.
- Explore hydrothermal habitats. (Potential for discovery of evidence of past and present life is greatly improved.)
- Search for present life.
- Explore evolution of Mars.

The primary short-term goal is the search for evidence of past life. If hydrothermal vents were found (none have been discovered as yet) the search would focus on such locations. The search for present life would "follow upon the discovery, by earlier orbiting or landed missions, that environments on present Mars have the potential to support life."

The theme of evolution of Mars would only be emphasized if the "currently favored hypotheses for the climatic history of Mars are incorrect. [If future missions] show that there remains no convincing evidence of wet conditions on ancient Mars involving standing bodies of water, as has been interpreted from orbital remote sensing to date, the program's current focus on the search for surface habitats would be given a significantly lower priority—unless, of course, liquid water is found on or near the surface of Mars today.[3] With this surprising discovery would come the attendant mystery of how the terrestrial planets evolved so very differently, given their gross similarity." The evolution of Mars would deal with the loss mechanisms and sinks for water and CO_2 over time, and inter-comparisons of the similarities and differences between the three terrestrial planets: Venus, Earth, and Mars.

In October 2004, more than 130 terrestrial and planetary scientists met in Jackson Hole, WY, to discuss early Mars.[4] The search for life on Mars is a central theme in their report. The word "life" occurs 119 times in their 26-page report, an average of almost five times per page. The Introduction to that report says: "Perhaps the single greatest reason scientists find this early period of martian geologic history so compelling is that its dynamic character may have given rise to conditions suitable for the development of life, the creation of habitable environments for that life to colonize, and the subsequent preservation of evidence of those early environments in

[2] *Mars Exploration Strategy 2009–2020*, Dan McCleese (ed.), Mars Science Program Synthesis Group, Jet Propulsion Laboratory, 2005.
[3] However, liquid water is not stable at Mars surface temperatures and pressures. Therefore, standing pools of liquid water cannot occur on or near the surface of Mars. Liquid water could conceivably exist deep under the surface where temperatures are higher, and it might be possible (but extremely unlikely) for liquid water under pressure to rarely flow up to the surface due to some underground event, where it would quickly freeze.
[4] News and Views: "Key Science Questions from the Second Conference on Early Mars: Geologic, Hydrologic, and Climatic Evolution and the Implications for Life," *Astrobiology*, Vol. 5, No. 6, 2005.

the geologic record." One of the three "top science questions related to early Mars" was stated to be "Did life arise on early Mars?" Later in the report it says: "The question of martian life embodies essentially three basic aspects. The first involves the possibility that Mars may have sustained an independent origin of life. The second involves the potential for life to have developed on one planet and be subsequently transferred to another by impact ejection and gravitational capture (i.e., panspermia). The third focuses on the potential for Mars to have sustained and evolved life, following its initial appearance." The report then admits that: "How life begins anywhere remains a fundamental unsolved mystery" and it further admits that: "The proximity of Mars and Earth creates an ambiguity as to whether Earth and Mars hosted truly independent origins of life. Meteoritic impacts like those that delivered martian meteorites to Earth might also have exchanged microorganisms between the two planets. Impact events were even more frequent and substantial in the distant geologic past, including a period of time after which life began on Earth. Thus we cannot be sure whether the discovery of life on Mars would necessarily constitute the discovery of a truly independent origin of life."

Since liquid water is considered to be a necessary (but not necessarily sufficient) requirement for life to evolve from inanimate matter, the Mars science community places great emphasis on seeking evidence of past action of liquid water on the surface (it cannot exist there under present conditions). The search for evidence of past conditions that may have supported life remains a central theme in exploration of Mars.

According to the Mars Exploration Program:

"The defining question for Mars exploration is: life on Mars? Among our discoveries about Mars, one stands out above all others: the possible presence of liquid water on Mars, either in its ancient past or preserved in the subsurface today. Water is key because almost everywhere we find water on Earth, we find life. If Mars once had liquid water, or still does today, it's compelling to ask whether any microscopic life forms could have developed on its surface. Is there any evidence of life in the planet's past? If so, could any of these tiny living creatures still exist today? Imagine how exciting it would be to answer, 'Yes!' "

Similarly, the main basis of NASA Exploration on other bodies in the solar system and beyond is the search for life.[5]

The emphasis on the search for life in the NASA community has swayed a number of competent and even prominent scientists to develop programs, papers, and reports to analyze, hypothesize, and imagine the possibility of liquid water and life on other planetary bodies, with prime emphasis on Mars—and these occasional musings have been blown out of proportion by the press. A not-so-subtle pressure weighs on Mars scientists to find implications for water and life in studies of Mars.

[5] For example, the "search for extraterrestrial intelligence" (SETI).

For example, a news release[6] in 2005 said:

"WASHINGTON—A pair of NASA scientists told a group of space officials at a private meeting here Sunday that they have found strong evidence that life may exist today on Mars, hidden away in caves and sustained by pockets of water.

The scientists, Carol Stoker and Larry Lemke of NASA's Ames Research Center, told the group that they have submitted their findings to the journal *Nature* for publication in May, and their paper currently is being peer reviewed. What Stoker and Lemke have found, according to several attendees of the private meeting, is not direct proof of life on Mars, but methane signatures and other signs of possible biological activity remarkably similar to those recently discovered in caves here on Earth. Stoker and other researchers have long theorized that the Martian subsurface could harbor biological organisms that have developed unusual strategies for existing in extreme environments. That suspicion led Stoker and a team of U.S. and Spanish researchers in 2003 to southwestern Spain to search for subsurface life near the Rio Tinto river—so-called because of its reddish tint—the product of iron being dissolved in its highly acidic water ... Making such a discovery at Rio Tinto, Stoker said in 2003, would mean uncovering a new, previously uncharacterized metabolic strategy for living in the subsurface. 'For that reason, the search for life in the Rio Tinto is a good analog for searching for life on Mars,' she said.

Stoker told her private audience on Sunday evening that by comparing discoveries made at Rio Tinto with data collected by ground-based telescopes and orbiting spacecraft, including the European Space Agency's Mars Express, she and Lemke have made a very strong case that life exists below Mars' surface.

The two scientists, according to sources at the Sunday meeting, based their case in part on Mars' fluctuating methane signatures that could be a sign of an active underground biosphere and nearby surface concentrations of the sulfate jarosite, a mineral salt found on Earth in hot springs and other acidic bodies of water like Rio Tinto that have been found to harbor life despite their inhospitable environments."

Another example[7] is an article on the Internet reporting on an interview with Steve Squyres, project scientist of the Mars Exploration Rovers mission. Two excerpts from this article are: "Squyres is reported to have said he hopes the rovers will answer two questions: Are we alone in the universe? and How does life come to be? Most importantly, they've found evidence that water once existed on Mars. And where there's water, there's life."

[6] *http://www.space.com/scienceastronomy/mars_life_050216.html*
(Exclusive: "NASA Researchers Claim Evidence of Present Life on Mars," Brian Berger, *Space News* Staff Writer, posted: 16 February 2005, 02:09 PM ET.)
[7] *http://www.berkshireeagle.com/fastsearch/ci_3289264*
"The Mars man—A surface scratched—Scientist behind planetary probes," Benning W. De La Mater, *Berkshire Eagle* Staff.

It is hard to believe that Squyres said this exactly. How can the press say such ridiculous stuff? What evidence is there that a planet that had liquid water at some time *must* have life? Can't we distinguish between necessary and sufficient? Water may be necessary for life, but is it sufficient for life? There is zero evidence that it is. And how can anyone in their right mind believe that the MER rovers will answer the question of how life forms? This is not science. It is pseudo-science in its worst form.

The press releases based on interviews with Carol Stoker and Steve Squyres are just the tip of the iceberg. The Internet is full of wild, unfounded assertions in frequent press releases attributed to prominent and accomplished space scientists.

At what point did science go from hypotheses to be proven by measurements, from conservative understated conclusions, and painstakingly validating theories before going public—to wild unproven assertions, to unfounded claims, and repeated press releases reporting fluff?

1.2   THE CURMUDGEON'S VIEW ON THE SEARCH FOR LIFE ON MARS

One of the great, unsolved puzzles in science is how life began on Earth. The prevailing view amongst scientists today seems to be that life forms fairly easily with high probability on a planet, given that you start with a temperate climate, liquid water, carbon dioxide, and perhaps ammonia, hydrogen, and other basic chemicals, and electrical discharges (lightning) to break up the molecules to form free radicals that can react with one another. In fact, Leslie Orgell, a prominent expert on the origin of life, said at a 1997 JPL seminar: "I am confident that somebody is going to do it [create life in the lab] in the next 10 or 20 years."

How can such ideas be promulgated in the science community? At least part of the answer seems to be due to the observation that "Some 4.6 billion years ago the planet was a lifeless rock; a billion years later it was teeming with early forms of life." The fact that life arose on Earth comparatively early in the history of the Earth is the foundation of the widely believed argument that life forms easily and with high probability—an argument that has no basis that I can discern. First of all, we don't know if life "began" on Earth or was transferred to Earth from another body. Second, we don't understand the process by which life was formed, so how can we be sure that the relatively early emergence of life on Earth is indicative of anything? There is no evidence or logic to suggest that if life arose, say, 3 billion years after the formation of the Earth (rather than 1 billion), this would require that the probability of life forming is lower than if it formed in 1 billion years. And even if this argument held, it would only be a factor of 3, whereas the innate probability of forming life must be a very large negative exponential.

If you can imagine a million planets distributed around stars in a billion galaxies, all of which have the usual requirements (temperate climate, liquid water, carbon dioxide, and perhaps ammonia, hydrogen, and other basic inorganic chemicals, and electrical discharges [lightning]) you will note that if life originates on any one of these, and evolution develops thinking beings, the people there will be aware of themselves—because they are there. In the words of Descartes: "I think, therefore I exist." Now suppose that the innate probability of life forming on such a planet is very small, and that it takes an extremely unlikely fortuitous conflux of chemical, electrical, and geological events to provide the necessary conduit for life to form from natural chemicals. Further suppose that out of those 1,000,000 planets, life arose only once on one planet. The people who evolved on that planet would then think that they were prototypical of other planets and that life abounds through the universe. We are aware of life because we are alive. We have zero information on whether life ever arose anywhere else.

Considering the complexity of life—even the simplest bacterium requires perhaps 2000 complex organic enzymes in order to function—the probability seems extremely small that starting with simple inorganic molecules, life would evolve spontaneously. Fred Hoyle's book[8] estimates this probability to be extremely minute. Hoyle goes on to claim that life originated elsewhere in the universe and was "seeded" to the Earth via interstellar dust grains. Gert Korthof provides a detailed criticism of many of the arguments in the Hoyle book that favor seeding of life from extraterrestrial sources.[9] Most of these criticisms appear to have merit. However, we still remain with the question of how life originated, whether on Earth or elsewhere. Hoyle, faced with the dilemma that the probability of life arising spontaneously is remotely small, postulated a quasi-religious view that there is "intelligent control" in the universe in which life was created by higher forces that we cannot understand. It is noteworthy that Robert Shapiro[10] in his 1986 book on the origin of life provides a humorous allegory of a seeker of the answer to the origin of life, who goes to a great guru in the Himalayas. Each day, the guru tells the seeker of another far-fetched "scientific" theory, and each day the seeker is not satisfied. Finally, on the last day, the guru reads the first page of the Book of Genesis[11] and the seeker concludes that this explanation is about as good as the "scientific" explanations.

The problem for all of the explanations of the origin of life from inanimate matter is that none of them holds up to even cursory scrutiny. The one thing that seems most likely is that the innate probability is very small, and given 1,000,000 planets in the universe with a climate that could theoretically support life, it is possible that only an extremely rare and fortuitous conflux of events led to the formation of life on one planet (or possibly a few). If life evolved on only one planet, we are the one—which we recognize by Descartes' logic. That being the case, the search for life on Mars appears to be doomed to failure.

Depending on how one phrases the fundamental questions, the entire thrust of exploration and research can be altered. For example, the ESA Cosmic Vision[12] poses

---

[8] *The Intelligent Universe*, Fred Hoyle, Michael Joseph, London, 1983, ISBN 0 7181 22984.
[9] *http://home.wxs.nl/~gkorthof/kortho47.htm*
[10] *Origins. A Skeptic's Guide to the Creation of Life on Earth*, William Heinemann, 1986, ISBN 0 4346 95203
[11] "In the beginning, . . ."
[12] *Cosmic Vision: Space Science for Europe 2015-2025*, ESA BR-247, October 2005.

one of the "four great questions": What are the conditions for planetary formation and the emergence of life?

This biases the entire framework toward the prevailing view that there exists a set (or sets) of conditions (temperature, pressure, atmospheric constituents, liquid water, energy input, ...) that, given sufficient time, will deterministically generate life from inanimate matter just as a matter of chemistry. This viewpoint has influenced (in my opinion warped) the entire Mars Exploration Program into a fruitless, bound-to-fail search for life on Mars, and has engendered many specious articles regarding the search for evidence of life.

The fact is that we do not even know if life started on Earth or was transmitted there from without. So it is not clear that life originates on planets at all. It may well be that evolution of life from inanimate matter is a very difficult, improbable, almost impossible process that requires many improbable sequential events to occur so that life only originated once in the universe, and we will never know where or how. The widespread belief that life will evolve deterministically in many locations in the universe where there are stars and planets with water and temperate climates does not appear to have any foundation.

That being said, there still remain good reasons to explore Mars. These include the following:[13]

> "But beyond the question of life, understanding the conditions that prevailed on early Mars is also likely to provide important clues with regard to how the Mars we see today came to be. In this respect, Mars may also provide critical insight into understanding the nature of the early Earth. As much as 40% of the martian surface is believed to date back to the Noachian, but this period is barely represented in the Earth's geologic record, as those few exposures that have been identified from that time are highly metamorphosed (i.e., with uncertain preservation of original texture and chemistry). Since Earth and Mars are Solar System neighbors, they undoubtedly shared certain early (pre-3.7 Ga) processes, and studies of Mars may provide essential clues for our home planet."

1.3 WHY SEND HUMANS TO MARS?—THE ENTHUSIAST'S VIEW

The typical rationale for exploration of the Moon or Mars is based on three themes: *science*, *inspiration*, and *resources*. Paul Spudis provided this basis for lunar exploration[14] but many of these same arguments have been applied to Mars by enthusiasts.

Robert Zubrin is a prominent advocate of Mars exploration and is founder and president of the Mars Society. In his article "Getting Space Exploration Right"[15] he

---

[13] News and Views: "Key Science Questions from the Second Conference on Early Mars: Geologic, Hydrologic, and Climatic Evolution and the Implications for Life," *Astrobiology*, Volume 5, Number 6, 2005.
[14] "Why We're Going Back to the Moon," P. Spudis, *Washington Post*, December 27, 2005.
[15] *The New Atlantis*, Spring, 2005, pp. 15–48.

expounds why he believes humans should explore Mars. In fact, his enthusiasm outreaches his common sense when he suggests that we could do this in one decade. Zubrin argues that "of all the planetary destinations currently within reach, Mars offers the most—scientifically, socially, and in terms of what it portends for the human future."

Zubrin echoes a theme that is fairly widespread in the science community, the belief that any planet where liquid water is flowing on the surface in sunlight will eventually spontaneously evolve life. And since there is considerable photographic and geological evidence that liquid water once flowed on Mars, Zubrin concludes: "So if the theory is correct that life is a naturally occurring phenomenon, emergent from chemical 'complexification' wherever there is liquid water, a temperate climate, sufficient minerals, and enough time, then life should have appeared on Mars."

This was based on his reasoning that "liquid water flowed on the surface of Mars for a period of a billion years during its early history, a duration five times as long as it took life to appear on Earth after there was liquid water here."

Zubrin contemplates finding "fossils of past life on its surface," as well as using "drilling rigs to reach underground water where Martian life may yet persist." He believes that there is great social value in the inspiration resulting from a Mars venture. Finally, he says: "the most important reason to go to Mars is the doorway it opens to the future. Uniquely among the extraterrestrial bodies of the inner solar system, Mars is endowed with all the resources needed to support not only life, but the development of a technological civilization ... In establishing our first foothold on Mars, we will begin humanity's career as a multi-planet species."

Zubrin has support from a good many Mars enthusiasts. (The goal of the Mars Society is "to further the goal of the exploration and settlement of the Red Planet.") They seem to believe that we can send humans to Mars "in ten years" and begin long-term settlements. Each year, the International Space Development Conference hosts a number of futurists who lay out detailed plans for long-term settlements on Mars. The Mars Society often describes settlements on Mars as the next step in the history of "colonization", and warns not to make the same mistakes that were made in colonizing on Earth. For example, the Oregon Chapter of the Mars Society says:[16]

> "When the initial settlements are set up, there will most likely be a few clusters of small settlements. As time goes on, they should spread out. The more spread out the developing townships are, the more likely they will develop their own culture. In the beginning, townships will be dependant [sic] upon each other for shared resources, such as food, water, fuel, and air. Once a more stable infrastructure is set up on Mars, then people should be encouraged to set up more isolated townships. In any area where colonization or expansion has occurred, one important item that cannot be ignored is the law. Some form of law will be needed on Mars. Looking at the system that was used in the old west, we can see that whoever enforces the law can have difficulty completing his job. The 'sheriffs' on Mars must be trustworthy individuals that the majority of people agree on.

[16] *http://chapters.marssociety.org/or/msoec1.html*

They should not be selected by the current form of politically interested members of society; this only encourages corruption. Instead, some sort of lottery system of volunteers should be allowed. As for the law itself, it should be set in place to guarantee all of the basic rights of everyone, from speech to privacy.

While these zealots are already concerned with establishing law and order on Mars, this humble writer is merely concerned with getting there and back safely.

The JSC Mars design reference mission ("DRM-1")[17] expounded at some length on the rationale for human exploration of Mars. In August 1992, a workshop was held at the Lunar and Planetary Institute in Houston, TX, to address the "whys" of Mars exploration. The workshop attendees identified six major elements of the rationale for a Mars exploration program and these are summarized below:

- Human evolution—Mars is the most accessible planetary body beyond the Earth–Moon system where sustained human presence is believed to be possible. The technical objectives of Mars exploration should be to understand what would be required to sustain a permanent human presence beyond Earth.
- Comparative planetology—The scientific objectives of Mars exploration should be to understand the planet and its history to better understand Earth.
- International Cooperation—The political environment at the end of the Cold War may be conducive to a concerted international effort that is appropriate, and may be required, for a sustained program.
- Technology advancement—The human exploration of Mars currently lies at the ragged edge of achievability. Some of the technology required to achieve this mission is either available or on the horizon. Other technologies will be pulled into being by the needs of this mission. The new technologies or the new uses of existing technologies will not only benefit humans exploring Mars but will also enhance the lives of people on Earth.
- Inspiration—The goals of Mars exploration are bold, are grand, and stretch the imagination. Such goals will challenge the collective skill of the populace mobilized to accomplish this feat, will motivate our youth, will drive technical education goals, and will excite the people and nations of the world.
- Investment—In comparison with other classes of societal expenditures, the cost of a Mars exploration program is modest.

DRM-1 then goes on to say: "In the long term, the biggest benefit of the human exploration of Mars may well be the philosophical and practical implications of settling another planet." DRM-1 mentions human history, migrations of people stimulated by overcrowding, exhaustion of resources, the search for religious or economic freedom, competitive advantage, and other human concerns. "Outside the area of fundamental science, the possibility that Mars might someday be a home for humans is at the core of much of the popular interest in Mars exploration. A human settlement on Mars, which would have to be self-sufficient to be sustainable,

[17] http://exploration.jsc.nasa.gov/marsref/contents.html

would satisfy human urges to challenge the limits of human capability, create the potential for saving human civilization from an ecological disaster on Earth (for example, a giant asteroid impact or a nuclear incident), and potentially lead to a new range of human endeavors that are not attainable on Earth."

DRM-1 then suggests that three considerations are important:

- Demonstrating the potential for self-sufficiency.
- Demonstrating that human beings can survive and flourish on Mars.
- Demonstrating that the risks to survival faced in the daily life of settlers on Mars are compatible with the benefits perceived by the settlers.

Another point of view was expressed by Rycroft.[18] He suggested "that the over-arching goal of space exploration for the 21st century should be to send humans to Mars, with the primary objective of having them remain there." The basis for this goal is to provide the human species with "a second base in the Solar System ... because, at some point in the future, the Earth may no longer be habitable." Rycroft points out that this could "arise because of a disastrous catastrophe on Earth. Civilization may self-destruct, or there may be a giant natural event rendering the Earth uninhabitable. Such possibilities include: overpopulation, global terrorism, nuclear war or accident, cyber technology war or accident, biological war or accident, occurrence of a super-virus, asteroid collision, geophysical events (e.g., earthquakes, tsunamis, floods, volcanoes, hurricanes), depletion of resources (e.g., oil, natural gas reserves), climate change, global warming and sea level rise, stratospheric ozone depletion, other anthropogenic abuses of Earth." He quotes M. Rees who said: "the odds are no better than 50–50 that our present civilization on Earth will survive to the end of the present century." In particular, the threats due to overpopulation, pollution, global warming, depletion of resources, and the worldwide expansion of Islamic terrorism leading to a World War III between the West and Islam are the most pressing issues.

While Rycroft properly emphasizes the seriousness of these threats, his suggested strategy of "colonization of Mars before the end of the 21st century" would only add to mankind's woes, not relieve the problems. If we cannot find a way to populate the Earth and live in harmony, how are we going to do it on Mars with its much harsher climate?

1.4   SENDING HUMANS TO MARS—THE SKEPTIC'S VIEW

It is nice to know that the Mars Society is concerned about establishing townships with law and order on Mars. However, before we even imagine "settlements" on Mars, there are shorter-term challenges involved in sending the first humans to Mars for preliminary exploration, and the costs and risks are very high. There are several

---

[18] "Space exploration goals for the 21st century," Michael J. Rycroft, *Space Policy*, 2006.

questions involved:

(1) What are the main goals of Mars exploration?
(2) What is the benefit/cost comparison for robotic vs. human exploration?
(3) What are the risks and challenges involved in attempting to send humans to Mars?

As we discussed in previous sections, the prevailing view in both scientific and futuristic circles is that the main reason to explore Mars is the search for life, which in turn requires a search for liquid water (past or present). Futurists and visionaries have imaginations that reach well beyond this initial phase to the point where human settlements are established for their "social, inspirational and resource value."

Even if we accept the unwarranted proposition that the search for life is central to exploration of Mars, the question then arises as to comparative costs and prospective results based on robotic vs. human exploration of Mars. It seems likely that the benefit/cost ratio will be far greater for robotic exploration.

Furthermore, since the search for life is likely to end in failure, the real value in exploring Mars is to understand why the three terrestrial planets—Venus, Earth, and Mars—turned out to be so different, presuming that they were initially endowed with similar resources. Venus has a thick atmosphere of carbon dioxide whereas Mars has very little atmosphere. There are theories for why this is so, but exploration of the planets is necessary to unravel the geological history of how this came about. Sending humans to Mars seems to be a very expensive and risky approach compared with robotic exploration.

In regard to the broader, visionary viewpoint expressed in DRM-1, the drive toward a sustained human presence beyond Earth appears to be premature by a few hundred years. Certainly, the presence of a handful of humans on Mars will not relieve the Earth of any of its pressures due to overpopulation, pollution, or resource depletion. Comparative planetology is a worthwhile goal but it is not clear that a human presence is needed to accomplish this. Surely, there are plenty of opportunities for international cooperation without sending humans to Mars? The conclusion that the investment required to send humans to Mars is "modest" is derived by comparing with larger societal expenditures. But when compared with traditional expenditures for space, it is huge. On the other hand, there may be merit to the claims that the new technologies or the new uses of existing technologies will not only benefit humans exploring Mars but will also enhance the lives of people on Earth, and the boldness and grandeur of Mars exploration "will motivate our youth, will drive technical education goals, and will excite the people and nations of the world." Here, it all boils down to the benefit/cost ratio, which seems likely to be low.

2

Planning space campaigns and missions

2.1 CAMPAIGNS

A campaign is a series of closely related space missions that sequentially contribute to fulfillment of overall campaign goals. In some cases, each mission in the campaign is distinct, and the main value to each subsequent mission provided by previous missions is the knowledge gained that may influence sites for subsequent missions, and validate instruments, flight technologies, or other mission design elements. This typically prevails in the case of robotic missions to Mars. For human missions to Mars, previous robotic missions will be necessary to validate new technologies on Mars prior to use by humans, and early human missions may leave behind facilities and infrastructure to be utilized by subsequent human missions.

For example, the Mars Exploration Program envisages a campaign of exploratory robotic missions to Mars in which each mission provides important insights as to where to go and what to look for in the next mission(s).[19] The current NASA lunar exploration initiative is an outline of a campaign, but, unfortunately, the campaign has not been defined very well except that it will begin with short-duration "sortie" missions that lead up to establishment of a lunar "outpost" with uncertain location and functions. In fact, initial planning has not even resolved many important aspects of the sortie missions or the Lunar Surface Access Module (LSAM), while almost all the focus seems to have been on the so-called Crew Exploration Vehicle (CEV). In this process, NASA seems to have lost sight of the overall campaign and how the pieces fit together. For example, although *in situ* resource utilization for producing oxygen for ascent propulsion remains a central theme for outposts, the elimination of oxygen as an ascent propellant suggests that different groups working on the lunar exploration initiative are not only not communicating, but are at cross-purposes.

---

[19] *http://mars.jpl.nasa.gov/missions/future/2005-plus.html*
http://mars.jpl.nasa.gov/missions/future/futureMissions.html

At the highest level, one would start with a set of goals that must be achieved by a campaign. One would define a set of hypothetical missions that are potential building blocks of campaigns. Then one would define a set of metrics to evaluate "missions" and use these as a starting point to select metrics for campaigns. Campaigns are assemblages of missions but the sequence of missions in a campaign is probabilistic:

- Each mission involves at least two possible outcomes with probabilities assigned.
- After Mission 1, do Mission 2A if Event A happens, or Mission 2B if Event B occurs.
- There are a number of possible outcomes for each campaign (each with a different series of missions, and differing cost, risk, and performance).

Alternative pathways for carrying out a campaign can be represented by a "tree-diagram" showing campaigns as paths through arrangements of sequentially arranged missions. A number of investigators have been studying approaches for seeking the best campaign (i.e., best sequence of missions) according to some figure of merit for the campaign. However, this is a complex subject and is beyond the scope of this discussion.

In order to make a wise choice for a campaign, the properties, characteristics, and requirements of the individual missions that make up a campaign need to be understood.[20]

2.2   PLANNING SPACE MISSIONS

In planning a space mission, the first thing to consider is why do we want to do it, and what do we hope to derive from the results? The next questions deal with the feasibility of the enterprise as expressed by the following questions:

- How much does it cost? Is it affordable?
- Is it technically (and politically) feasible?
- How safe is it, and what is the probability of failure?
- Can we launch (and possibly assemble in space) the required vehicles?

It is generally very difficult to arrive at even approximate answers to these questions without investing a great deal of effort in analysis. Furthermore, there are typically a number of architectural variations in space vehicles, their sequencing, their phasing, and their destinations that can be utilized to carry out such a space mission. These alternate variations are referred to as "mission architectures" or simply "architectures". It would require very large financial resources and considerable time and effort to carry out detailed analyses of each potential architectural option. Therefore, the typical approach utilized by NASA involves an initial analysis to compare

[20] *Architecting Space Exploration Campaigns: A Decision-Analytic Approach*, Erin Baker, Elisabeth L. Morse, Andrew Gray, and Robert Easter, IEEEAC Paper #1176 (2006).

architectural options, and from this a short list of favored architectures can be identified that should be examined more thoroughly. For rough initial analysis, there is widespread use of the "initial mass in low Earth orbit" (IMLEO) as a rough measure of mission cost, and since IMLEO is typically calculable it is used as a surrogate for mission cost. This is based on the notion that, in comparing a set of alternative potential missions to carry out a desired goal, the amount of "stuff" that you need to transport to LEO is proportional to the cost. IMLEO is the total mass initially in LEO, but it does not specify how this total mass is partitioned into individual vehicles. The mass of the largest vehicle in LEO dictates the requirements for Launch Vehicle capability (how much mass a Launch Vehicle must lift in "one fell swoop"—unless on-orbit assembly is employed). Thus, the initial planning of space missions and preliminary selection of the mission architecture depend on two main things: (1) IMLEO, and (2) the required Launch Vehicle and number of launches.

It is important to understand that the requirements for space missions are dominated by the requirements for accelerating vehicles to high speeds. Unlike an automobile, which has a large crew compartment and a small gas tank, a spacecraft typically has large propellant tanks and a relatively small crew compartment. A space mission is composed of a series of propulsion steps, each one of which may have a good deal more propellants than payload. Each propulsion step requires accelerating not only the payload, but also the propellants reserved for later acceleration steps. As a result, most of IMLEO is typically propellants, not payload. The mass of propellants delivered to LEO for getting from here to there (and back) is then the determining factor in whether a space mission is feasible and affordable. And this is embodied in the magnitude of IMLEO, which is mainly made up of propellants, not payload.

2.3 ARCHITECTURES

In a simple space mission, a single spacecraft may be placed atop a Launch Vehicle and sent on its way to its destination in space. In such a case, there is no need to discuss "architectures". However, in more complex space missions, particularly in human exploration missions, a number of alternative approaches can be conceptualized for scheduling and phasing the launches, rendezvous, assemblies and disassemblies, descents and ascents, and other operations involved in the overall mission. Each one of these overall mission designs is referred to as a "mission architecture" (or simply "architecture"), and characterizing, evaluating, and comparing alternative architectures constitutes a major portion of early planning of complex space missions.

As a current example, the mission architecture adopted in 2005–6 by the NASA Exploration Systems Architecture Study (ESAS) for lunar sortie[21] missions is shown

[21] A "sortie" mission is a short-term mission with a minimal payload that is designed to prove out the functionality of the various space systems and operations, and obtain a limited amount of useful data. It is typically a precursor to longer term missions to establish an "outpost".

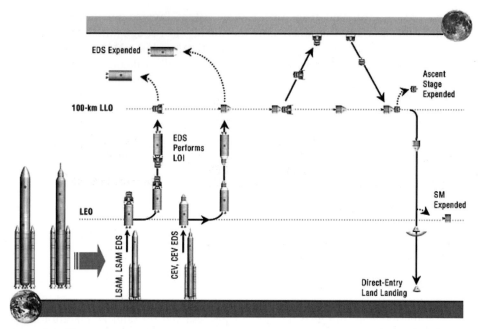

Figure 2.1. The NASA ESAS architecture for lunar sortie missions. Vehicle symbols are defined in the text. "LOI" stands for lunar orbit insertion. [From NASA "ESAS Report," 2005.]

in Figure 2.1. In this architecture, the following vehicles are defined:

- EDS = Earth Departure System. (This is a propulsion system consisting of propellant tanks, plumbing, rockets, and propellants to send the assembled system from LEO on a path toward the Moon.)
- LSAM = Lunar Surface Access Module (sum of Descent Stage [DS], Ascent Stage [AS], and Habitat [H]). The Ascent and Descent Stages are propulsion systems consisting of propellant tanks, plumbing, rockets, and propellants. The Habitat is the capsule in which the crew resides during transit between lunar orbit and the lunar surface as well as for several days on the surface. The Descent Stage transports H + AS to the lunar surface. The Ascent Stage transports H back to lunar orbit to transfer the crew to the waiting CEV.
- CEV = Crew Exploration Vehicle (sum of Service Module [SM] and Crew Module [CM]). The Crew Module houses the crew in transit from Earth to Moon and back, and the Service Module provides the Crew Module with a propulsion system to return to Earth, as well as other support systems.

There are two Launch Vehicles: the Cargo Launch Vehicle (CaLV) to launch cargo (EDS + LSAM) to low Earth orbit (LEO), and the Crew Launch Vehicle (CLV) to launch the crew (in the CEV) to LEO. The architecture requires a rendezvous and assembly of the EDS + LSAM and the CEV into a single unit in LEO, and a

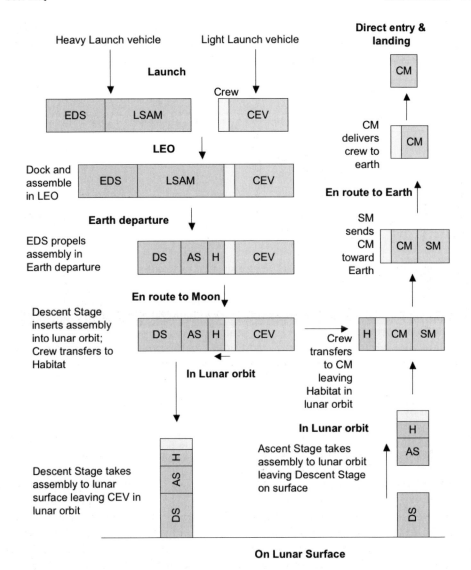

Figure 2.2. State–step representation of ESAS lunar sortie architecture. The entire mission consists of a set of alternating states and steps.

disassembly in lunar orbit in which the CEV remains in lunar orbit while the LSAM descends to the surface (and the EDS is discarded). The Ascent Stage + Habitat ascends to lunar orbit from the lunar surface, performs a rendezvous with the CEV, and transfers the crew to the CEV. The CEV then returns to Earth with the crew.

The ESAS architecture is more fully described in Figure 2.2. It can be seen that there are a variety of vehicles and steps involved in the mission.

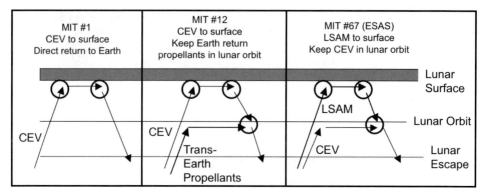

Figure 2.3. Three of the architectures considered by MIT for landing on the Moon. [Based on a presentation at JPL by P. Wooster, March 2, 2005.]

This is only one of many architectures that are conceivable. Some of these conceptual architectures would eliminate the rendezvous in Earth orbit, or the disassembly and rendezvous in lunar orbit. In fact, an MIT study examined large numbers of alternative lunar architectures.[22] As the MIT study points out, "Initial Mass in Low Earth Orbit (IMLEO) is typically used as a top-level screening criterion in architecture selection; however, ... additional factors must be taken into account in determining the preferred architecture." These include cost and safety.

The MIT study concluded that the ESAS architecture shown in Figures 2.1 and 2.2 is not optimum. Figure 2.3 illustrates three partial architectures for lunar landing as defined by the MIT study. In the architecture on the left, the CEV goes directly to the lunar surface and returns to Earth without rendezvous and crew transfer in lunar orbit. The architecture in the center of this figure is a variant in which propellant tanks for Earth return are kept in lunar orbit so they don't have to be carried down and up from the lunar surface. The architecture on the right is the ESAS architecture involving rendezvous and crew transfer on the way in and the way out from the Moon. The key issue is whether the CEV should remain in lunar orbit or whether it should land on the lunar surface. If the CEV is landed on the Moon, it will be heavier because it will require extra capabilities. However, this will eliminate the need to develop a second Habitat and it will eliminate several significant operations in lunar orbit. MIT identified other benefits as well. The MIT study concluded that architecture #1 in Figure 2.3 was superior to the ESAS architecture, which maximizes the number of vehicles needed and the number of in-space operations. I happen to agree with MIT, but that is not the issue here. The point being made here is that human exploration missions can typically be implemented with a number of conceptual architectures, and choosing an optimum approach requires a great deal of analysis and judgment.

---

[22] "Crew Exploration Vehicle Destination for Human Lunar Exploration: The Lunar Surface," Paul D. Wooster, Wilfried K. Hofstetter, and Edward F. Crawley, *Space 2005, 30 August–1 September 2005, Long Beach, California*, AIAA 2005-6626.

For human missions to Mars, many of the same options occur as in human missions to the Moon. There are alternative options in Earth launch, Earth orbit assembly, Earth departure, Mars orbit insertion, Mars orbit operations, Mars descent and ascent, and Earth return steps. Each architecture involves a serial set of steps leading from departure to return.

2.4   A MISSION AS A SEQUENCE OF STEPS

Space missions can be described in terms of a series of *states* connected by *steps*:

- A *state* is a condition of relative stability and constancy (coasting, remaining in orbit, operating on the surface, etc.)
  —Each state is characterized by a set of vehicle masses.

- A *step* is a condition of change (fire a rocket, jettison excess cargo, transfer crew between vehicles, etc.)
  —Each *step* is characterized by propulsion parameters or other relevant data that characterize the dynamic operations that take place during the step.

The *states* and *steps* matrix provides a simple means of summarizing the major elements of a space mission. Unless all the *states* and *steps* of a mission are defined, it is not possible for an independent party to verify that the mission is feasible. The current culture in NASA space mission description utilizes long reports (typically many hundreds of pages long) in which ferreting out *state–step* information is time-consuming and frustrating, and often impeded by missing or ill-defined data. It is typically difficult to summarize, understand, and recreate the *states* and *steps* in complex multi-element missions from current NASA and ESAS study reports.

Table 2.1 shows a *state–step* table for a simple hypothetical mission involving transfer of a payload that weighs 10 mass units to Mars orbit and return from Mars orbit to Earth. Actually, the process begins on the launch pad with launching to LEO. However, the launch process is typically considered separately, and the mission design presumes that launch has taken place and initial assets are located in LEO. Therefore, the first *state* is defined to be in LEO.

The *steps* are given in the first row. The *states* at the start and end of each step are given in rows 8 and 9. In order to characterize a *step*, a considerable amount of data must be specified in rows 2–6 (but it is not necessary for the reader to assimilate all of this). The most important of these is Δv, the change in velocity imparted to a vehicle by firing a rocket.

$$\Delta v = v(\text{after step}) - v(\text{before step})$$

The greater that Δv is, the more propellant needs to be used in the rocket burn to

Table 2.1. *State–Step* table for a hypothetical mission to Mars orbit and return.

Row	Step $\Rightarrow$	LEO to TMI	TMI to circular Mars orbit	Mars orbit to TEI	TEI to Earth
1	Δv (km/s)	3.9	2.5	2.4	*
2	Propulsion system	LOX–LH$_2$	LOX–CH$_4$	LOX–CH$_4$	*
3	Propulsion specific impulse (sec)	450	360	360	*
4	Rocket equation exponential	2.421	2.031	1.974	*
5	Propulsion stage % of propellant mass	10	12	12	*
6	Entry system % of delivered mass	0.0	0.0	0.0	70
7					
8	*State at start*	*LEO*	*TMI*	*Mars orbit*	*TEI*
9	*State at end*	*TMI*	*Mars orbit*	*TEI*	*Earth*
10					
11	Entry system mass	0.0	0.0	0.0	7.0
12	Payload mass	10.0	10.0	10.0	10.0
13	Propellant mass	146.0	44.7	18.8	
14	Stage mass	14.6	5.4	2.3	
15	Total mass at start	248.7	88.1	38.0	17.0
16	Total mass at end	88.1	38.0	17.0	10.0
17					
18	*LEO mass/orbit payload mass*	6.5			
19	*LEO mass/round trip mass*	24.9			

* Note that the Earth entry step typically uses an aeroshell, not propulsion, so no Δv is specified in the Earth entry column. It has been assumed here arbitrarily that the entry system mass is 70% of the mass delivered to Earth.

achieve the required Δv. Some typical values of Δv are provided in the second row of Table 2.1.[23]

The columns in Table 2.1 represent the *steps* in the mission to transfer a 10-unit mass to Mars orbit, and return it to Earth. Each *step* is a transition from an initial *state* to a final *state* (rows 8 and 9 in the table). The mass prior to and after each *step* is shown in rows 15 and 16. At first, it may seem like working backwards, but the pathway to estimate IMLEO for a space mission is not to start at the launch pad or LEO, but rather to begin at the destination, and then work backwards to arrive at an estimate of the initial mass needed in LEO to send the various vehicles to their destinations. Thus, Table 2.1 is generated by starting in the far right column, and working backwards toward the left. The final mass in any column is the initial mass in the column to its right.

The first *step* is Earth departure (or, as it is referred to by professionals, trans-Mars injection [TMI]), in which a rocket (or rockets) is fired to send the vehicle out of LEO on its way toward Mars. This vehicle would cruise toward Mars for typically 6

[23] One meter/second is equivalent to 2.24 miles per hour. A Δv of 4,000 m/s is equivalent to 8,960 mph.

to 9 months, carrying propellants and propulsion stages for subsequent firings. The second *state* is a quiescent cruise toward Mars (TMI). A few minor mid-course trajectory corrections may be needed along the way, introducing additional *steps*, but these are minor and are not included in the table. On arriving in the vicinity of Mars, a significant retro-rocket firing *step* is needed to slow the vehicle down by 2,500 m/s in order to insert it into Mars orbit. To return toward Earth, yet another rocket firing is needed to provide the 2,400 m/s needed to escape Mars orbit and head toward Earth. At Earth, an aeroshell entry system is used instead of a retro-propulsion system. Finally, the "gear ratios" are given in rows 18 and 19. These gear ratios give the ratio of initial mass in LEO (IMLEO) to the mass delivered to Mars orbit, or to the mass that undergoes a round trip to Mars. It will be noted that it requires about 25 mass units initially in LEO to send one mass unit on a round trip to Mars orbit. Similar (though more complex) tables can be used to describe missions to the surface of Mars with multiple vehicles.

It is not expected that the reader can fully digest Table 2.1. A greater under-standing of such tables will be provided later in this book. However, several aspects of this table are worth discussing even at this early point. In this table, the ultimate payload is an undefined mass of 10 units (it doesn't matter whether it is 10 kg, 10 tons, or whatever, because everything else scales to this mass.)

The IMLEO is 248.7 mass units. This IMLEO is constituted from a number of components, as shown in Figure 2.4. These include propulsion for trans-Mars injection, propulsion for Mars orbit insertion, propulsion for trans-Earth injection

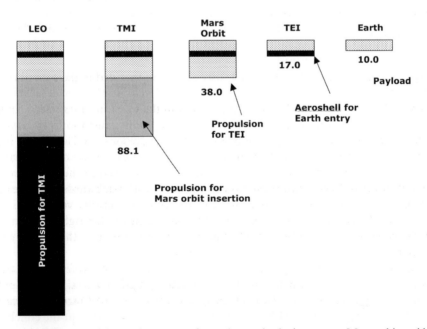

Figure 2.4. Sequence of decreasing masses for each *state* in the journey to Mars orbit and back. Each column represents a *state*, and the transition from one column to the next column is a *step*.

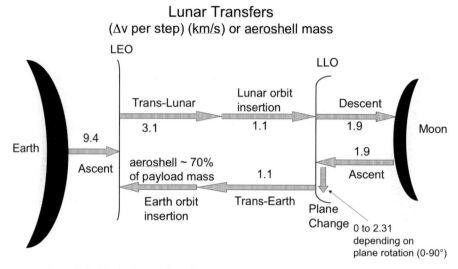

Figure 2.5. Typical set of Δv for each step in lunar sortie missions (km/s).

(TEI), and finally an aeroshell for Earth entry. The mass on the launch pad is roughly 20 times IMLEO, or about 5,000 mass units. Thus it takes roughly 5,000 mass units on the launch pad to transfer 10 mass units on a round trip to Mars orbit. These diagrams show that most of what we launch into space is propulsion systems, and the payload is typically a very small fraction of the total mass.

The key conclusion to be drawn is that it takes a great deal of IMLEO to send a spacecraft to Mars and back. The "gear ratios" define the ratio of IMLEO to the payload mass that is transported through space.

In any mission design, the first and foremost thing that is needed is the set of Δv for all the mission *steps*. Estimates of Δv for various *steps* in Mars or lunar missions can be made by standard trajectory analysis. These values of Δv will depend to some extent on other factors. For example, in lunar missions, the values of Δv vary depending on whether global access, anytime return, and use of "loitering" in the lunar vicinity is employed. For Mars missions, the Δv varies with launch opportunity and the desired trip time to Mars (shorter trip times require higher Δv and therefore more propellants). Some typical sets of Δv for lunar and Mars missions are shown in Figures 2.5 and 2.6.

2.5   WHAT'S AT THE DESTINATION?

Figure 2.7 shows one particular model for a human mission to Mars. There are three major deliveries: (1) the crew to Mars surface, (2) the cargo to Mars surface, and (3) the Earth Return Vehicle (ERV) to Mars orbit. In this scenario, at the conclusion of the Mars surface phase, the crew ascends to rendezvous with the ERV in Mars orbit, and transfers to the ERV for the return trip to Earth. Based on the values of Δv

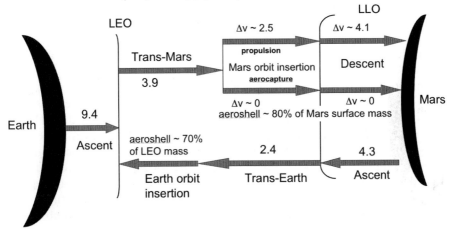

Figure 2.6. Typical set of Δv (km/s) for each step in Mars missions (km/s). Two options are shown for orbit insertion and descent: one using propulsion, and the other using aero-assist.

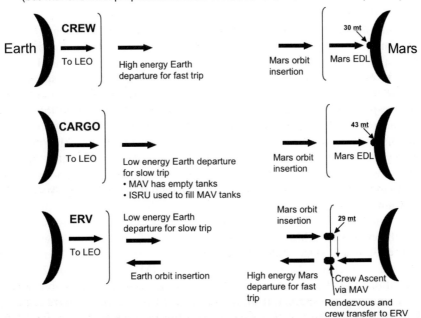

Figure 2.7. Vehicles delivered to destinations according to JSC's Mars Design Reference Mission 3.0. The ERV is the Earth Return Vehicle and the MAV is the Mars Ascent Vehicle.

for each step, one can work backwards to estimate the propellant mass required for each step along the way, and thereby eventually arrive at the initial mass in LEO (IMLEO). The reason to work backwards is that each previous step must provide the propellants needed to accelerate propellants needed for future steps, and this cannot be known if one starts at the beginning. In addition, the required lifting capability of the Launch Vehicle on the Earth launch pad can be derived from the mass in LEO of the heaviest vehicle involved.

Estimation of the masses of vehicles at their destinations requires extensive, detailed analysis, and is beyond the scope of this book. However, previous studies have provided some insights into what these are likely to be. They include Habitats, life support systems, radiation shielding, mobility systems on the surface, power systems, communication systems, propulsion systems, and various other engineering systems to support the mission. Figure 2.7 provides one estimate of vehicle masses at their destinations; however, several optimistic assumptions were made regarding life support, *in situ* resource utilization, and Habitat masses in this study. Actual required masses are likely to exceed those shown in Figure 2.7. We shall subsequently comment on this in greater detail.

2.6   WHAT'S IN LOW EARTH ORBIT

As we have discussed, the initial mass that must be lifted from the Earth's surface to LEO (IMLEO) is a rough measure of the mission cost, and is widely used to compare mission architectures at early stages of planning to help identify the optimum architecture at an early stage in planning. As we have indicated in the previous section, one starts with the vehicles required at the ultimate destinations (in the case of Figure 2.7 this would be three vehicles: the Crew Lander, the Cargo Lander, and the Earth Return Vehicle) and works backwards to estimate for each step how much propellant mass is needed to accelerate these vehicles through the various *steps* with their corresponding values of Δv. After adding up the masses of all the propulsion stages and propellants needed to take the various spacecraft through their various operations and steps in the missions, one ends up with a total mass that must be sent on its way toward Mars from LEO. The IMLEO is the sum of this mass that must be sent on its way to Mars plus the mass of the Earth Departure System—a propulsion system to impart an impulse to the spacecraft to send it out of LEO on its way toward Mars. Using current propulsion systems based on hydrogen–oxygen propellants, it requires roughly 1.8 mT of propulsion system in LEO dedicated to send 1 mT of spacecraft on its way to Mars. Of the 1.8 mT of propulsion system, roughly 0.2 mT is the dry propulsion stage (tanks, plumbing, and thrusters) and 1.6 mT is liquid propellants. Therefore, the initial mass in LEO (IMLEO) is about 2.8 times the aforementioned 1 mT mass sent on its way toward Mars. Once this propulsion system has fired and sent the spacecraft on its way toward Mars, the liquid propellants are used up and the dry stage is jettisoned in space since there is no further use for it. For example, if a 41 metric ton (mT) vehicle must be sent on its way toward Mars, this requires an IMLEO of about $2.8 \times (41 + 9) = 140$ mT, of which $\sim$41 mT is the payload, $\sim$9 mT is the dry propulsion stage, and $\sim$81 mT is the liquid propellants. The ratio of IMLEO to payload sent toward Mars is $140/41 \sim 3.4$.

For large masses that must be sent to Mars, the initial mass in LEO (IMLEO) is considerably larger than the mass sent toward Mars due to the factor of $\sim$3.4.

2.7 WHAT'S ON THE LAUNCH PAD?

A Launch Vehicle is a rocket used to transport the basic space vehicle to LEO. Launch vehicles are like olives. They come in three sizes: *large*, *giant*, and *jumbo*. Jumbo Launch Vehicles have not been used since Apollo used the Saturn V for Moon missions. A Launch Vehicle is made up of mostly propellants to lift a relatively small payload to LEO. It typically requires about 20 mT on the launch pad to lift 1 mT of payload to LEO. The 20 mT is composed of roughly 1 mT of payload, 2 mT of structure and propulsion stages, and about 17 mT of propellants.

Therefore, it takes about $20 \times 3.4 = 68$ mT on the launch pad to send 3.4 mT to LEO, which, in turn, can send 1 mT of payload on its way in transit toward Mars.

The Heavy-Lift Launch Vehicle being developed by NASA for lunar missions will lift about 125–150 mT to LEO and will weigh about 2,500 mT on the launch pad.

2.8 IMLEO REQUIREMENTS FOR SPACE MISSIONS

One of the first and most important requirements in planning space missions is making a preliminary estimate of IMLEO. One must then devise a scheme for transporting this mass to LEO, either in one fell swoop with a large Launch Vehicle, or possibly in several launches followed by rendezvous and assembly in LEO.

To estimate IMLEO for any conceptual Mars mission architecture, we have seen that the following steps are needed:

(1) Estimate the payload masses of vehicles that must be transported through various journeys.
M_1 = mass transported to Mars orbit
M_2 = mass transported to Mars surface
M_3 = mass transported to Mars orbit and returned to Earth
M_4 = mass transported to Mars surface and returned to Earth

(2) Using models for Δv for the various steps, estimate the gear ratios for each of the four journeys.
G_i = gear ratio for ith journey ($i = 1, 2, 3,$ or 4)

(3) Estimate IMLEO $= G_1 M_1 + G_2 M_2 + G_3 M_3 + G_4 M_4$
IMLEO dictates the scale of the mission, the size of the required Launch Vehicle, and leads to a rough estimate of probable cost. This is the most basic and important thing to do at the outset of planning a space mission.

3

Getting there and back

While there are many challenges involved in planning human missions to Mars, the problems involved in launching, transporting, landing, and returning large masses from these bodies appear to be perhaps the most formidable of these hurdles.

3.1 PROPULSION SYSTEMS

In a round-trip human mission to Mars or the Moon, there are a number of steps that must typically be traversed:

- Launch to LEO.
- Trans-Mars injection from LEO (Earth orbit departure).
- Mars orbit insertion.
- Entry, descent, and landing at Mars.
- Ascent to Mars orbit.
- Trans-Earth injection (Mars orbit departure).
- Entry, descent, and landing at Earth.

Each of these steps is a major undertaking, requiring significant amounts of propellants and propulsion stages, and/or aeroshells. Because these steps are sequential, they are leveraged so that the overall delivered fractional payload is a product of the fractional payloads for individual steps. For example, on an overall basis, it appears to require roughly 200 metric tons (mT) on the launch pad to land 1 mT on Mars.

Sending spacecraft out to various solar system destinations requires firing rockets to change the velocity of the spacecraft. When departing from a sizable body such as the Earth, there is a large penalty involved in escaping from the Earth's gravity pull. If there is an atmosphere (as there is on Earth) there is also a penalty due to atmospheric drag. When approaching a sizable body such as Mars, there is a significant velocity change involved in slowing down the spacecraft so it does not fall into the planet in a high-speed collision. Thus, departing from and landing on planetary bodies are two of the most demanding steps that require acceleration with rockets. In addition, changing course and in-flight maneuvers also require acceleration with a rocket, albeit with a much smaller Δv.

3.1.1   Propellant requirements for space transits

In a typical space transit the rocket "burns" and, after the burn is completed, the empty propulsion system is jettisoned so it does not have to be carried along for future steps (see Figure 3.1).

The performance of a rocket is determined by three parameters:

Specific impulse (I_{SP})—This is basically the exhaust velocity of the expelled propellants (divided by $g = 9.8 \text{ m/s}^2$). It is a measure of the effectiveness of the propulsion system because, for any fixed amount of propellant expelled by the rocket, the momentum imparted to the payload is proportional to the rocket exhaust velocity.

Here are some typical values of I_{SP} for rocket systems: space-storable mono-propellant (hydrazine), solid propellant, storable bi-propellant, and LOX/LH$_2$, respectively:

$$I_{SP} \text{ (Mono)} = 225 \text{ s}$$

$$I_{SP} \text{ (Solid)} = 290 \text{ s}$$

$$I_{SP} \text{ (Bi-Prop)} = 315\text{–}325 \text{ s}$$

$$I_{SP} \text{ (LOX/LH}_2) = 450\text{–}460 \text{ s}$$

The specific impulse of the LOX/LH$_2$ chemical propulsion system used for departure from LEO is typically about 450 s corresponding to a rocket exhaust velocity of about

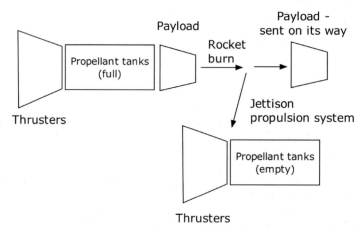

Figure 3.1. Model of a rocket burn.

4,400 m/s (9,860 miles per hour). Other chemical propulsion systems have lower exhaust velocities.

The specific impulse that can be achieved with any propellant combination in a rocket depends on a number of subordinate factors, such as the thrust level, the mixture ratio, the propellant feed (pressure or pump fed), whether the exhaust is into a vacuum or an atmosphere, and the nozzle design. When mission planners select values of specific impulse for various propellant mixtures in rockets, they rarely specify these details. The values of specific impulse selected for any propellant combination vary considerably from study to study, with values tending to cluster into two groups for each combination (optimistic and pessimistic). Furthermore, it is interesting that JPL propulsion experts tend to be pessimistic, whereas ISRU advocates and JSC mission designers tend to be optimistic.

For methane/oxygen, JSC DRM-1 and DRM-3 used 374 s, and Mars Direct used 373 s. Other optimists have used values as high as 379 s. At the present time, 360 s seems to be a popular figure used by several less optimistic groups. However, a detailed analysis,[24] using the rigorous JANNAF procedure and assuming a parabolic wall nozzle, determined the theoretical specific impulse of the oxygen–methane propellant combination for a range of mixture ratios and thrust levels. The theoretical performance (including kinetic, two-dimensional, and boundary layer losses) was reduced by 2% to account for an assumed 98% combustion efficiency (i.e., vaporization and mixing efficiency). The results are plotted in Figure 3.2 (see color section).

As can be seen from Figure 3.2, even for large rockets with an ideal mixture ratio (about 3.1), the highest predicted specific impulse for methane–oxygen approaches 3,400/9.8 or 347 s. However, as Thunnisen et al. point out, if an ISRU system based on hydrogen and the Sabatier-electrolysis process is used to produce methane and oxygen in a 1:4 mixture ratio, the most favorable initial mass in LEO is achieved with this higher mixture ratio even though the specific impulse may be around 337 s. Additional analysis of the methane–oxygen rocket was made in 2004.[25] This study performed a two-dimensional kinetics analysis for a thrust level of 450 N. They found a maximum specific impulse of 348 s at a mixture ratio of 3.0.

For liquid hydrogen/liquid oxygen rockets, the optimists typically use values of I_{SP} that range from 460 to 470 s, whereas the pessimists have used values as low as 420 s. In an expanded report,[26] the Thunnissen team reported the specific impulse of a

[24] *A 2007 Mars Sample Return Mission Utilizing In-Situ Propellant Production*, Daniel P. Thunnissen, Donald Rapp, Christopher J. Voorhees, Stephen F. Dawson, and Carl S. Guernsey, American Institute of Aeronautics and Astronautics, AIAA 99-0851 (1999).

[25] *Advanced Space Storable Propellants for Outer Planet Exploration*, Daniel P. Thunnissen, Carl S. Guernsey, Raymond S. Baker, and Robert N. Miyake, American Institute of Aeronautics and Astronautics, AIAA-2004-3488.

[26] *Space Storable Propulsion—Advanced Chemical Propulsion Final Report*, Carl Guernsey, Ray Baker, Robert Miyake, and Daniel Thunnissen, Jet Propulsion Laboratory, JPL D-27810, February 17, 2004.

hydrogen–oxygen rocket to be roughly flat at 440 s for mixture ratios from 3 to 5 with a fall-off at higher mixture ratios (430 s at $MR = 5$ and 420 s at $MR = 6$).

There seems to be a mismatch between the figures being bandied about by some optimistic mission planners, and figures compiled by propulsion experts.

Dry mass of the propulsion system—because the dry propulsion system (thrusters, propellant tanks, plumbing, ...) must be accelerated along with the payload, the dry propulsion system acts as a drag on the acceleration produced by the rocket. For the LOX/LH$_2$ propulsion used for departure from LEO, a rough guess is that the dry propulsion system weighs perhaps 10–12% of the propellant mass. This dry propulsion system is typically jettisoned after the burn, although in some cases it may be retained for further burns.

Change in velocity produced by the rocket $(\Delta v = v(\text{final}) - v(\text{initial}))$—This is the net effect of firing the rocket. The combination of the payload and the propulsion system are accelerated to a new velocity, during which the propellant tanks are gradually being emptied as the rocket burns.

The propellant requirement increases exponentially as Δv increases, and decreases as the specific impulse is increased. For each rocket burn in the overall mission, Δv is a critical parameter that determines how much propellant is required. Ultimately, the mass that must be delivered from Earth to space in mostly propellants, and the mass of propellants depends on the accumulation of Δv in the various mission steps.

3.1.2   The rocket equation

A critical element of any space mission involves transfers of space vehicles from one place to another. Most such transfers are achieved by operating propulsion systems that send effluents out the back of the vehicle, propelling the vehicle forward by Newton's law of action and reaction. The basic theory here is the law of conservation of momentum. To achieve a change in velocity of the space vehicle, an appropriate amount of propellant products must be expelled out the back of the vehicle. The so-called rocket equation allows one to calculate the amount of propellant required to apply a change in velocity Δv to a space vehicle.

We define the following quantities:

M_S = spacecraft mass (metric tons [mT])
M_R = mass of dry rocket (including structure, storage tanks, plumbing, thrusters, etc.) used to accelerate the spacecraft (mT)
M_P = mass of propellants initially stored in the rocket prior to rocket burn (mT)
m_P = mass of propellants instantaneously stored in the rocket at any point during the burn
$M_T = M_{\text{initial}}$ = total mass before rocket burn = $M_S + M_R + M_P$

v_E = exhaust velocity at which rocket effluents are expelled from the back of the rocket (km/s)

v_S = velocity of the spacecraft (km/s)

g = Earth's gravity acceleration = 0.0098 (km/s^2)

I_{SP} = specific impulse of rocket = v_E/g (s)

Δv = change in speed of the (spacecraft + rocket system) (km/s)

If we consider a small differential amount of propellant burned in the rocket, conservation of momentum requires that at any stage in the burn,

$$(M_R + M_S + m_P)\,dv_S = v_E\,dm_P$$

This differential equation can be integrated:

$$dv_S = -v_E\,dm_P/(M_R + M_S + m_P)$$

If we integrate v_S from its starting value to its final value, we obtain Δv_S on the left side and

$$v_E \log\{M_T/(M_R + M_S)\}$$

on the right side. This can be rearranged to provide one form of the so-called "rocket equation":

$$M_T/(M_R + M_S) = \exp\{(\Delta v)/(gI_{SP})\} = \text{"}q\text{"}$$

Other forms of the rocket equation are:

$$M_P/(M_R + M_S) = q - 1$$

$$M_T/M_P = q/(q - 1)$$

For any spacecraft of mass M_S, the object is typically to use as efficient a rocket as possible to minimize the propellant mass, M_P, required to achieve a given Δv. This is accomplished by using propellants with the highest available specific impulse. However, the "dry mass" of the rocket (M_R) is a detriment because it must be accelerated along with the spacecraft and I_{SP} is not the sole determinant of the best performance (see Figure 3.1). In addition, the volume of the propellant tanks is another important parameter that affects the viability of such rockets.[27]

In many cases, we have a spacecraft of known mass, M_S, and we desire to estimate the required amount of propellants to deliver the spacecraft to its destination. We may not know the dry rocket mass very accurately but there are usually "rules of thumb" for estimating this as a function of propellant mass. It is reasonable to assume that a relationship of the form:

$$M_R = A + KM_P$$

---

[27] A figure of merit for propulsion systems that takes several factors into account was defined in *Propulsion: Its Role in JPL Projects, the Status of Technology, and What We Need to Do in the Future*, C. Guernsey and D. Rapp, Jet Propulsion Laboratory, D-5617, Pasadena, CA, August 8, 1988; also see *Advanced Space Storable Propellants for Outer Planet Exploration*, Daniel P. Thunnissen, Carl S. Guernsey, Raymond S. Baker, and Robert N. Miyake, American Institute of Aeronautics and Astronautics, AIAA 2204-3488.

is adequate, and in some cases even the constant A has been neglected. If we can use the approximation:

$$M_R = KM_P$$

then using the form of the rocket equation:

$$M_P/(M_S + M_R) = q - 1$$

we have

$$M_P = (q-1)KM_P + M_S(q-1)$$

and

$$\frac{M_P}{M_S} = \frac{(q-1)}{1 - K(q-1)}$$

Now, using the other form of the rocket equation:

$$M_T/(M_R + M_S) = q$$

we find that

$$M_{\text{initial}} = qM_S + qM_R$$

Substituting (KM_P) for (M_R), we obtain:

$$\frac{M_{\text{initial}}}{M_{PL}} = \left(\frac{q}{1 - K(q-1)} \right)$$

Using this equation, we can estimate the initial mass required to deliver a payload through an acceleration of Δv using a rocket with specific impulse I_{SP}.

For Earth departure using LOX/LH$_2$ propulsion, $q \sim 2.22$ for a Δv of 3.61 km/s. If we assume the parameter $K \sim 0.12$, the ratio turns out to be:

$$\frac{M_{\text{initial}}}{M_S} = 2.50$$

Thus, only 40% of the mass in LEO can be sent as payload on its way toward Mars. It should also be noted that the Δv of 3.61 km/s corresponds to an idealized value for a rather slow transit to Mars. If it is desired to send a spacecraft to Mars more quickly (as, for example, in sending a human crew) the required value of Δv could increase to over 4 km/s, and this ratio would increase to about 2.77. In that case, only ~36% of the mass in LEO could be injected as payload on a path toward Mars. These estimates are merely examples. In actuality, the Δv varies considerably with launch opportunity and launch date within an opportunity.

Upon arriving at Mars, if we use a bi-propellant propulsion system to retard the spacecraft to 2.09 km/s, we find that $q \sim 1.93$. Assume in this case that $K \sim 0.15$, and we obtain:

$$\frac{M_{\text{initial}}}{M_S} = 2.16$$

as the ratio of mass approaching Mars to the payload inserted into Mars circular orbit at 300 km altitude. The ratio of mass in LEO to payload mass inserted into Mars orbit is then estimated roughly as $2.50 \times 2.16 = 5.4$ for a slow trip and $2.77 \times 2.16 = 5.98$ for a fast trip.

If, instead, we insert into an elliptical orbit with a Δv of 0.822 km/s, and use aerobraking thereafter to put the spacecraft into a circular orbit, the value of $q \sim 1.294$ and

$$\frac{M_{\text{initial}}}{M_{PL}} = 1.345$$

and the ratio of mass in LEO to payload mass inserted into Mars orbit is then $2.50 \times 1.345 = 3.36$.

All of the above deals with a single rocket burn. It must be noted that when large velocity changes are required, staging provides great benefits. If staging is used, as is typical for Launch Vehicles, then the above formulas apply to each stage, with the value of M_R decreasing for each successive stage. For ascent from planets, the situation is more complicated due to the rotation of the planet, as well as effects of gravity loss and drag loss.

It should also be noted that in all cases it is impossible to burn all of the propellant stored in the propellant tanks. There is always a residual amount that remains in the tanks. In the above equations, M_P should be the amount of propellant that is actually burned, while the excess unburned propellant should be added to the rocket mass.

3.1.3   Dry mass of rockets

In early high-level planning of space missions, the estimated dry mass of the rocket is usually assumed to be proportional to the propellant mass:

$$M_R = KM_P$$

Thunnisen *et al.* (*loc. cit.*) estimated the elements that contribute to the constant "K" in some detail for planetary spacecraft. They included seven types of valves, transducers, sensors, filters, flow controllers, lines and fittings, thrusters, ullage systems, support structures, and propellant tanks. Unfortunately, they did not specifically report estimates of "K" but, instead, used implicit values of "K" to estimate payloads that could be sent to outer planets for various propellant combinations. It is particularly noteworthy that the masses of propellant tanks depend on the propellant volume, and low-density propellants require higher tankage mass than more dense propellants.

Various groups have assumed values of K from 0.05 to over 0.20 for various applications. The consensus seems to be that for space-storable propulsion systems, K decreases as the propulsion system size increases. For small robotic planetary spacecraft, K is expected to be in the range 0.12 to 0.15, but for large, human-scale propulsion, it is likely to be smaller, perhaps around 0.10. However, use of cryogenic propellants will increase K, either by adding insulation and bulk for passive systems, or by adding power and cryocoolers (as well as insulation) for active systems.

The 2005 design of the Lunar Surface Access Module (LSAM) involved the masses shown in Table 3.1. As this table shows, the ratio of dry propulsion system to propellant mass is 0.15 for the descent system and 0.25 for the ascent system. Since

Table 3.1. Masses for the LSAM ascent and descent systems.

	LSAM ascent system		LSAM descent stage	
	Mass* (2005)	Mass** (2007)	Mass (2005)	Mass (2007)
1.0 Capsule/support structure	1,025	1,147	1,113	2,214
2.0 Protection	113	113	88	88
3.0 Propulsion	893	718	2,362	2,761
4.0 Power	579	1,205	468	486
5.0 Control	0	0	92	92
6.0 Avionics	385	385	69	69
7.0 Environment	896	1,152	281	284
8.0 Other	382	382	640	715
9.0 Growth	855	1,020	1,023	1,342
10.0 Non-cargo	834	153	1,033	2,498
11.0 Cargo	0	0	2,294	500
12.0 Non-propellant	131	173	486	659
13.0 Propellant (burned)	4,715	6,238	25,105	30,319
Dry mass		6,123		8,051
Inert mass	5,962	6,276	9,464	11,049
Wet mass (sum 1 to 13)	10,809	12,687	35,055	42,027

Propulsion				
Propellants	CH_4/O_2	NTO/MMH	CH_4/O_2	H_2/O_2
Main engine	94		527	
RCS thrusters	155			
CH_4/O_2 tanks	485		1,758	
Structure	44		0	
Growth	179		472	
Plumbing and pressurization tanks	159		77	
Total	1,116		2,834	
Residual propellant and pressurant	188		650	
Propellant boil-off	0		384	
Total for propulsion	188		1,034	
Propellants burned (B)	4,715		25,105	
Dry propulsion system	1,116		2,834	
Residual propellant and pressurant	188		1,034	
Total propulsion system (A)	1,304		3,868	
A/B	0.28		0.15	
$A/(A+B)$	0.22		0.13	
RCS propellant	172			
RCS thrusters	155			
RCS tanks	0			
Total RCS	327			
Propellants w/o RCS (C)	4,543			
Propulsion w/o RCS (D)	1,149			
C/D	0.25			
$C/(C+D)$	0.20			

* Figures from NASA reports c. 2005.
** Figures from "MoonHardware22Feb07_Connolly.pdf" on *NASA Watch* website.

the descent system utilizes roughly five times as much propellant, there is clearly an economy of scale.

In the absence of more definitive information, a value of $K \sim 0.12$ for large cryogenic propulsion systems in space is recommended.

For descent and ascent propulsion, higher values appear to be appropriate. New data released in 2007 are also included in this table for comparison.

3.2 TRAJECTORY ANALYSIS

As we have shown, a space mission involves a series of propulsion steps, each of which requires a mass of propellant to be burned. The properties of propulsion systems are known. Thus, the major factors that determine the propellant mass requirements, and ultimately the mission initial mass in LEO (IMLEO), are the Δv for all the steps in the mission.

3.2.1 Rocket science 101

Trajectory analysis provides detailed estimates for Δv for the steps in the mission. However, high-fidelity trajectory analysis is quite complex, and we shall present only a simplified approximate treatment for (1) departure from Earth orbit toward Mars (trans-Mars injection), (2) Mars orbit insertion, and (3) departure from Mars orbit (Trans-Earth injection).[28]

3.2.1.1 Constants of motion

We will be concerned with spacecraft that are in Earth orbit or Mars orbit. Therefore, we deal with a spacecraft of mass m moving under the influence of a planet of mass M. The radial force that attracts the spacecraft to the planet is GMm/r^2 where r is the distance of the spacecraft from the center of the (spherical) planet and G is the gravitational constant 6.6742×10^{-11} m^3/s^2/kg. This force can be derived from a potential energy $[-GMm/r]$. We denote the product (GM) as μ.

In any unperturbed trajectory of the spacecraft under influence by the planet, energy and angular momentum are the two key constants of motion (they are conserved).

In general, v is the magnitude of the velocity at the radius r. The orbital energy is the sum of the kinetic energy plus the potential energy. However, we shall deal with the *specific orbital energy*, ε, that is just the energy per unit spacecraft mass.

$$\varepsilon = \frac{v^2}{2} - \frac{m}{r}$$

In the following, ρ is the specific orbital angular momentum (i.e., per unit spacecraft

---

[28] This section was written by Dr. Mark Adler of JPL and the author made small modifications to his treatment.

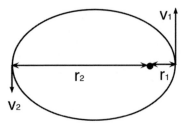

Figure 3.3. Flight path at periapsis showing flight path is perpendicular to radius.

mass) and γ is the flight path angle at radius r (i.e., γ is the angle that the velocity vector deviates from normal to the radius vector). By convention, γ is positive when the object is moving away from the body, and γ is negative when it is moving towards the body. In a circular orbit, γ is always zero. Then $v\cos[\gamma]$ is the velocity component normal to the radius vector, and the specific angular momentum is simply:

$$r = vr\cos(\gamma)$$

Note that this is actually the *magnitude* of the angular momentum, which is a vector. From here on, we will refer to *energy* and *angular momentum* with the understanding that they are *specific* (i.e., per unit spacecraft mass), and that the angular momentum is a magnitude.

  In the derivations here, we will also make use of Kepler's first law, which states that an orbit is an ellipse with the central body at one focus of the ellipse. Next, consider a spacecraft in orbit about a planet as shown in Figure 3.3. In this figure, r_1 is the periapsis, r_2 is the apoapsis, a is the semi-major axis, and b is the semi-minor axis. Note that, as per the usual convention, the γ shown in Figure 3.4 is considered negative, since the velocity vector with the direction shown is inside the normal to the radius vector (i.e., heading towards the body).

3.2.1.2   *Energy of an orbit*

Consider a closed, elliptical orbit with periapsis radius r_1 and apoapsis radius r_2, as shown in Figure 3.4. At periapsis we have velocity v_1 and, at apoapsis, velocity v_2.

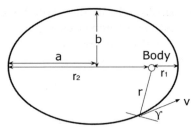

Figure 3.4. Spacecraft in orbit about a planet at a focus of an ellipse. Periapsis is the distance of closest approach (far right) and apoapis is the maximum distance (far left).

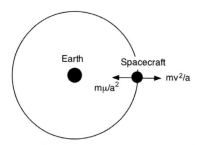

Figure 3.5. Force balance for a circular orbit.

Note that at both positions the flight path angle γ is zero (i.e., the flight path is normal to the radius vector). So we have:

$$r = v_1 r_1 = v_2 r_2$$

We now want to determine the energy of a spacecraft in a closed orbit about a planet in terms of μ (that characterizes the planet) and a (that characterizes the orbit). For the special case of a circular orbit, the inward attractive force is $(m\mu/a^2)$ and the outward centrifugal force is mv^2/a (see Figure 3.5).

If we set the two forces equal, we find

$$\varepsilon = mv^2/2 - \mu/a = -\mu/(2a)$$

In the more general case of an elliptical orbit, by setting the energy at apoapsis to the energy at periapsis and using the fact that angular momentum is conserved (same at both apoapsis and periapsis), it can be shown after some considerable algebraic manipulation that the energy of a spacecraft in a closed orbit is given by the same expression:

$$\varepsilon = -\frac{\mu}{(2a)}$$

As we showed in the previous section, the spacecraft energy ε is negative for bound orbits. However, if ε is positive, then the spacecraft is not bound to the planet, and it can fly off to large distances where r becomes ∞.

$$\varepsilon = \frac{v_\infty^2}{2}$$

v_∞ is the approach and departure velocity of a hyperbolic trajectory that passes by the body (once). The v_∞ and closest approach radius, r_{CA}, define that trajectory. The energy is defined by v_∞, and the angular momentum is defined by v_∞ and r_{CA}. The velocity at closest approach v_{CA} can be calculated from the energy:

$$\varepsilon = \frac{v^2}{2} - \frac{\mu}{r} = \frac{v_\infty^2}{2}$$

$$v_{CA} = \sqrt{v_\infty^2 + \frac{2\mu}{r_{CA}}}$$

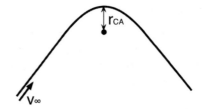

Figure 3.6. Typical hyperbolic trajectory.

Since the velocity at closest approach is normal to the radius, we have:

$$r = v_{CA}r_{CA} = \sqrt{v_\infty^2 r_{CA}^2 + 2\mu r_{CA}}$$

A typical hyperbolic trajectory is shown in Figure 3.6.

3.2.1.3   *Measured values for the Sun, Earth, and Mars*

Here are some actual values of μ and distances that allow us to derive results:

  – One astronomical unit:

$$1\,\text{AU} = 1.49597870691 \times 10^8 \text{ km};$$

  – GMs of the bodies:

$$\mu_{\text{Sun}} = 1.32712440018 \times 10^{11} \text{ km}^3/\text{s}^2$$
$$\mu_{\text{Earth}} = 398600.44 \text{ km}^3/\text{s}^2$$
$$\mu_{\text{Mars}} = 42828.3 \text{ km}^3/\text{s}^2$$

  – GM of the Earth–Moon system for more careful escape calculations (left as an exercise for the reader):

$$\mu_{\text{E-M}} = 403503.24 \text{ km}^3/\text{s}^2$$

  – The semi-major axis of Earth and Mars orbits around the Sun:

$$a_{\text{Earth}} = 1.00000261 \text{ au}$$
$$a_{\text{Mars}} = 1.52371034 \text{ au}$$

  – Equatorial radii of Earth and Mars:

$$r_{\text{Earth}} = 6378.14 \text{ km}$$
$$r_{\text{Mars}} = 3396.2 \text{ km}$$

  – Representative low-orbit altitudes:

$$z_{\text{Earth}} = \sim 200 \text{ km}$$
$$z_{\text{Mars}} = \sim 300 \text{ km}$$

3.2.1.4   *Escaping the influence of a planet*

For a spacecraft in a hyperbolic orbit, moving past a planet, the influence of the planet persists out to considerable distances because of the $1/r$ in the potential energy. In this section, we deal briefly with the time required to effectively depart from the planetary influence to the point where the spacecraft velocity approaches v_∞. Of course, you never reach v_∞, but you can get pretty close. Consider Figure 3.7, which plots velocity vs. distance for a typical escape from Earth, where we have arbitrarily selected v_∞ to be 3 km/s. (It can be any value.) Note that when the spacecraft is under the gravitational attraction of the Earth, it is speeded up considerably compared with its velocity at infinity.

Although one never quite gets to v_∞, we can arbitrarily take $(1.05 v_\infty)$ as a marker point where the spacecraft has almost left the influence of Earth (to within about 95%).

We find the distance at which the spacecraft velocity is $0.95 v_\infty$ by using the conservation of energy between two points (one at the point where $v = 0.95 v_\infty$, and one at ∞). Thus:

$$\frac{v_\infty^2}{2} = \frac{v^2}{2} - \frac{\mu}{r}$$

$$\frac{v_\infty^2}{2} = \frac{(1.05 v_\infty)^2}{2} - \frac{\mu}{r}$$

Using the values:

$$\mu = 3.986 \times 10^5 \text{ km}^3/\text{s}^2$$

$$v_\infty = 3 \text{ km/s}$$

we can solve for

$$r = 8.642 \times 10^5 \text{ km} = 135.5 \text{ Earth radii } (6{,}378 \text{ km})$$

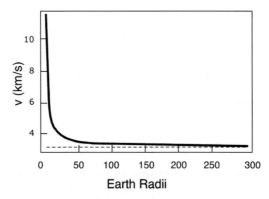

Figure 3.7. Velocity vs. distance from Earth for escape from Earth on a hyperbolic trajectory. The dotted line is the asymptotic value of v_∞.

A rough estimate of the time required to reach this point can be made by assuming that the velocity is always 3 km/s. The result is 8.642×10^5 km/ $(3 \text{ km/s}) = 2.88 \times 10^5$ s $= 3.33$ days. However, the actual time will be 2.9 days because the spacecraft is moving faster than 3 km/s, particularly when it is close to Earth.

This is small compared with the orbital period of Earth about the Sun of about 365 days. Hence, departure from Earth takes place rapidly compared with the Earth's motion about the Sun and we can treat the Earth as approximately stationary during Earth departure by a spacecraft. The spacecraft can be assumed to reach v_∞ relative to Earth while the Earth is fixed relative to the Sun.

3.2.1.5   Earth and Mars solar orbit velocities

For any orbit the energy can be expressed as:

$$\varepsilon = -\frac{\mu}{(2a)} = \frac{v^2}{2} - \frac{\mu}{r}$$

For a circular orbit, r is always equal to a, so the above reduces to:

$$v^2 = \frac{\mu}{a}$$

Earth's velocity around the Sun (ignoring eccentricity) is:

$$v_{\text{Earth}} = \sqrt{\frac{\mu_{\text{Sun}}}{a_{\text{Earth}}}} = 29.7847 \text{ km/s}$$

For Mars (ignoring its greater eccentricity):

$$v_{\text{Mars}} = \sqrt{\frac{\mu_{\text{Sun}}}{a_{\text{Mars}}}} = 24.1291 \text{ km/s}$$

3.2.1.6   Hohmann transfer to Mars

The most energy-efficient transfer from Earth orbit to Mars orbit is half of a solar orbit whose periapsis just touches the orbit of the body closer in (Earth), and whose apoapsis just touches the orbit of the body farther out (Mars) as shown in Figure 3.8.

This shows that $\Delta v_{\text{out}} = 2.945$ km/s and $\Delta v_{\text{in}} = 2.649$ km/s. We assume circular orbits for Earth and Mars, and we will assume that they are in the same plane. Our object here is to estimate how much velocity change Δv_{Hper} must be imparted to the spacecraft in Earth orbit to put it on the Hohmann trajectory toward Mars; and how much velocity change Δv_{Hapo} must be imparted to the spacecraft on the Hohmann trajectory to put it into a Mars orbit at Mars. The transfer orbit velocity vector where it touches the Earth and Mars orbits will be in the same direction as the Earth and Mars orbital velocities. Therefore, we can simply subtract the orbital velocities at each point from the velocity at each point on the Hohmann trajectory to get the Δv required at each end—the Δv at the Earth end to get on the trajectory heading for Mars, and the Δv at the Mars end to match Mars' orbital velocity about the Sun.

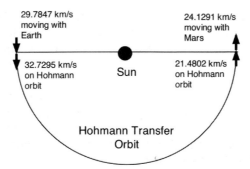

Figure 3.8. Transfer from moving with the Earth to Hohmann, moving along Hohmann, and then to moving with Mars. This shows that $\Delta v_{\text{out}} = 2.945$ km/s and $\Delta v_{\text{in}} = 2.649$ km/s.

Note that these values of Δv are calculated on the basis that the spacecraft would go from Earth's orbit around the Sun to Mars' orbit around the Sun without the planets actually being there. However, when the presence of the planets is taken into account, the required Δv's will change due to the forces exerted by the planets. However, we will need these velocities for the pure orbital change (with the planets neglected) in order to estimate more realistic Δv's for the spacecraft transfers when the planets are present (in Section 3.2.1.9). Therefore, we proceed with this unrealistic calculation as a step along the way toward a more realistic estimate.

To compute the energy of the transfer orbit we note that a, the semi-major axis of the Hohmann transfer orbit, is the sum of the semi-major axes for the Earth and Mars about the Sun:

$$a = a_{\text{Earth}} + a_{\text{Mars}}$$

and the energy of the spacecraft in the Hohmann transfer orbit is:

$$\varepsilon_H = -\frac{\mu}{2a} = -\frac{\mu_{\text{Sun}}}{a_{\text{Earth}} + a_{\text{Mars}}}$$

$$\varepsilon_H = -351.517 \left(\frac{\text{km}^2}{\text{s}^2} \right)$$

From this, we can calculate the velocities at periapsis and apoapsis in the Hohmann transfer orbit:

$$v_{\text{Hper}} = \sqrt{2 \left(\varepsilon_H + \frac{\mu_{\text{Sun}}}{a_{\text{Earth}}} \right)}$$

$$v_{\text{Hper}} = 32.7295 \text{ km/s}$$

$$v_{\text{Hapo}} = \sqrt{2 \left(\varepsilon_H + \frac{\mu_{\text{Sun}}}{a_{\text{Mars}}} \right)}$$

$$v_{\text{Hapo}} = 21.4802 \text{ km/s}$$

The Δv's for the spacecraft transfers can now be calculated by subtracting v_{Earth} from v_{Hper} and subtracting v_{Mars} from v_{Hapo}. The results for transfer out of Earth orbit (about the Sun) into the Hohmann transfer orbit and from the Hohmann orbit to the Mars orbit (about the Sun) are:

$$\Delta v_{out} = v_{Hper} - v_{Earth} = 32.7295 - 29.7847 = 2.945 \text{ km/s}$$

$$\Delta v_{in} = v_{Mars} - v_{Hper} = 24.1291 - 21.4802 = 2.649 \text{ km/s}$$

As previously stated, these do not take into account the presence of the planets, and these calculations are only made as inputs to more realistic estimates to be made in a subsequent section. It requires an increase in velocity to transfer from the Hohmann transfer orbit to a motion in the Mars orbit about the Sun, neglecting the presence of the planet Mars. In a subsequent section, we will show that in the presence of the gravitational field of Mars, the spacecraft speeds up and in order to insert the spacecraft into an orbit about Mars, the spacecraft must actually be slowed down.

3.2.1.7 *Earth and Mars low-orbit velocities*

In the previous section, we calculated the values of Δv for (a) departing from the synchronous motion of the Earth about the Sun to the Hohmann transfer orbit, and (b) from the Hohmann transfer orbit to a motion synchronous with the motion of Mars about the Sun. What we'd really like to know is how much Δv we'll need to get from a spacecraft rotating about the Earth in low Earth orbit (LEO) to a spacecraft rotating about Mars in low Mars orbit (LMO). First, we'll need the low-orbit velocities. Again using:

$$v^2 = \frac{\mu}{a}$$

for a circular orbit, we find for LEO and LMO:

$$v_{LEO} = \sqrt{\frac{\mu_{Earth}}{r_{Earth} + z_{Earth}}}$$

$$v_{LEO} = 7.7843 \text{ km/s}$$

$$v_{LMO} = \sqrt{\frac{\mu_{Mars}}{r_{Mars} + z_{Mars}}}$$

$$v_{LMO} = 3.4040 \text{ km/s}$$

These are the velocities of the spacecraft relative to the planets when it is in low orbit about a planet.

3.2.1.8 *Earth escape*

We're going to transfer from low Earth orbit to a hyperbolic escape trajectory to get on the transfer orbit to Mars. That hyperbolic trajectory will have a v_∞ equal to the Δv needed to get from Earth's orbit to the transfer orbit. Here is why: As explained previously, we can consider the process of hyperbolic escape from Earth and reaching v_∞ to be an almost instantaneous event with respect to the Earth and transfer solar orbits. The v_∞ then is simply the velocity of the spacecraft relative to Earth. (Of course, we arrange for the direction of that relative velocity to line up with the Earth's orbital velocity.) Then the sum of the Earth's orbital velocity and v_∞ for the new orbit's velocity at that point is the velocity of the spacecraft relative to the Sun in the Hohmann transfer orbit. So if the v_∞ magnitude is equal to Δv_{out}—*voilà*, we end up on the transfer orbit.

Launch vehicle performance for Earth escape missions is usually quoted in terms of a quantity called "C_3" for historical reasons; C_3 is simply the square of v_∞. Therefore, for transfer to Mars, we obtain

$$C_{3M} = \Delta v_{out}^2 = 8.6719 \ (km/s)^2$$

For real trajectories, typical Mars C_3's can range from 8.67 to values as high as $16 \ (km/s)^2$ or more. These differences are due to the fact that the assumptions above are not quite true—that is, the Earth's and especially Mars' orbits are not circular, and the orbits are not in the same plane.

Now we compute the velocity of the hyperbolic trajectory at the altitude of the low Earth orbit:

$$\frac{v_\infty^2}{2} = \frac{v^2}{2} - \frac{\mu}{r}$$

$$v_{esc} = \sqrt{(\Delta v_{out}^2) + \frac{2\mu_{Earth}}{r_{Earth} + z_{Earth}}}$$

$$v_{esc} = 11.3957 \ km/s$$

We will assume that the plane of the low Earth orbit has been carefully matched to line up with the outgoing trajectory. That is the job of the Launch Vehicle mission designers when selecting the parking orbit. Then, we can simply subtract the low Earth orbit velocity from the escape velocity to get the actual spacecraft Δv to get from LEO to the Mars transfer orbit:

$$\Delta v_{inject} = v_{esc} - v_{LEO} = 11.3957 - 7.7843 = 3.611 \ km/s$$

3.2.1.9   Mars orbit insertion—part 1

The treatment of Mars orbit insertion is very much analogous to Earth departure. On arrival at Mars, the approach v_∞ at Mars is the relative velocity between Mars and the transfer orbit, Δv_{in}. The velocity of that hyperbolic trajectory at the altitude of low Mars orbit is:

$$v_{app} = \sqrt{(\Delta v_{in}^2) + \frac{2\mu_{Mars}}{r_{Mars} + z_{Mars}}}$$

Noting that $\Delta v_{in} = 2.649 \ km/s$ and $z_{Mars} = 300 \ km$, we find that $v_{app} = 5.4947 \ km/s$.

Again we arrange the arrival trajectory to match the plane of the desired orbit, so that we can just subtract the low Mars orbit velocity from the approach velocity to get the spacecraft Δv required for insertion into a path that follows Mars in its orbit about the Sun:

$$\Delta v_{\text{insert}} = v_{\text{LMO}} - v_{\text{app}} = -2.0907 \text{ km/s}$$

The spacecraft must be slowed down by 2.0907 km/s in order to put it into a circular orbit about Mars at an altitude of 300 km. This assumes, correctly, that Earth and Mars are going in the same direction around the Sun. If Mars were going in the opposite direction, then this approach velocity would be much higher!

3.2.1.10   *Summary of transfers to and from Hohmann orbit*

The Earth and Mars are traveling about the Sun in the same direction. Neglecting the eccentricity of their orbits (assuming circular orbits—which is rather a crude approximation for Mars), the velocities of the planets in their paths about the Sun are:

$$v_{\text{Earth}} = 29.7847 \text{ km/s}$$

$$v_{\text{Mars}} = 24.1291 \text{ km/s}$$

The Earth is moving faster than Mars, and this is what makes it so difficult to carry out a round trip to Mars. By the time the spacecraft has reached Mars, Earth has moved around the Sun and is unapproachable for return without waiting a long period for Earth to re-emerge on the Mars side of the Sun.

If we consider a simple Hohmann transfer orbit to go from Earth to Mars, assuming that the spacecraft is initially moving with the Earth in Earth's orbit about the Sun and ends up moving with Mars in Mars' orbit about the Sun, the appropriate velocities of the Hohmann transfer orbit are:

$$v_{\text{Hper}} = 32.7295 \text{ km/s}$$

$$v_{\text{Hapo}} = 21.4802 \text{ km/s}$$

at periapsis and apoapsis, respectively.

Thus, in the simple model where the spacecraft is not in orbit about the planets, but moving in unison with the planets, the Δv for transfer from moving with the Earth to the Hohmann orbit is:

$$\Delta v_{\text{out}} = v_{\text{Hper}} - v_{\text{Earth}} = 32.7295 - 29.7847 = 2.945 \text{ km/s}$$

and the Δv for transfer from the Hohmann orbit to moving with Mars is:

$$\Delta v_{\text{in}} = v_{\text{Mars}} - v_{\text{Hper}} = 24.1291 - 21.4802 = 2.649 \text{ km/s}$$

This is illustrated in Figure 3.8.

However, we are really interested in the transfer from LEO to Hohmann, and from Hohmann to LMO. We note that the velocities of a spacecraft in LEO or LMO relative to the planet are:

$$v_{LEO} = 7.7843 \text{ km/s}$$

$$v_{LMO} = 3.4040 \text{ km/s}$$

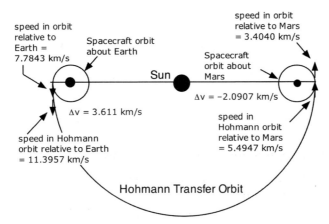

Figure 3.9. Transfer from Earth orbit to Hohmann orbit, and transfer from Hohmann orbit to Mars orbit.

(The reason that the velocity is greater in LEO is due to the higher mass of Earth, necessitating a higher velocity to generate a centrifugal force to balance the inward gravitational attractive force.)

We have the velocity of the spacecraft in LEO. This must be subtracted from the velocity in the Hohmann transfer orbit at LEO in order to obtain the velocity change that must be imparted to the spacecraft to go from LEO to the Hohmann orbit in LEO,

$$\Delta v_{\text{inject}} = v_{\text{esc}} - v_{\text{LEO}} = 11.3957 - 7.7843 = 3.611 \text{ km/s}$$

Similarly, on arrival at Mars, we need the difference between the velocity in low Mars orbit and the velocity in the Hohmann orbit at LMO, to obtain the velocity change that must be imparted to insert the spacecraft into LMO.

$$\Delta v_{\text{insert}} = v_{LMO} - v_{\text{app}} = 2.0907 \text{ km/s}$$

See Figure 3.9.

3.2.1.11 Orbital period

To calculate the period of an orbit, we will derive (and use) Kepler's second law, which states that equal area is swept in equal time. Consider a triangle with one end at the central attractive body and the other two ends separated by an infinitesimally small distance on the ellipse, dx, as shown in Figure 3.10. The area of that triangle to first order in dx is:

$$dA = \frac{r\,dx\,\cos(\gamma)}{2}$$

where γ is the flight path angle. The velocity at radius r gives us the rate of change of x:

$$v = \frac{dx}{dt}$$

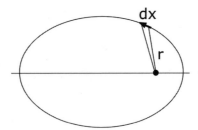

Figure 3.10. Diagram for deriving Kepler's second law.

Recalling the definition of the angular momentum, ρ, we see that the areal sweep rate is given simply by:

$$\frac{dA}{dt} = \frac{rv\cos(\gamma)}{2} = \frac{\rho}{2}$$

Since

$$A = \pi ab$$

the time required to complete one circuit is $A/(dA/dt)$ or:

$$T = 2\pi a \frac{b}{\rho}$$

If we go back to Figure 3.1 and apply conservation of energy and conservation of angular momentum at points 1 and 2 in the figure (at apoapsis and periapsis) we can express the angular momentum in terms of r_1 and r_2. The steps are as follows:

$$\rho = r_1 v_1 = r_2 v_2$$

$$\varepsilon = \frac{v_1^2}{2} - \frac{\mu}{r_1} = \frac{v_2^2}{2} - \frac{\mu}{r_2}$$

$$\frac{1}{2}(v_1^2 - v_2^2) = \mu\left(\frac{1}{r_1} - \frac{1}{r_2}\right)$$

$$\frac{1}{2}\left(\frac{\rho^2}{r_1^2} - \frac{\rho^2}{r_2^2}\right) = \mu\left(\frac{1}{r_1} - \frac{1}{r_2}\right)$$

$$\frac{\rho^2}{2} = \mu\left(\frac{r_1 r_2}{r_1 + r_2}\right)$$

Having expressed the angular momentum in terms of r_1 and r_2, we can then proceed to express the period in terms of orbit parameters. Solving for ρ and inserting into the expression for the period, T, we obtain:

$$T = 2\pi ab \sqrt{\left(\frac{r_1 + r_2}{2\mu r_1 r_2}\right)}$$

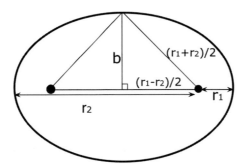

Figure 3.11. Right triangle formed by semi-minor axis and half of a constant-length cord from the foci to the ellipse.

Noting that $a = r_1 + r_2$, we obtain:

$$T = 2\pi\sqrt{\frac{a^3 b^2}{\mu r_1 r_2}}$$

A geometric characteristic of the foci of an ellipse is that the sum of the lengths of the two lines from any point on the ellipse to the two foci is a constant. (In fact, an ellipse can be defined as the locus of points that satisfy this criterion.) We use this property and apply the Pythagorean theorem to compute the semi-minor axis b in terms of r_1 and r_2. In Figure 3.11, we note that the horizontal lines are of length r_1 and r_2. Therefore, the horizontal line in the right triangle has length $(r_2 - r_1)/2$. We then construct the right triangle and note the height is b. Since the length of the hypotenuse of the right triangle is the same as the length of the other slanted line, their sum is $r_1 + r_2$. Therefore, the length of the hypotenuse is $(r_2 + r_1)/2$. Now, if the Pythagorean theorem is applied to the right triangle in Figure 3.9, we obtain:

$$b^2 + \left(\frac{r_2 - r_1}{2}\right)^2 = \left(\frac{r_2 + r_1}{2}\right)^2$$

from which, we can show that

$$b^2 = r_1 r_2$$

and this leads to Kepler's second law:

$$T = 2\pi\sqrt{\frac{a^3}{\mu}}$$

3.2.1.12 Mars orbit insertion—part 2

In Section 3.2.1.9, we showed that the approach velocity of a spacecraft in a Hohmann transfer orbit is 5.4947 km/s when it is at 300 km altitude above Mars. We also showed that to insert the spacecraft into a circular orbit of altitude 300 km requires imparting a Δv of -2.0907 km/s to the spacecraft to reduce its velocity down to the velocity of a spacecraft in a circular orbit at 300 km altitude: 3.4040 km/s.

We can now contrast what would happen if we imparted a lesser negative Δv to the spacecraft than 2.0907 km/s. First, note that if we impart no Δv to the spacecraft, it will just fly past Mars and go out into space. Next we determine the minimum negative Δv to capture the spacecraft into an extreme elliptical orbit about Mars. We deal with the limiting case where you go to an energy of zero—right on the borderline of being captured at all. We compute the velocity of that almost-orbit at a periapsis coincident with low Mars orbit at an altitude of 300 km:

$$\varepsilon = \frac{v^2}{2} - \frac{\mu}{r} \sim 0$$

$$v_{\text{Zero}} = \sqrt{\frac{2\mu_{\text{Mars}}}{r_{\text{Mars}} + z_{\text{Mars}}}}$$

The result is:

$$v_{\text{Zero}} = 4.8140 \text{ km/s}$$

and therefore the velocity change imparted is

$$\Delta v = v_{\text{Zero}} - v_{\text{app}} = 4.8140 - 5.4947 = 0.6807 \text{ km/s}$$

This implies that if the spacecraft is slowed down by less than 0.6807 km/s, it will not be captured. If Δv imparted to the spacecraft is just 0.6807 km/s, it will be captured into an extremely elongated elliptical orbit. As the absolute value of Δv is increased above 0.6807 km/s, the size (and period) of the elliptical orbit decrease, until at a Δv of -2.0907 km/s the spacecraft enters a circular orbit at 300 km altitude with a velocity of 3.4040 km/s.

Consider, for example, an intermediate value of Δv to place the spacecraft in an elliptical orbit with a period of 48 hours. We have previously derived the relation between period and semi-major axis:

$$T = 2\pi \sqrt{\frac{a^3}{\mu}}$$

We can use this relation to calculate that $a_{\text{P48}} = 31{,}878$ km.
    For any orbit,

$$\varepsilon = -\frac{\mu}{(2a)} = \frac{v^2}{2} - \frac{\mu}{r}$$

In our case,

$$r = r_{\text{Mars}} + z_{\text{Mars}}$$

where $z_{\text{Mars}} = 300$ km, and the velocity v_{P48} is at the distance of closest approach. Thus, we can solve for the velocity that must be achieved to put the spacecraft into a 48-hour orbit:

$$v_{\text{P48}} = 4.6723 \text{ km/s}$$

and the change in velocity needed to do this is:

$$\Delta v_{\text{P48}} = 4.6723 - 5.4947 = 0.8223 \text{ km/s}$$

So, it can be seen that when a spacecraft approaches Mars, it can be inserted in a wide variety of orbits, depending on the Δv that is applied to slow down the approaching

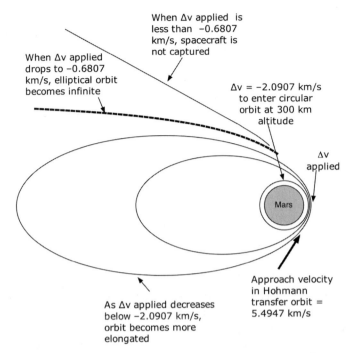

Figure 3.12. Dependence of Mars orbit on Δv applied to retard spacecraft.

spacecraft. If a Δv as large as 2.0907 km/s is applied at an altitude of 300 km, the spacecraft will enter a 300 km circular orbit. If Δv is between 0.6807 km/s and 2.0907 km/s the spacecraft will enter an elongated elliptical orbit whose period will be lower as Δv increases. See Figures 3.12 and 3.13.

Clearly, it requires more propellant to insert a spacecraft into a circular orbit. In order to reduce the propellant requirement for Mars orbit insertion into a circular orbit, the process of aerobraking was devised. In this process, the spacecraft is initially placed into a ~48-hour elliptical orbit ($\Delta v \sim 0.8$ km/s) and it is gradually slowed down by atmospheric drag to a smaller orbit. Some small further propulsive corrections are needed, but the overall Δv is about 1.0 km/s as opposed to all-propulsive orbit insertion that requires a Δv of about 2 km/s.

3.2.1.13   Summary

According to the simple models developed in previous sections, we have found the following:

- The Δv required for departure from Earth orbit toward Mars along a Hohmann trajectory is estimated to be 3.61 km/s.
- The Δv required for insertion into a ~300 km circular orbit from a Hohmann trajectory is estimated to be 2.09 km/s.

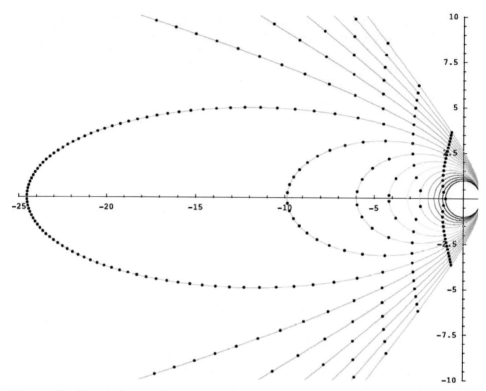

Figure 3.13. Hyperbolic and elliptical trajectories with linearly varying velocities at the closest approach that possess the same distance of closest approach. The outermost hyperbola is a flyby with no engine firing, and each curve inside represents a linear increase in applied velocity at the distance of closest approach. For the seventh inward curve, orbit insertion is achieved to a highly elliptical orbit. The innermost orbit is circular. The dots are linear time-ticks, and widely spaced ticks represent fast-moving spacecraft.

- The Δv required for capture of a spacecraft into an elongated elliptical orbit from a Hohmann trajectory is estimated to be > 0.68 km/s. As the Δv increases above 0.68 km/s, the eccentricity of the resultant orbit decreases, until at $\Delta v \sim 2.09$ km/s it becomes circular. Capture into a 48-hour orbit requires a Δv of about 1.0 km/s.

These estimates are idealized and the actual data will differ from these estimates, depending on the specific launch date.

3.2.2   Mars mission duration and propulsion requirements

In this section, we describe a general overview of Mars trajectories, based on a simplified model assuming circular orbits, two-body interactions, and planar trajectories. The actual situation, while similar in many respects to the simplified model, differs in some important details that will be discussed in Section 3.2.3.

The Earth circulates around the Sun in its orbit at the rate of 360° per Earth year, whereas Mars circulates at the rate of 191° per Earth year. Both planets move in the same direction around the Sun. As we showed in the previous section, the lowest energy trajectory from LEO to Mars is the so-called Hohmann trajectory that is tangent to the Earth's path at departure, and tangent to the Mars path at arrival, since the thrust imparted to the spacecraft is always in the direction of spacecraft motion. This trajectory was illustrated in Figure 3.9.

Similarly, a Hohmann transfer trajectory can be defined for return from Mars to Earth.

Wertz provides excellent descriptions of these trajectories.[29] According to Wertz's model, the overall Δv for transfer from LEO to Mars orbit is 5.6 km/s and the Δv for return to Earth orbit is also 5.6 km/s. Note that our simplified treatment produced the value $3.6 + 2.1 = 5.7$ km/s for this figure, which is close enough for approximate analysis.

According to Wertz, the Hohmann traverses to and from Mars each take 259 days and the spacecraft spends 454 days at Mars while it waits for a propitious time to return. Because the Earth moves about the Sun more rapidly than Mars does, the timing of the Hohmann orbit requires that Mars must be leading the Earth at spacecraft launch, and the spacecraft arrives at Mars tangent to the Mars orbit when the Earth is on the other side of the Sun. From this configuration, it takes roughly a year and a quarter before the relative positions of Mars and the Earth are positioned just right for a similar return flight from Mars. The total elapsed time for a Hohmann departure and return to LEO in Wertz's model is roughly 2.7 years. These models are based on assumed circular orbits and two-body calculations. Exact calculations will vary somewhat from opportunity to opportunity, but the rough general outlines given here will be retained.

The Hohmann-type orbit is an ideal way to send cargo to Mars because the propellant requirement is a minimum and the relatively long trip time ($\sim$265 days) is not generally problematic for cargo. The return flight from Mars to Earth is very similar with Mars once again leading the Earth when departure from Mars orbit is begun so that the Earth "catches up" with the spacecraft moving inward toward the Earth's orbit.

It should be noted that because of the requirement that the two planets, Earth and Mars, be positioned appropriately for a transfer between them, realistic opportunities for transfers to Mars occur for limited periods (several weeks) every 26 months. That means that sequential departures for Mars must be spaced by approximately 26-month periods.

---

[29] "Rapid Interplanetary Round Trips at Moderate Energy," James R. Wertz, *International Astronautics Federation Congress, October 4–8, 2004, Vancouver, BC, Canada*, Paper No. IAC-04-Q.2.a.11 and Paper No. IAC-03-Q.4.06; also "Interplanetary Round Trip Mission Design," James R. Wertz, *54th International Astronautical Congress, September 29–October 3, 2003, Bremen, Germany*; also *http://www.smad.com/ie/ieframessr2.html*

Table 3.2. Effect of increasing Δv on trip time, time on Mars, and mission duration for Mars "long-stay" missions. The overall Δv is for a round trip from LEO. (Wertz, *loc. cit.*)

Overall Δv (km/s)	Years on Mars	Mission duration (years)	Days on Mars	Average trip time (years)
11	1.25	2.70	456	265
12	1.37	2.72	500	246
13	1.50	2.70	548	219
14	1.58	2.68	577	201
15	1.64	2.65	599	184
18	1.70	2.64	621	171
20	1.75	2.62	639	159
25	1.84	2.61	672	141
30	1.90	2.60	694	128
35	1.97	2.58	719	111
40	2.03	2.56	741	97
45	2.08	2.55	759	86

If a more powerful propulsion system is used, generating a higher Δv than the minimum (the Hohmann), the trip time can be reduced, and the staying time on Mars is increased. In this case, the thrust is not along the spacecraft path, as the purpose of the extra propulsive thrust is to shorten the trip from Earth to Mars. The return trip is again the inverse of the outbound trip. This type of trajectory would presumably be used for sending the crew to Mars and returning them to Earth. By increasing the Δv above that required for the Hohmann orbit, the trip time to (and from) Mars can be reduced substantially. However, one must still wait for the proper orientation of the two planets prior to return from Mars, and therefore the elapsed time from Earth departure to ultimate Earth return will not change much as Δv is increased. Hence, as Δv is increased, the trip times to and from Mars are decreased, but the length of stay on Mars is increased, so that the overall mission duration does not change much. Estimates provided by Wertz are given in Table 3.2. These values will actually vary somewhat with mission opportunity, but will generally remain in the same "ballpark" as the predictions of the simple model.

All of the trajectories in Table 3.2 are based on the Earth making one extra revolution around the Sun compared with the spacecraft. Generally, as we apply more Δv, we can leave later and arrive earlier relative to a given opposition (opposition is when Mars and Earth are closest). For a given Mars opposition, if we can arrive quickly enough to spend a brief period on Mars and still catch a trip back to Earth before the Earth has moved too far away, then the trip will be relatively short in total duration. Thus, we can imagine "short-stay" missions to Mars in which high-energy trajectories are used for fast transfers to and from Mars, and the stay on Mars is limited to a maximum of a few weeks. Wertz provides estimates of requirements for short-stay missions as summarized in Table 3.3.

It can be seen that the total round-trip time is reduced from about 2.7 years to typically less than half that amount for "short-stay" missions. However, the Δv must

Table 3.3. Characteristics of Mars "short-stay" missions. (Wertz, *loc. cit.*)

Overall Δv (km/s)	Total years elapsed	Days on Mars	Days to reach Mars	Days to return
35	1.06	7	112	268
38	1.12	38	106	264
40	1.14	55	102	259
45	1.25	117	94	245
50	1.30	147	88	240
55	1.37	193	81	226
60	1.41	223	75	217

be increased to a high value, and the propellant requirements increase exponentially with Δv. Another problem is that perihelion occurs near or inside the orbit of Venus on the return flight. The impact on the mission (in terms of radiation and thermal input) is significant. Wertz has made the point that these models are based on all-propulsive mission steps, and use of aerocapture could reduce the overall Δv require-ments, although the masses of the aeroshells would have to be taken into account. However, the Δv will remain very high, and that, coupled with the horrendous stresses introduced in the return flight by spending a good deal of time in high radiation, high thermal, zero-g environments, as well as the uncertainty that the crew can do anything useful in 20–30 days on Mars ("plant the flag and footprints" mode), makes this approach appear unattractive. Yet, JSC continues to consider such approaches based on logic that is difficult to grasp. Robert Zubrin has suggested that there are three possible reasons for advocating the short-stay scenario:

- You want to develop an exotic propulsion system.
- You don't understand the problem.
- You want to kill the Program.

3.2.3 More realistic models

Trajectory analysis for transits to and from Mars is a complex subject. JPL has carried out realistic detailed analyses of requirements and options for transfers to and from Mars for each mission opportunity (26-month spacings) over the next 30 years or so.

Each mission opportunity includes a "window" of about 2 months during which departure from LEO toward Mars can be implemented with reasonable propulsion requirements. Outside this window, propulsion requirements tend to rise signifi-cantly. Within this ~2-month window, one can select a specific departure date and a Δv imparted for Earth departure. For the higher values of Δv, the trajectory to Mars always remains inside the Mars orbit and the trip times to Mars are minimal (typically 150–250 days, dependent on departure date and Δv). Shorter trips are considered necessary for crew transfers to Mars to reduce exposure to zero-gravity and radiation.

For the lower values of Δv, the trajectories pass outside the Mars orbit and curve back in to rendezvous with Mars (these are called "type 2 trajectories"). These lower energy trajectories may require from 300 to over 400 days for transit to Mars. They may be appropriate for cargo transfers to Mars.

The dependence of Δv and trip time on mission opportunity is significant. Fast trips of 150–200 days' duration may require a Δv of 3.8 km/s to 4.4 km/s for departure from LEO depending on the launch date, and slow trips of 300–400 days' duration may require a Δv of 3.6 km/s to 4.0 km/s for departure from LEO depending on the launch date. Similar variations govern the return flight from Mars.

Since it may typically require a Δv of up to 2.4 km/s for Mars orbit insertion, a more realistic overall Δv for transfer from LEO to Mars orbit is somewhat higher than the prediction of the simple model for minimum energy transfer given in the previous section as 5.6 km/s.

3.3   EARTH TO LOW EARTH ORBIT

In 1999, Dr. Michael Griffin (current NASA Administrator) presented a set of lectures at the University of Wisconsin entitled: "Heavy Lift Launch for Lunar Exploration." He defined the requirements for a Launch Vehicle (LV) for human missions to the Moon. He suggested a Heavy-Lift LV capable of delivering >100 mT to LEO is "highly desirable if not mandatory" in order to send a ~50 mT vehicle on its way toward the Moon.

Bob Zubrin[30] has shown how the LV requirement for lunar missions depends on the mission architecture (number of launches and rendezvous). He described how one can break down the load to be sent to the Moon into four, two, or one package(s) requiring four, two, or one separate launch(es) with assembly in space for the multiple launch cases. He also compared direct return from the surface of the Moon with lunar orbit rendezvous on return. For a four-launch mission, he estimated that the largest single package that must be delivered to LEO is roughly 33 metric tons (mT). For a single-launch direct mission, he estimated that the requirement is about 120 mT. Recent NASA planning for lunar missions appears to require Launch Vehicles that can deliver 125–150 mT to LEO. Similar considerations can apply to Mars missions except that the delivered masses are higher and the propulsion requirements are greater.

In 2004–2005, the Orbital Sciences Corporation carried out a series of economic-based trade studies that concluded that major benefits result from development of a heavy-lift launch system to support crewed lunar and Mars missions.[31]

[30] *Review of NASA Lunar Program Architectures*, Robert Zubrin, Pioneer Astronautics, Zubrin@aol.com, January 10, 2005.
[31] AIAA-2005-2515.

Table 3.4. Cost per kg delivered to LEO according to Zubrin (*loc cit.*). The Lockheed HLV is projected and has not yet been built. It is not clear whether these include amortized development costs.

Vehicle	LEO delivery (kg)	Cost ($M)	kg/$M	$/kg
Pegasus XL	443	13.5	32.8	30,488
Taurus	1,320	19.0	69.5	14,388
Delta IV Medium	8,600	82.5	104.2	9,597
Delta IV Medium plus	13,600	97.5	139.5	7,168
Delta IV Heavy	25,800	155.0	166.5	6,006
Atlas IIAS	8,618	97.5	88.4	11,312
Atlas IIIB	10,718	97.5	109.9	9,099
Atlas V 400	12,500	82.5	151.5	6,607
Atlas V 500	20,050	97.5	205.6	4,864
Lockheed HLV (projected)	*150,000*	*300*	*500*	*2,000*

Robert Zubrin[32] discussed the economics of Launch Vehicles and pointed out that using the basic figure of merit ($ per kg delivered to LEO), larger Launch Vehicles (LVs) are more cost-effective. He provided the data in Table 3.4.

JSC[33] and others[34] have carried out paper studies of hypothetical human missions to Mars that are called "design reference missions" (DRMs). One of the most important conclusions we can draw from the work done so far on Mars DRMs is the fact that the scale of vehicles and operations for a human mission to Mars is innately huge. The challenge in launching massive vehicles, delivering them to Mars, and returning safely from Mars remains as the major impediment to human exploration of Mars. There is also the problem of packaging the mass into elements that can be transported by conceivable Launch Vehicles. In this connection, volume limitations are also likely to be important. In some scenarios, assembly in Earth orbit is required, but the specific pieces to be assembled and the methods for accomplishing this have not been described to any significant degree.

JSC's DRM-1 estimated that a capability to deliver 240 mT to LEO was needed for human missions to Mars if no in-space assembly was used. They suggested four possible options for an LV as shown in Table 3.5.

Because of the cost and difficulty in developing a 240-ton-class Launch Vehicle, consideration was given by DRM-1 to the option of launching several hardware

---

[32] Robert Zubrin, *loc. cit.*
[33] DRM-1: *Human Exploration of Mars: The Reference Mission of the NASA Mars Exploration Study Team*, Stephen J. Hoffman and David I. Kaplan, eds., Lyndon B. Johnson Space Center, Houston, TX, July 1997, NASA Special Publication 6107 DRM-3: NASA/SP–6107–ADD, Reference Mission Version 3.0; addendum to the *Human Exploration of, Mars: The Reference Mission of the NASA Mars Exploration Study Team*, Bret G. Drake (ed.), Lyndon B. Johnson Space Center, Houston, TX, 1998. Link: *http://exploration.jsc.nasa.gov/marsref/contents.html*
[34] *Mars Direct* led by Robert Zubrin, and the *Mars Society DRM* led by James Burke.

Table 3.5. JSC DRM-1 Launch Vehicle concepts.

Option	Payload mass (mT) to 360 km circular orbit	Key technology assumptions
1	179	Modified Energia core with eight Zenith-type strap-on boosters. New upper stage using a single Space Shuttle Main Engine (SSME).
2	209	New core stage based on Space Transportation System (STS) external tank and SSMEs. Seven new strap-on boosters each use a single RD-170 engine. New upper stage using a single SSME.
3	226	New core stage based on STS external tank and four of the new Space Transportation Main Engines. Four strap-on boosters each with a derivative of the F-1 engine used on the first stage of the Saturn V. New upper stage using a single SSME.
4	289	New vehicle using technology derived from the Saturn V launch vehicle. Boosters and first stage use a derivative of the F-1 engine, and the second stage uses a derivative of the J-2 engine.

elements to LEO using smaller vehicles, assembling (attaching) them in space, and then sending the assembled system on the outbound trajectory to Mars. This smaller Launch Vehicle (with a 110 to 120 mT payload capability to LEO) would have the advantage of more reasonable development costs and is within the envelope of capability demonstrated by the unmodified U.S. Saturn V and Russian Energia programs. However, this smaller Launch Vehicle introduces several potential difficulties to the mission scenario. The most desirable implementation using this smaller Launch Vehicle is to simply dock the two elements in Earth orbit and immediately depart for Mars. To avoid boil-off losses in the departure stages (assumed to use liquid hydrogen as the propellant), all elements must be launched from Earth in quick succession, placing a strain on existing launch facilities and ground operations crews. Assembling the Mars vehicles in orbit and loading them with propellants just prior to departure may alleviate the strain on launch facilities, but the best Earth orbit for Mars missions is different for each launch opportunity, so a permanent construction and/or propellant storage facility in a single Earth orbit introduces additional constraints. DRM-1 summarized the situation as follows: "A 240-ton payload-class launch vehicle is assumed for the Reference Mission. However, it is beyond the experience base of any space-faring nation. While such a vehicle is possible, it would require a significant development effort for the launch vehicle, launch facilities, and ground processing facilities; and its cost represents a considerable fraction of the total mission cost. The choice of a launch vehicle remains an unresolved issue for any Mars mission."

JSC's DRM-3 was an addendum to DRM-1 that dealt with the problem of the need for a large mass in LEO by making two important changes: (a) they "scrubbed" the vehicle mass estimates from DRM-1 and succeeded in reducing the required mass in LEO from around 240 mT to about 160–170 mT (although justification for these mass reductions remains unclear), and (b) they postulated a Shuttle-derived LV that could lift 80–85 mT to LEO, and employed on-orbit assembly to achieve the required mass in LEO. This new LV concept was designated as the "Magnum" to differentiate it from the numerous other past Launch Vehicle studies. JSC's "Dual Landers" DRM subsequently raised the Magnum capability to 100 mT Earth-to-orbit (ETO).

Robert Zubrin's Mars Direct DRM hypothesized an advanced Launch Vehicle designated as an "Ares", that was optimized for Earth escape missions. The Ares would be able to send about 47 mT *en route* toward Mars based on lifting about 121 mT to LEO.[35] Mars Direct also mentions a more conservative LV that can deliver 40 mT to trans-Mars injection (TMI) based on 106 mT to LEO. The Mars Direct LV weighed in at around 2,300 mT at takeoff.

The Mars Society Mission (MSM) defined a family of Launch Vehicles with second stages for TMI, scaled to the mass of vehicles involved. The largest Launch Vehicle in the family was capable of sending 55 mT on its way toward Mars. For the largest launch system envisaged by MSM, the mass at takeoff from the Earth was ~2,450 mT. This configuration used a launch system with a dry mass of 264 mT and 1,976 mT of propellants to reach Earth orbit. The upper stage for TMI had a dry mass of 18 mT and used 187 mT of propellants. The delivered mass to LEO was estimated to be about 150 mT.

The 2005 NASA lunar Exploration Systems Architecture Study (ESAS) discussed LV options at some length. The ESAS architectures utilize separate Launch Vehicles for the crew and the cargo for lunar missions, and it is almost certain that the same principle will be applied by NASA to Mars missions. Indeed, the brief description of Mars human missions in the ESAS report[36] utilizes this approach. The two ESAS LVs are:

- *Crew Launch Vehicle (CLV)*—sends the crew to orbit with a capability of delivering about 20–25 mT of payload to LEO.
- *Cargo Launch Vehicle (CaLV)*—sends cargo to orbit with a capability of delivering about 125 mT of payload to LEO.[37]

The ESAS report asserts that a 100–125 mT to LEO LV will suffice for Mars missions (although the evidence for this claim is flimsy).

---

[35] This was based on an optimistic specific impulse of 465 s for Earth departure propulsion.

[36] http://www.spaceref.com/news/viewsr.html?pid=19094

[37] Actually, this LV sends about 145 mT to LEO, but about 20 mT is tied up in a propulsion stage, leaving only about 125 mT available for payload. This propulsion stage is a carry-over from a sub-orbital burn used to place the system into LEO and the same propulsion system is used for Earth departure. Had NASA staged these two propulsion systems, they would have saved a considerable amount of mass by jettisoning the dry propulsion system for orbit insertion prior to Earth departure, but this would have necessitated investment in two propulsion systems, and this would be more costly.

Based on the aforementioned DRMs and the 2005 ESAS study, it appears that NASA is limiting its cargo Launch Vehicle capability to a takeoff mass of roughly 2,500 mT that can transfer perhaps 125–150 mT of useful payload to LEO. It is noteworthy that a 2007 NASA release[38] increased the Ascent and Descent Vehicle masses, so presumably the Launch Vehicles will be expanded accordingly.

3.4 DEPARTING FROM LEO

3.4.1 The Δv requirement

The Δv requirement for trans-Mars injection (TMI) from LEO can be described in the following way.[39] It is desired to send a spacecraft from LEO on its way toward Mars (or the Moon) with a velocity v_∞ after it leaves the Earth.

The orbital speed of a spacecraft in a circular orbit in LEO is

$$v_{orb} = (GM_E/R_{LEO})^{1/2}$$

where G is the gravitational constant, M_E is the mass of the Earth, and R_{LEO} is the radius of the low Earth orbit (measured from the center of the Earth).

If we use a rocket to impart additional speed to the spacecraft to escape from Earth orbit, the speed of the spacecraft must exceed the escape speed of the Earth:

$$v_{esc} = (2GM_E/R_{LEO})^{1/2}$$

The speed imparted to the spacecraft after departing from the Earth's gravitational influence is v_∞ and, for historical reasons, we define:

$$C_3 = (v_\infty)^2$$

The total kinetic energy acquired by the spacecraft in LEO is called E_{tot}. This energy is the sum of the escape energy that is just enough to release the spacecraft from orbit with zero speed, plus the kinetic energy at infinity.

$$E_{tot} = (1/2)m_{SC}[v_{esc}^2 + v_\infty^2]$$

where m_{SC} is the mass of the spacecraft.

Thus, the total speed acquired by the spacecraft is:

$$v_{tot} = [C_3 + v_{esc}^2]^{1/2}$$

The Δv requirement for trans-Mars injection (TMI) (or trans-lunar injection [TLI]) from LEO is the difference between the total speed acquired and the initial speed while in LEO. Thus:

$$\Delta v = v_{tot} - v_{orb}$$

---

[38] "MoonHardware22Feb07_Connolly.pdf" on *NASA Watch* website.
[39] I am indebted to Dr. Mark Adler of JPL who instructed me on this section.

Since $GM_E \sim 398{,}600 \ \text{km}^3/\text{s}^2$, and $R_{LEO} = (6{,}378 + H)$ km,

$$\Delta v = [C_3 + 7.972 \times 10^5/(6{,}378 + H)]^{1/2} - [3.986 \times 10^5/(6{,}378 + H)]^{1/2}$$

For any altitude H (km) and C_3 (km/s)^2, the value of Δv can be calculated. For TMI, appropriate values of C_3 vary with launch opportunity (approximately every 26 months). At each launch opportunity, a map can be constructed that plots arrival date vs. departure date, in which contours of constant C_3 are shown and lines of constant trip time to Mars are also shown. A typical example of such a "pork-chop" plot is shown in Figure 3.14 (see color section) for the 2022 launch opportunity.

It can be seen that the diagram splits into two sections. The upper section of contours is for so-called Type II trajectories in which the spacecraft (on its way to Mars) goes more than $180°$ about the Sun, and the lower section of contours pertains to Type I trajectories where the spacecraft goes less than $180°$ about the Sun. The dividing line between these is the Hohmann trajectory that goes exactly $180°$ around the Sun. While simple models based on circular orbits and planar trajectories (see Section 3.2) predict that the Hohmann $180°$ trajectory has the lowest energy requirement, this is not the actual case. The lowest energy trajectory with a reasonable launch window for cargo delivery in 2022 is a Type II trajectory located in the small central ellipse of the upper set of contours with a $C_3 \sim 14.5$ (km/s)^2, and a trip time of about 400 days. However, this trip time can be reduced to $\sim$350 days if C_3 is increased to $\sim$17 (km/s)^2. The best fast trajectory for crew transport with a reasonable launch window is a Type I trajectory near the center of the innermost ellipse in the lower contours with a C_3 of about 22 (km/s)^2 with a trip time of about 175 days.

3.4.2 Mass sent toward Mars

The calculated values of Δv with $H \sim 200$ km are:

Lowest energy cargo delivery (400 days)	3.86 km/s
Cargo delivery in 350 days	3.97 km/s
Crew delivery in 175 days	4.18 km/s

As we have said, these pertain to only one launch opportunity and there is considerable variation from opportunity to opportunity. Review of the "pork-chop" plots for opportunities 2009 through 2026 provides the data shown in Table 3.6. However, no account was taken of arrival time, and constraints to arrive during daylight with a communication link to Earth could further increase Δv requirements beyond those listed in Table 3.6. It can be shown that as the trip time is reduced by applying a higher propulsive impulse, the payload fraction of mass in LEO decreases.

The mass sent toward Mars as a percentage of mass in LEO for an LH$_2$/LOX trans-Mars injection propulsion system with $I_{SP} = 450$ s and dry stage mass $= 13\%$ of propellant mass is given in Table 3.7.

Table 3.6. Characteristics of Earth departure steps for fast and slow trips to Mars at various launch opportunities.

Year	Lowest energy Type II trip			Low energy Type II trip			Fast Type I trip			Fastest Type I trip		
	C_3 (km/s)²	Trip time (days)	Δv (km/s)	C_3 (km/s)²	Trip time (days)	Δv (km/s)	C_3 (km/s)²	Trip time (days)	Δv (km/s)	C_3 (km/s)²	Trip time (days)	Δv (km/s)
2009	11	325	3.71	12	300	3.76	23	175	4.22			
2011	10	300	3.67	11	275	3.71	20	175	4.10			
2013	10	325	3.67	13	275	3.80	15	175	3.89			
2016	9	300	3.63				12	175	3.76	15	150	3.89
2018	14	280	3.84				12	175	3.76	15	150	3.89
2020	18	400	4.01	23	340	4.22	16	175	3.93	20	150	4.10
2022	14.5	400	3.86	17	350	3.97	22	175	4.18			
2024	13	350	3.80	16	320	3.93	22	200	4.18	28	175	4.43
2026	11	300	3.71	12.5	275	3.78	17.5	200	3.99	23	175	4.22

Table 3.7. Mass sent toward Mars as % of mass in LEO for a trans-Mars injection propulsion system with $I_{SP} = 450$ s and dry stage mass $= 13\%$ of propellant mass.

Year	Lowest energy Type II trip	Low energy Type II trip	Fast Type I trip	Fastest Type I trip
2009	36	35	30	
2011	36	35	31.5	
2013	36	35	32.5	
2016	37		35	32.5
2018	34	35	32.5	
2020	32.5	30	33	31.5
2022	34	32.5	30	
2024	35	33	30	28
2026	36	35	32.5	30

3.4.3 Nuclear thermal rocket for TMI

The baseline propulsion system used for trans-lunar (or trans-Mars) injection is a LOX/LH$_2$ propulsion stage atop the Launch Vehicle. For transfers from LEO toward the Moon (or Mars) using LOX/LH$_2$ propulsion, roughly 55% (or 65%) of the mass in LEO is required for the propellant and the propulsion stage, and ~45% (or 35%) of the mass in LEO consists of payload that is sent on its way to the Moon (or Mars).[40] For Mars, as we have seen, the actual propulsion requirements depend on several factors (e.g., the specific launch opportunity and the desired duration of the trip to Mars). One can either use a lower-energy trajectory with a trip time of typically 300–400 days that requires less propellant (appropriate for cargo transfer), or a higher energy trajectory that uses more propellant with a trip time of typically 170 to 200 days (appropriate for crew transfer). LOX/LH$_2$ is the most efficient form of chemical propulsion that is available. The technology for use of LOX/LH$_2$ propulsion for Earth departure is fairly mature, having been used on the Space Shuttle for years.

Despite the fact that LOX/LH$_2$ is the most efficient form of chemical propulsion, the requirement that ~3 mass units in LEO are required to send 1 mass unit on its way toward Mars is a major factor in driving up the IMLEO for Mars missions. To partly mitigate this onerous requirement, JSC mission planners have proposed using a form of exotic propulsion instead of chemical propulsion for Earth departure in their design reference missions (DRMs). In DRM-1 and DRM-3,[41] a nuclear thermal rocket (NTR) was conjectured. More recently, the ESAS report continues to place great importance on the NTR.[42] Use of an NTR instead of chemical propulsion for

---

[40] Therefore, it requires about 2.8 mass units in LEO to send 1 mass unit on its way to Mars.
[41] These are available at *http://exploration.jsc.nasa.gov/marsref/contents.html*
[42] The ESAS architecture does not address the Mars phase in detail, but it says "it is recognized that traditional chemical propulsion cannot lead to sustainable Mars exploration with humans. Nuclear Thermal Propulsion (NTP) is a technology that addresses the propulsion gap for the human Mars era."

Table 3.8. Estimated payload reduction to circular orbit vs. altitude.

Altitude (km)	Circular orbit payload	% Reduction from 200 km
250	3,220	1.9
490	2,960	9.8
750	2,740	16.5
1,000	2,590	21.1
1,250	2,495	24.0

trans-Mars (or lunar) injection from LEO would significantly increase the payload fraction in LEO by doubling the specific impulse from 450 s to 900 s, but the increase in payload fraction will be limited by the dry mass of the NTR system including the nuclear reactor and hydrogen storage.

DRM-1 assumed a hydrogen propellant mass of 86 mT and an NTR dry mass[43] of 28.9 mT (dry mass ~34% of propellant mass). DRM-3 used a hydrogen propellant mass of 45.3 mT and a 23.4 mT NTR dry mass (dry mass ~52% of propellant mass). Robert Zubrin's "Mars Direct" provides a formula that would have indicated more optimistic estimates of 20 mT and 12 mT for the NTR dry mass of these systems. Further, it is not clear how the large amount of hydrogen propellant would be stored. The volumes of hydrogen implied by these masses are 123 and 65 cubic meters.

Since there is considerable uncertainty regarding the NTR dry mass percentage, we can treat this in terms of a parameter:

$$(\text{dry mass}) = K(\text{propellant mass})$$

where K is an unknown parameter that might be somewhere between 0.2 and 0.6.

In addition to the problem of the NTR dry mass, another problem in the use of the NTR is the fact that, for safety reasons—both real and imagined, it is likely that public policy will require that the NTR be lifted to a higher Earth orbit before it is turned on. However, this requires that the Launch Vehicle burn more propellants, and therefore the net benefit of using the NTR will be reduced compared with firing it up in LEO. In fact, the ESAS report indicates that it would be lifted to 800–1,200 km altitude rather than the typical starting point of ~200 km altitude of LEO. The estimated reduction in payload lifted to various Earth orbit altitudes is given in Table 3.8. For example, the mass lifted to 1,000 km altitude is about 80% of that which could be lifted to 200 km altitude. Table 3.9 shows the fraction of mass originally in a 200 km LEO that can be sent on a fast trajectory to Mars in 2022. Using chemical propulsion and departing from 200 km LEO, 31% of the mass can be sent toward Mars. The percentage for NTR propulsion depends on K and the altitude of start-up. At the likely conditions of $K \sim 0.5$ and start-up at 1,250 km, the NTR produces a very small improvement (33%). This small improvement would come at an enormous cost to develop, test, and validate the NTR. The benefit/cost ratio is

---

[43] Here, the term "dry mass" refers to the sum of the masses of the nuclear reactor and rocket assembly plus the empty hydrogen propellant tanks.

Table 3.9. Fraction of total mass in 200 km LEO that can be sent toward Mars. These figures pertain to a fast flight at the 2022 opportunity.

Altitude at start-up (km)	Chemical propulsion	NTR with various K				
		0.2	0.3	0.4	0.5	0.6
200	0.31	0.55	0.51	0.47	0.43	0.40
1,000	—	0.45	0.42	0.39	0.36	0.33
1,250		0.42	0.39	0.36	0.33	0.30

clearly unfavorable, but JSC has resolutely included the NTR in its plans for DRM-1, DRM-3, and in 2005–6 the ESAS report.

In a NASA presentation in late 2006, it was proposed to utilize the NTR not only for Earth departure, but also for Mars orbit insertion and Earth return from Mars orbit. This would entail storing many tens of mT of hydrogen for up to two years in space. Exactly how this would be accomplished was not revealed.

3.4.4   Solar electric propulsion for orbit raising

One of the major requirements for propellants in any space exploration venture is for departing from the gravitational influence of the Earth. While most mission concepts call for an Earth departure propulsion stage to fire up in LEO, another conceivable alternative is to use solar electric propulsion to raise the orbit of the spacecraft from LEO to a high Earth orbit, and thereby greatly reduce the propulsion requirements for Earth departure from this high orbit that is outside most of the Earth's gravitational influence. The question then arises as to the requirements for the solar electric propulsion system, and whether such a system is feasible and desirable.

In a series of design reference mission studies for human missions to Mars, JSC had to deal with the inevitable problem of sending large masses to Mars, which required much larger masses in LEO. During the late 1990s, their studies utilized a nuclear thermal rocket (NTR) for Earth departure from LEO to reduce the IMLEO. However, in the "Dual Landers" study (c. 2000) they eliminated the NTR and used solar electric propulsion (SEP) for orbit raising instead. It is not clear why this change was made, but presumably it was because of uncertainties in the political viability of firing up an NTR in LEO, as well as concerns about the cost of developing the NTR.

In the Dual Landers concept, a "tug" powered by solar electric propulsion is used to lift space vehicles from LEO to an elongated elliptical Earth orbit because trans-Mars injection with chemical propulsion requires far less propellant from this orbit than it would from LEO. The energy that would have been used for departure from LEO using chemical propulsion is mostly replaced by solar energy that drives the electric propulsion system used to raise spacecraft to a high orbit. The documentation of the Dual Landers mission is sparse, and it is difficult to appraise the feasibility of

Table 3.10. Estimated masses of SEP orbit-raising system for 50 mT payload. (Woodcock, *loc. cit.*)

Up trip time	240 d
Return trip time	61 d
Array area	2,000 m^2
Payload accommodation mass	5,000 kg
Array mass	2,500 kg
Thruster mass	1,000 kg
PPU and cabling mass	2,000 kg
Propellant tank mass	2,060 kg
Structure mass	4,483 kg
Inert mass	19,255 kg
Return cutoff mass	19,255 kg
Up propellant	31,394 kg
Return propellant	7,841 kg
Unusable propellant	1,962 kg
Total propellant	41,196 kg
Total initial mass in LEO	110,451 kg

the masses used in this mission concept. The slow spiraling out of the SEP tug (several months required for transfer) creates time delays and operational scheduling difficulties. Because the SEP tug drags the vehicles slowly through the radiation belts, the crew would have to wait until the Trans-Habitat Vehicle reached HEO before using yet another vehicle, a fast "crew taxi" to rendezvous with the Trans-Habitat Vehicle.

Woodcock[44] describes a hypothetical SEP system for orbit raising of heavy loads. This reference utilizes a payload of 50 mT driven by a 500 kW solar electric propulsion system with a specific impulse of 2,000 s. The trip time (up) is 240 days and (down) is 60 days. The critical parameters of the propulsion system were estimated as:

Thruster mass	2 kg/kW
Power processing mass	4 kg/kW
Array mass	143 W/kg = 250 W/m^2
Array areal density	1.8 kg/m^2

These are ambitious figures, but may possibly be achievable some day. The masses of various elements are summarized in Table 3.10. The required amount of Xe propellant per transfer is 41.2 mT. According to estimates on the Internet, world production of Xe is presently 10×10^6 liters/yr = 53 mT/yr. Thus, one transfer would require approximately the present annual world production of Xe. Furthermore, Xe

[44] "Controllability of Large SEP for Earth Orbit Raising," Gordon Woodcock, *40th AIAA/ ASME/SAE/ASEE Joint Propulsion Conference and Exhibit, July 11–14, 2004, Fort Lauderdale, FL*, American Institute of Aeronautics and Astronautics, AIAA 2004-3643.

presently costs about \$10/liter, so the cost of Xe for one orbit transfer could be \$100M. While it may be possible to increase world production significantly, recent articles on anesthesiology suggest difficulties. The viability of the SEP tug concept depends critically on use of a hypothetical high-efficiency lightweight solar array that is likely to be difficult to develop, and lightweight propulsion components. Furthermore, it seems unlikely that the required amount of xenon propellant could be obtained, and, if obtainable, whether the cost would be affordable. Radiation would gradually diminish the efficiency of the solar arrays with each passage through the radiation belts. The total cost of the system includes the SEP tug, the mission operations involved, and the Fast Transit Vehicle to take the crew up to high orbit for rendezvous with the Trans-Habitat Vehicle. Because of the long time required for transfer, several of these "tugs" might be needed.

At this point, it seems unlikely that such a scheme would be viable.

3.5   MARS ORBIT INSERTION

Based on the "pork-chop" plots, it is estimated that planetocentric approach velocities (v_∞) to Mars can vary from 2.5 to 3.8 km/s for slow trips and up to 7.2 km/s for fast trips, depending on the launch opportunity. This will affect the Δv required for propulsive orbit insertion as well as the aeroshell requirements for aerocapture at Mars.

A very rough estimate can be made of the effect of entry velocity on aeroshell requirements for aerocapture at Mars.[45] The following simplifying assumptions are adopted:

1. It is assumed that there is no backshell. This requires an aeroshell large enough to prevent the wake from extending around the shoulder of the aeroshell and impinging on the vehicle being protected. This avoids the need to guess how the backshell would change with entry velocity, which is very complicated. This assumption is made only for purposes of calculation. In reality, it is likely that a backshell will be required.
2. The vehicle has a typical medium-to-high ballistic coefficient—for example, 100 to several hundred kg/m^2 (i.e., it is not an inflatable).
3. The thickness and therefore the areal density of the thermal protection material is made proportional to the total heat load. This ignores non-linear effects.
4. The total heat load is proportional to the vehicle energy dissipated by the aerocapture process. This assumption is a significant oversimplification due to (1) increased radiative heating at higher entry velocities, and (2) the likelihood of use of different strategies to balance heat rate against heat load at higher entry velocities by (a) adjustment of the entry flight path angle, (b) adjustment of the flight path with lift, and (c) greater use of reradiating energy, etc. Perhaps even

---

[45] The author is indebted to Mark Adler of JPL for instruction on this section.

more significantly, a completely different ablative material might be used for different entry velocities.

5. The structural mass of the aeroshell is proportional to the mass of the thermal protection material.
6. Ballast is not required to effect positive separation of the heat shield. (If it was, one could convert ballast mass to thermal protection material mass without penalty.)

We are interested in the energy that must be dissipated to inject an incoming spacecraft into Mars orbit by aerocapture. To perform this calculation, we first calculate the energy of the spacecraft after it is in Mars orbit, and subtract that energy from the energy of a spacecraft when it is approaching Mars from a distance. Although this calculation is oriented toward aerocapture, it is also relevant to estimation of the Δv for propulsive orbit insertion.

First, we calculate the energy of the spacecraft while it is in Mars orbit. The equation of force balance while a spacecraft is in circular orbit is

$$mv_O^2/R = GMm/R^2$$

or

$$v_O^2 = GM/R$$

where $m =$ spacecraft mass
$M =$ planet mass
$G =$ gravitational constant
$R = r_{PL} + H$
$r_{PL} =$ planet radius
$v_O =$ the orbital velocity, and
$H =$ altitude of orbit.

While in a circular orbit, the energy of the spacecraft is

$$E_O = mv_O^2/2 - GMm/R = -\tfrac{1}{2}GMm/R$$

for a circular orbit, and for an elliptical orbit, it is:

$$E_O = -GMm/(R_A + R_P)$$

where the subscripts A and P refer to apoapsis and periapsis, respectively.

When the spacecraft is distant from the planet and approaching Mars, the energy is

$$E_\infty = mv_\infty^2/2$$

where v_∞ is the approach velocity of the spacecraft before it reaches the gravitational influence of the planet.

Thus, the change in energy due to the aerocapture process into elliptical orbit is

$$\Delta E = E_\infty - E_O = m\{v_\infty^2/2 + GM/(R_A + R_P)\}$$

Some important constants are:

$$G = (6.6742 \pm 0.0010) \times 10^{-20} \text{ km}^3 \text{ kg}^{-1} \text{ s}^{-2}$$

$$M_{\text{Mars}} = 0.642 \times 10^{24} \text{ kg}$$

$$M_{\text{Earth}} = 5.97 \times 10^{24} \text{ kg}$$

$$r_{\text{Mars}} = 3,400 \text{ km}$$

$$r_{\text{Earth}} = 6,380 \text{ km}$$

For Mars, $GM = 42,800 \text{ km}^3 \text{ s}^{-2}$. Therefore

$$\Delta E = m\{v_\infty^2/2 + 42,800/(R_A + R_P)\}$$

As an illustration, suppose a spacecraft is captured into an orbit with altitude $H = 300$ km at apoapsis, and $H = 50$ km at periapsis (a likely scenario). Then $(R_A + R_P) \sim 7,150$ km and

$$\Delta E = m\{v_\infty^2/2 + 6.0\}$$

As stated at the beginning of this section, planetocentric approach velocities (v_∞) to Mars can vary from 2.5 to 3.8 km/s for slow trips and up to 7.2 km/s for fast trips, depending on the launch opportunity. Over this range, the change in energy due to the aerocapture process into elliptical orbit (ΔE) varies from 9.1 m to 31.9 m, a dynamic range of ~ 3.5. The heating rate is dependent on the energy dissipated in capture, and this simplistic calculation indicates that the requirements for the aero-shell could vary significantly from opportunity to opportunity, and from cargo delivery to crew delivery within the same opportunity.

In the jargon of entry, descent, and landing (EDL) technology, the "approach velocity" (v_∞) is the velocity of the spacecraft approaching a planet outside the gravitational influence of the planet. If the spacecraft enters the Mars environment (direct entry), it will speed up because it is in the gravitational well of the planet. This velocity that it reaches at an altitude of 125 km is called the "entry velocity" (v_E).

When the spacecraft is distant from the planet, the energy is $mv_\infty^2/2$. In direct entry, the spacecraft proceeds directly in toward Mars. When it reaches ~ 125 km altitude, it may be construed as "entering" the atmosphere, and the "entry velocity" can be computed at this altitude. At this point, conservation of energy requires that

$$mv_E^2/2 - GMm/R = mv_\infty^2/2$$

Therefore

$$v_E^2 = v_\infty^2 + 2GM/R$$

At 125 km altitude on Mars, this reduces to

$$v_E^2 = v_\infty^2 + 24 \text{ (km}^2/\text{s}^2)$$

As noted previously, based on the "pork-chop" plots, it is calculated that planetocentric approach velocities (v_∞) to Mars can vary from 2.5 to 3.8 km/s for slow trips and up to 7.2 km/s for fast trips, depending on the launch opportunity. Thus, the entry velocity is likely to range from 5.5 km/s to 8.7 km/s. The effect of entry velocity on heating rate and EDL mass is difficult to quantify but it is likely to be important.

Future vehicles for human missions to Mars will require "pinpoint landing" to land within some radius of previously deployed surface assets. The requirements for precision in this regard have not been determined, but something less than 100 m seems to be appropriate. In carrying out these steps, there is a maximum deceleration limit that is necessary to maintain crew health and performance during the aero-braking maneuver. JSC's DRM-3 estimated this to be $5g$. The cargo lander can presumably be allowed to undergo higher accelerations. The propulsion requirements for pinpoint landing also need to be folded in.

For propulsive Mars orbit insertion, we may regard the process as the inverse of Mars orbit departure. The Δv for Mars orbit departure is estimated in Section 3.7 and tabulated in Table 3.12.

3.6   ASCENT FROM THE MARS SURFACE

Propellant mass requirements for ascent from Mars surface to Mars orbit depend on (1) mass of the ascent system, (2) propulsion system used, and (3) orbit to which transfer is made for rendezvous. According to the rocket equation, the propellant requirement is determined by:

$$M_P = M_D * \{\exp(\Delta v/(9.8 * I_{SP})) - 1\}$$

where $M_P =$ propellant mass;
         $M_D =$ mass of capsule + dry propulsion system + crew;
         $I_{SP} =$ specific impulse (sec).

For methane–oxygen an optimistic guess for I_{SP} is 360 s.

Δv for ascent to a circular orbit is $\sim$4,300 m/s and to an elliptical orbit it is $\sim$5,600 m/s. Thus, for ascent to circular or elliptical orbits, M_P/M_D is 2.38 or 3.89, respectively. That makes a huge difference. ISRU advocates using an elliptical orbit, thus saving on Δv for Mars orbit insertion (MOI) at the expense of higher ascent propellant requirements (supplied by ISRU). Anti-ISRU advocates using a circular orbit to minimize ascent propellants and accept a higher propellant requirement for MOI. Finally, we need estimates for ascent dry mass. Three JSC DRMs have made such estimates and these are summarized in Table 3.11. However, note that—as Table 3.1 shows—the ratio of ascent stage to ascent propellants for the lunar Ascent Vehicle is about 0.25, and the estimates provided by DRM-1 and DRM-3 for the ascent stage mass appear to be optimistic.

The Mars Direct DRM and the Mars Society DRM went directly from the Mars surface back to Earth, and the Ascent Vehicle thereby incurred the need for life support for >6 months, a more significant Habitat, additional Δv for Earth return, and an entry system for use at Earth. Mars Direct estimated the ascent propellant mass to be 96 mT and the Mars Society DRM estimated 136 mT.

Table 3.11. Ascent systems from JSC DRMs. Masses are in kg.

	JSC DRM-1	JSC DRM-3	JSC Dual Landers
Crew Module	5,500	4,829	2,066
Ascent propulsion stage	2,550	4,069	3,716
Crew	600	600	600
Other			
Total dry mass	8,650	9,500	6,380
Destination	Elliptical Mars orbit	Elliptical Mars orbit	Circular Mars orbit
Propellant mass from source	26,000	39,000	15,600
Propellant mass calculated herein	33,650	37,000	15,200
Stage mass/Propellant mass	0.08	0.11	0.24

3.7 TRANS-EARTH INJECTION FROM MARS ORBIT

The procedure used in Section 3.4.1 can be used here to estimate Δv for trans-Earth injection from Mars orbit, except that the mass and radius of Mars must be used instead of Earth. The Δv requirement for trans-Earth injection (TEI) from a circular Mars orbit is

$$\Delta v = v_{\text{tot}} - v_{\text{orb}}$$

Since $GM_E \sim 42,650 \text{ km}^3/\text{s}^2$, and $R_{LMO} = (3,397 + H)$ km,

$$\Delta v = [C_3 + 8.53 \times 10^4/(3,397 + H)]^{1/2} - [4.265 \times 10^4/(3,397 + H)]^{1/2}$$

For any altitude H (km) and C_3 (km/s)^2, the value of Δv can be calculated. The values in Table 3.12 are for a 300 km circular Mars orbit.

3.8 EARTH ORBIT INSERTION

According to "pork-chop" plots, the Earth-centric velocity (v_∞) at which the returning spacecraft encounters Earth varies from 3.2 to 4 km/s for slow cargo trips, and up to 5–10 km/s for the fastest crew return trips, depending on the opportunity. This could require very different aeroshell masses, depending on entry velocity. The formula derived previously for Mars:

$$v_E^2 = v_\infty^2 + 2GM/R$$

when applied to Earth return at 125 km altitude reduces to

$$v_E^2 = v_\infty^2 + 123 \text{ (km}^2/\text{s}^2)$$

At the high end of v_∞ (about 10 km/s), the entry velocity is about 15 km/s.

Table 3.12. Characteristics of fast and slow trips to Earth from Mars orbit at various return opportunities.

Year	Lowest energy trip			Low energy trip			Fast trip			Fastest trip		
	C_3 (km/s)²	Trip time (days)	Δv (km/s)	C_3 (km/s)²	Trip time (days)	Δv (km/s)	C_3 (km/s)²	Trip time (days)	Δv (km/s)	C_3 (km/s)²	Trip time (days)	Δv (km/s)
2020	16	260	2.85				13	185	2.61	18	150	3.01
2022	13	282	2.61	18	250	3.01	20	175	3.17	27	150	3.68
2024	10	300	2.35				18	200	3.01	24	175	3.46
2026	8.5	325	2.22	10	300	2.35	18	200	3.01	25	175	3.54

3.9 GEAR RATIOS

3.9.1 Introduction

Almost any conceivable human mission to Mars involves transfer of some assets to
Mars orbit, and some assets to the Mars surface. A widely accepted surrogate for
estimating mission cost for human missions to Mars is the required initial mass in low
Earth orbit (IMLEO). This, in turn can be calculated by estimating how much mass
must be delivered to Mars orbit (M_{MO}) and how much mass must be delivered to the
Mars surface (M_{MS}), and multiplying each figure by its appropriate "gear ratio":
mass required in LEO to deliver one mass unit to Mars orbit or the Mars surface.

The overall gear ratio for transfer from LEO to Mars orbit or surface can be
subdivided into a product of subordinate gear ratio factors for each step along the
way. Gear ratios for propulsive steps can be estimated from the rocket equation. Gear
ratios will be estimated for various mission steps using several chemical propulsion
systems. For aero-assisted orbit insertion and entry, descent, and landing, the recent
models developed by B. Braun and the Georgia Tech Team were used to estimate
entry system masses (see Section 4.6).

The gear ratio for propulsive transfer from LEO to Mars orbit depends on
whether the orbit is elliptical or circular. The gear ratio is greater for a circular orbit.
However, when ascending from the Mars surface to rendevous in Mars orbit, the gear
ratio is considerably higher for the elliptical orbit.

When nuclear thermal propulsion (NTP) is utilized for Earth departure, two
important factors are the minimum altitude allowed for start-up and the propulsion
system dry mass fraction. We have parameterized these and estimated gear ratios for
a range of values. If NTP can be fired up in LEO, and if the dry mass fraction is as low
as 0.2 to 0.3, the use of NTP in place of LOX/LH$_2$ for Earth departure can reduce
IMLEO by ~40%. However, if the NTP must be raised to an altitude of over
1,000 km prior to start-up, and if the dry mass fraction is ~0.5, the benefit of using
NTP diminishes to almost nothing (see Table 3.9).

In situ resource utilization (ISRU) on Mars for propellant production can reduce
the landed mass (M_{MS}) and, if an elliptical orbit is used, it can also reduce the
propellant mass required for orbit insertion and Earth return. Mass savings from
use of ISRU are estimated to be significant.

Finally, a recent (2006) NASA Mars mission plan was analyzed in terms of gear
ratios and it was found that the estimated value of IMLEO was ~1,600 metric tons
(mT), whereas the NASA claim is that IMLEO = 446 mT. NASA mass estimates for
human missions to Mars appear to be overly optimistic.

3.9.2 Gear ratio calculations

All previous NASA design reference missions for human missions to Mars involved
transfer of some assets (mass = M_{MO}) to Mars orbit, and some assets to the Mars
surface (mass = M_{MS}). The optimal choice of the Mars orbit depends on (1) whether
in situ resource utilization (ISRU) is used to produce ascent propellants from indi-

genous resources, and (2) the relative values of M_{MO} that dictate propellant requirements for orbit insertion and departure, and the mass of the Ascent Capsule that determines the mass of propellants needed for ascent. If an elliptical orbit is used, the propellant requirements for orbit insertion and orbit departure are lower, but the propellant requirement for ascent to orbit is much higher. If ISRU is utilized, the elliptical orbit is certainly preferred because the ascent propellants are produced on Mars and need not be delivered from Earth. If ISRU is not utilized, the circular orbit may be preferable if the Ascent Capsule is sufficiently massive, because the disadvantage of high ascent propellant mass would outweigh the benefits of ease of orbit insertion and departure. However, if the Ascent Capsule has sufficiently low mass compared with the mass of assets delivered to Mars orbit, the elliptical orbit would still be preferred for the non-ISRU case because the benefits of ease of orbit insertion and departure would outweigh the penalty of delivering a greater amount of ascent propellants from Earth.

G_{MO} = Mass required in LEO to deliver 1 mass unit of payload to Mars circular orbit

G_{MS} = Mass required in LEO to deliver 1 mass unit of payload to the Mars surface

Thus, for assets transported to Mars orbit:

$$\text{IMLEO}_{MO} = M_{MO}G_{MO}$$

and for assets transported to the Mars surface:

$$\text{IMLEO}_{MS} = M_{MS}G_{MS}$$

and the total IMLEO is:

$$\text{IMLEO} = \text{IMLEO}_{MO} + \text{IMLEO}_{MS}$$

Simple algorithms are derived for estimating IMLEO for any combination of values of M_{MO} and M_{MS} for several propulsion systems and scenarios.

In recent NASA scenarios for human missions to Mars, transport from LEO to the Mars surface proceeds through the intermediate step of inserting into Mars circular orbit prior to descent and landing.[46] While it is true that the Mars Direct[47] and Mars Society[48] design reference missions utilized direct return from Mars (without rendezvous in Mars circular orbit), and the MIT Study (Wooster *et al.*, *loc cit.*) indicated that direct return has advantages, recent NASA documents

---

[46] Anonymous (2005), "Exploration Systems Architecture Study (ESAS)," NASA-TM-2005-214062, *www.sti.nasa.gov*, November 2005; Anonymous (2006), "Project Constellation presentation on Mars mission architectures," attributed to D. Cooke.

[47] *Practical Methods for Near-Term Piloted Mars Mission*, R. M. Zubrin and D. B. Weaver, American Institute of Aeronautics and Astronautics, AIAA-93-20898, 1993; *Mars Direct: A Simple, Robust, and Cost Effective Architecture for the Space Exploration Initiative*, R. M. Zubrin, D. A. Baker, and O. Gwynne, American Institute of Aeronautics and Astronautics, AIAA-91-0328, 1991.

[48] *A New Plan for Sending Humans to Mars: The Mars Society Mission*, Christopher Hirata, Jane Greenham, Nathan Brown, Derek Shannon, and James D. Burke, California Institute of Technology, 1999, Jet Propulsion Laboratory, informal report.

(Anonymous, *loc cit.*) suggest that this approach is very unlikely to be adopted by NASA. If we define the gear ratio for descent and landing from Mars circular orbit to be:

G_{DL} = Mass required in Mars orbit to deliver 1 mass unit to the Mars surface

then:

$$G_{MS} = G_{DL}G_{MO}$$

The gear ratio for transport from LEO to Mars orbit is:

$$G_{MO} = G_{ED}G_{MOI}$$

where G_{ED} = mass required in LEO to implement Earth departure of 1 mass unit to a trans-Mars trajectory;
G_{MOI} = mass required in a trans-Mars trajectory to insert 1 mass unit into Mars circular orbit;

and requirements for (minor) mid-course corrections are neglected.

The gear ratio for transfer from approaching Mars to the Mars surface is:

$$G_{ML} = G_{MOI}G_{DL}$$

3.9.3   Gear ratio for Earth departure

The gear ratio for Earth departure to a trans-Mars trajectory was discussed in Section 3.4. The required value of Δv varies with launch opportunity, launch date within any opportunity, and the desired trip time to Mars. A fast "Type 1" trajectory for crew transfers may require up to about 4,200 m/s whereas a slower low-energy "Type 2" trajectory for cargo transfers may require perhaps 3,900 m/s (these are representative values). Estimates of Δv are taken from JPL internal reports showing "pork-chop" plots of Δv for various trip times for various departure dates. The specific impulse of a LOX/LH$_2$ departure stage may be taken as roughly 450 seconds. Guernsey *et al.* (*loc. cit.*) estimated that the ratio of dry propulsion system mass to propellant mass, $K \sim 0.11$ for space-storable propellants. Larson and Wertz (1999) indicate that for cryogenic Earth departure propulsion systems, $K \sim 0.11$. However, for longer-term use, we expect K to be higher so we have estimated K to be about 0.12 for Mars space transfers. With these values, we find that for fast crew departures, $G_{ED} \sim 3.2$ and for slow cargo departures, $G_{ED} \sim 2.9$. The impact of using nuclear thermal propulsion for Earth departure was discussed in Section 3.4.3.

3.10   LEO TO MARS ORBIT

The simplest (but not necessarily best) method for Mars orbit insertion involves retrofiring of a rocket to slow down the spacecraft approaching Mars until it falls into orbit (see Section 3.2.1.9). As the spacecraft approaches Mars, application of the retro-rocket places the spacecraft into an elongated elliptical orbit. Further propulsion burns can modify this orbit and transform it to a ~400 km circular orbit that is convenient for many purposes in remote sensing of Mars from orbit. The total Δv for insertion into a circular orbit is typically about 2.4 to 2.5 km/s (depending on launch date and trajectory) of which about two-thirds is to establish the elliptical orbit and roughly one-third is for circularization of the elliptical orbit.

There are several possibilities for the propulsion system used for Mars orbit insertion. These are:

(1) *Space-storable propellants* (NTO and MMH). These propellants are liquids at room temperature and can be stored on a spacecraft without requiring cryogenic tanks. They are relatively dense and therefore minimize the required volume. However, their relatively modest specific impulse (about 320 s) requires significant amounts of propellant.

(2) *Cryogenic LOX and methane*. These propellants produce a higher specific impulse (about 360 s), but they must be stored at around 100 K ($-173°$C). This, in turn, requires some allowance for boil-off from a passive storage system during the 6 to 9 month transit to Mars, or, possibly, use of an active refrigeration system to prevent boil-off (with its attendant mass penalty, complexity, and risk).

(3) *Cryogenic LOX and LH₂*. These propellants produce a specific impulse of about 450 s, but there is a challenge in storing hydrogen at ~30 K ($-243°$C) by either accepting some boil-off during the 6 to 9 month transit to Mars or using a larger active refrigeration system to prevent boil-off. Either way, this will add a mass penalty, complexity, and risk. It should also be noted that the use of LH_2 introduces some issues regarding volume because of its low density.

The effect of boil-off during the transit from LEO to Mars is to increase the required mass in LEO by the amount of propellant (plus additional mass for a larger dry propulsion system) that boils off.

There are two other alternatives to propulsive orbit insertion at Mars: "aerobraking" and "aerocapture." Aerobraking is a very gradual process that has the advantage that small reductions in spacecraft velocity are achieved by drag of the solar arrays in the outer atmosphere, so no additional mass for a heat shield is necessary. However, the process is slow and takes several months of intensive management. Aerocapture is very rapid, but has the disadvantages that a heavy heat shield is needed and the resultant g-forces may be high.

Aerobraking requires prior use of retro-propulsion to insert the spacecraft into an elongated elliptical orbit. Aerobraking then employs atmospheric drag to reduce orbit energy and thereby decrease the long axis of the ellipse by repeated passes through the atmosphere near the short axis of the ellipse. Thus, aerobraking saves the propellant needed to convert the initial elliptical orbit to a circular orbit. The primary drag surface is typically the solar array panels. Aerobraking has been used on a number of missions at Earth, Venus, and Mars. Despite advances in aerobraking automation, aerobraking remains a human-intensive process that requires 24/7 maintenance for several months with close cooperation between navigation, spacecraft, sequencing, atmosphere modeling, and management teams. Based on previous mission experience, it is estimated that—using aerobraking—the ratio of mass in transit to Mars to the mass injected into Mars orbit drops to about 1.6 (the entry system has a mass of roughly $0.6/1.6 \sim 38\%$ of mass injected into orbit) and the ratio of mass in LEO to mass injected into Mars orbit drops to $\sim$4.5. If the Mars orbit is an elliptical orbit, aerobraking is not applicable.

Aerocapture begins with a shallow approach angle to the planet. Descent into the relatively dense atmosphere is sufficiently rapid that the deceleration causes severe heating that requires a heat shield (aeroshell). The trajectory of the inbound spacecraft is bent around Mars and as the spacecraft exits the atmosphere, the heat shield is jettisoned and a propulsive maneuver is performed to raise the short axis of the resulting elliptical orbit. The entire operation is short-lived and requires the spacecraft to operate autonomously while in the planet atmosphere. Later, a final propulsive maneuver is used to adjust the orbit to make it circular. Demands placed on the vehicle depend greatly on the specifics of the planet being approached and the mission. Key variables include atmospheric properties, desired orbit insertion geometry, interplanetary approach accuracy, entry velocity, and vehicle geometry. Aerocapture has been proposed and planned several times, and was even partly designed for the MSP 2001 Orbiter, but it has never actually been implemented. Aerocapture was considered for the MSP 2001 Orbiter, but, finally, it was decided to use initial propulsive capture followed by aerobraking to a circular orbit. A critically important unknown for aerocapture is the expected value of the ratio of mass in transit toward Mars (approach mass) to the mass injected into Mars orbit. This ratio was 1.7 for the relatively small MSP 2001 Orbiter design. This would imply that $1.0/1.7 \sim 60\%$ of the mass approaching Mars can be placed into Mars orbit as payload. For human-scale payloads, significant technical challenges are involved (see Section 4.6).[49] Based on models generated by Braun and co-workers, it appears that for human-scale payloads, roughly 60% of the mass approaching Mars can be inserted into Mars orbit by means of aerocapture.

JSC's design reference mission DRM-1 assumed that the mass inserted into Mars orbit was 85% of the approach mass. JSC's DRM-3 mentions a number of unpublished studies that led to estimates near 85%, so they also adopted that figure. This appears to be quite optimistic. Unfortunately, none of the putative JSC studies were published, so there is no way to check them.

The appropriate mass ratio (approach mass/payload mass into orbit) for Mars orbit insertion using aerocapture is considerably lower than when propulsion is used. This implies that more payload mass can be inserted into Mars orbit for a given fixed mass approaching Mars when aero-assist is used. However, aero-assist may not be easily implemented on human-scale missions. Aerocapture is quick, but it may lead to unacceptably high g-forces if deceleration is too rapid, and, more importantly, scaling

---

[49] *Entry, Descent, and Landing Challenges of Human Mars Exploration*, G. Wells, J. Lafleur, A. Verges, K. Manyapu, J. Christian, C. Lewis, and R. Braun, AAS 06-072; "Mars Exploration Entry, Descent and Landing Challenges," R. D. Braun and R. M. Manning, *Aerospace Conference, 2006 IEEE, March 2006*.

up the aeroshell to very large loads needed for human missions may place requirements on the Launch Vehicle shroud diameter that are difficult to fulfill. Development of aerocapture systems for large human-size payloads will require a long,
arduous, and expensive program (see Section 4.6.5).

Tables 3.13 and 3.14 provide rough estimates of relevant gear ratios for fast and
slow traverses to Mars orbit.

3.11   LEO TO THE MARS SURFACE

The total Δv for descent from a Mars circular orbit is estimated to be roughly
4.1 km/s. Just as in the case of Mars orbit insertion, the simplest (but not necessarily
best) method for Mars descent involves retro-firing of a rocket to de-orbit the spacecraft from orbit around Mars, with continual retro-firing to allow it to descend until it
lands gently on the surface. As in the case of Mars orbit insertion, use of higher
performance propellants tends to improve mass ratios, but boil-off acts in the
opposite direction to increase mass ratios. Descent from Mars orbit using aero-assist
is a complex process. It involves (a) initially firing a retro-rocket to de-orbit the
spacecraft and cause it to drop, (b) use of an aeroshell for thermal protection as
the spacecraft descends, (c) deployment of a parachute(s) for further slowing of the
descending spacecraft, (d) jettisoning the heat shield, and (e) final retro-propulsive
landing. Braun and co-workers have estimated the mass requirements for aero-
assisted descent to the surface from Mars orbit for human-scale payloads (see
Section 4.6).

Estimates of mass required in LEO to land 1 mass unit of payload on the surface
of Mars are given in Tables 3.15 and 3.16.

Using space-storable propellants requires over 38 mass units in LEO to deliver
1 mass unit to the Mars surface. Considering that the mass requirements for human
missions are many tens of metric tons (mT) to the Mars surface, this approach
appears to require untenably high IMLEO. Use of higher performance propellants
reduces the mass in LEO considerably. However, even with LOX/LH_2, it requires
about 1,800 mT in LEO to land 100 mT on Mars. That would require about 12
launches with the largest conceivable Heavy-Lift Launch Vehicle (150 mT to
LEO). Clearly, aero-assisted descent is the only hope for a feasible human mission
to Mars, but even this approach requires about six launches of a Heavy-Lift Launch
Vehicle to deliver 100 mT to Mars.

3.12   IMLEO FOR MARS MISSIONS

3.12.1   Chemical propulsion and aero-assist

Using the estimates for G_{MO} and G_{MS} from previous tables, we can infer what
the required IMLEO is for any Mars mission that involves delivery of mass M_{MO}
to Mars circular orbit and another mass M_{MS} to the Mars surface. No specification is

Table 3.13. Gear ratios for transfer from LEO to circular or elliptical Mars orbit with propulsive orbit insertion.

	Earth departure				Mars orbit insertion					Overall gear ratio	
	From ⇒ LEO							Trans-Mars	Trans-Mars	LEO	LEO
	To ⇒ Trans-Mars							Circular orbit	Elliptical orbit	Circular orbit	Elliptical orbit
Trip	I_{SP}	Δv	G_{ED}	Propellants	I_{SP}	Circular orbit Δv	Elliptical orbit Δv	G_{MOI}	G_{MOI}	G_{MO}	G_{MO}
Fast	450	4.2	3.20	LOX + LH$_2$	450	2.4	1.2	1.89	1.36	6.05	4.36
Fast	450	4.2	3.20	LOX + CH$_4$	360	2.4	1.2	2.24	1.48	7.17	4.73
Fast	450	4.2	3.20	NTO + MMH	325	2.4	1.2	2.46	1.54	7.87	4.94
Slow	450	3.9	2.92	LOX + LH$_2$	450	2.4	1.2	1.89	1.36	5.52	3.98
Slow	450	3.9	2.92	LOX + CH$_4$	360	2.4	1.2	2.24	1.48	6.54	4.31
Slow	450	3.9	2.92	NTO + MMH	325	2.4	1.2	2.46	1.54	7.18	4.50

Table 3.14. Gear ratios for transfer from LEO to circular Mars orbit using aerocapture for MOI.

Trip	I_{SP}	Δv	Earth departure From ⇒ LEO To ⇒ Trans-Mars G_{ED}	MOI Trans-Mars Mars circular orbit G_{MOI}	Overall gear ratio LEO Mars circular orbit G_{MO}
fast	450	4200	3.20	1.67	5.34
slow	450	3900	2.92	1.67	4.88

Table 3.15. Gear ratios for transfer from LEO via circular Mars orbit to the Mars surface using propulsion for orbit insertion, descent, and landing.

Trip	Earth departure From ⇒ LEO To ⇒ Trans Mars G_{ED}	Propellants	Mars orbit insertion Trans-Mars Mars orbit G_{MOI}	LEO Mars orbit G_{MO}	I_{SP}	Δv	Mars descent and landing Mars orbit Mars surface G_{DL}	Trans-Mars Mars surface G_{ML}	Overall gear ratio LEO Mars surface G_{MS}
Fast	3.20	LOX + LH$_2$	1.89	6.05	450	4,100	3.11	5.88	18.8
Fast	3.20	LOX + CH$_4$	2.24	7.17	360	4,100	4.34	9.72	31.1
Fast	3.20	NTO + MMH	2.46	7.87	325	4,100	5.29	13.01	41.6
Slow	2.92	LOX + LH$_2$	1.89	5.52	450	4,100	3.11	5.88	17.2
Slow	2.92	LOX + CH$_4$	2.24	6.54	360	4,100	4.34	9.72	28.4
Slow	2.92	NTO + MMH	2.46	7.18	325	4,100	5.29	13.01	38.0

Table 3.16. Gear ratios for transfer from LEO via circular Mars orbit to Mars surface using full aero-assist for orbit insertion, descent, and landing.

Trip	I_{SP}	Δv	Earth departure From ⇒ LEO To ⇒ Trans-Mars G_{ED}	Mars entry, descent, and landing Trans-Mars Mars surface G_{ML}	Overall gear ratio LEO Mars surface G_{MS}
Fast	450	4,200	3.20	3.6	11.52
Slow	450	3,900	2.92	3.6	10.51

made as to what those masses are or what functions they perform. The net result is shown in Table 3.17 using propulsion for orbit insertion and descent, and Table 3.18 using aero-assist for orbit insertion and descent. Thus, for example, if the mission required delivery of $M_{MO} = 80$ mT to Mars circular orbit and $M_{MS} = 100$ mT to the surface, IMLEO would be estimated to be 3,684 mT using propulsion and 1,579 mT using aero-assist. Clearly, aero-assist has a major beneficial effect compared with propulsion.

We may regard Table 3.18 as a baseline for IMLEO for human missions to Mars. It is based on LOX/LH$_2$ propulsion for Earth departure, and aero-assisted orbit insertion, descent, and landing at Mars using the Georgia Tech models for aero-assist masses.

3.12.2   Use of Nuclear Thermal Propulsion

Table 3.19 provides the gear ratios for Earth departure using nuclear thermal propulsion. These gear ratios depend on the dry mass fraction and the altitude for start-up. These can be simply converted to a "benefit factor" (β)—the smaller, the better—that multiplies any data in Tables 3.17 or 3.18 to obtain what IMLEO would be if NTP replaced chemical propulsion for Earth departure. A set of values of β is given in

Table 3.17. Estimated IMLEO (mT) using a fast trip to Mars and LOX/CH$_4$ propulsion for orbit insertion and EDL at Mars for any combination of masses delivered to orbit and delivered to surface.

		M_{MO} =Mass sent to Mars circular orbit (mT)				
		20	40	60	80	100
M_{MS} =	20	765	909	1,052	1,196	1,339
Mass sent	40	1,387	1,531	1,674	1,818	1,961
to Mars	60	2,009	2,153	2,296	2,440	2,583
surface	80	2,631	2,775	2,918	3,062	3,205
(mT)	100	3,253	3,397	3,540	3,684	3,827

Table 3.18. Estimated IMLEO (mT) using a fast trip to Mars and full aero-assist at Mars for any combination of masses delivered to orbit and delivered to surface.

		M_{MO} =Mass sent to Mars orbit (mT)				
		20	40	60	80	100
M_{MS} =	20	337	444	551	658	764
Mass sent	40	568	674	781	888	995
to Mars	60	798	905	1,012	1,118	1,225
surface	80	1,028	1,135	1,242	1,349	1,456
(mT)	100	1,259	1,366	1,472	1,579	1,686

Table 3.20. As an example, if $K \sim 0.5$ and start-up occurs at 1,250 km, the NTR produces a modest improvement in mass sent toward Mars ($\beta \sim 0.94$). This small improvement would come at an enormous cost to develop, test, and validate NTP.

Recent NASA documents indicate that the latest NASA plan for human missions to Mars utilizes NTP for Mars orbit insertion and Earth return, in addition to Earth departure.

Use of NTP for human-scale missions typically requires many tens of mT of hydrogen as a propellant. Previous JSC design reference missions assumed the use of NTP only for Earth departure, but NTP was not considered for Mars orbit insertion (which occurs 6 to 9 months after Earth departure) or Earth return (which occurs more than two years after Earth departure) because of presumed difficulties in storing hydrogen for such long periods.

However, a NASA presentation delivered in 2006 proposed a Mars mission architecture in which the crew transfer involves use of NTP for Earth departure, Mars orbit insertion, and Earth return. Cargo transfer uses NTP only for Earth departure. This is illustrated in Figure 8.2 (see color section). With this mission model, NASA provided an estimate that the value of IMLEO for an entire Mars mission is 446 mT, as illustrated in Figure 8.3 (see color section).

We can estimate data for a replacement for Tables 3.17 and 3.18 using NTP for both Earth departure and Mars orbit insertion. In doing this, however, we must take into account several factors. One is the effect of start-up altitude and dry mass fraction on G_{ED}, as expressed in Table 3.19. As we stated previously, the ESAS report has indicated that Earth rendezvous prior to Earth departure to Mars would occur in the altitude range 800–1,200 km.

We can also estimate G_{MOI} using NTP as shown in Table 3.21.

Table 3.19. Estimated values of G_{ED} for Earth departure using NTP.

Altitude at start-up of NTR	NTR $K = 0.2$	NTR $K = 0.3$	NTR $K = 0.4$	NTR $K = 0.5$	NTR $K = 0.6$	NTR $K = 0.7$
200 km	1.82	1.96	2.13	2.33	2.56	2.78
750 km	2.17	2.38	2.56	2.78	3.03	3.33
1,250 km	2.44	2.56	2.78	3.03	3.33	3.70

Table 3.20. Benefit factor (β) that multiplies data in Tables 3.15 or 3.16 to obtain IMLEO when NTP is substituted for chemical propulsion for Earth departure. Smaller (β) are better.

Altitude at start-up of NTR	NTR $K = 0.2$	NTR $K = 0.3$	NTR $K = 0.4$	NTR $K = 0.5$	NTR $K = 0.6$	NTR $K = 0.7$
200 km	0.56	0.61	0.66	0.72	0.79	0.86
750 km	0.67	0.74	0.79	0.86	0.94	1.03
1,250 km	0.76	0.79	0.86	0.94	1.03	1.15

Table 3.21. Estimated values of G_{MOI} for Mars orbit insertion, using NTP.

Orbit	NTR $K = 0.2$	NTR $K = 0.3$	NTR $K = 0.4$	NTR $K = 0.5$	NTR $K = 0.6$	NTR $K = 0.7$
Circular	1.40	1.45	1.50	1.56	1.62	1.68
Elliptical	1.18	1.20	1.22	1.24	1.26	1.28

We could combine Tables 3.19 and 3.21 to obtain an estimate for G_{MO}, the overall ratio from LEO to Mars orbit, but a few caveats are in order first. In transporting many tens of mT of hydrogen to Mars, one can either accept some boil-off and start with a larger tank that is only partly filled on arrival at Mars, or an active cooling system can be used for zero boil-off, at a cost of some mass for power, cryo-cooling, controls, radiator, etc. In either case, it adds to the propulsion system dry mass, thus increasing K. In addition, it is clear that the same basic power system and propulsion system will be used for Mars orbit insertion as was previously used for Earth departure. But, for Earth departure the Δv was around 4,200 m/s whereas for Mars orbit insertion, it is only 2,400 m/s resulting in a lesser amount of hydrogen propellant being used. But, if the same power/propulsion system is used with less propellant, the inevitable result is that K will be greater for Mars orbit insertion than it was for Earth departure. The combination of these two factors suggests that K will be quite a bit larger for Mars orbit insertion than it was for Earth departure. Without detailed analysis it is impossible to know which value of K in Table 3.21 to combine with any value of K in Table 3.19. However, just as a hypothetical example, if the effective K for Earth departure is, say, 0.5 and the start-up altitude is 1,000 km, and the K for Mars elliptical orbit insertion is 0.7, the overall G_{MO} would be:

$$G_{MO} \sim 2.90 \times 1.28 = 3.71$$

But, if one is going to go to the trouble (and expense) of transporting many tens of mT of hydrogen to Mars as propellant for NTP, it is worth comparing this scenario with one where a much smaller amount of hydrogen (plus an appropriate amount of oxygen) is transported to Mars for Mars orbit insertion using LOX/LH$_2$ chemical propulsion. If chemical propulsion is used for Earth departure and LOX/LH$_2$ is also used for Mars orbit insertion, Table 3.13 shows that

$$G_{MO} = 4.36$$

If NTP is used for Earth departure (at 1,000 km and $K = 0.5$) and LOX/LH$_2$ is used for Mars orbit insertion, Table 3.20 indicates that this would decrease to

$$G_{MO} = 4.36 \times 0.9 = 3.92$$

Obviously, we have only cited these values as illustrations. We don't have credible estimates for K in NTP in Earth departure, and we are even less informed on K at Mars.

Nevertheless, it does appear that the benefit of using NTP downstream of Earth departure is minimal. Using NTP for both Earth departure and Mars orbit insertion,

G_{MO} is around 4, and using NTP for Earth departure and full aero-assist for EDL, G_{MS} is around 10. That being the case, if one has to land 121 mT on the Mars surface and insert roughly 100 mT into Mars orbit, the likely requirement for IMLEO would be

$$\text{IMLEO} = 10 \times 121 + 4 \times 100 = 1{,}600\ \text{mT}$$

That is quite a bit more than the estimate of 446 mT given by NASA in Figure 8.3.

3.12.3   Use of ISRU

If *in situ* resources are utilized (ISRU) on Mars for propellant production, the absolute value of M_{MS} may be reduced. Furthermore, it is quite certain that if ISRU is used to produce ascent propellants from indigenous resources on the Mars surface, the ERV would be inserted into an elongated elliptical orbit, thus reducing the Δv required for orbit insertion from about 2,400 m/s for a circular orbit to about 1,200 m/s (and therefore also reducing the propellant requirement). The appropriate gear ratios for transfer to an elliptical orbit are provided in Table 3.13.

In addition, the required Δv for departure from the elliptical orbit ought to be about half of that required for departure from a circular orbit (also ~2,400 m/s), resulting in additional mass savings. Although this would increase the Δv required for ascent from the surface to orbit, resulting in a significant increase in the mass of ascent propellants, the increased amount of propellants required for ascent to an elliptical orbit can be provided by ISRU. Thus, the benefits of ISRU for propellant production are threefold: (1) reduction in landed mass (M_{MS}), (2) reduction in propellant mass for Mars orbit insertion of the ERV, and (3) reduction in propellant mass for Mars orbit departure. Furthermore, because the Δv required for Mars orbit insertion into an elliptical orbit would be significantly reduced, use of aerocapture could potentially be eliminated and propulsive orbit insertion would likely be quite competitive. Aero-assisted descent would nevertheless be retained. We shall assume that all propulsive transfers near Mars are accomplished with methane/LOX propellants at $I_{SP} \sim 360$ s and we assume a fast transfer from LEO to Mars. However, it should be noted that an elliptical orbit might be beneficial even if ISRU is not used, because the benefits of easy orbit insertion and departure would likely outweigh the disadvantage of heavier ascent propellants if (a) the Ascent Capsule is sufficiently light weight compared with assets placed in Mars orbit, and (b) aero-assisted entry, descent, and landing is used.

Additional reductions in M_{MS} from production of life support consumables by ISRU are difficult to quantify, but are likely to be significant.

According to Table 3.13, the overall gear ratio for transfer from LEO to an elliptical orbit is 4.73. The overall gear ratio for transfer from LEO to the Mars surface via propulsive insertion into an elliptical orbit is estimated by multiplying 4.73 by G_{DL}, based on aero-assisted descent, which is equal to 2.16. Thus, the overall gear ratio for transfer from LEO to the Mars surface via an elliptical orbit is estimated to be $2.16 \times 4.73 = 10.2$. This is actually lower than using aerocapture into a circular orbit as the intermediate step.

The reduction in landed mass produced by use of ISRU depends on a number of aspects of mission design. In the JSC DRMs, the Ascent Vehicle together with a nuclear power system and an ISRU plant were sent to Mars 26 months prior to crew departure from Earth. These were used to produce propellants for ascent, and the tanks of the Ascent Vehicle would be filled prior to crew departure from Earth. This allowed use of a relatively small ISRU plant to function 24/7 for about a year with about the same power requirement as will eventually be needed when the crew subsequently arrives. As a result, the same power system is used for ISRU prior to crew departure, and for crew support after the crew arrives, and the mass of the nuclear power system is not attributed to the ISRU installation. The required mass of ascent propellants depends on the masses of the capsule to hold the crew during ascent and rendezvous, as well as the mass of the ascent propulsion system. The total Δv required for ascent to an elliptical orbit is about 5,600 m/s and we assume use of an LOX/CH$_4$ propulsion at an I_{SP} of 350–360 s. The value of "q" is 4.891. We use the formula presented in Section 3.1.2:

$$\frac{m_P}{m_{PL}} = \left(\frac{q - 1}{1 - K(q - 1)} \right)$$

where m_P is the propellant mass, and m_{PL} is the payload delivered to orbit. For ascent, the payload is the capsule that transports the crew. It should be carefully noted that when $[K(q - 1)]$ approaches unity, the denominator approaches zero and you cannot lift any payload with that propulsion system. Obviously, the value of K has a strong impact on the ascent propellant requirement. Since $q = 4.891$, when K approaches 0.257, the denominator goes to zero and no payload can be lifted. While DRM-1 and DRM-3 assumed $K \sim 0.10$, preliminary planning data for lunar ascent from unpublished Project Constellation documents indicate that $K > 0.2$. The value of m_P/m_{PL} is 6.37 when K is 0.1, but rises to 17.5 when K is 0.2. Based on their estimates of m_{PL} and using their assumed value $K \sim 0.10$, JSC derived values of m_P for ascent of 26 mT in DRM-1 and 39 mT in DRM-3. Clearly, the value of K for ascent has a highly leveraged effect on the required mass of ascent propellants.

The effect of use of ISRU on IMLEO is treated in further detail in Sections 5.3.3 and 6.6.4.

4

Critical Mars mission elements

4.1 LIFE SUPPORT CONSUMABLES

As we have seen in Chapter 3, the problems involved in launching, transporting, landing, and returning large masses from Mars present formidable challenges. However, other challenges exist in sending humans to Mars. These include life support (consumables and recycling), mitigation of radiation, and low-gravity effects, providing abort options, utilization of indigenous planetary resources, as well as human factors.[50]

4.1.1 Consumable requirements (without recyling)

Life support during the three major legs of a Mars human mission (transit to Mars, surface stay, and return to Earth—and also descent and ascent) poses major challenges for human missions to Mars.[51] The estimated total consumption of consumables for a crew of six for a round trip to the surface of Mars far exceeds 100 metric tons (mT) and may be as much as ~200 mT. This could require an IMLEO of over 2,000 mT, or roughly 13 launches with a Heavy-Lift Launch Vehicle just for life support consumables—if neither recycling nor use of indigenous water from Mars were used. Clearly, life support is a major mass driver for human missions to Mars, and recycling and possibly use of indigenous Mars water resources are necessary elements of any rational plan to make such missions feasible and affordable.

---

[50] See: *http://www.ingentaconnect.com/content/asma/asem/2005/00000076/A00106s1/art00012*
[51] See: *http://www.marsjournal.org/contents/2006/0005/files/rapp_mars_2006_0005.pdf*

Life support, as defined by the NASA Advanced Life Support (ALS) Project, includes the following elements:

- Air supply.
- Biomass production.
- Food supply.
- Waste disposal.
- Water supply.

Each of these elements participates in a comprehensive overall *Environmental Control and Life Support System* (ECLSS) that maximizes recycling of waste products. These systems are complex and highly interactive.

Consumption requirements are summarized in Table 4.1. These estimates were derived from ALS reports but further refinement is needed, particularly for water.

In order to characterize ECLSS for a human mission to Mars, a first step would be to catalog the inventory of consumables that are needed for each leg of the trip to support a crew of six, assuming no ECLSS is utilized. One would tabulate how much food, water (various qualities), oxygen, atmospheric buffer gas, and waste disposal materials are needed, first on a per-crew-member-per-day basis, and then for the whole stay for a crew of say, six. Unfortunately, this basic information is not presented in any of the ALS reports. Therefore, we have estimated these data based on Table 4.1. The resultant gross life support consumptions for a human mission to Mars are summarized in Table 4.2 (p. 89) assuming no use is made of recycling or indigenous resources. The total mass consumed over all mission phases is 201 mT. The appropriate "gear ratios" to estimate IMLEO for each phase are also listed,

Table 4.1. Estimated consumption requirements for long-term missions.

Item	Requirements [kg/(person-day)]
Oral hygiene water	0.37
Hand/face wash water	4.1
Urinal flush water	0.5
Laundry water	12.5
Shower water	2.7
Dishwashing water	5.4
Drinking water	2.0
Total water	*27.6*
Oxygen	1.0
Buffer gas (N_2?)	3.0
Food	1.5
Waste disposal materials	0.5

based on estimates in Chapter 3. The total IMLEO for all phases is over 2,000 mT, or roughly 13 launches just for life support. Clearly, this would be impractical and unaffordable without recycling.

4.1.2   Use of recycling systems

The NASA Advanced Life Support (ALS) Program is actively working on processes and prototype hardware for ECLSS that involves recycling vital life support consumables so as to reduce the mass that must be brought from Earth.

Not only must the ECLSS provide the gross requirements for these elements, but it must also monitor trace contaminants and reduce them to an acceptable level. NASA life support data are reported in two segments. One segment is claimed to be "state of the art" based on "the International Space Station (ISS) Upgrade Mission" and the other segment is for an "Advanced Life Support" (ALS) system that is based on advanced technologies currently under development within NASA. It is claimed that only technologies included in the assessments have (as a minimum) been taken to the breadboard test stage. The ALS reports provide numerical estimates for mass and power requirements of ECLSS systems. However, the connection between the baseline data in the reports and actual experience with the ISS is very difficult to discern. It is not clear how much experimental data underlie the tables, and how many data are estimated from modeling. Nor is it clear whether these systems are reliable for the long transits and surface stays of Mars missions. The longevity and mean time between failures of these systems has not yet been reported. It is noteworthy that all of the mass estimates provided by the ALS are from the research arm of NASA, and they do not include allowances for margins, redundancy, or spares.

Most of the available reports provide system estimates of masses for the various elements of the ECLSS. The basic element masses are listed, as well as "equivalent system masses" (ESMs) that include additional mass to account for the required power system, thermal system, and human oversight requirements associated with operation of the LSS. However, we do not use equivalent system masses herein.

The major elements that are recycled are air and water. For each recycle system, the mass of the physical plant needed to supply the consumables must be estimated, as well as the recovery percentages for the air and water systems. From the recovery percentage, one can calculate the size of the back-up cache needed for replenishment of lost resources during recycling. Then, for each of the air and water systems, five quantities would be reported for each mission leg:

(1) The total mass of the resource needed for a crew of six over the duration of the mission leg (M_T).
(2) The mass of the physical plant (M_{PP}).
(3) The recovery percentage (R_P) (percent of used resource that is recovered in each cycle).

(4) The mass of the back-up cache needed for replenishment of losses in recycling:

$$M_B = (100 - R_P)M_T/100$$

(5) Total mass of the ECLSS that supplies M_T of resource during the mission leg: (sum of physical plant + back-up cache)

$$M_{LS} = M_{PP} + M_B$$

A useful figure of merit is the ratio M_T/M_{LS} that specifies the ratio of the mass of resource supplied to the total mass of the ECLSS system. The larger this ratio, the more efficient the ECLSS is.

In addition to these performance estimates, the reliability and longevity of such systems should be discussed, and additional mass provided for margins, spares, and redundancy, as needed.

Finally, the potential impact of utilizing indigenous water on Mars for surface systems should be considered and incorporated into plans as appropriate.

Only water and air are susceptible to recycling, whereas the elements biomass production, food supply, waste disposal, and thermal control are not recycled.

We hypothesize a mission with a crew of six, 180-day transits to and from Mars, and a 600-day stay on Mars. The requirements for non-recycled elements were estimated in Table 4.2. Using JSC estimates for recycling, we obtain Table 4.3. The total mass delivered for each mission phase is the sum of ECLSS masses for air and water, plus food and waste disposal materials, although no recycling is assumed for the short-duration ascent and descent steps. The total required IMLEO for all mission phases is 570 mT. But this is an optimistic estimate, and it is far from certain that the recovery percentages listed in Table 4.3 can be achieved for fail-safe applications to Mars over a 2.7-year mission.

A recent report[52] indicates that: "Experience with Mir, International Space Station (ISS), and Shuttle, have shown that even with extensive ground checkout, hardware failures occur. For long duration missions, such as Mir and ISS, orbital replacement units (ORUs) must be stored on-orbit or delivered from Earth to maintain operations, even with systems that were initially two-fault tolerant. Long surface stays on the Moon and Mars will require a different method of failure recovery than ORU's." This might add to the required back-up cache and/or require some spares or redundant units that would double (or more) the mass of the system. Obviously, long-term testing is needed here. Sanders and Duke (*loc. cit.*) emphasize the need for ISRU as a back-up for an Environmental Control and Life Support System (ECLSS), pointing out the unreliability of ECLSS. It is also interesting that the ALS appears to be rather cautious regarding the potential for widespread indigenous Mars water resources to impact life support on Mars, despite the fact that this impact could potentially be a major benefit in mass reduction and safety. Admittedly, acquisition of such water resources will require sophisticated machinery and there are

---

[52] *ISRU Capability Roadmap Team Final Report*, J. Sanders (JSC) and M. Duke (Colorado School of Mines) (eds.), Informal Report, March 2005.

Table 4.2. Gross life support requirements for a human mission to Mars without recycling or *in situ* resource utilization for a crew of six (in metric tons).[53]

Mission phase ⇒	Transit to Mars	Descent	Surface stay	Ascent	Earth return
Duration (days)	180	15	600	15	180
Water	29	2	100	2	29
Oxygen	1.1	0.3	4	0.3	1.1
Food	1.6	0.1	5.4	0.1	1.6
Waste disposal materials	0.6		1.8		0.6
Buffer gas	3.3	0.9	12	0.9	3.3
Total consumed	*36*	*3*	*123*	*3*	*36*
"Gear ratio" for IMLEO	3	9.3*	9.3*	76	18
Required IMLEO mass	*108*	*28*	*1,144*	*228*	*656*

* According to Table 3.16, the current best estimate for this gear ratio is around 11:1 but that estimate includes a 30% margin. With a 15% margin, the gear ratio drops to 9.3 and we use this more optimistic figure here.

concerns regarding planetary protection. Nevertheless, this aspect would seem to deserve more attention in ALS activities.

Use of indigenous water on Mars may provide significant mass savings as well as great risk reduction. It is conceivable that the entire water supply needed while the crew is on the surface could be supplied from near-surface indigenous resources, thus eliminating water recycling altogether on the surface of Mars. In addition, the entire surface oxygen supply could be provided as well. The only commodity needing recycling on Mars would then be atmospheric buffer gas. See Appendix C for a discussion of Mars water resources.

It is hoped that in the future the ALS will:

(1) Concentrate on systems with very high reliability for long durations rather than systems with very high recovery percentages. For Mars, an LSS with 90% recovery and 99.8% reliability would be far more valuable than one with 99.8% recovery and 90% reliability.
(2) Provide clearer delineation of data sources with particular emphasis on which data are based on experiment, and how long the experiments lasted.
(3) Consider use of the widespread near-surface water resources on Mars.

[53] It is noteworthy that NASA reports on life support consumables do not provide the data for "total consumed" but only present their estimates for the mass of an ECLSS system to supply these (unstated) requirements. The only place in the literature where I could find estimates of the gross requirements was in a University of Houston SICSA report (see Section 4.5.5.2). The SICSA estimate for water is about the same as that given in Table 4.2 but the SICSA estimate for food is about 80% greater than that given in Table 4.2.

Table 4.3. Required IMLEO for life support consumables using JSC estimates for ECLSS.

Mission phase ⇒	Transit to Mars	Descent	Surface stay	Ascent	Earth return
Duration (days)	180	15	600	15	180
Water requirement = M_T	29	2	100	2	29
Water ECLSS plant mass	1.4		4.1		1.4
Water ECLSS recovery %	>99		94		>99
Water ECLSS back-up cache mass	0.3		6.3		0.3
Total water ECLSS mass = M_{LS}	1.7	2	10.4	2	1.7
M_T/M_{LS} ratio	17	1	10	1	17
Air requirement = M_T	4	0.9	12	0.9	4
Air ECLSS plant mass	0.5		1.3		0.5
Air ECLSS recovery %	83		76		83
Air ECLSS back-up cache mass	0.7		2.9		0.7
Total air ECLSS mass = M_{LS}	1.2	0.9	4.2	0.9	1.2
M_T/M_{LS} ratio	3	1	3	1	3
Food	1.6	0.15	5.4	0.15	1.6
Waste disposal materials	0.5	0.05	1.8	0.05	0.5
Total mass delivered to Mars	*5.0*	*3.1*	*21.8*	*3.1*	*5.0*
"Gear ratio" for IMLEO	3	9.3*	9.3*	76	18
Required IMLEO mass with *ECLSS*	*15*	*29*	*203*	*235*	*90**

* According to Table 3.16, the current best estimate for this gear ratio is around 11:1, but that estimate includes a 30% margin. With a 15% margin, the gear ratio drops to 9.3 and we use this more optimistic figure here.

4.1.3   Life support summary

In summary, we can draw the following conclusions regarding life support consumables:

- Although there are half a dozen elements involved in life support, water is by far the greatest factor in determining the mass requirements for life support.
- If no recycling or use of indigenous resources is used, the requirements for consumables in a human mission to Mars would be about 200 mT, which, in turn, would require over 2,000 mT for IMLEO. Such a scenario would be prohibitive and is totally impractical.
- If recycling is used for air and water, and ALS mass estimates for ECLSS based on ISS experience are adopted, the total mass brought from Earth decreases from 201 mT to 38 mT, and IMLEO decreases from over 2,000 mT to about 570 mT.

This would still require several launches solely for life support consumables and recycling plants.

- The connection between ALS estimates of ECLSS system masses and actual ISS experience has not been divulged. Therefore, the experimental basis for ALS estimates is unclear.
- The ALS reports do not discuss longevity, reliability, and mean time between failures of ECLSS. It is not clear how transferable ISS data are to Mars missions where ECLSS must function without failure for 2.7 years.
- The estimates provided herein do not include allowances for margins, redundancy, or spares.
- Use of indigenous Mars water resources has the potential to eliminate the need for recycling of water and oxygen on the surface of Mars.

4.2 RADIATION EFFECTS AND SHIELDING REQUIREMENTS

Radiation in space poses a threat to humans embarked on missions to the Moon or Mars.[54] There is considerable uncertainty as to the biological effects of various levels of radiation exposure, and how much exposure should be permitted in deep space. In the absence of anything better, standards for LEO are typically extrapolated to the very different conditions encountered in deep space. A number of studies compare the "point estimates" of levels of radiation doses in space with allowable doses, including the effects of various forms of shielding. However, considerable uncertainty is involved in understanding the health effects of radiation doses. As a result, any given point estimate of radiation dose may suggest a predicted mean biological impact, but there is a wide range of biological impacts (lesser or greater than the mean) that could be attributed to that exposure. Recently, there has been a shift in emphasis from "point estimates" to 95% confidence intervals. If 95% confidence is demanded, the biological impact will typically be three to four times greater than that associated with the point estimate. Recent reports issued by NASA as well as the Exploration Systems Architecture Study (ESAS) have estimated the radiation effects for some lunar and Mars mission scenarios. However, radiation effects and the effectiveness of shielding remain uncertain. Preliminary estimates indicate that radiation effects will be at least a serious problem for Mars missions and may be a "show stopper".

4.2.1 Radiation sources

From the standpoint of radiation protection for humans in interplanetary space, the two important sources of radiation for lunar and Mars missions are:

- The heavy ions (atomic nuclei with all electrons removed) of galactic cosmic rays (GCRs).
- Sporadic production of energetic protons from large solar particle events (SPEs).

[54] *http://www.marsjournal.org/contents/2006/0004/files/rapp_mars_2006_0004.pdf*

Galactic cosmic radiation consists of the nuclei of the chemical elements that have been accelerated to extremely high energies outside the solar system. Protons account for nearly 91% of the total flux, alpha particles account for approximately 8%, and HZE (high charge and energy for $Z > 3$) particles account for less than 1% of the total flux. Even though the number of HZE particles is relatively small, they contribute a large fraction of the total dose equivalent. At Solar Maximum conditions, GCR fluxes are substantially reduced, producing a dose of roughly half of that produced by the Solar Minimum GCR flux.

The constant bombardment of high-energy GCR particles delivers a lower steady dose rate compared with large solar proton flares which can on occasion deliver a very high dose in a short period of time (on the order of hours to days). The GCR contribution to dose becomes more significant as the mission duration increases. For long-duration missions, the GCR dose can exceed allowable annual and career limits. In addition, the biological effects of the GCR high-energy and high-charge particles are not well understood and lead to uncertainties in biological risk estimates. In transit to Mars, shielding can provide benefits, but these benefits are limited and they are costly. For systems on the surface of Mars, the planet itself blocks out one-half of the GCR, the atmosphere removes additional GCR, and regolith can hypo-thetically be used to provide additional shielding. However, as more shielding is applied one obtains diminishing returns because high-energy particles are very pene-trating, and because secondaries are produced by impact with shields. Furthermore, the Habitats described in Section 4.5 do not appear to be compatible with use of regolith for shielding.

Solar particle events (SPEs) occur when a large number of particles, primarily protons, move through the solar system. These events happen during periods of increased solar activity and appear to correspond to large coronal mass ejections. Large SPEs are rare and last only a matter of hours or days. In the last 50 years, we have typically had only one or two major SPEs per 11-year solar cycle.

The largest SPEs observed in the past were the February 1956, November 1960, August 1972, and the October 2003 events. The largest flares recorded since August 1972 occurred in the months of August through October 1989. The magnitude of the October 1989 flare was on the same order as the widely studied August 1972 event. The addition of the three 1989 flare events, which occurred within 3 months of each other, can provide a fairly realistic estimate of the flare environment that may be encountered during missions taking place in the 3 or 4 years of active Sun conditions (Solar Maximum). The greatest concern for SPE exposure is that the 30-day exposure limit might be exceeded. Regolith is a fairly good shield for SPE radiation.

There are also smaller, more frequently occurring solar particle events, through-out a solar cycle. These events are not considered here since the shielding designed to reduce the GCR dose and a large solar particle event dose to within acceptable limits will dominate the shield design calculations.

4.2.2   Definitions and units

The units used to describe the radiation effects on humans are sometimes confusing. The following definitions are most useful:

- *Absorbed dose* measures energy absorbed by a target per unit mass of the target. The fundamental unit is *1 Rad = absorption of 100 ergs of energy per gram of material.*
- *Equivalent dose* modifies the absorbed dose to account for the estimated biological effect of the absorbed energy: *1 Sievert = 1 Rad × (weighting factor)* where the weighting factor amplifies or reduces the absorbed dose based on the estimated biological effect on tissue.

Most of the data reported herein are for equivalent dose based on estimates of biological impact on blood-forming organs (BFOs). Wherever the simple word "dose" appears, it should be interpreted as equivalent dose for BFO.

All of the above units may have prefixes with $c = 1/100$ and $m = 1/1,000$ so that (for example) 1 cSv = 0.01 Sv.

4.2.3 Radiation effects on humans and allowable dose

4.2.3.1 Allowable exposure

Because the biological effects of exposure to space radiation are complex, variable from individual to individual, and may take years to show their full impact, definition of allowable exposure will always include some subjectivity. Aside from the difficulty in quantifying the biological impacts of exposure to radiation in space, there is also subjectivity in defining how much risk is appropriate.

Presently, there are no guidelines for allowable radiation exposure in deep space. A common assumption is to use LEO guidelines as a first approximation for deep space. The standards presently adopted by the National Council on Radiation Protection and Measurement (NCRP) for low Earth orbit (LEO) are based on the "point estimate" for the levels of radiation that would cause an excess risk of 3% for fatal cancer due to this exposure. These guidelines are summarized in Tables 4.4 and 4.5. It should be noted that if the mortality rate is 3% then the morbidity rate is probably more like 4.5%. A point estimate is an estimate of what appears to be the most probable result even though there is uncertainty in the results.

Table 4.4. Recommended organ dose equivalent limits for all ages from NCRP-98 (1989) and repeated by NCRP-132 (2001). "BFO" = blood-forming organ.

Exposure interval	BFO dose equivalent (cSv)	Ocular lens dose equivalent (cSv)	Skin dose equivalent (cSv)
30-day	25	100	150
Annual	50	200	300
Career	See Table 4.5	400	600

Table 4.5. LEO career whole-body effective dose limits (Sv) from NCRP-132 (2001).

Age	25	35	45	55
Male	0.7	1.0	1.5	2.9
Female	0.4	0.6	0.9	1.6

4.2.3.2   Radiation effects on humans

Most of the data and understanding of radiation effects relates to X-ray and gamma-ray exposure, and relatively little is known about continuous low dose rate heavy-ion radiation. From the standpoint of radiation protection for humans in interplanetary space, the heavy ions (atomic nuclei with all electrons removed) of galactic cosmic rays (GCRs) and the sporadic production of energetic protons from large solar particle events (SPEs) must be considered.

Radiation exposure limits have not yet been defined for missions beyond LEO. For LEO operations, in addition to a federally mandated obligation to follow the ALARA principle of keeping exposure as low as reasonably achievable, NASA adopted and OSHA has approved the radiation exposure recommendations of the National Council on Radiation Protection and Measurements (NCRP) contained in NCRP Report No. 98 (1989). This report contains monthly, annual, and career exposure limits in dose equivalents. The career limits were based on equivalent doses to blood-forming organs, and not on effective dose to the entire body. About 12 years later, the NCRP recommended new exposure limits contained in NCRP Report No. 132 (2001) (see Tables 4.4 and 4.5). These allowable doses are based on the belief that they lead to an excess risk of 3% for fatal cancer due to this exposure.

For high-energy radiation from GCR and solar proton flares, the dose delivered to the vital organs is the most important with regard to latent carcinogenic effects. This dose is often taken as the whole-body exposure and is assumed equal to the blood-forming organ (BFO) dose. When detailed body geometry is not considered, the BFO dose is conservatively computed as the dose incurred at a 5-cm depth in tissue (can be simulated by water). A more conservative estimate for the skin and eye dose is made using a 0-cm depth dose. Dose equivalent limits are established for the short-term (30-day) exposures, annual exposures, and career exposure for astronauts in LEO. Short-term exposures are important when considering solar flare events because of their high dose rate. Doses received from GCR on long-duration missions are especially important to annual limits and total career limits. Long-term career limits vary with the age and gender of the individual.

Current thinking seems to favor use of the LEO limits as guidelines for deep-space mission exposures, principally because computation of conventional exposures based on linear energy transfer (LET) in a target medium by flux of ionizing radiation may be performed with little ambiguity. However, the basis for radiation damage to mammalian cellular systems by continuous low dose rate heavy-ion radiation (galactic cosmic rays—GCRs) is related to LET in an indirect and complex fashion.

For a given ionizing particle species and energy, cell damage is highly variable for different cell types.

An interesting study was reported by Cohen (2004) in which the biological impact of 1 GeV iron particles was measured by counting chromosomal aberrations in lymphocyte cell samples ranging from 150 to 3,000 cells per dose. He found that in comparing polyethylene shields with carbon shields, the polyethylene produced a lower dose but a greater biological effect than carbon.

Recently, Hada and Sutherland[55] investigated the levels and kinds of multiple damage, called damage clusters, produced in DNA by high-energy radiation beams. Damage clusters are dangerous because they can cause genetic mutations and cancers, or they can be converted to double-strand breaks. They found that protons produced a spectrum of cellular damage very similar to the pattern caused by high-energy iron ions and other heavy charged particles. These results cast doubt on the extrapolation of radiation effects from X-ray and gamma-ray exposure to energetic proton exposure.

These studies introduce uncertainty and doubt into all estimates of the biological impact of deep-space radiation.

4.2.4   Confidence intervals and point estimates

It is conventional for most analysts to generate point estimates of radiation dose for various scenarios and then compare these with the allowable exposures in Tables 4.4 and 4.5. However, Cucinotta et al.[56] have analyzed the uncertainty in predictions of *risk of exposure-induced death* (REID) and they have shown that the uncertainties in the point estimates of REID are large. Figure 4.1 shows a schematic plot of probability vs. REID. The most probable value of REID is the one corresponding to the apex of the curve. A confidence interval is the range of REID (below and above the most probable) for which one can assign a probability level. For example, in Figure 4.1, confidence intervals are shown schematically for 30%, 60%, and 95% confidence. In order to have 95% confidence that one encompasses the true range of REID, the confidence interval widens significantly. To assure that one has 95% confidence in the estimate of REID, one must use the highest REID in the interval, indicated by point "B".

Cucinotta et al. adopted the 95% confidence interval (CI) REID as a basis for evaluating radiation risk, and this leads to biological risks that are typically a factor of 3 (or more) higher than point estimates. Therefore, when various investigators calculate point estimates of dose equivalent, a rough expedient would be to multiply these point estimates by a factor of ~3.5 to obtain a higher value (representing a supposed 95% CI) that can be compared with allowable levels from Tables 4.4 and 4.5.

[55] "Spectrum of Complex DNA Damages Depends on the Incident Radiation," M. Hada and B. M. Sutherland, *Radiation Res.*, Vol. 165, 223–230, 2006, doi:10.1667/RR3498.1
[56] *Managing Lunar, Radiation Risks, Part I: Cancer, Shielding Effectiveness*, Francis A. Cucinotta, Myung-Hee Y. Kim, and Lei Ren, NASA/TP-2005-213164, 2005.

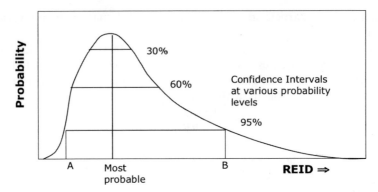

Figure 4.1. Schematic diagram of probability vs. REID. The most probable equivalent dose is the vertical line. To be 95% confident that the equivalent dose is included within a range, the range from point A to point B must be included. Thus, for 95% confidence, the maximum equivalent dose is at point B, and that is likely to be three to four times greater than the most probable REID.

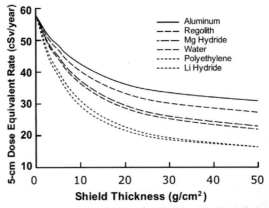

Figure 4.2. Point estimates of 5-cm depth dose for GCR at Solar Minimum as a function of areal density for various shield materials.[57]

4.2.5   Radiation in space

A number of point estimates of radiation doses in various locations with various levels of shielding are presented in the literature. Figure 4.2 shows the effect of shielding on GCR.

Moderate amounts of shielding are beneficial in reducing the lower energy components of GCR, but, as more shielding is added, diminishing returns result

---

[57] "Analysis of Lunar and Mars Habitation Modules for the Space Exploration Initiative (SEI)," L. C. Simonsen, Chapter 4 in *Shielding Strategies for Human Space Exploration*, J. W. Wilson, J. Miller, A. Konradi, and F. A. Cucinotta (eds.), NASA Conference Publication 3360, December, 1997.

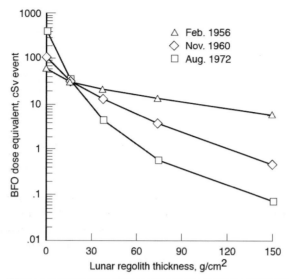

Figure 4.3. Point estimates of BFO dose equivalent as a function of lunar regolith thickness for three large SPEs.[58]

due to the penetrating power of high-energy components as well as formation of secondaries.

Figure 4.3 shows the effect of regolith thickness in reducing SPE radiation.

4.2.6   Radiation levels in Mars missions

A Mars surface mission is likely to involve ~400 days of round-trip transit in space, plus ~600 days on the surface of Mars. The data presented herein are abstracted from a review of the literature provided by D. Rapp.[59]

Transits to Mars

The point estimate of the GCR dose equivalent amounts to about 30 cSv with 10 g/cm^2 of aluminum shielding (about 4 cm of aluminum thickness) for a transit to Mars during Solar Minimum. Comparing with Table 4.4, this would be somewhat less than the annual allowable level. However, it is likely that a computation of the risk of exposure-induced death (REID) based on the point estimate of dose equivalent would yield wide error bars. As we have discussed, a rough expedient to estimate the maximum biological impact within the 95% confidence interval (CI) is to multiply the point estimate by about 3.5. This would increase the dose equivalent to ~100 CSv—that is, double the allowable annual dose for BFO.

---

[58] Simonsen, *loc. cit.*
[59] "Radiation Effects and Shielding Requirements in Human Missions to Moon and Mars," D. Rapp, *http://www.marsjournal.org/contents/2006/0004/files/rapp_mars_2006_0004.pdf*

Because crew transits to Mars are likely to require about 200 days in space each way, the probabilities of encountering major SPEs during Solar Maximum for each leg of the round trip are approximately the same as those for a 180-day stay on the Moon. During Solar Maximum, occurrence of a 4X 1972 SPE is about 1.2% probable and occurrence of a 1X 1972 SPE is about 10% probable for each leg. For the round trip the corresponding figures are 2.4% and 20%. Even with 10 or 20 g/cm^2 of aluminum shielding, the point estimate of dose equivalent for a 1X 1972 SPE would be comparable with the 30-day limit in Table 4.4. However, if the 95% CI REID were calculated, the REID would rise well above the nominal 3% that is the basis for Table 4.1.

Dose on Mars

The effect of the Mars atmosphere is significant in reducing radiation levels compared with space. The point estimate of the GCR dose equivalent on the surface of Mars is estimated to be roughly 0.06 cSv/day. However, neither aluminum nor regolith shielding is very effective at reducing this. Over the course of a year, the accumulated point estimate of the GCR dose equivalent is about 22 cSv, and this is somewhat less than the annual allowable in Table 4.4. However, if, as before, the REID is computed with 95% CI, it is likely that the biological effect would be excessive. As we have discussed, a simplistic approach is to multiply the point estimate by 3.5. In this case, the annual dose would rise to 70 cSv, and this exceeds the annual allowable of 50 cSv from Table 4.4. For a 600-day stay on Mars, the cumulative 95% CI dose is about 125 cSv. This would exceed the career allowable dose for most females and younger males.

The effect of the Mars atmosphere on SPE radiation is significant. One point estimate for the 1956 SPE dose equivalent inside a Habitat on Mars using regolith shielding is about 10 cSv/event. Doses would be lower for other major SPEs. Nevertheless, if we multiply the point estimate by ~3.5 to roughly estimate the 95% CI biological impact, it would exceed the 30-day allowable dose by a moderate amount. The probability of encountering a major SPE in a 600-day surface stay during Solar Maximum would be about 3.6% for occurrence of a 4X 1972 SPE and about 30% probable for occurrence of a 1X 1972 SPE.

Mars mission summary

In this section, we follow the results of Cucinotta et al.[60] For a Mars surface mission involving 400 days of transit in space plus 600 days on the surface, they estimate the total exposure (dose equivalent) assuming protection afforded by 5 or 20 g/cm^2 of aluminum wall (5 or 20 g/cm^2 corresponds to about 2 cm or 8 cm thickness of aluminum) at Solar Minimum and Solar Maximum. Their estimated dose equivalents are due to a sum of GCR and SPE sources. The risk of exposure-induced death (REID) based on any computed dose equivalent is also given, but the uncertainties in

---

[60] *Managing Lunar, Radiation Risks, Part I: Cancer, Shielding Effectiveness*, Francis A. Cucinotta, Myung-Hee Y. Kim, and Lei Ren, NASA/TP-2005-213164, 2005.

Table 4.6. Estimated risks associated with round-trip human missions to Mars.

When?	Thickness of aluminum shielding (g/cm^2)	Indicated dose equivalent (Sv)	Indicated REID (%) (point estimate)	95% confidence interval for risk of exposure-induced death (REID) (%)
Solar Minimum	5	1.07	5.1	1.6–16.4
	20	0.96	4.1	1.3–13.3
Solar Maximum	5	1.24	5.8	2.0–17.3
	20	0.60	2.9	0.9–9.5

REID are large. The 95% confidence interval (CI) for REID varies over a wide range. To be assured of 95% confidence, one must choose the highest REID. The results are shown in Table 4.6.

Note that the estimated dose equivalents are slightly higher at Solar Maximum because the higher dose from an assumed SPE more than makes up for the fact that the GCR is lower during Solar Maximum. Also note that shielding is more effective against lower energy SPE than against higher energy GCR, and that is why shielding is more effective at Solar Maximum.

For 95% confidence, the risk of exposure-induced death varies from 9.5% to 17.3% depending on timing and shielding.

4.2.7   Radiation summary

The subject of radiation effects in human missions to Mars is fraught with uncertainty. There is uncertainty as to how much biological impact is tolerable. The connection between dose equivalent and biological impact contains considerable uncertainty. For any given point estimate of dose equivalent, the uncertainty in biological impact may be roughly treated by arbitrarily multiplying the dose equivalent by a factor of about 3.5 prior to comparing with allowable dose equivalent. This would provide a crude measure of the 95% CI biological impact. When this is done, the exposures involved in human missions to Mars appear to be excessive. Further study of these effects is needed.

4.3   EFFECTS OF MICROGRAVITY

4.3.1   The Whedon–Rambaut review of low-*g* effects

According to Whedon and Rambaut (W&R):[61] "One of the major effects of prolonged weightlessness seen in long-duration space flights has been an extended loss of

---

[61] "Effects of long-duration space flight on calcium metabolism: Review of human studies from Skylab to the present," G. Donald Whedon and Paul C. Rambaut, *Acta Astronautica*, Vol. 58, 59–81, 2006.

bone from the skeleton. The principal characteristics of this loss were shown in the metabolic studies carried out on the Skylab flights of 1, 2 and 3 months in 1973 and 1974. These studies now provide the background for a comprehensive review of the considerable number of subsequent calcium studies in humans during space flights from that time until the present [2005]. Because of the close similarities in pattern and degree between space flight and bed rest in effects on calcium metabolism, relevant long-term human bed rest studies have been included. An analysis is presented of the bone calcium loss data with respect to degree, duration and significance, as well as relative failure of reversibility or recovery following flights."

W&R observe from Skylab that: (1) in one month in space, there is an increase in urinary calcium to nearly double pre-flight levels that remaining elevated thereafter; (2) negative calcium balances yield an average loss of 5 g/month or 0.4%/month of total body calcium; (3) a significant bone loss by densitometry was observed in the lower extremities of three astronauts; (4) there was a progressive decrease in net intestinal calcium absorption.

W&R conclude that "The pattern of the rise in urinary calcium excretion in Skylab was strikingly similar in the two lack-of-weight-bearing conditions: bed rest and space flight in proportion to the extent of physical inactivity. The degree of calcium loss during the Skylab missions varied considerably from astronaut to astronaut and was substantial but not as great as that found during horizontal bed rest ... While this rate of loss does not seem to be large with regard to the whole skeleton, it is sufficient to account for a regional loss which occurred in the lower extremities ... Furthermore, demineralization of the weight-bearing lower extremities in subjects at bed rest and patients with paralytic poliomyelitis has been estimated to be at least five times that of the whole skeleton."

Studies of patients subjected to very long bed rests (>1 year) showed that, despite administration of drugs like bisphosphonate etidronate (EHDP) and a vigorous treadmill and bicycle ergometer program, calcium loss continued for over a year, and this is likely to occur in space as well.

W&R also discuss recovery upon return to Earth: "Of the three Skylab astronauts who lost bone in flight only one showed recovery, observed at 90 days postflight; ... all three showed additional bone loss 5 years later, ranging from ~3.4% to ~5.6% from their 90 day post-flight values."

W&R quote a paper that claims that: "... assuming bone is lost at the same rate during 2-year flights as in 3-months flights, the localized lower limb bone loss might be so great as to present an immediate threat of fracture to astronauts either at landing on Mars or on returning to Earth."

Finally, W&R conclude that: "... studies of recovery of bone lost during space flight show that it is either slow, incomplete or not at all and highly variable from one astronaut to another."

A side-issue for microgravity is that "[renal] stone formation of the calcium phosphate variety often presents a major complication to diseases involving immobilization. This requires a 'need to maintain astronauts' fluid intake so that their daily urine volumes will be kept above 2 liters. This requirement for high volume of fluid intake and output, unless it can be modified by more attention to other procedures, may pose a serious engineering problem (providing enough fluid intake volumes) for very long flights, especially to Mars. The formation of a kidney stone not only endangers the health and safety of the crew-member but also imperils the success of the mission."

Finally, it is concluded that "current exercise countermeasures applied during exposure to microgravity have not been completely successful in maintaining or restoring impaired musculoskeletal and cardiovascular functions."

4.3.2 Artificial gravity

Artificial gravity for human missions has been discussed by a number of system engineers.[62]

They point out that serious concerns are continuing regarding the human physiological effects of long-duration microgravity exposure, including loss of bone mineral density, skeletal muscle atrophy, and orthostatic hypertension. Current countermeasures are "deemed ineffective (in particular with respect to bone mineral density loss)."

Paloski raises the following generic questions:

- How much artificial gravity is needed to maintain physiological function/ performance?
- What additional countermeasures would be required to supplement artificial gravity?
- What are the acceptable and/or optimal ranges for radius and angular velocity of a rotating space vehicle or centrifuge?

Paloski also raises these specific artificial gravity research questions:

- What are the physiological thresholds for effective gravitational force?
- What minimum and/or optimum g-force should be used during transit?
- Would artificial gravity be required on the lunar or Martian surface?
- What are the untoward physiological consequences of rotational artificial gravity?
- What are the physiological limits for angular velocity, g-gradient, etc.?
- What duty cycle is optimal?

These are good questions. So far, we don't have answers. In conceptualizing artificial gravity systems, the first question that arises is whether it is necessary to establish $1g$ or whether partial gravity would be sufficient. It is argued that no data are available on the physiological effects of "hypogravity" and that "acquiring this data would

[62] *Artificial Gravity for Human Exploration Missions*, K. Joosten (NASA-JSC), NEXT Briefing, July 16, 2002; *Human Mars Exploration Mission Architectures and Technologies*, J. Connolly and K. Joosten (NASA-JSC), January 6, 2005; *Artificial Gravity for Exploration Class Missions?* W. H. Paloski, (NASA-JSC) September 28, 2004.

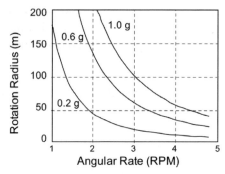

Figure 4.4. The vertical scale is rotation radius and the horizontal scale is rpm. Each curve represents a level of acceleration in units of "*g*". The uppermost curve corresponds to 1*g*. A design point at 4 rpm and radius = 56 m is selected to achieve 1*g*. [Abstracted from *Artificial Gravity for Human Exploration Missions*, Ken Joosten, NASA NEXT Status Report, July 16, 2002.]

likely be difficult, time-consuming, and expensive." Therefore, most plans for artificial gravity assume that 1*g* is the goal. Based on room-rotation studies, it is claimed that a crew can adapt to a rotating room at 4 rpm, so rotation levels have been set at ≤ 4 rpm.[63] While it is true that a rotating body will produce an outward acceleration that can be used to create artificial gravity, it is also true that, if the person is not perfectly stationary, there are cross-coupling effects that can create difficulties. Short-radius centrifugation is a potential countermeasure to long-term weightlessness. Unfortunately, head movements in a rotating environment induce serious discomfort, loss of stability, motion sickness, and subjective illusions of body tilt.[64] However, these were reduced with adaptation.

In order to achieve a centrifugal acceleration of 1*g* at 4 rpm, the minimum rotation radius is 56 meters. Figure 4.4 shows the dependence of acceleration on rotation radius and rpm.

Three conceptual artificial gravity configurations were described by JSC. Greatest emphasis was placed on the "fire-baton" approach as shown in Figure 4.5. In this configuration, the Habitat is counterweighted by the reactor/power conversion systems, and the entire vehicle rotates. Various analyses were claimed to have been performed on trajectory analysis, dynamics, structures, power, propulsion, habitation configuration, and other systems architecture issues.

Joosten's 2002 NEXT presentation discusses various aspects of artificial gravity. However, this discussion is built on a mission concept that involves an 18–24 month round trip to Mars with 18 months as a goal, and 3 months dwell in the Mars system.

---

[63] It is not clear how to obtain reports on these room rotation studies, and therefore it is difficult to verify this claim.
[64] "Artificial Gravity: Head Movements During Short Radius Centrifugation," Laurence R. Young, Heiko Hecht, Lisette E. Lyne, Kathleen H. Sienko, Carol C. Cheung, Jessica Kavelaars, *Acta Astronautica*, Vol. 49, No. 3-10, 215–226, 2001.

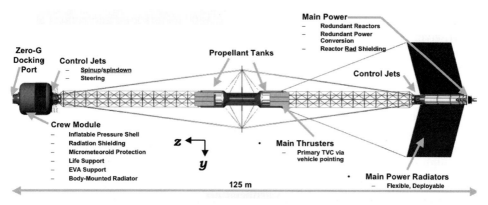

Figure 4.5. "Fire-baton" concept for artificial gravity. The Habitat is at one end and the nuclear reactor is at the other end. The entire structure rotates about a central axis. [Abstracted from *Artificial Gravity for Human Exploration Missions*, Ken Joosten, NASA NEXT Status Report, July 16, 2002.]

On the return flight, the spacecraft had to pass within ~0.4 AU of the Sun. The IMLEO was stated to be "less than 200 mT". It also included 5–12 MW nuclear electric propulsion (NEP) and efficient ECLSS for air and water. All of this was planned for a 2018 mission! Considering that the assumptions regarding NEP performance and availability, IMLEO, and mission duration appear to be impractical, it is difficult to treat the artificial gravity aspect of this presentation with great credulity. The joint Joosten–Connolly presentation of 2005 seems to simply repeat material from the 2002 presentation, and one must wonder whether any technical progress at all was made between 2002 and 2005.

Mars Direct[65] argued that artificial gravity can be provided to the crew on the way out to Mars by tethering off the burnt-out Earth departure upper stage and spinning up at 1 rpm. In the Mars Direct plan "a tether is used to create artificial gravity on the Earth-to-Mars leg only. Shortly after the habitat is injected onto its trans-Mars trajectory the upper stage separates from the bottom of the crew habitat and maneuvers around to the other side of the habitat and begins moving away, pulling the tether off the habitat's roof as it goes. Once the stage has drawn out the full length of the tether it fires its reaction control thrusters and accelerates tangentially. This gradually pulls the tether taught and begins to create artificial gravity for both the habitat and the stage. When the stage reaches a relative velocity of 400 m/s its engines stop and all remaining cryogenic propellants are dumped. (The cryogenic propellants powered the main engines that put the vehicle on the trans-Mars trajectory, whereas the reaction control engines, using hydrazine, performed the tether

---

[65] *Scientific American*, March, 2000; Addendum II: Mars Direct—A Practical Low-Cost Approach to Near-Term Piloted Mars Missions, *http://www.iaanet.org/p_papers/add2.html*; and *Mars Direct: A Simple, Robust, and Cost Effective Architecture for the Space Exploration Initiative*, Robert M. Zubrin, David A. Baker and Owen Gwynne, AIAA-91-0328.

extension and spin-up maneuver.) The tether is 1500 meters long and rotates at one RPM, giving the crew an acceleration of 0.38 Earth-g (one Mars-g). The Habitat is connected to the tether with a pyrotechnically releasable end fitting. This allows the tether to be rapidly dropped in the event of the onset of any unanticipated irremediable tether dynamic modes. Since the tether links the Habitat to only the burnt out upper stage, and not a mission critical item, the tether can be dropped and the mission continued in a zero-gravity mode.

"As a matter of routine, the tether is dropped shortly before the habitat begins its approach towards Mars aerocapture." Evidently, Mars Direct allows for the possibility that the tether system may not work and the crew may have to endure zero gravity all the way to Mars. The artificial gravity system is described as "not mission critical". Also note that no artificial gravity is used on return from Mars.

Mars Direct goes on to say: "Having a rotating spacecraft traveling through interplanetary space presents many design challenges: How are maneuvers performed? How are communications maintained between Earth and the spacecraft? How is power gathered from the sun using solar arrays? How will navigation sensors view stars, Mars and its moons?" Arguing by analogy to missions in the past that used spinning spacecraft, Mars Direct concludes that this can easily be achieved. On the other hand, Mars Direct seems to assume that everything is easy. The tethered artificial gravity system conjectured by Mars Direct might work if enough billions are poured into it. Time will tell.

According to the Mars Society DRM,[66] an artificial gravity system was deemed necessary for the Mars Society DRM's outbound Habitat flight to "(1) minimize bone loss and other effects of freefall; (2) reduce the shock of deceleration during Mars aerobraking; and (3) have optimal crew capabilities immediately upon Mars landing. Experience with astronauts and cosmonauts who spent many months on Mir suggests that if the crew is not provided with artificial gravity on the way to Mars, they will arrive on another planet physically weak." They go on to say that "unless a set of countermeasures that can reduce physiological degradation in microgravity to acceptable levels is developed, the only real alternatives to a vehicle that spins for artificial gravity are futuristic spacecraft that can accelerate (and then decelerate) fast enough to reach Mars in weeks, not months. To save on mass, the Mars Society DRM uses an artificial gravity system with the habitat counterbalanced by a burned-out Earth departure stage, as in Mars Direct. The Mars Society DRM created a Mars-like gravity at 3 RPM with a 125 m truss."

Use of Mars gravity was a compromise between desired fitness of the crew and mass budget concerns stemming from a larger truss. A truss connection between the Habitat and the burnt-out upper stage was chosen over a tether because the truss had (1) a much lower risk of failure when impacted by a micrometeorite; (2) no risk of snag; (3) less energy stored in the tension of the connecting structure that could be potentially damaging if released.

---

[66] *A New Plan for Sending Humans to Mars: The Mars Society Mission*, California Institute of Technology, 1999, Contributors: Christopher Hirata, Jane Greenham, Nathan Brown, and Derek Shannon, Advisor: James D. Burke, Jet Propulsion Laboratory.

It seems likely that, technically, mitigation of zero-g health effects can be achieved to some degree with some form of artificial gravity. However, considerably more research needs to be done on physiological effects vs. design parameters (rotation speed, rotation radius, ...), prospective designs for trusses and configurations of artificial gravity systems including packaging and deployment in space, methods and procedures for steering the structure, and realistic cost estimates for development, validation, and implementation. In doing this, artificial gravity needs to be separated and divorced from the outlandish schemes of short-stay missions with close approaches to the Sun using NEP. It is not clear what the penalty will be for utilizing artificial gravity, but deployment or construction of the system shown in Figure 4.5 that is bigger than a football field, is likely to add significantly to mission complexity, risk, and cost.

Another question is whether artificial gravity is needed for both legs of the trip to Mars (outbound and return) or as Mars Direct had it, only outbound.

A remaining unknown is the effect of ~550 to 600 days on the surface of Mars in a gravity field about 0.38 of that on Earth. If artificial gravity is needed on the surface, that might add very significantly to mission mass, complexity, cost, and risk.

4.3.3   NASA plans for low-g effects

NASA plans for lunar and Mars human missions are outlined in the so-called "ESAS report"[67] encompassing 758 pages of descriptions of lunar and Mars mission architectures and vehicles. It appears from this report that NASA has no plans at all to develop artificial gravity technology. The ESAS report says:

"The Mars Exploration DRM employs conjunction-class missions, often referred to as long-stay missions, to minimize the exposure of the crew to the deep-space radiation and zero-gravity environment while, at the same time, maximizing the scientific return from the mission. This is accomplished by taking advantage of optimum alignment of Earth and Mars for both the outbound and return trajectories by varying the stay time on Mars, rather than forcing the mission through non-optimal trajectories, as in the case of the short-stay missions. This approach allows the crew to transfer to and from Mars on relatively fast trajectories, on the order of 6 months, while allowing them to stay on the surface of Mars for a majority of the mission, on the order of 18 months."

Thus, the NASA approach seems to be that if they can shorten the trips to and from Mars to 6 months each, they plan to just live with the zero-g effects that ensue over these durations. While they do mention that the 1988 OEx study recommended development of artificial gravity systems, there is no evidence that NASA has done this.

---

[67] *http://www.spaceref.com/news/viewsr.html?pid=19094*

4.4 ABORT OPTIONS AND MISSION SAFETY

4.4.1 Abort options and mission safety in ESAS lunar missions

The word *"abort"* occurs 314 times in the ESAS report. Evidently, NASA is keenly aware of the dangers involved in human missions into deep space and has made plans and allowances for "aborting" missions if serious problems arise. Nevertheless, there are occasions when a major system failure may not allow a mission abort and a mission failure results. The consequences of a major mission failure are categorized at two levels: loss of mission (LOM), and loss of crew (LOC), with LOC clearly being the more serious of the two.

For lunar missions, the ESAS report estimated probabilities of LOM and LOC occurrences in all stages of the missions including:

- Cargo launch to LEO.
- Crew launch to LEO.
- LEO propulsion burns (suborbital, circularization, and trans-lunar injection).
- LEO dockings.
- Lunar orbit insertion.
- Lunar orbit un-docking.
- Lunar descent.
- Lunar surface operations.
- Lunar ascent.
- Lunar docking.
- Lunar orbital operations.
- Trans-Earth injection.
- Earth return.
- Earth entry and descent.

They did this for nine different conceptual lunar mission architecture variations. The overall probability of LOM varied from about 5.5% to 7.5%, and was about 5.8% for the selected lunar architecture. The overall probability of LOC varied from about 1.6% to 2.5%, and was about 1.6% for the selected lunar architecture. The details of these calculations are contained in Appendix 6D of the ESAS report, which was not released to the public, so there is no way to review the details.

The safety of lunar missions is greatly enhanced by the facts that:

- While in lunar orbit, the crew can elect to return to Earth in the CEV at any time.
- While on the lunar surface, the crew can elect to ascend to rendezvous with the CEV in lunar orbit at any time. (The NASA Constellation design activities that followed the ESAS report provide a few insights into mission abort requirements for lunar human missions. A basic requirement is that "in the event of an abort from the lunar surface, return of crew to the Earth's surface will take no longer than 5 days—independent of orbital alignment.")

- In any instance where the life support recycling system fails, there will always be sufficient resources in the back-up cache for a quick return to Earth (requirement is return in 5 days).

It should be emphasized that an escape route is always present from lunar orbit or the lunar surface in case there is a subsystem failure.

However, Mars is a very different story. Although all of the abort modes during launch and LEO operations that were designed for the lunar missions should be applicable to Mars missions as well, abort options downstream of LEO become problematic for Mars missions. As we discussed in Section 3.2.2, the Earth moves away from Mars during the six-to-nine month traverse from LEO to Mars, and if on arrival at Mars a problem develops the propulsion requirements for immediate return are huge. *Hence, it becomes necessary to remain in the Mars vicinity for a considerable length of time (typically 500–600 days, depending on specific launch date) before attempting to return.* The duration of a Mars mission ($\sim$200 days getting there, 500–600 days at Mars, and $\sim$200 days getting back) requires that iron-clad, fail-safe life support systems function for a total of more than 2.5 years, which is far more demanding than any requirement for lunar missions where a 5-day return to Earth can be implemented at any time.

The present ESAS plan for Mars human missions involves the crew descending to the surface to conduct an exploratory short-term stay and, after perhaps 30 days, they will make a "go/no-go decision" for the surface stay mission. If surface systems and operations are functional and satisfactory, they will elect to remain on the surface for the full 500–600 days. If not, they will return to orbit. However, if they return to orbit, the consequences are:

- Life support for an additional 500–600 days must be added to the Earth Return Vehicle to support the crew in orbit for that period, thus driving up the mission mass and cost. The requirements for fail-safe life support for a crew of six for up to 600 days will be very significant. (This would not be needed if the crew remained on the surface for the intended period.)
- The crew will be exposed to zero-*g* for the additional 500–600 days in orbit unless artificial gravity is implemented in Mars orbit—which seems very unlikely.
- The crew will be exposed to excessive radiation for the additional 500–600 days in orbit, having lost the benefits of (a) shielding by the planet from below, (b) shielding provided by the atmosphere, and (c) possible use of regolith piled on top of the Habitat as shielding.
- The psychological effects of sequestering the crew into a small Habitat in space for this length of time are likely to be very debilitating.[68]

Although NASA continues to promulgate this approach, it is clearly not viable and, furthermore, it defeats the whole basis for going to Mars—which is to explore the

[68] *http://www.ingentaconnect.com/content/asma/asem/2005/00000076/A00106s1/art00012;*
http://www.anacapasciences.com/publications/Astrolabe.pdf

surface *in situ*. The point is that NASA must do the preliminary work to assure that all systems are utterly reliable and fail-safe, which includes many un-crewed robotic demonstration missions.

4.4.2   The Mars Society Mission

So far, NASA has not fully evaluated risk or dealt with abort options for Mars missions. The only attempt to deal with risk in Mars missions was carried out briefly in the Mars Society Mission (MSM) concept.[69] This DRM claimed that: "a politically and scientifically viable mission with unified support has yet to be realized. The MSM resolves this problem by addressing the safety and scientific shortcomings of the JSC's DRM-3 and Mars Direct."

The MSM began by including a launch escape system for launch failures; this was not included in DRM-3 or Mars Direct, but it has been incorporated into current ESAS plans for lunar missions and would almost certainly be incorporated into any future NASA plan for Mars. The MSM also introduced increased redundancy for maximum safety. In the MSM plan, the outbound Habitat is connected to the burned-out upper stage of the Launch Vehicle via a truss to create artificial gravity (0.4g). A Crew Return Vehicle (CRV) transports the crew to the Habitat. The CRV accompanies the outbound Habitat module in close proximity, "and in the event of habitat failure would be able to support the crew until arrival on Mars or Earth. After a 612-day surface stay, both the Mars Ascent Vehicle (MAV) and Earth Return Vehicle (ERV) will accompany the crew during the return to Earth. If either ERV or MAV fails, or Mars orbital rendezvous does not take place, either component could return the crew."

In the MSM, the CRV escort tags along with the Habitat on the trip to Mars to provide the crew with a back-up spacecraft that can keep them alive in the event of a failure of a critical system on the Habitat. One instance in which the CRV would be used is failure of the Habitat life support. The Habitat life support system was designed with a small back-up cache, assuming 98% closed loop cycling. This implies that a 2% back-up cache is provided along with the recycling plant. Therefore, the Habitat life support system can malfunction without endangering the crew for up to 2% of the 900-day potential operational lifetime—18 days total. If the malfunction cannot be repaired, the 18-day allowance provides time to spin down the Habitat, dock with the CRV, and transfer the crew. However, the MSM assumed that 98% cycling, fail-safe, long-life, life-support systems can be placed on both the Habitat and the CRV, and that each life support system weighs a mere ~3 mT. As can be seen from Table 4.3, the actual mass of a such a life support system would be much greater, and achieving the required lifetime of ~2.7 years will be very difficult.

The MSM also introduced other innovations for mission safety. One was utilization of free-return trajectories. This is a complex topic that will not be discussed here.

---

[69] *A New Plan for Sending Humans to Mars: The Mars Society Mission*, California Institute of Technology, 1999, Contributors: Christopher Hirata, Jane Greenham, Nathan Brown, and Derek Shannon, Advisor: James D. Burke, Jet Propulsion Laboratory.

Another innovation was utilization of the fact that, in addition to the large amount (11.8 mT) of hydrogen brought to Mars for *in situ* resource utilization (ISRU) production of ascent propellants, the Habitat in the MSM carried a relatively small amount (211 kg) of liquid hydrogen to the Mars surface for use in ISRU production of life support consumables. This was not even close to sufficient for supporting the crew in its ~600-day stay on the surface, but it provided a small amount of life support consumables in case of an emergency. Although it was not essential for the MSM, inclusion of this hydrogen and a small ISRU plant for generating oxygen and water was an important safety feature, as it allowed the crew to produce 1.9 mT of water and 0.1 mT of oxygen, "sufficient for the crew to survive for 19 days on the surface of Mars running on open-loop life support, even if they do not land next to their MAV and cargo lander. The water and oxygen produced could also be used to support the crew on the Mars surface for 630 days in the event that the life support system loses efficiency and can only achieve 97% closure for water and oxygen loops." However, if the crew does not land near the MAV, they are not going to ever leave Mars, and keeping them alive for 19 more days would seem to be an effort in futility. Furthermore, I would contend that if the survival of the crew depends on 630 days' operation of a 98% life support system (or 97% as the case may be) this should be fully proven by preliminary test, and not left to chance.

The MSM emphasizes that "when sending humans to Mars, . . . risk estimation is an inexact science that is made even more approximate by the fact that many of the relevant systems do not yet exist." Although MSM introduced a launch escape system and a CRV back-up to the outbound Habitat, they were concerned that "Mars aerocapture is performed at high speed. Unfortunately, at the present time, we can say very little about the reliability of planetary aerocapture fifteen years in the future; any numerical estimates would be very speculative." They go on to say that "an absolute [risk analysis] would require knowledge of the exact systems to be used; since a mission architecture does not include such specific components, a bottoms-up risk analysis of this kind was impossible. Absolute risk estimation would also require analysis of factors such as radiation . . ."

4.4.3   Abort options conclusions

It appears that, having traveled to Mars, there are no opportunities for immediate return, and the full mission duration must be implemented. Unlike lunar missions, in which space systems can be validated with short-term "sortie" missions prior to estab-lishment of an "outpost," all Mars missions require a round-trip duration >2.5 years.

However, utilization of "free-return trajectories" may allow an immediate departure from Mars with a 2-year return flight in zero-g with full radiation exposure. This would also require that the Earth Return Vehicle possess sufficient life support capability for this duration. This does not seem to be a practical solution, but further study might uncover new options.

On arrival in Mars orbit, a decision could be made to not descend to the surface if signals were received from below that indicates a major system failure (e.g., the nuclear reactor failed). However, the consequences of this are (a) the crew would have to wait in Mars orbit for the full term originally accorded to the landed mission, enduring zero-g and high-radiation exposure, (b) the Earth Return Vehicle would have to possess a massive life support capability for this additional duration, and (c) the psychological effects of such confinements would be very deleterious.

There are a number of potential mission architectures for human missions to Mars and each of these has somewhat different implications for abort options and safety. All of the early Mars DRMs (JSC's DRM-1 and DRM-3, Mars Direct, and MSM) sent a large infrastructure to the surface of Mars 26 months prior to crew departure. Central to this infrastructure was a nuclear reactor power plant, an Ascent Vehicle with empty propellant tanks, and an *in situ* resource utilization (ISRU) plant capable of filling the Ascent Vehicle's propellant tanks by reacting hydrogen (brought from Earth) with indigenous carbon dioxide. If the tanks of the Ascent Vehicle were filled prior to crew departure from Earth, that would add an element of mission safety. For this mission architecture, it is essential that the crew link up with the landed infrastructure after descent, or they will have no means of returning from Mars. Thus, pinpoint-landing accuracy is a necessary requirement for these mission concepts. If they do not link up with the landed infrastructure, the remaining options depend on the ability of the landed Habitat to provide life support for 26 months on the surface to keep the crew alive until a rescue mission can be sent. But, since crew descent takes place $\sim$7–8 months after crew departure, only about 18–19 months remain before the rescue mission must be launched. Therefore, we may regard pinpoint landing and crew link-up to the landed infrastructure as a critical necessity for these architectures. Having linked up to the landed infrastructure, the option would exist to ascend at almost any time if something went wrong with the landed assets. However, such ascension would terminate in rendezvous with the Earth Return Vehicle waiting in Mars orbit, and transfer of the crew to the Earth Return Vehicle. They would then have to wait there for the full term originally accorded to the landed mission, enduring zero-g and high radiation exposure. In addition, the Earth Return Vehicle would have to possess sufficient life support capability for this duration.

In the current ESAS architecture for human missions to Mars, the plan is to emulate the mission architecture used for lunar missions. In this respect, the Ascent Vehicle is placed directly above the Descent Vehicle in a single structure called the "Surface Access Module". Using this system, the Ascent Vehicle cannot be landed 26 months before the crew, but instead the crew and the Ascent Vehicle would land as a unit. This prevents ISRU from filling propellant tanks prior to crew departure, and since ISRU, if it is used, would be used in parallel with crew occupancy on the surface, ISRU would add to mission risk, rather than reduce it. It seems likely that. with this architecture, ISRU would be eliminated and the Ascent Vehicle would be transported to Mars fully loaded with ascent propellants (in analogy with lunar missions). However, cryogenic propellants would certainly be needed on Mars and boil-off might be a severe problem for the long durations involved. One advantage of the ESAS architecture for human missions to Mars is that in case descent did not link up with previously landed infrastructure, the Ascent Vehicle (fully loaded with propellants) could immediately ascend. However, as before, such ascension would terminate in rendezvous with the Earth Return Vehicle waiting in Mars orbit, and transfer of the crew to the Earth Return Vehicle.

Clearly, in human missions to Mars everything has to work. Therefore, an unprecedented level of validation and verification of all mission systems would be required prior to actually sending humans to Mars. This would entail multiple unmanned precursors sent to Mars over a period of years at increasing levels of scale and complexity. NASA has never shown a proclivity to carry out such a long-drawn-out, expensive program with robotic precursors, although the Apollo project might qualify.

4.5 HABITATS

4.5.1 DRM-1 Habitats

JSC's DRM-1 chose common Habitats for transfer to and from Mars, for descent and ascent, and for the surface stay. The descent/ascent Habitat provided redundancy to the Surface Habitat for the surface stay. Because seven separate Habitats would be required to support three crews sent successively to Mars, unification of the design provides great cost advantages. While it may not be feasible to use a common design for all of the components that make up a Habitat, the significant elements—such as the pressure vessel (both primary and secondary structure), electrical distribution, hatches, and docking mechanisms—lend themselves to a common approach. However, DRM-1 said: "A significant amount of work still remains on definition and design of interior details of the habitats that will become part of future efforts associated with Mars mission planning. Study team members were not unanimous in the choice of a common habitat for space transit, for landing on the surface, and for surface habitation. Some argued that, due to the different requirements, a common design was not in the best interest of the mission."

In DRM-1 the crew was transported to Mars in a Transit/Surface Habitat that was identical to the Surface Habitat/Laboratory deployed robotically on a previous mission. Although a smaller Habitat might suffice for a crew of six during the approximately 6 months of transit time, designing the Habitat so that it can be used during transit and on the surface provides redundancy during the longest phase of the mission and reduces the risk to the crew.

Each Habitat consists of a structural cylinder 7.5 meters in diameter and 4.6 meters long with two elliptical end caps (overall length of 7.5 meters). The available volume is $90\,\mathrm{m}^3$ per crew-member. The internal volume is divided into two levels oriented so that each "floor" is a cylinder 7.5 meters in diameter and approximately 3 meters in height. The primary and secondary structure, windows, hatches, docking mechanisms, power distribution systems, life support, environmental control, safety features, stowage, waste management, communications, airlock function, and crew egress routes are identical to the other Habitats (the Surface Habitat/Laboratory and the Earth-return Habitat).

Table 4.7. DRM-1 Mars Transit/Surface Habitat

Subsystem	Subsystem mass (mT)	Consumables subtotal (mT)	Dry mass Subtotal (mT)
Physical/chemical life support	6.00	3.00	3.00
Crew accommodations	22.50	17.50	5.00
Health care	2.50	0.50	2.00
Structures	10.00	0.00	10.00
EVA	4.00	3.00	1.00
Electrical power distribution	0.50	0.00	0.50
Communications/information	1.50	0.00	1.50
Thermal control	2.00	0.00	2.00
Spares/growth/margin	3.50	0.00	3.50
Science	0.90	0.00	0.90
Crew	0.50	0.50	0.00
Total estimate	*53.90*	*24.50*	*29.40*

The Mars Transit/Surface Habitat contains the required consumables for the Mars transit and surface duration of approximately 800 days (approximately 180 days for transit and approximately 600 days on the surface) as well as all the required space-craft systems for the crew during the 180-day transfer trip. Table 4.7 provides a breakdown of the estimated masses for this particular Habitat, although it is not entirely clear what "crew accommodations" includes and how "physical/chemical life support" relates to it. It is not a simple matter to separate the physical Habitat from the life support system.

Once on the surface of Mars, this Transit/Surface Habitat is physically connected to the previously landed Surface Laboratory, doubling the usable pressurized volume (to approximately 1,000 cubic meters) available to the crew for the 600-day surface mission. This configuration is illustrated in Figure 4.6 with the first of the Transit Habitats joined to the previously landed Surface Habitat/Laboratory. However, it is clear that allocations of mass for life support were optimistic compared with Table 4.3.

4.5.2   DRM-3 Habitats

DRM-3 removed the first Habitat that was sent by DRM-1 to Mars 26 months prior to crew departure, thus eliminating Surface Habitat redundancy. To make up for this loss of conservatism, they adopted a speculative approach to Habitat design based on inflatable structures. However, they provided no information on the inflatable Habitat design, but they simply reduced the Habitat mass estimated from DRM-1 (see Figure 4.7). Comparisons of DRM-1 and DRM-3 masses are given in Tables 4.8 and 4.9.

Figure 4.6. Dual connected Habitats in DRM-1. This design does not seem to lend itself to shielding by piling regolith over it. [Reproduced from *Human Exploration of Mars: The Reference Mission of the NASA Mars Exploration Study Team*, Stephen J. Hoffman and David I. Kaplan (eds.), Lyndon B. Johnson Space Center, Houston, TX, July 1997, NASA Special Publication 6107.]

4.5.3   Dual Landers Habitat

In keeping with the JSC policy of improving Mars missions by reducing system masses (in PowerPoint), the Dual Landers mission further reduced masses compared with DRM-3. They utilized inflatable Habitats, but details are not provided (see Figure 4.8). A mass table is given in Table 4.10.

4.5.4   The "MOB" design

4.5.4.1   Basic design

In 2003, an analysis of requirements and design for a Mars Surface Habitat was prepared by a group of students at the University of Colorado[70] under the instruction of Dr. David Klaus. The starting point for the work was JSC's DRM-1 and DRM-3.

[70] *Martian Habitat Design, Mars Or Bust (MOB)*, University Of Colorado, Boulder, Aerospace Engineering Sciences, ASEN 4158/5158, December 17, 2003.

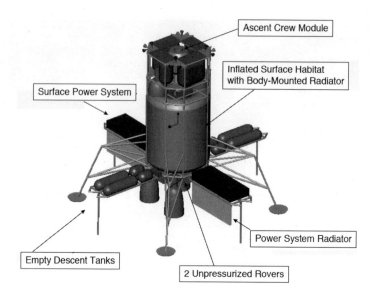

Figure 4.7. DRM-3 Habitat with Ascent Vehicle mounted above it. This design does not seem to lend itself to shielding by piling regolith over it. [Reproduced from Addendum to the *Human Exploration of Mars: The Reference Mission of the NASA Mars Exploration Study Team*, Bret G. Drake (ed.), Lyndon B. Johnson Space Center, Houston, TX, 1998, NASA/SP-6107-ADD, Reference Mission Version 3.0.]

Table 4.8. Comparison of Piloted Crew Lander masses of scrubbed DRM-1 with DRM-3 (kg).

	DRM-1	Scrubbed DRM-1	DRM-3	Difference
Habitat element		33,657	19,768	−13,889
P/C LSS	6,000	3,000	4,661	1,661
Crew accommodations	22,500	16,157	11,504	−4,653
EVA equipment	4,000	1,000	969	−31
Comm/info management	1,500	1,500	320	−1,180
Power distribution	500	500	275	−225
Thermal	2,000	2,000	500	−1,500
Structure	10,000	5,500	1,039	−4,461
Crew	500	500	500	0
Health care	2,500			
Spares	3,500	3,500	0	−3,500
Science	900			
3 kWe PVA/RFC		1,700	1,700	0
Unpressurized rovers (3)		440	500	60
EVA consumables		2,300	2,300	0
Crew + EVA suits		1,300	0	−1,300
Total payload mass	*53,900*	*39,397*	*24,268*	*−15,129*

Table 4.9. Comparison of Earth Return Vehicle masses of scrubbed DRM-1 with DRM-3 (kg).

	DRM-1	SDRM-1 = Scrubbed DRM-1	DRM-3	DRM-3 -SDRM-1
Habitat element		31,395	21,615	−9,781
P/C LSS	6,000	2,000	4,661	2,661
Crew accommodations	22,500	13,021	10,861	−2,160
Health care	2,500			
EVA equipment	4,000	500	485	−155
Comm/info management	1,500	1,500	320	−1,180
Power distribution	500	500	275	−225
30 kWe PVA power		2,974	2,974	0
Thermal	2,000	2,000	500	−1,500
Structure	10,000	5,500	1,039	−4,461
Science equipment	900	900	500	−400
Crew	500			
Spares	3,500	2,500	0	−2,500
Total payload mass	*53,900*	*31,395*	*21,615*	*−9,920*

Some of the top-level requirements were:

• Habitat must support crew of six.
• Habitat must support crew for 600 days without re-supply.
• Habitat must be deployed 2 years before first crew arrives.
• All systems in Habitat must have a minimum operational lifetime of 15 years.
• All mission critical systems must have two-level redundancy.
• All life critical systems must have three-level redundancy.
• There will be communication satellite(s) in orbit around Mars.

The general layout of the Mars Habitat is shown in Figure 4.9.

The layout of the Habitat was designed with the following considerations in mind: orientation, volume allocations, use of curved wall space, ease of access, noise proximity, systems proximity, center-of-mass considerations, and radiation shielding. Figures 4.10 and 4.11 show the floor plans for the upper and lower floors of the Habitat. Each floor is 2.25 m from floor to ceiling and the floors are only 6.2 m wide, due to radiator storage and the curvature of the Habitat.

Each crew-member has a $5\,m^2$ of area ($11.25\,m^3$ of volume) in a private bedroom. There is a wardroom on one side of the personal rooms with a galley and storage, and the area that is vacated by the airlock on the other side would also be used as a wardroom and exercise room. The area near the galley would provide space for the entire crew to eat together, and either space can be used for crew meetings and recreation. The bottom floor is primarily used as a lab. The space vacated by the airlocks on the bottom floor would also be used for the lab. There is storage on both

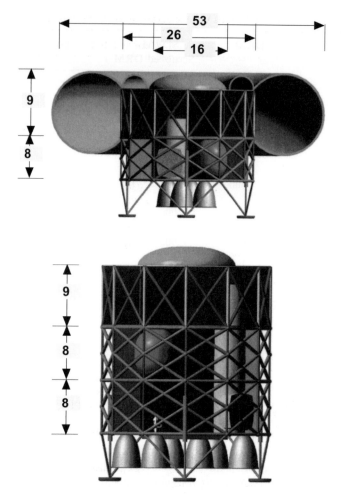

Figure 4.8. Dual Landers landed vehicles (dimensions in feet). Upper figure is the Surface Habitat. Lower figure is the Descent/Ascent Vehicle. The landing systems are the same for both vehicles. [Based on NASA JSC *Dual Landers* study, 1999.]

sides of this floor, as well as a safe haven and a medical/dental suite. The safe haven houses the command/control/communication computers and can protect the crew during large solar events.

The orientation of the cylindrical Habitat was chosen to be on its side, providing greater stability and ease of movement in the Habitat.

The leg supports elevate the Habitat to keep it thermally insulated from the Martian surface and to make the underside of the Habitat more easily accessible for repairs. They must support the entire mass of the Habitat on Mars. They also must keep the Habitat stable during a Martian windstorm. This design does not appear to be compatible with use of regolith for radiation shielding.

Table 4.10. Masses of Dual Landers Habitat and Crew Landers (kg).

Item	Habitat Lander	Crew Lander
Power system	5,988	4,762
Avionics	153	153
Environmental control and life support	3,949	1,038
Thermal management	2,912	527
Crew accommodations	3,503	728
EVA systems	1,174	1,085
ISRU	165	0
Mobility	0	1,350
Science	830	301
Structure	4,188	3,015
Margin (15%)	1,775	1,438
Food	6,840	360
Crew	0	558
Total payloads and systems	*32,652*	*15,314*

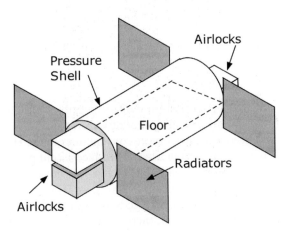

Figure 4.9. MOB concept for Habitat. [Based on *Martian Habitat Design, Mars Or Bust (MOB)*, Informal Report, University of Colorado, December 17, 2003.]

The MOB Habitat has a pressurized volume of $616\,m^3$, an unused volume of $211\,m^3$, an overall mass of $\sim$68,000 kg, and a maximum power consumption of 43 kWe.

4.5.4.2   ISRU

An ISRU system arrives with the cargo lander 450 days prior to the departure of the crew from Earth. The ISRU is required to produce a total of 3,900 kg of nitrogen, 4,500 kg of oxygen, and 23,000 kg of water before the crew arrives as per the DRM-1.

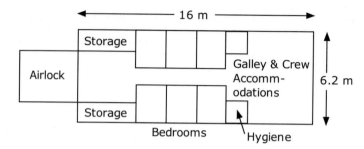

Figure 4.10. MOB plan for upper floor of Habitat. [Based on *Martian Habitat Design, Mars Or Bust (MOB)*, Informal Report, University of Colorado, December 17, 2003.]

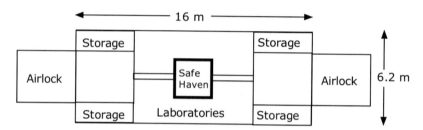

Figure 4.11. MOB plan for lower floor of Habitat. [Based on *Martian Habitat Design, Mars Or Bust (MOB)*, Informal Report, University of Colorado, December 17, 2003.]

This results in an estimated production rate of consumables of 8.39 kg/day of nitrogen, 9.68 kg/day of oxygen, and 49.46 kg/day of water. During the crewed mission, the ISRU interface will need to transfer these consumables to the Habitat. This ISRU system is distinct from an ISRU system used to produce ascent propellants. Note that the ISRU system provides all the breathable air needed, but only about 20% of the water needed to support a crew of six for ~600 days. Therefore, in this design, it would appear at first glance that no air needs to be recycled, but water requires recycling at ~80% efficiency.[71] However, there is a contradiction built into the report because the ECLSS design seems to be based on the assumption that there is no ISRU, and furthermore there is a top-level requirement which states: "ISRU cannot be initially relied upon." The ISRU system seems to be regarded as a back-up for an ECLSS that does not fundamentally rely on ISRU.

4.5.4.3   Habitat mass

The mass, power, and volume breakdown for the MOB Habitat is shown in Table 4.11. Note that the power associated with ISRU is merely for maintenance and transfer of fluids. Power required for ISRU performed prior to Habitat occupancy

---

[71] See Table 4.2 for estimates of air and water requirements for human Mars missions.

Table 4.11. Mass, power and volume breakdown of MOB Habitat. ["Martian Habitat Design, Mars Or Bust (MOB)," Informal Report, University of Colorado, December 17, 2003.]

Subsystem	Mass (kg)	Power (kW)	Volume (m^3)
ISRU	330	0.5	0.7
Structures	21,000	N/A	150.0
Power distribution and storage	2,000	11.3	4.1
ECLSS*	*31,200	9.6	84.0
Thermal control	5,000	2.0	14.0
Mission ops/crew accommodation	6,000	11.8	46.0
Command, control, and communication	530	1.9	0.3
Robotics/automation	880	0.5	0.6
EVA systems	970	5.0	105.0
Unused interior volume			211
Grand totals	*67,910*	*42.6*	*616*
Habitat without ECLSS	*36,710*	*33.0**	

* It is difficult to understand how the figure 31,200 kg for ECLSS was derived.

was not specified by MOB, nor could this have been done since the ISRU system was not clearly specified.

4.5.4.4   Habitat structure

The mass breakdown for the Habitat structure was estimated as shown in Table 4.12.

The entire Habitat is protected from radiation by an aluminum–lithium shield covering the outside of the structure. Radiation shielding at the level of 15–20 g/cm^2 (150–200 kg/m^2) was used. It was assumed that radiation shielding would not be necessary on the bottom of the Habitat (on Mars). The water storage tanks are located on the top of the Habitat and the stored water will provide some protection along with the 5 mm of aluminum from the pressure shell and tank. With the density of water (1,000 kg/m^3) and the density of aluminum (2,710 kg/m^3), it can be seen the total amount of protective shielding is already at 190 kg/m^2, which eliminates the need for any extra material on the top of the Habitat. Storage is placed on the sides of the Habitat, which will provide some radiation shielding as well. The average density of all the stored objects was assumed to be 500 kg/m^3 with an average thickness of about 30 cm (averaged over the mission duration). There is also about 3 mm of aluminum from the structure to provide protection. These materials provide a protective shielding of 158 kg/m^2, which means there is an extra 32 kg/m^2 of aluminum needed on the sides of the Habitat to have an equal amount of radiation shielding. This equates to 1.2 cm of extra aluminum on the sides. The mass of the necessary additional shielding is 2,700 kg + an additional 25% margin.

Table 4.12. MOB Habitat structural mass breakdown. [Based on "Martian Habitat Design, Mars or Bust (MOB)," Informal Report, University of Colorado, December 17, 2003.]

Component	No.	Unit mass (kg)	Design mass (kg)#	Unit volume (m³)	Volume (m³)#
Pressure shell	1	2,000	2,500	0.73	0.91
Radiation shielding	1	2,700	3,375	0.8	1
Safe haven	1	3,800	4,750	1.4	1.75
Top floor structure	1	1,360	1,700	263	2.5
Bottom floor structure	1	1,360	1,700	263	2.5
Center truss	1	520	650	10	12.5
Chassis	1	400	500	7.5	9.4
Wheels	8	50	500	0.24	2.4
Leg supports	6	5	38	0.12	0.9
Secondary floors	2	40	100	0.52	1.3
Secondary walls	30	5.5	225	0.075	2.8
Airlock structure	3	1,000	3,750	0.2	0.75
Radiator supports	4	80	400	0.5	2.5
Supports for other subsystem components	1	350	438	10	12.5
Totals			*20,625*		*114*

Includes 25% allowance for volume growth.

4.5.4.5 ECLSS

A great deal of effort went into estimating the mass and power requirements for the ECLSS system. ECLSS serves two purposes: (1) maintenance of the life support environment including trace contaminant control, and (2) recycling of consumables. The ECLSS subsystem was separated into four subsystems: atmosphere, water, waste, and food. The four subsystems were then integrated into one functional system.

The cabin was maintained at 3.1 psia partial pressure (pp) O_2, with an acceptable range of 2.83 to 3.35 psia pp O_2. The total pressure was regulated to 10.2 psia. The thermal control system maintains relative humidity between 25% and 70% and temperature between 18.3°C and 26.7°C. Carbon dioxide removal needs to offset the metabolic production rate of 0.85 kg CO_2/person/day in addition to technology products that interact with the crew cabin atmosphere.

The discussion of ECLSS systems in the MOB report is somewhat garbled and rather incomprehensible. It is very difficult to follow the logic. A rough interpretation is provided in Table 4.13. The totals are similar but not quite the same as those given in Table 4.11. The logic behind these figures is somewhat obscure. The water figures seem low whereas food, waste, and air seem high.

Table 4.13. MOB estimates for ECLSS requirements. [Based on "Martian Habitat Design, Mars Or Bust (MOB)," Informal Report, University of Colorado, December 17, 2003.]

Element	Mass (kg)	Volume (m³)	Power (kW)
Food—supply cache	11,100	31.7	
Food processing and storage equipment	3,030	5.3	2.2
Food total	*14,130*	*37*	*2.2*
Water—supply cache	6,480	6.5	
Water processing and storage equipment	890	3.3	4.2
Water total	*7,370*	*9.8*	*4.2*
Air—supply cache	5,370	5.6	
Air processing and storage equipment	5,900	16.6	3.9
Air total	*11,270*	*22.2*	*3.9*
Waste—supply cache			
Waste processing and storage equipment	280	2.1	0.4
Waste total	*280*	*2.1*	*0.4*
Grand Total	*33,050*	*71.1*	*10.7*
From Table 4.11	*31,200*	*84*	*9.6*

4.5.5   Sasakawa International Center for Space Architecture

4.5.5.1   *Habitat configurations*

The Sasakawa International Center for Space Architecture (SICSA) has published a number of reports on studies of Habitats for space missions.

An early study[72] reviews pre-1988 space Habitats. The following excerpt provides a good introduction:

"Human factors planning must consider means to maintain the psychological and physical well being of the crews under isolated and confined circumstances. Interior areas should be as comfortable and attractive as possible, emphasizing flexibility and convenience. Equipment design should reflect a good understanding of changes in body posture, leverage, procedures and other conditions imposed by zero- or reduced gravity. Variety is important to prevent boredom and depression. Means to change and personalize the appearance of interior areas and incorporate color and interest into the surroundings will be helpful. Menus should offer the widest practical range of choices, emphasizing enjoyment as well

[72] "Living in Space: Considerations for Planning Human Habitats Beyond Earth, *Sasakawa Outreach*, Vol. 1, No. 9, October/December, 1988 (Special Information Topic Issue).

as nutrition. Schedules and accommodations should encourage exercise, recreation and social activities to help free time pass pleasantly. People require time and places for private leisure. Sleeping quarters, for example, should be environments conducive to reading, listening to music and other solitary activities, incorporating devices to avoid intrusions of objectionable sounds, odors and other disturbances. Means to maintain good hygienic conditions are extremely important. Since space Habitats are closed systems, microbial growth can occur and spread rapidly, potentially causing human infections and foul odors. Problem areas and surfaces should be accessible and should be designed to facilitate easy cleaning.

The absence of gravity in orbiting Habitats strongly affects most human activities. For example, in microgravity, the directions of 'up' and 'down' are established by the interior layout of the facilities, not by the orientation with respect to Earth. People can move freely in all directions. Therefore, ceilings, walls and floors can all serve as functional work areas. Since most interior surfaces are likely to be used as push-off places when people float from one area to another, switches and fragile items such as lighting elements should be protected. Sharp corners that can cause injuries when bumped should be avoided. Anchorage devices are needed to hold people in place while they are performing stationary tasks, and to secure loose items that will otherwise float away. Skylab crews had cleats attached to their shoes that were inserted into triangular grid openings in floors for this purpose. Storage systems should be designed to keep contents from escaping when opened. Body posture is altered significantly under weightless conditions. Without gravity to compress the spinal chord, the human torso elongates a few inches, but is not as stiffly erect as on Earth. Sitting in standard chairs is uncomfortable because without gravity, people need to constantly tense stomach muscles to keep their bodies bent. Accordingly, tables and other work surfaces should be raised to crouching heights of users since chairs are not needed. Tabletops can be tilted since items placed on top must always be secured to keep them from drifting away. Rigorous exercise regimes are necessary to help offset physical deconditioning effects of prolonged weightlessness. Life in zero-gravity leads to loss of muscle mass and weakened heart-lung systems. Bones leach calcium and become more brittle. Blood and other body fluids which normally collect in the legs under the pull of gravity collect in the chest and head, causing swollen faces, nasal congestion and occasional shortness of breath."

A more recent study[73] discusses two options for configuration (see Figure 4.12).

"A 'bologna-slice configuration' [left side of Figure 4.12] offers advantages for large diameter modules. This configuration is only acceptable for modules with diameters >15 m. Smaller dimensions will limit sight lines, creating claustrophobic conditions. Equipment racks and other elements located around the

[73] "Mars Habitat modules: Launch, scaling and functional design considerations," L. Bell and G. D. Hines, *Acta Astronautica*, Vol. 57, No. 1, 48–58, July 2005.

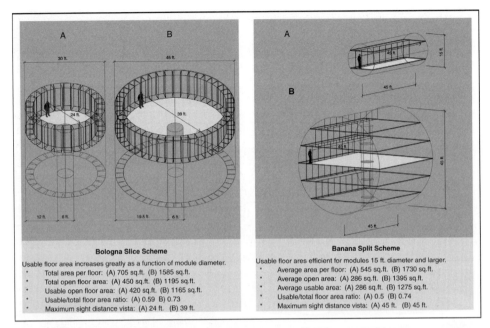

Bologna Slice Scheme

Usable floor area increases greatly as a function of module diameter.
* Total area per floor: (A) 705 sq.ft. (B) 1585 sq.ft.
* Total open floor area: (A) 450 sq.ft. (B) 1195 sq.ft.
* Usable open floor area: (A) 420 sq.ft. (B) 1165 sq.ft.
* Usable/total floor area ratio: (A) 0.59 B) 0.73
* Maximum sight distance vista: (A) 24 ft. (B) 39 ft.

Banana Split Scheme

Usable floor ares efficient for modules 15 ft. diameter and larger.
* Average area per floor: (A) 545 sq.ft. (B) 1730 sq.ft.
* Average open area: (A) 286 sq.ft. (B) 1395 sq.ft.
* Average usable area: (A) 286 sq.ft. (B) 1275 sq.ft.
* Usable/total floor area ratio: (A) 0.5 (B) 0.74
* Maximum sight distance vista: (A) 45 ft. (B) 45 ft.

Figure 4.12. Layout/configuration options for cylindrical modules. ["Mars Habitat modules: Launch, scaling and functional design considerations," L. Bell and G. D. Hines; pers. commun., Larry Bell, Sasakawa International Center for Space Architecture (SICSA), April 18, 2007.]

Figure 4.13. SICSA logistic module concept. ["Mars Habitat modules: Launch, scaling and functional design considerations," L. Bell and G. D. Hines; pers. commun., Larry Bell, Sasakawa International Center for Space Architecture (SICSA), April 18, 2007.]

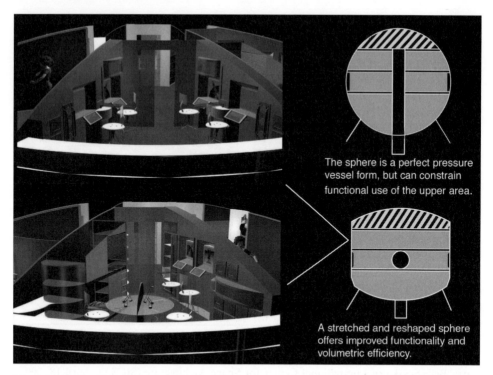

The sphere is a perfect pressure vessel form, but can constrain functional use of the upper area.

A stretched and reshaped sphere offers improved functionality and volumetric efficiency.

Figure 4.14. Considerations influencing Habitat module geometry. ["Mars Habitat modules: Launch, scaling and functional design considerations," L. Bell and G. D. Hines; pers. commun., Larry Bell, Sasakawa International Center for Space Architecture (SICSA), April 18, 2007.]

internal perimeter will reduce open areas, and vertical circulation access ways between floors will further diminish usable space. A broad base provides a wide footprint for landing stability, which is further enhanced by a short height that lowers the center of gravity. Tall, narrow modules would have a great tendency to topple over on a rocky and possibly hilly terrain. A 'banana-split approach' (right side of Figure 4.12) can be applied for both small and large diameter modules. A longitudinal floor orientation can serve satisfactorily for modules with diameters >5 m. Unlike the bologna-slice circular arrangements, the rectangular spaces offer considerable versatility to accommodate diverse equipment and functional arrangements. Long modules with horizontal layouts will, however, present special landing problems for planetary surface applications. Modules that are less than 5 m in diameter will be incompatible with efficient equipment rack design and layout due to the small-radius curvature of the walls. Beyond the 5 m diameter minimum, the modules are scaled in increments based upon the desired number of floors, allowing for a 3 m floor-to-floor height."

The SICSA team selected 15 m modules for the Mars surface and 5 m modules for transit to and from Mars. The SICSA team preferred the "banana-split" configuration, but they modified it as shown in Figure 4.14.

4.5.5.2 *Logistic and crew support*

SICSA estimated a requirement of 28.3 kg (7.5 gal) of water per person, per day. For their 500-day mission with eight crew-members[74] this would require a gross total of 113,200 kg (30,000 gal).[75] SICSA assumed an average recycling efficiency of 90%,[76] implying that 1/10 of 113,200 kg of water had to be brought to Mars as a back-up cache, but they added 10% margin to that for a water cache of 12,500 kg (12.5 m^3). Food consumption was estimated to be 2.3 kg/day per person, or ~9,200 kg for a 500 day surface stay.

Equipment needed to support food preparation, toilet, hygiene, and other basic Habitat functions is included. An ecologically closed life support system is essential to conserve water. Each module had a solar radiation shelter that used stored water for shielding. Pressurized compartments that provide passageways between adjacent surface modules also serve as airlocks that are large enough to accommodate EVA suits and tools. Habitable facilities would be located at a safe distance from landing/launch areas. All habitable elements were designed to be joined together to provide interconnecting pressurized interior passageways. The connections would be sealed to enable isolation of modules that can possibly pose hazards to others due to fires or other emergency circumstances. These sealed connectors will also serve as airlocks to support access and egress for surface crews.

4.5.6 Inflatable Habitats

Although inflatable Habitats have been mentioned in studies and press releases, precious little detailed data have been provided.

A published article in 1998 briefly described JSC efforts in inflatable Habitats:[77]

"The design is compiled of a metallic central core with a flexible composite outer shell that is cylindrical and has toroidal ends. The inflatable structure is packaged around the central core to decrease its volume for launch, then inflated, on-orbit, to its approximate 7.6 m diameter by 9.1 m length (Figure 4.15).

... The inflatable structure is a series of material layers that perform numerous functions including: gas retention, structural restraint, micrometeoroid/orbital debris impact protection, thermal protection, and radiation protection. Gas retention is achieved by a doubly redundant bladder assembly. Structural

---

[74] SICSA selected a crew of eight rather than six because of the diversity of roles and skills that they conjectured would be required.

[75] It is noteworthy that in their reports on ECLSS, NASA does not provide the total amount of resources used, but, rather, only the estimated mass of the ECLSS system. The SICSA report was the only place in the literature where this figure was cited.

[76] Use of a figure of 90% recycling for water is very conservative, as various DRMs have typically used estimates of ~98%.

[77] *Inflatable Composite Habitat Structures for Lunar and Mars Exploration*, D. Cadogan, J. Stein, and M. Grahne, IAA-98-IAA.13.2.04.

Figure 4.15. 30-foot diameter TransHab Test Unit. [*Inflatable Composite Habitat Structures for Lunar and Mars Exploration*, D. Cadogan, J. Stein, and M. Grahne, IAA-98-IAA.13.2.04, 1998.]

restraint is achieved by a series of Kevlar webbings that are interwoven and indexed to one another to form a shell. The webbings are terminated to pins that are mounted to the ends of the metallic core structure. The webbings are sized to withstand 101 kPa internal pressure loads with a factor of safety of 4 over ultimate. This yields a structure that must withstand approximately 2,232 kg/cm maximum stress in the hoop direction and 893 kg/cm stress in the longitudinal direction. Other materials, such as Vectran, are also being considered for the webbings in the later configurations. Micrometeoroid and orbital debris (MMOD) impact protection is accomplished by a series of woven 1.5 mm thick Nextel layers separated by foam spacers to create a multi-hull structure. This structure was tested for hypervelocity particle impact by JSC and found to provide greater protection than the current Space Station design. Testing revealed that 1.8 cm particles traveling at 7 km/sec would not penetrate the structure. This provided an improvement over the current limit of 1.3 cm particles at 7 km/sec for the Space Station in a roughly equivalent mass system. The Nextel layers were coated with polyethylene to enhance their stability. The polyethylene also provides a significant amount of radiation protection. Thermal protection is accomplished by a series of metallized films on the exterior of the assembly that reflect radiation.''

An Internet article on the JSC inflatable TransHab appeared in 2000 based on an interview with George Parma of JSC. It said:

> "TransHab provides greater protection against space debris than metal … The skin is over a foot thick and made of almost 24 layers. Starting from the outside, tightly woven white Beta cloth protects TransHab from erosion from the 'sand-blast' effect of atomic oxygen. Insulation is provided by blankets and Mylar. Next comes debris protection, consisting of multiple layers of Nextel, between layers of open cell foam. Any particle aimed at the walls would 'shatter' as it hits, causing it to lose energy as it penetrates deeper into the layers. The shell also contains a 'restraint layer' of Kevlar, that holds the shape of the module, with air held in pressure bladders made of an air-tight material layered with Kevlar. The inside wall is made of fireproof Nomex cloth, that protects the bladders from scuffs or scratches. Even if space debris did manage to penetrate the TransHab wall, it's interesting to note that the module would not 'burst' like a balloon. It would leak, but not pop. The reason for this is that the pressure difference between the interior of TransHab and space is only about 10 lbs per square inch.
>
> A number of tests have been run on TransHab to investigate its durability. In 1998, a test module with a diameter of 23 feet was inflated in the huge pool used by astronauts for EVA training at the Neutral Buoyancy Laboratory at JSC. The shell was found to hold pressure four times as great as the Earth's atmosphere at sea level. Later that year, the full-scale module, with a diameter of 27 feet, was inflated in JSC's vacuum chamber, to simulate the vacuum of space.
>
> The amount of air required to inflate TransHab is 3 times that required for an existing Space Station module. Sufficient supplies will therefore need to be taken up along with the deflated module.
>
> TransHab might serve as a prototype habitation for a Mars mission or the establishment of a lunar base. The only adaptation required for these types of environments would be the addition of some lightweight floors, as both Mars and the Moon are subject to gravity. He feels TransHab could be ideal for taking on a long-duration mission, due to the fact that it can be transported in a compressed state and inflated when it reaches its intended location."

Details on these tests do not seem to be available to the public. Since year 2000, almost no data or information have been released. Inflatable Habitats may provide great benefits to Mars missions, but the technology does not seem to have evolved past the "gee whiz" press release phase.

4.6 AERO-ASSISTED ORBIT INSERTION AND ENTRY, DESCENT, AND LANDING

4.6.1 Introduction

Most of the material in this section was abstracted from presentations prepared by Professor Robert Braun for his students at Georgia Tech.

It is amply demonstrated in this book that the most important technology needed to enable human missions to Mars is efficient aero-assist technology for Mars orbit insertion and for descent to the surface from Mars orbit. Aero-assist technology would reduce IMLEO by typically more than 1,000 mT as compared with use of chemical propulsion for entry, descent, and landing (EDL).

We define the following aero-assist operations:

Direct entry is a direct flight into the planet's atmosphere from hyperbolic approach without first injecting into Mars orbit. The entry vehicle can be passive (ballistic) or actively controlled. The passive vehicle is guided prior to atmospheric entry and proceeds into the planet's atmosphere as dictated by the vehicle shape and the atmosphere. An actively controlled direct entry vehicle may maneuver autonomously while in the atmosphere to improve landed location, or modify the flight environment. Direct entry was successfully performed on Viking, Apollo, Shuttle, Pioneer–Venus, Galileo, Mars Pathfinder and MER missions.

In direct entry, the spacecraft is pointed toward Mars at some flight path angle, and it gradually speeds up as it enters the gravitational attraction of Mars. As it impinges on the upper atmosphere, drag forces will cause intense heating on the frontal surfaces of the spacecraft causing it to slow down. To protect the spacecraft, a heavy aeroshell loaded with thermal protection material is placed in front of the spacecraft. In addition, a backshell is likely to be needed to prevent hot gases from flowing into the spacecraft behind the aeroshell. Although very intense heating takes place, the heat transferred to the vehicle is limited due to the short duration of the deceleration period. The disadvantages of this method are that high deceleration loads are encountered, and the weight penalty of the thick aeroshell. The accelerations encountered in direct entry are likely to be excessive for human landings but might be acceptable for cargo.

Table 4.14. Sequence for direct entry.

Step	Altitude (km)	Time from previous step	Comment
Separate entry system from cruise stage	12,000	0	Begin entry process
Reach outer edge of Mars atmosphere	131	~30 min	Begin deceleration
Attain peak heating rate	37	~70 sec	
Endure peak g-load	30	~10 sec	About 15–20 g's
Deploy parachute	8	~80 sec	
Heat shield jettison	—	—	—
Land	0	~2 min	Terminal descent includes several steps

Propulsion/Aerobraking is a process for inserting a spacecraft into a low Mars orbit (typically a circular orbit). If aerobraking is not employed, and an incoming space-craft is inserted into Mars orbit using only chemical propulsion, orbit insertion would take place in two steps:

(1) Insert into an elongated elliptical orbit ($\Delta v \sim 1.7$ km/s).
(2) With burns at apoapsis and periapsis, reduce the orbit to a circular orbit of perhaps 400 km altitude ($\Delta v \sim 0.8$ km/s).

Aerobraking is a technique for eliminating the second step, thereby reducing Δv for orbit insertion from ~ 2.5 km/s to about 1.7 km/s. Aerobraking is a relatively low risk maneuver that consists of repeated dips into an atmosphere to generate drag and lower velocity. Large performance margins are maintained to accommodate signifi-cant atmospheric variability. Generally, the total heat flux and peak temperatures are low enough to fly without a thermal protection system. The primary drag surface is typically the solar array panels, and the maximum heating rate (typically ~ 0.6 W/cm^2) is dictated by the temperature limit of the solar array (typically $\sim 175°$C). Aerobraking has been used on a number of missions at Earth, Venus, and Mars. Despite advances in aerobraking automation, aerobraking remains a human-intensive process that requires 24/7 maintenance for several months with close cooperation between navigation, spacecraft, sequencing, atmosphere modeling, and management teams. The time-sequence of aerobraking is shown in Figure 4.16.

Aerocapture is a maneuver that takes advantage of a planet's atmosphere to slow a spacecraft to orbital capture velocities and results in orbit insertion in a single pass. Descent into the atmosphere causes sufficient deceleration and heating to require a massive heat shield (aeroshell). The trajectory of the inbound spacecraft is bent around Mars as shown in Figure 4.17. As the spacecraft exits the atmosphere, the heat shield is jettisoned and a propulsive maneuver is performed to raise the periapsis. The entire operation is short-lived and requires the spacecraft to operate

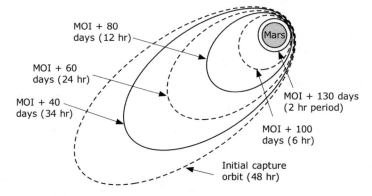

Figure 4.16. MGS aerobraking process. [Based on *NASA GSFC* website.]

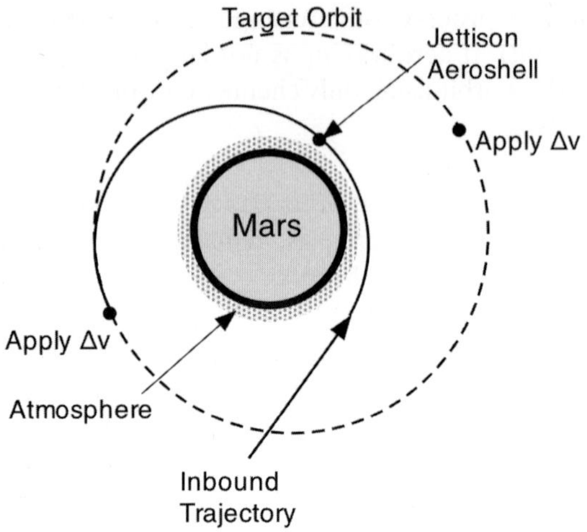

Figure 4.17. Schematic sequence of events in aerocapture.

autonomously while in the planet atmosphere. Later, a final propulsive maneuver is used to adjust the periapsis.

In one design[78] an elongated elliptical orbit with a major axis of ~10,000 km and an eccentricity of ~0.34 was generated by initial aerocapture. Subsequent maneuvers to raise the periapsis to a safe 400 km altitude (while reducing the apoapsis to ~3,800 km) require a Δv of about 250 m/s.

Aerocapture has been proposed and planned several times, and was even developed for the MSP 2001 Orbiter, but it has never actually been implemented. A change in the MSP 2001 Orbiter was made to propulsive capture followed by aerobraking. Nevertheless, it is instructive to review the design for the MSP 2001 Orbiter. In this design, the minimum altitude (~50 km) was reached about two minutes after entry into the atmosphere. The maximum deceleration was reached at that point (about 4.5 Earth g). Exit from the atmosphere occurred at about 9 minutes after minimum altitude, or 11 minutes after entry. Various masses associated with this system are listed in Table 4.15.

A comparison of aerocapture and aerobraking options for the MSP 2001 Orbiter is given in Table 4.16. It should be noted that the mass of the entry system for aerocapture was about 60% of the mass inserted into Mars orbit. There is little advantage mass-wise for aerocapture over aerobraking. However, aerobraking has the disadvantage of taking several months whereas aerocapture is accomplished in a few days.

---

[78] *Sizing of an Entry, Descent, and Landing System for Human Mars Exploration*, John A. Christian, Grant Wells, Jarret Lafleur, Kavya Manyapu, Amanda Verges, Charity Lewis, and Robert D. Braun, Georgia Institute of Technology, preprint, 2006.

Table 4.15. Some characteristics of the MSP 2001 Orbiter aerocapture system (as of the end of Phase B of the mission).

Maximum deceleration	$\sim$4.4g's at 48 km altitude
Minimum altitude	$\sim$48 km
Time duration at altitude <125 km	$\sim$400 sec
Mass prior to Mars entry	544 kg
Heat shield mass	122 kg
Backshell mass	75 kg
Total entry system mass (includes all structures and mechanisms)	197 kg
Entry system mass/Mass prior to Mars entry	0.36
Entry system mass/Mass placed into orbit	0.6
Approach mass/Mass placed into orbit	1.66
Post-aerocapture periapsis-raise maneuver propellants	
(400 km circular orbit)	20 kg
Payload mass in Mars orbit	327 kg

Table 4.16. Comparison of aerocapture and aerobraking options for the MSP 2001 Orbiter.

Component	Aerocapture system mass (kg)	Propulsive capture/ aerobraking system mass (kg)
Launch mass (wet)	647	730
Cruise stage	71	71
Cruise propellant	32	32
Entry system	197	0
Propellant at Mars	20	321
Payload mass	327	377
Payload mass ratio	0.51	0.52

4.6.2   Challenges for aero-assist technology

This section is abstracted from papers by Robert Braun and his students at Georgia Tech, and Rob Manning of JPL.[79]

As Braun and Manning point out, the United States has successfully landed five robotic systems on the surface of Mars. These systems all had landed mass below

---

[79] *Entry, Descent,and Landing Challenges of Human Mars Exploration*, G. Wells, J. Lafleur, A. Verges, K. Manyapu, J. Christian, C. Lewis, and R. Braun, AAS 06-072; "Mars Exploration Entry, Descent and Landing Challenges," R. D. Braun and R. M. Manning, *Aerospace Conference, 2006 IEEE, March 2006*; *Sizing of an Entry, Descent, and Landing System for Human Mars Exploration*, John A. Christian, Grant Wells, Jarret Lafleur, Kavya Manyapu, Amanda Verges, Charity Lewis, and Robert D. Braun, Georgia Institute of Technology, preprint, 2006.

600 kg, had landed footprints on the order of 100's of km, and landed at sites below −1 km MOLA elevation due to the need to perform entry, descent, and landing operations in an environment with sufficient atmospheric density. For human missions to Mars, vehicles of mass 40 mT or greater may have to be landed. According to NASA mission concepts, payloads will be aerocaptured into Mars orbit prior to descent to the surface. Since several payloads must be landed at the same site, pinpoint landing to within 10–100 m is also required. Compared with robotic landers, human missions require a simultaneous two orders of magnitude increase in landed mass capability, four orders of magnitude increase in landed accuracy, and EDL operations that may need to be completed in a lower density (higher surface elevation) environment.

As of 2006, robotic exploration systems engineers were struggling with the challenges of increasing landed mass capability to 1 mT, while reducing landed errors to tens of km, and landing at a site as high as +2 km MOLA elevation.

4.6.2.1   Atmospheric variations

Relative to the Earth, the Mars atmosphere is thin, approximately 1/100th in atmospheric density. As a result, Mars entry vehicles tend to decelerate at much lower altitudes and, depending upon their mass, may never reach the subsonic terminal descent velocity of Earth aerodynamic vehicles. Because hypersonic deceleration occurs at much lower altitudes on Mars than on the Earth, the time remaining for subsequent EDL events is often a concern. On Mars, by the time the velocity is low enough to deploy supersonic or subsonic parachutes, the vehicle may be near the ground with insufficient time to prepare for landing. Atmospheric variability across a Mars year limits our ability to develop a common EDL system. In addition, significant atmospheric dust content (a random occurrence) increases the temperature of the lower atmosphere, reducing density and requiring conservatism in the selection of landing site elevation. The Mars EDL challenge is exacerbated by the bi-modal Mars surface elevation, where fully half of the surface of Mars has been out of reach of past landers due to insufficient atmosphere for deceleration. To date, all successful Mars landings have been to surface sites with elevation less than −1.3 km MOLA. Coupling Mars' low atmospheric density with the mission requirements for deceleration has led to entry systems designed to produce a high hypersonic drag coefficient. One such system, the Viking-era 70° sphere cone aeroshell, has been used on every U.S. mission to the surface of Mars.

4.6.2.2   Aeroshell shape and size

The major challenge in developing aero-assist systems for Mars is the thinness of the atmosphere. In order to achieve rapid deceleration at sufficiently high altitudes that enough altitude remains for final descent with parachutes, the aerodynamic forces must be made as large as possible compared with the inertial force of the mass being decelerated. The ratio of inertial force to aerodynamic force is defined by the so-called ballistic coefficient:

$$\beta = m/(C_D A) \quad [\mathrm{kg/m^2}]$$

where $m =$ vehicle mass;
  $C_D =$ vehicle drag coefficient;
  $A =$ reference area, typically defined by maximum diameter.

A low ballistic coefficient vehicle will achieve lower peak heat rate and peak deceleration values by decelerating at a higher altitude in the Mars atmosphere. This system
will be characterized by more timeline margin for the subsequent descent and
landing events. To reduce the ballistic coefficient, systems engineers tend towards
the largest aeroshell diameters possible, where this diameter is generally limited by
physical accommodation within Launch Vehicle and/or integration and test
facilities. For robotic missions, the Delta II class Launch Vehicles can accommodate
aeroshell diameters to 2.65 m. At higher cost, the Atlas V class launch system can
accommodate aeroshell diameters as large as 5 m.

In order to achieve a fairly low value of $\beta = 64\,\mathrm{kg/m^2}$, the Viking mission
adopted an entry vehicle design described as a 70° sphere-cone aeroshell (see
Figure 4.18). The Viking program developed the aeroshell, the SLA-561V forebody
thermal protection material, and the supersonic disk-gap band parachute. With some
modification, these three EDL components have formed the backbone of all Mars
EDL architectures since. To date, β for Mars missions has ranged from 63 to 94 kg/
m^2.

In order to land a higher mass on Mars, it may be necessary to utilize entry
vehicles with higher values of β. However, this reduces the altitude available for
parachute deployment, and a point is reached where β is too large to permit safe
landing. Furthermore, when β is increased, higher elevations on Mars become less
accessible. Braun and Manning suggest that $\beta \sim 150\,\mathrm{kg/m^2}$ may be "the largest β one
can imagine for robotic Mars systems over the next several decades." In order to

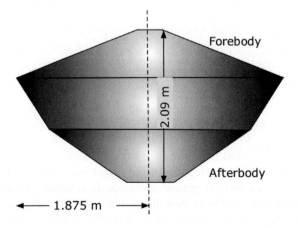

Figure 4.18. Typical design for a 70° sphere-cone aeroshell.

prevent β from growing beyond acceptable levels as m increases, the ratio m/A must be maintained fairly constant. However, m tends to increase proportionally to the volume of the entry vehicle, which is proportional to the cube of a standard dimension, whereas the area grows as the square of the standard dimension. Therefore, the packing density within the aeroshell must be reduced as m grows. This implies that large increases in area will be needed to accommodate large values of m.

Braun and Manning suggest that, to increase the landed mass on Mars, one alternative is to decrease β and enter with a very large hypersonic decelerator. An inflatable entry aeroshell is a logical option. They indicate that blunt body entry vehicles with β as low as $25\,\mathrm{kg/m^2}$ would have the advantage of eliminating the need for a separate supersonic parachute. These systems would simply require a drogue for stabilization and a large subsonic parachute or propulsive decelerator. However, the aeroshell would be very large (for a 1 mT lander, the aeroshell would have to be about 11.5 m in diameter). Full-scale testing of such systems at Earth under Mars-like conditions would be required (at high altitude and hypersonic speeds).

4.6.2.3   Supersonic parachute

The terminal velocity of a Mars entry system is generally greater than a few hundred m/s. This is too high an impact velocity for a lander. As a result, all previous and currently planned EDL architectures deploy a supersonic parachute to increase the descent β and slow the vehicle to subsonic speeds before too much altitude is lost. Besides the added drag, the parachute also provides sufficient vehicle stability through the transonic regime and the marked increase in descent β allows for positive separation of the aeroshell forebody (heat shield), a critical step in the preparation for landing.

All of the U.S. Mars landing systems used parachute systems derived directly from the Viking parachute development program. In 1972, high-altitude, high-speed qualification tests of the Viking parachute in Earth's atmosphere were successfully conducted in Mars relevant conditions. Due to the expense of these tests, similar tests have not been attempted since. Instead, all subsequent Mars EDL systems including those planned in the foreseeable future rely on inflation qualification by similarity to the Viking design and focus on parachute strength qualification through lower-cost subsonic and static testing. The Viking project selected a disk-gap-band (DGB) parachute, that consists of a canopy, a small gap, and a cylindrical band. This system had a 16 m diameter. The Mars Science Laboratory has baselined a slightly larger diameter supersonic parachute than the one flown on Viking (19.7 m diameter). To land greater masses on Mars, a new high-altitude supersonic parachute qualification program will be needed that is likely to be expensive and time-consuming.

Once subsonic conditions are achieved, a larger parachute that is less expensive to qualify can be deployed to reduce the velocity further, as well as potentially provide added time for lander reconfiguration and sensing events. For large-mass landed systems, such staged parachute systems may provide compelling system benefits to outweigh their inherent complexity and risk.

4.6.2.4   *Landing accuracy*

All previous Mars landers flew ballistic (non-lifting) entries, and as such had no means of exerting aerodynamic control over the atmospheric flight path. As such, the design landing footprints were relatively large (80 to 200 km downrange ellipse major axis). The addition of improved approach navigation can likely reduce this somewhat (to about 60 km), in order to improve lander accuracy further, a real-time hypersonic guidance algorithm to autonomously adjust its flight within the Mars atmosphere to achieve closer proximity to a desired landing site. The MSL mission will take the first major step toward performing precision landing at Mars, combining aeromaneuvering technology with improved approach navigation techniques to set down within 10 km of a specified target. Such an advance is possible as a result of improved interplanetary navigation techniques and the qualification for flight of a lifting aeroshell configuration directed by an autonomous atmospheric guidance algorithm that controls the aeroshell lift vector during the high dynamic-pressure portion of atmospheric flight.

4.6.3   Entry, descent, and landing requirements for human missions to Mars

4.6.3.1   *Initial results*

Human missions to Mars will require delivery of vehicles to Mars orbit, delivery of cargo to the surface, and delivery of the crew to the surface. While direct entry might prove to be viable for delivering cargo to the surface, NASA appears to have a strong preference for a two-step process: initial insertion into Mars orbit followed by de-orbiting and descent to the surface. Crew transfer to the surface is clearly earmarked for the two-step process. Therefore, direct entry will not be considered further in connection with human missions to Mars.

The masses of vehicles involved in human missions to Mars are likely to be greater than 30 mT and possibly larger than 60 mT. If chemical propulsion were to be used for these transfers, the required IMLEO would increase by several thousand mT, making the cost and complexity of such a mission prohibitive. Therefore, use of full aero-assist has been assumed by all DRMs. Using aero-assist to inject such large masses into Mars orbit, or to transfer them to the surface, poses major technical challenges.

Pinpoint landing will be required to perhaps 10 m to 100 m accuracy. The reliability of entry systems will have to be extraordinarily high. Environmental control during entry will be very demanding, in regard to both temperature and g-forces.

All deliveries to Mars will begin with aerocapture into Mars orbit. Aerobraking is not considered to be appropriate because of the long duration of the process.

Braun *et al.*[80] carried out simulation models of the orbit insertion and descent processes. For the orbit insertion step, they assumed a $5g$ limit to be the maximum tolerable deceleration for short periods by a crew of de-conditioned astronauts.

---

[80] *Entry, Descent, and Landing Challenges of Human Mars Exploration*, G. Wells, J. Lafleur, A. Verges, K. Manyapu, J. Christian, C. Lewis, and R. Braun, R., AAS 06-072.

A study of aerocapture trajectories was performed to determine bounding entry velocities that would allow for at least a 1° aerocapture entry corridor width into a 400 km circular orbit around Mars. This was done to accommodate a navigation requirement of ±0.5° on flight path angle at arrival. For a vehicle with lift/drag ratio $L/D = 0.3$, this corridor width limit yields an entry velocity constraint between 6 and 8.8 km/s that corresponds to a wide range of interplanetary trajectory options. Therefore, they concluded that an entry vehicle with an L/D of 0.3 would be sufficient for Mars aerocapture.

After aerocapture, a vehicle delivering humans or cargo to the surface must perform entry, descent, and landing. As in the case of aerocapture, a 5g acceleration limit was assumed. Braun *et al.* estimated the final altitudes of the initial descent trajectories as a function of mass with varying end conditions of Mach 2, 3, and 4. These results define how much altitude is available to provide descent deceleration via parachutes or propulsion or some combination. They considered two options: (1) a single heat shield used for both aerocapture orbit insertion and descent, and (2) a dual heat shield set in which one shield is used for aerocapture orbit insertion and jettisoned prior to descent using the second shield.

The single heat shield approach is appealing in its simplicity. However, Braun and Manning[81] raise three concerns about this approach. First, since the TPS must be sized for the harsher aerocapture environment, the vehicle performs its entry-from-orbit with a more massive, higher ballistic coefficient heat shield than would nominally be required for descent, exacerbating heating and deceleration concerns. This also depresses the altitude at which the vehicle has slowed to its Mach 3 or 4 transition altitude. Second, if the vehicle does not jettison the aerocapture heat shield following the aerocapture maneuver, it must be designed to withstand a large amount of heat soaking back into the vehicle structure from the thermal protection system (TPS). Finally, a third challenge results from the fact that the vehicle, having been inserted into Mars orbit, must operate functionally prior to descent. It is not clear how to accommodate power, thermal, orbit-trim propulsion, communications, and other spacecraft function needs from within the confines of the aeroshell, which implies that the backshell (and possibly the forebody) of the vehicle must allow openings for items such as solar arrays, radiators, engines, thrusters, and antenna.

The alternate approach for the TPS configuration would be to use separate, nested heat shields for aerocapture and EDL. This provides the benefit of jettisoning the hot aerocapture TPS immediately following the aerocapture maneuver, and allows the use of much lighter TPS for entry, thus minimizing the vehicle's ballistic coefficient for that maneuver. The disadvantage of this approach is that packaging two nested heat shields on the vehicle requires a means of securing the primary heat shield to the structure and separating it without damaging the secondary heat shield, and likely results in an overall mass penalty.

Analysis of entry/descent systems was made for 10 m and 15 m diameter heat

---

[81] "Mars Exploration Entry, Descent and Landing Challenges," R. D. Braun and R. M. Manning, *Aerospace Conference, 2006 IEEE, March 2006.*

shields, using single or dual heat shields, and using propulsive descent with or without assistance from a 30 m parachute.

The mass of the entire entry, descent, and landing (EDL) system was computed by combining the estimated masses of each of the major EDL subsystems (heat shield, main propulsion system, reaction control system [RCS], backshell, and parachute). Additionally, an EDL dry mass margin of 30% was included.

Propulsion   The primary descent propulsion system was assumed to be a liquid bipropellant engine using a LOX/CH_4 propulsion system with an assumed specific impulse of 350 s. The total mass of propellant was calculated by integrating the equations of motion during powered descent.

Parachute   It was found by analysis that if the parachute is sufficiently light, it can reduce the overall EDL mass. However, if it is too heavy, the EDL mass is minimized with all-propulsive descent with no parachute. The mass of the parachute system is theorized to be somewhere between 1% and 8% of the total entry mass. In each case, the parachute mass percentage was varied, and the overall EDL mass was compared with and without the parachute. From this, the maximum allowable parachute mass percentage was determined for which the parachute produced a net benefit compared with all-propulsive descent.

Heat shield   The mass of the heat shield is due to (1) the amount of ablative TPS material required, and (2) the mass of the underlying structure. The TPS mass was estimated by calculating the required thickness due to estimated heating rates. In the single heat shield scenario, where the same heat shield is used for Mars aerocapture and entry, the total required thickness was estimated as the sum of the required thickness for each individual stage. The underlying heat shield structure was estimated as 10% of the total entry mass.

Backshell   The backshell mass was estimated as 15% of the total entry mass. It is common in human mission designs for the aeroshell to be integral with vehicle's primary structure because studies have shown that this design results in a net increase in the payload delivered to the Martian surface.

Reaction Control System   The vehicle's RCS configuration was assumed to use the propellant combination of monomethyl hydrazine and nitrogen tetroxide, with a specific impulse of approximately 289 seconds. The propellant tanks and propulsion systems were sized using conventional methods.

An initial study compared scenarios that use an aeroshell for descent down to ~Mach 3, and then propulsively decelerate without the use of parachutes for the ultimate descent to the surface. Parachutes were not included in this initial assessment due to the large uncertainty in their mass. This initial comparison was used to identify the most promising heat shield strategy (single vs. dual), lift-to-drag ratio of the entry vehicle, and Mach number for the final gravity turn to point vertically downward. The parameters that were varied were:

- Entry vehicle $L/D = 0.3$ or 0.5.
- Heat shield: single or dual, 10 m or 15 m diameter.
- Final gravity turn: at Mach 2 or Mach 3.
- Total mass at Mars entry (sum of entry system + payload): 20 mT to 100 mT.

The figure of merit chosen was the percent of total entry mass assigned to the entry system (the payload percent of total entry mass is 100 minus this percentage). *It is important to note that the best system has the lowest figure of merit.* In general, it was found that use of a single heat shield resulted in better figures of merit as compared with use of dual heat shields. For the set of runs covering all the parameters, the figure of merit varied from a low of 65% to a high of 94%. For total entry masses less than 60 mT, the best figure of merit averages to about 67%, for a 15 m heat shield and an $M = 2$ gravity turn. This indicates that a payload of one-third of the mass approaching Mars can be delivered to the surface in this mass range (the maximum payload meeting this condition is 20 mT). For higher total entry masses, use of a single heat shield with $L/D = 0.3$ or 0.5 with an $M = 3$ gravity turn led to similar results. The figure of merit for total entry masses from 60 to 100 mT for this design was 70%, indicating a maximum payload of 18 mT to 30 mT over this range of total entry masses.

Using this scenario as a point of departure, the potential advantages of a parachute system (up to 30 m diameter) were assessed. Examination of these results showed that the parachute-augmented system is superior as long as the parachute mass does not exceed ~3% of the total entry mass and is deployed at Mach 3. Figure 4.19 shows the figure of merit as a function of total entry system mass for several parachute mass percentages.

Noting that the mass percentage of the parachute is likely to be in the 3% to 4% range, it is a toss-up as to whether the parachute provides any benefit. Assuming an entry mass of 60 mT, a comparison of the mass breakdowns for both scenarios is illustrated in Figure 4.20. In this example, the system equipped with a parachute (3% of total entry system mass) is capable of delivering a payload mass fraction of 0.33 to the surface, while the propulsive-only option can only deliver a payload mass fraction of 0.29.

Braun *et al.* conclude: "Because of expected landed mass requirements, it is concluded that Mars human exploration aerocapture and EDL systems will have little in common with current and next-decade robotic systems. As such, significant technology and engineering investment will be required to achieve the EDL capabilities required for a human mission to Mars. Technology advances that require further analysis include aerocapture, ... high Mach aerodynamic deceleration concepts other than parachutes, supersonic propulsive descent capabilities, and thermal protection and structural concepts for large-diameter aeroshell systems."

Braun *et al.* did not divide the EDL system into separate parts for aerocapture into orbit and descent to the surface. It seems likely that the majority of the heat shield and aeroshell masses are attributable to orbit insertion whereas the majority of propulsion mass (and, of course, the parachute) would be attributable to descent. Based on Figure 9.6, it would appear that the total EDL system mass can be allocated

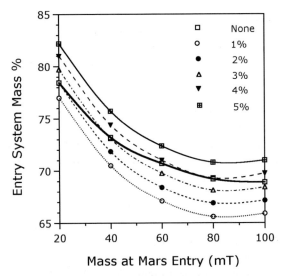

Figure 4.19. Percentage of total mass at entry consumed by EDL-related systems for various mass fractions. These data correspond to a vehicle diameter of 15 m and 0.3L/D. [Based on data from *Entry, Descent, and Landing Challenges of Human Mars Exploration*, G. Wells, J. Lafleur, A. Verges, K. Manyapu, J. Christian, C. Lewis, and R. Braun, AAS 06-072.]

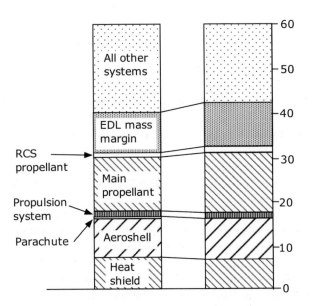

Figure 4.20. Mass breakdowns for 15 m diameter, 60 mT, single heat shield vehicle with and without a parachute (3% of total entry mass), 0.3L/D. "All other systems" implies the payload that is either 20 or 17.4 mT. [Based on data from *Entry, Descent, and Landing Challenges of Human Mars Exploration*, G. Wells, J. Lafleur, A. Verges, K. Manyapu, J. Christian, C. Lewis, and R. Braun, AAS 06-072.]

roughly 56% to aerocapture orbit insertion and 44% to descent and landing. Since the total EDL system mass was estimated to be roughly 70% of a large mass approaching Mars, the mass requirements for entry systems are estimated as follows. For a 100 mT mass approaching Mars, about 30 mT can be landed on Mars. The remaining 70 mT consists of two subordinate entry systems, with ~40 mT allocated to aerocapture orbit insertion and ~30 mT allocated to a descent and landing system. Thus, a 40 mT entry system can place 60 mT into Mars orbit, but 28 mT of the mass in orbit represents the descent and landing system. We conclude that, for large masses, the entry system for orbit insertion is estimated to be roughly $40/60 \sim 67\%$ of the mass inserted into orbit, and the entry system for orbit insertion, descent, and landing is about $70/30 = 230\%$ of the landed payload mass.

4.6.3.2   Updated results: 2006

In a more recent study,[82] the Georgia Tech group pointed out that most of the EDL challenges center around the heat shield design. They used carbon–carbon (rather than fiber-form) as the thermal protection material and found that a single heat shield provided better performance. However, they estimated that fabrication limitations will prevent any one carbon–carbon panel from exceeding approximately 5 m in its largest dimension, and that the carbon–carbon sections must therefore be paneled to cover larger diameter vehicles, resulting in numerous seams along the windward side of the vehicle. After entry, the protected descent engines must be exposed. They suggest two options: (1) jettison the heat shield prior to engine ignition and (2) open doors in the heat shield to expose the engines. They report significant challenges in both of these options. Other options involving either engines that deploy off the side or top of the vehicle or facing towards the leeward side of the vehicle were dismissed as being too complex or risky.

They also found that cargo unloading on the Martian surface imposes configuration challenges. The first problem is unloading the cargo from a high platform above the surface. Due to the landing gear and the size of the engines and propellant tanks, it is unlikely that the cargo will be very close to the Martian surface. A second problem is the question of whether the backshell is dropped prior to landing. In landing a crewed vehicle or one with a pressurized payload, significant mass, volume, and complexity penalties may be incurred by separating the backshell from the pressure vessel and primary structure. The study therefore used a backshell that was integral to the vehicle's primary structure and was not discarded. Descent engine configuration also presented a number of interesting challenges for a vehicle of this shape.

The study included 10 m and 15 m diameter entry vehicles. The heat shield consisted of the thermal protection system (TPS) material and the underlying structure. The carbon–carbon was assumed to be 1 mm thick (density of 1890 kg/m^3) and

---

[82] *Sizing of an Entry, Descent, and Landing System for Human Mars Exploration*, John A. Christian, Grant Wells, Jarret Lafleur, Kavya Manyapu, Amanda Verges, Charity Lewis, and Robert D. Braun, Georgia Institute of Technology, preprint, 2006.

the thickness of the fiber-form insulation (density of 180 kg/m^3) was chosen such that the temperature at the bondline remained below 250°C at all times. The underlying heat shield structure was estimated as 10% or 15% of the total entry mass for the 10 m or 15 m diameter vehicles, respectively.

An important question is whether it is feasible to use parachutes in the descent sequence for human-scale Mars missions. The study concluded that "material strength and stability concerns led to a limit on the parachute diameter of 30 m." All of the cases investigated in the study pushed up against this limit and required a 30 m diameter parachute. A parachute-sizing tool was developed to estimate parachute mass.

The backshell and primary vehicle structure was assumed to constitute 25% of the vehicle dry mass.

It was assumed that the propulsion system used during the powered flight portion of the EDL sequence utilized LOX/CH$_4$ with a specific impulse of ~350 seconds at a mixture ratio of 3.5.

As in previous work by this group, a mass contingency of 30% was applied to the entry vehicle dry mass. See Table 4.17.

The figure of merit for the study was the *payload fraction* = ratio of landed payload mass to initial mass (sum of masses of payload plus entry system). They compared results for $L/D = 0.3$ and $L/D = 0.5$ entry vehicles, and they found that the lower L/D can land slightly larger payloads on the Martian surface. Results are shown in Figure 4.21.

It can be noted from Figure 4.21 that the payload fraction is higher using the smaller diameter entry vehicle. However, it was impractical to use the smaller entry vehicle for initial mass greater than 60–80 mT (landed payload mass ~18–20 mT). Using the larger entry vehicle (15 m diameter), an initial mass of ~100 mT can land 25 mT of payload for a payload fraction of ~0.25. In JSC DRMs, it was assumed that the payload fraction was ~0.7.

Another important consideration is how much volume is available within the aeroshell. The study concluded that pressurized volumes will be ≤60 m^3 and 287 m^3 in the 10 m and 15 m diameter entry vehicles, respectively. A significant portion of this pressurized volume will be taken up by cargo, crew accommodations, and other subsystems, resulting in a habitable volume that is noticeably lower. The study concluded that while such a small capsule might be allowable for the short descent and ascent transits, it would be very inadequate as a long-term Habitat on the surface. They concluded that a dedicated Habitat is needed for a human Mars mission. However, a rigid Habitat of sufficient size cannot fit within the capsule shape. An inflatable Habitat would likely be necessary.

4.6.4 Precision landing

Every Mars DRM prepared so far, and just about any conceivable Mars mission concept one can imagine, requires precision pinpoint landing, including landing payloads at the same site that were launched at different launch opportunities (separated by 26, 52, ... months). At this point, it is not certain what level of

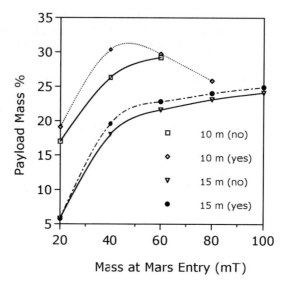

Figure 4.21. Payload delivered to the Martian surface as percentage of initial mass for 10 m and 15 m diameter entry vehicles with $L/D = 0.3$. Yes and no refer to use of a parachute. [Based on data from *Sizing of an Entry, Descent, and Landing System for Human Mars Exploration*, John A. Christian, Grant Wells, Jarret Lafleur, Kavya Manyapu, Amanda Verges, Charity Lewis, and Robert D. Braun, Georgia Institute of Technology, preprint, 2006.]

Table 4.17. Mass breakdown for 15 m diameter, 60 mT entry vehicle for $L/D = 0.3$. [Based on data from: *Sizing of an Entry, Descent, and Landing System for Human Mars Exploration*, John A. Christian, Grant Wells, Jarret Lafleur, Kavya Manyapu, Amanda Verges, Charity Lewis, and Robert D. Braun, Georgia Inst. of Technology, preprint, 2006.]

Element	Mass % without parachute	Mass % with parachute
Payload	21.4	23.2
Heat shield	17.7	17.7
Backshell and structure	14.2	14.3
Propulsion system	2.8	2.5
Parachute		0.4
Command, control, and communication	0.8	0.7
Power	2.2	2.2
Landing gear	1.8	1.9
Margin	17.1	17.1
Descent propellant	13.7	11.7
RCS and OMS propellant	7.9	7.9

landing accuracy will be required. That will depend upon the mobility of the surface elements that must be interconnected on Mars. However, it seems likely that the requirement will be for landing site accuracy to within about 10 m to 100 m. It is possible that the requirement may turn out to be that the aim point must be controllable absolutely to within 10 m, and the actual landing must be within 10 m of the aim point (relatively).

As discussed previously, it is claimed that the MSL mission will take the first major step toward performing precision landing at Mars, combining aeromaneuvering technology with improved approach navigation techniques to set down within 10 km of a specified target. The needs for human missions then involve scaling up the size of landed payloads by perhaps a factor of ~50, and improving the landing accuracy by about a factor of 1,000. It is not clear at this point what the requirements might be to achieve such demanding requirements, nor is it clear whether an approach for accomplishing this has been formulated.

Paper studies[83] have analyzed approaches for achieving pinpoint landing on Mars for robotic science landers. The EDL process relevant to pinpoint landing is divided into four phases: (1) approach, (2) hypersonic entry, (3) parachute phase, and (4) powered descent. The approach phase starts at the last propulsive maneuver before atmospheric entry and ends at entry interface (typically ~125 km, for Mars). The hypersonic entry phase starts at entry interface and ends at deployment of the parachute (or, if there is more than one parachute, the first parachute). The parachute phase starts at deployment of the first parachute and ends at ignition of the descent engines. The conditions required for successful parachute deployment in the thin Martian atmosphere (high enough dynamic pressure and low enough Mach number) are achieved at altitudes of typically 10 km or less above the surface of Mars. Parachute deployment must occur high enough to assure enough time for deceleration to a soft landing and completion of all required events before landing. This is especially difficult at higher landing site elevations and in seasons of low atmospheric density, where the atmosphere is thinnest and the demands on the system are greatest. The powered descent phase starts at ignition of the descent engines and ends at touchdown. The terminal velocity of a parachute in the thin Martian atmosphere is significantly higher than in Earth's atmosphere. Descent rates are typically in the ~50 m/s range at Mars with parachutes of reasonable size, requiring the use of thrust to slow the descent enough to permit a successful soft landing.

The factors that contribute to these position uncertainties at landing include uncertainties in navigating the spacecraft to the desired entry point in the atmosphere, atmospheric modeling uncertainties, uncertainties in vehicle aerodynamic coefficients, map-tie error, and wind drift. The estimated landing accuracy of Mars landers to date has steadily improved, from ~150 km of the target for Mars Pathfinder to ~90 km for Mars Polar Lander (not actually flown) to ~35 km for the Mars Exploration Rovers.

[83] *Systems for Pinpoint Landing at Mars*, Aron A. Wolf, Claude Graves, Richard Powell, and Wyatt Johnson, AAS 04-272.

The improvements made so far in landing accuracy have mainly been due to improved approach navigation. All previous landers flew unguided ballistic trajectories during EDL. Various optical-imaging improvements to approach navigation have been developed and will be used on the Phoenix and MSL missions with a goal of achieving a landing error less than about 10 km.

A rough estimate is that using the best possible approach navigation system, the landing error ellipse would be about 30 km × 3 km if there were no further introduction of uncertainties (as, for example, by winds) or any further trajectory corrections during entry and descent. It does not seem likely that the landing error ellipse could be reduced further by merely attempting to improve approach navigation. It is also estimated that using the current state of the art, additional uncertainties involved in descent propulsion and parachute drift (without applying new technology for correction) would increase the actual landing ellipse to about 38 km × 5 km.

In order to reduce the landing error further, active control is required during the hypersonic entry, parachute, and powered descent phases. Control during the hypersonic entry phase, during which the entering vehicle is slowed down to ~Mach 3 to permit deployment of a supersonic parachute, requires two innovations: (a) an entry vehicle with a moderate lift/drag ratio to enable steering the vehicle by changing its angle of attack, and (b) an optical means of observing the position of the vehicle relative to known features on the surface. These would be compared with onboard stored maps (previously taken from orbiters) to locate the position and speed of the vehicle while it is in hypersonic entry, thus allowing commands to be made to the lifting entry vehicle to steer it toward the desired landing point. Wolf *et al.* suggest that if there were no further errors due to descent propulsion uncertainty and wind drift of the parachute, the combination of best possible approach navigation plus advanced feature mapping with a vehicle having $L/D \sim 0.20$ to 0.25 could yield a landing error of several tens of meters. However, wind drift is likely to produce an uncertainty in landing of about 5 km and variations in descent propulsion could add another 0.5 to 1 km. These uncertainties can be greatly reduced by using descent imaging with feature recognition based on stored onboard maps. To correct for these factors, it is necessary that the parachute be deployed at a sufficiently high altitude, and there must be additional propellant added to the descent propulsion system with which to apply control authority to steer the descending lander in the final minute or two of descent.

In principle, pinpoint landing can be achieved in the following way. Using a best possible advanced approach navigation system, but no subsequent controls, the landing error ellipse is likely to be about 38 km × 5 km. By comparing imaging with stored surface maps (originally taken from orbit) during hypersonic entry, and using a lifting vehicle, most of the error that would otherwise occur during entry can be removed. This would still leave a landing error of >5 km due to uncertainties from winds and descent propulsion. Most of this can be removed using descent imaging in combination with onboard stored feature maps. In practice, the requirements for developing and validating such capabilities for large vehicles present serious technical and financial challenges.

A very recent study[84] reaffirms that the Mars Science Laboratory mission scheduled for 2009 launch will use guidance during hypersonic entry to improve this to
~10 km. To achieve pinpoint landing (within 100 m) for future missions, ways of
addressing the remaining error sources (approach navigation, wind drift, and map-tie
error) must be found. This reference claims that 100 m landing precision can be
achieved with the reference system design indicated in the study, but a great deal
more work appears to be required.

4.6.5   Development, test, and validation program

A NASA Roadmap Team prepared a roadmap for EDL technology in 2005.[85]
The team represented the majority of the planetary aero-EDL (AEDL) experience
in the U.S. This team met several dozen times and held three 3-day workshops.
There was universal agreement that today's robotic AEDL systems will not scale
up to human-scale AEDL, and we do not know which alternative path will ultimately
succeed at human scale. The Roadmap Team discussed and reviewed the many
challenges, but one challenge rose above the rest: we do not know how to get from
Mach 4 down to subsonic speeds with sufficient remaining altitude. Possible
approaches include supersonic propulsive decelerators, Rose Bowl size supersonic
parachutes, and inflatables. In addition, there is a need to understand the deceleration
effects that impact physiological performance on astronaut's duties during and just
after aerocapture and entry, as well as a need to better characterize atmosphere
variations—winds, density, dust—and how robust the human-scale AEDL system
will be to these variations.

The Roadmap Team envisaged a development and demonstration program for
AEDL that would last about 25 years. The first ~12 years would be divided into three
parallel tracks as summarized in Tables 4.18 and 4.19.

The cost and duration of the EDL development and validation program will be
determined by the number, scale, and sophistication of precursor tests, particularly
when tests must be carried out at Mars. Obviously, maximum use should be made of
testing in the Earth's upper atmosphere where densities can be comparable with those
near the surface of Mars, but such tests have limitations, and ultimately some testing
will have to be done at Mars. The NASA EDL roadmap is not particularly clear on
this aspect. It does say that: "full-scale AEDL Flight Tests can and should be done at
Earth (in order to get fast turn around between multiple tests)." In regard to the
question of whether a "full-scale validation flight test" is needed at Mars, the NASA
EDL roadmap says: "Not, specifically, but the AEDL community is very uncomfortable with the notion of the very first full scale AEDL being piloted. The full-scale
unpiloted AEDL advance cargo mission that immediately precedes the human land-

[84] *Trades for Mars Pinpoint Landing*, Aron A. Wolf, Jeff Tooley, Scott Ploen, Mark Ivanov,
Behcet Acikmese, and Konstantin Gromov, IEEEAC Paper #1661 (2006).
[85] "Aerocapture, Entry, Descent and Landing (AEDL) Capability Evolution toward Human-
Scale Landing on Mars," Rob Manning, *Capability Roadmap #7: Human Planetary Landing
Systems*, March 29, 2005.

Table 4.18. Brief summary of EDL roadmap (first half). [Based on: *Aerocapture, Entry, Descent and Landing (AEDL) Capability Evolution toward Human-Scale Landing on Mars*, Rob Manning, Capability Roadmap #7: Human Planetary Landing Systems, March 29, 2005.]

Track	Years ~1–4	Years ~5–8	Years ~9–12
Human-scale Mars AEDL development	Decide what could work: AEDL architecture assessment	Decide what to baseline: at Earth sub-scale AEDL component development and architecture evaluation flight testing	Validate AEDL models: scaled Mars AEDL validation flight(s)
Validate Mars models	Surface models	Atmosphere models via Mars missions	
Utilize Mars Robotic Overlap Technology	Pin point landing radar, terrain-relative navigation, terminal guidance, hazard avoidance sensors		

Table 4.19. Brief summary of EDL roadmap (second half).

Track	Years ~13–16	Years ~17–20	Years ~21–25
Human-scale Mars AEDL development	Continue to validate AEDL Models: scaled Mars AEDL Validation flight(s)		
	Develop and qualify full-scale hardware: Earth-based full-scale development and at-Earth flight test program		
		Fly first Mars human-scale demonstration missions (>40 mT landing capability)	
Validate Mars models	Continue atmosphere models via Mars missions		
Utilize Mars Robotic Overlap Technology	AEDL communication and navigation support		

ing could do the trick." Instead, the NASA EDL roadmap says that AEDL Flight Tests at perhaps 1/10 scale are needed at Mars. It may be possible to establish scaling laws from large tests in Earth orbit, and use these laws to infer full-scale performance at Mars from tests at ~1/10 scale at Mars. However, this seems unlikely to this writer. The systems that require validation are not merely entry, descent, and landing, but pinpoint landing as well. It may well be that cargo landings may provide sufficient validation, but at least two such landings with demonstration of an established infrastructure on Mars would likely be required prior to crew delivery to Mars.

The NASA EDL roadmap envisages a ~25-year development and validation process. No cost estimate was provided, but it seems likely that the cost might be >$10B.

validation, but at least two, such landing, with demonstration of an established infrastructure on Mars would likely be required prior to crew delivery to Mars. The NASA EOL roadmap envisages a 20 year development and validation process, but cost estimate was provided, but it seems likely that the cost might be $10B.

5

In situ utilization of indigenous resources

5.1 VALUE OF ISRU

In situ resource utilization (ISRU) on the Moon or Mars is an approach for converting indigenous resources into various products that are needed for a mission. By utilizing indigenous resources, the amount of material that must be brought from Earth is reduced, thus reducing IMLEO. ISRU has the greatest value when the ratio:

$$R = \frac{[\text{mass of products supplied by ISRU to mission}]}{[\text{mass of the ISRU system brought from Earth}]}$$

is large. Thus, in order for ISRU to have net value, it is essential that the mass of the ISRU system (i.e., the sum of the masses of the ISRU plant, the power plant to drive it, and any feedstocks brought from Earth) must be less than the mass of products produced and used by the mission. If $R \gg 1$, then in comparing the IMLEO for two similar missions—one using ISRU, and the other not using ISRU—the IMLEO using ISRU will be lower. This comparison of IMLEO with and without ISRU will provide one measure of the "value" of ISRU. However, from a broader point of view, one should compare total investments (rather than IMLEO) with and without ISRU. In this regard, the investment in ISRU includes the costs of (a) prospecting to locate and validate the accessibility of indigenous resources, (b) developing and demonstrating capabilities to extract indigenous resources, (c) developing capabilities for processing indigenous resources to convert them to needed products, and (d) any ancillary requirements specifically dictated by use of ISRU (e.g., possibly a nuclear power system). The cost saving using ISRU is the investment that is eliminated by reducing IMLEO for as many launches as the ISRU system serves. If this saving is greater than the investment required, then ISRU has net value for a mission or set of related missions. If one only compares ISRU system mass with ISRU product mass, one may in some cases conclude (incorrectly) that ISRU is beneficial, when in actuality it adds to the overall mission cost. Typically, NASA plans human missions without the use of ISRU, and then considers tacking on ISRU rather late in the campaign as an embellishment. This limits the putative benefits of ISRU because all vehicles are sized without utilizing the benefit of ISRU.

In the short run (next ~40 years) the main products that might be supplied to human missions by ISRU are:

- Propellants for ascent, thus reducing the mass of propellants that are brought from Earth. Propellants could also be used for intra-surface transportation.
- Life support consumables (water and oxygen). It is possible that, while on the surface, the usual life support cyclic system could be eliminated using ISRU to provide these commodities.

In addition, in principle, regolith can be piled up on top of a Habitat for radiation shielding. However, it remains to be seen whether NASA designs for Surface Habitats allow for use of regolith as shielding from radiation. In view of existing designs for Habitats it seems very unlikely that lunar and Mars Habitats will be compatible with regolith radiation shielding (see Section 4.5).

In the longer run, a wider range of products could possibly be produced, as the industrial and electronic revolution is transferred from Earth to extraterrestrial bodies. Unfortunately, there is no clear path leading from where we are now to such an ultimate utopia.

Although we are primarily interested in Mars, it is worthwhile to first examine NASA plans for lunar ISRU because NASA seems to be intent on concentrating its resources and efforts on lunar ISRU for the next couple of decades.

5.2   LUNAR ISRU

5.2.1   Introduction

The NASA ESAS report views lunar exploration as a stepping-stone to Mars. This report uses the term "extensibility" or "extensible" 62 times (mainly in regard to selecting lunar mission technology that can also be used for Mars missions). One of the important "stepping-stones" is demonstration of ISRU technology. The term "ISRU" occurs 106 times in the ESAS report. A few choice quotes are given below:

- "The ESAS architecture has two primary goals for lunar exploration. The first is developing and demonstrating the capabilities needed for humans to go to Mars and the second is lunar science. ISRU is a blend of science and the development of exploration capabilities. Specific requirements for ISRU will change based on what future lunar robotic probes may discover on the surface, but the benefits of reduced logistics and extended mission durations associated with ISRU are highly desirable."
- "ISRU: Technologies for 'living off the land' are needed to support a long-term strategy for human exploration."

- "The [lunar] lander's propulsion system is chosen to make it compatible with ISRU-produced propellants and common with the CEV SM propulsion system."
- "The SM propulsion for performing major CEV translational and attitude control maneuvers is a . . . propulsion system . . . using LOX and Liquid Methane (LCH$_4$) propellants. This propellant combination was selected for its relatively high I_{SP}, good overall bulk density, space storability, non-toxicity, commonality with the LSAM, and *extensibility* to In-Situ Resource Utilization (ISRU) and Mars, among other positive attributes."

5.2.2   Potential products of ISRU

Most discussions of lunar ISRU seem to assume that resources are readily available, and they proceed to emphasize processing, while minimizing logistics (prospecting, excavating regolith, regolith transport, deposition into and removal of regolith from reactor, dumping waste regolith, etc.) However, the quantity and composition of end products provides the entire basis for considering the use of lunar ISRU, and for setting the requirements for lunar ISRU systems. Therefore, we begin here with the potential end products.

5.2.2.1   *Ascent propellants*

In the initial NASA ESAS architecture, the propulsion system for ascent from the Moon was based on $CH_4 + O_2$ propellants in order that ISRU-generated oxygen from the Moon could be used. Although methane had to be brought from Earth, it provided an implicit connection to Mars ISRU by using oxygen as the oxidizer for ascent. Later, when the realities of cost and schedule to develop $CH_4 + O_2$ propulsion system became clearer, this ascent propulsion system was dropped in favor of space-storable propellants that are incompatible with lunar ISRU. Yet, NASA continued to claim that ISRU was a major part of the lunar exploration program!

However, the entire 2005–2006 architecture is being re-engineered. If the final architecture returns to using oxygen as an ascent propellant, that oxygen can potentially be provided by lunar ISRU. In the original 2005–2006 architecture, the plan was to have two ascents per year from the outpost, each requiring about 4 metric tons (mT) of oxygen, for an annual need of roughly 8 mT. It is not clear what fuel would be used in conjunction with the oxygen. If it is methane, it will have to be brought from Earth. If it is hydrogen, it could conceivably be produced from polar ice (but not from equatorial regolith). In a NASA release in February, 2007,[86] NASA persists in using space-storable propellants for ascent, which are incompatible with ISRU.

Since the "gear ratio" (mass in LEO/mass on surface) for polar outposts is about 4:1, the potential mass saving in LEO is ~16 mT per launch. However, because the Launch Vehicles were designed without lunar ISRU, they will remain unaffected by inclusion of lunar ISRU. Hence the benefit of lunar ISRU will be an ability to deliver extra infrastructure payload (~4 mT) to the outpost with each launch (but rather

---

[86] "MoonHardware22Feb07_Connolly.pdf" on *NASA Watch* website.

late in the campaign—probably beginning in the late 2020s). (Even this minor benefit disappears if NASA persists in its present plan to use space-storable propellants [NTO/MMH] for ascent, thus eliminating oxygen as an ascent propellant.) The "value" of the $\sim 4\,mT$ increase in payload delivery per launch using lunar ISRU can be estimated by noting that over a period of years, with continual infrastructure deliveries to the outpost, a cargo delivery launch might be eliminated once every several years with small incremental increases in mass delivered by each launch.

In addition, if "abort to orbit" remains a requirement for descent, then there is no possibility of providing ascent propellants using ISRU since the incoming LSAM must possess ascent propellants.

5.2.2.2 *Life support consumables*

Oxygen requirements depend on crew activity, but an average value is about 1 kg per crew-member (CM) per day. Water requirements have been estimated by JSC to be about 27 kg/CM-day (see Table 4.1 for requirements).

To support a crew of four during one year, we require $4 \times 1 \times 365 = 1{,}460\,kg \sim 1.5\,mT$ of O_2, and $4 \times 27 \times 365 \sim 40\,mT$ of water.

It is likely that an Environmental Control and Life Support System (ECLSS) will be used to recycle these resources, thus greatly reducing mass requirements. JSC has estimated the mass of ECLSS systems. Using ISS experience as a basis, JSC estimated the mass and power requirements of ECLSS systems for a crew of six on Mars for 600 days (see Section 4.1). We can scale this to a crew of four for 365 days to estimate the mass of an ECLSS system for the Moon. For each resource (oxygen or water) there is a system mass and back-up cache mass to replenish losses. The estimated masses (in kg) of lunar ECLSS systems based on JSC estimates are summarized as:

System	Physical plant mass	Back-up cache mass	Total mass
Oxygen ECLSS	510	380	890
Water ECLSS	4,500	2,700	7,200

Even though lunar ISRU might supply the required amounts of oxygen and water, environmental control will still be required. An oxygen-only lunar ISRU system would reduce the mass of the ECLSS by a small amount and it is probably not worth integrating ISRU-produced oxygen to ECLSS. It appears that the ECLSS for lunar outposts might require replenishment of about $3\,mT$ of water per year. A lunar ISRU system that produces water could provide this replenishment, but it is likely that it would hardly be worth it since the reduction in ECLSS mass would be only a few mT. If the lunar ISRU system entirely replaced the ECLSS for water, the mass saving might be greater, but the ISRU system would be sizable since it would have to supply the entire required 40 mT of H_2O per year.

Exactly how a water-based lunar ISRU system would be integrated to an ECLSS remains to be determined. There might be some mass benefits, but they appear to be modest. If the ECLSS works as well as NASA hopes, there may not be much benefit to joining the lunar ISRU and ECLSS systems.

5.2.2.3   *Propellants delivered to LEO from the Moon*

For a typical Mars-bound vehicle in LEO prior to trans-Mars injection, about 60% of the total mass in LEO consists of $H_2 + O_2$ propellants for trans-Mars injection. If Mars-bound vehicles could be fueled in LEO with H_2 and O_2 delivered from the Moon, then only the remaining 40% of the total vehicle wet mass would need to be delivered from Earth to LEO. The other 60% would be provided from lunar resources. For example, a Mars-bound vehicle that weighs, say, 250 metric tons in LEO, would include about 150 mT of propellant for trans-Mars injection. If fueled by hydrogen and oxygen from the Moon, the mass that would have to be lifted from Earth to LEO would only be about 100 mT instead of 250 mT. This would have a huge impact on the feasibility of launching large Mars-bound vehicles.

The question that we must deal with is: How feasible is it to transfer water (and then by electrolysis produce $H_2 + O_2$) from the Moon to LEO? If this process is efficient, the scheme of supplying propellants to LEO from the Moon may be less costly than launching the propellants from Earth. If the transfer process is very inefficient, it is likely to be less costly to simply deliver propellants to LEO from Earth.

It is implicitly assumed here that accessible water ice can be exploited on the Moon. If that is not the case, this entire concept becomes moot. Furthermore, the process may become untenable if the transfer vehicle masses are too high. If these vehicles are too heavy, all the water ice excavated on the Moon would be used to produce $H_2 + O_2$ to deliver the vehicles, and ultimately no net transfer of water to LEO would be feasible. Therefore, it is necessary to examine the details of the transfer process and estimate what percentage of water excavated on the Moon can be transferred to LEO. The percentage of water mined on the Moon that can be transferred from the Moon to LEO for fueling Mars-bound vehicles can be estimated as discussed in Section 5.4.

The figure of merit is the net percentage of water mined on the Moon that can be transported to LEO for use by Mars-bound vehicles. As this percentage increases, the cost of transporting water to LEO from the Moon becomes more favorable. Section 5.4 provides detailed calculations of the efficiency of transport of water from the Moon to LEO. My best estimate is that most of the water excavated on the Moon is used up in transferring the tankers to LEO, and almost no net water is transferred to LEO. On the other hand, if these tanker vehicles can be made much less massive, such transfer might become feasible.

5.2.2.4   *Propellants delivered to lunar orbit for descent (and ascent)*

Whereas the amount of oxygen required for ascent from the Moon is a rather puny ~4 mT, the amount of oxygen required for descent is over 20 mT.[87] If oxygen (and less importantly hydrogen as well) can be delivered to lunar orbit for fueling Moon-bound descent vehicles, the potential payoff from lunar ISRU would be much higher than if lunar ISRU were used only for ascent propellants. The gear ratio for delivery of payloads to lunar orbit from LEO is roughly 2.5. Therefore, lunar ISRU generation of oxygen as a descent propellant would save >55 mT in LEO. The combination of lunar ISRU-provided ascent and descent propellants (hydrogen + oxygen) would save more than 80 mT in LEO, and this mass saving is likely to increase in the forthcoming revised ESAS architecture if vehicles become more massive. The concept would then be as follows.

NASA would begin by establishing an outpost in a shadowed polar area of the Moon to excavate regolith, extract water, and to some extent, electrolyze water and store hydrogen and oxygen. This would have to be done robotically without crew participation. Is this possible? Who knows?

NASA would design and implement a tanker system for transferring water from the surface of the Moon to lunar orbit, and establish a filling station in lunar orbit to electrolyze water and fill tanks on incoming vehicles with hydrogen and oxygen. This tanker system would act as a shuttle to move back and forth between the lunar surface and lunar orbit, carrying full tanks on the way up and empty tanks on the way down. The percentage of water extracted on the lunar surface that can be delivered to lunar orbit (after providing propellants for descent of the empty tanker) is estimated in Table 5.21 (p. 202). This percentage depends on the inert mass of the tanker vehicle that is represented as a constant times the total mass. The percentage is 39% if the constant is 0.1, but drops to only 1% if the constant is 0.3.

Incoming LSAM vehicles on their way to the surface of the Moon would carry empty ascent and descent tanks, and would be fueled in lunar orbit prior to descent. In case of an unexpected problem, the crew could return to Earth in the CEV and never descend in the LSAM.

This system works (at least on paper) after it is established, but how does it get established? If NASA must send crew-members to the surface to establish the outpost and set up the tanker/refill system, then we are back to "square one" because NASA must send the LSAM with full descent and ascent tanks prior to the establishment of the outpost and the tanker/refill system. The potential equivalent mass saving in LEO is over 80 mT per launch (see Figure 5.8, p. 202).

However, as in the case of lunar ISRU providing only ascent propellants, this >80 mT reduction will not be realized in terms of reduced Launch Vehicle capability if lunar ISRU is adopted by NASA as an afterthought late in the campaign.

---

[87] These propellant masses are based on 2005–2006 data. In February 2007, a NASA release indicated that oxygen will not be used for ascent, and over 30 mT of LOX/LH$_2$ will be used for descent. *MoonHardware22Feb07_Connolly.pdf* on *NASA Watch* website.

5.2.2.5 *Regolith for radiation shielding*

Use of regolith piled on top of Habitats for radiation shielding is probably a legit-
imate potential use of *in situ* resources, but the requirements and benefits require
further study. It is not clear that current Habitat designs and plans for installing them
on the lunar surface are compatible with regolith shielding, nor is the requirement for
moving the regolith known. In fact, if existing designs for Habitats are examined (see
Section 4.5) it seems very unlikely that lunar and Mars Habitats will be compatible
with use of regolith for radiation shielding.

5.2.2.6 *Visionary concepts*

Visionaries and futurists[88] have identified six rationales for going back to the Moon:

(1) *Expansion of humans into space*—the quest for expansion. This theme reaches
    back to the days of exploration of the Earth by seafaring explorers and extrap-
    olates this forward into a parallel era of space exploration and colonization. This
    ideology is not really part of ISRU, but has more to do with the whole rationale
    for exploring space, as discussed in Section 1.3.
(2) *Providing energy to the Earth*. This includes:
    - *Solar Power Satellites (SPSs)*—located in geostationary Earth orbit. Large
      arrays would be assembled in GEO and located above the cities or regions
      that needed power. The arrays would be hundreds of square kilometers in
      dimension. Typical designs for these arrays have yielded masses of 50,000
      metric tons (mT) of material for each 2 GW of power capability. It is claimed
      that these satellites could be manufactured on the Moon. The environmental
      impacts, cost, and risk in use of such systems needs to be investigated by
      skeptics, rather than advocates.
    - *Lunar Power System (LPS)* is similar in principle to the SPS, but consists of
      large PV arrays at both limbs of the Moon beaming energy to Earth by
      microwaves. Silicon solar cells would be manufactured on the Moon from
      regolith, and power beamed to Earth. The environmental impacts, cost, and
      risk in use of such systems need to be investigated by skeptics, rather than
      advocates.
    - ^3He—in support of nuclear fusion on Earth. The Moon's regolith stores low
      ($\sim$25 ppb) but ubiquitous concentrations of ^{3}He that undergoes fusion reac-
      tions with deuterium and with itself. ^{3}He is quite rare on the Earth, but has
      been implanted in the surface of the Moon by the solar wind. It has been
      estimated that 40 mT of ^{3}He could provide current U.S. energy needs for a
      year. However, this would require processing more than a billion mT of
      regolith. Furthermore, it is not clear whether nuclear fusion can be made to
      work in a practical manner even with a ready supply of ^{3}He.

---

[88] *Development of the Moon*, M. Duke, L. Gaddis, G. J. Taylor, and H. Schmitt, *http://
www.lpi.usra.edu/lunar_resources/developmentofmoon.pdf*

(3) *The industrialization of space*—various types of production facilities might eventually be located in space as profit-making ventures or to relieve environmental pressures on Earth. Most investigations of material processing in space to date have been aimed at producing new and unique materials by taking advantage of the micro-gravity environment. Little research has been aimed at production in space, because the cost of space transportation is so high that only products that sell at prices several times the cost/mass of space transportation can be considered. One study concluded that a space manufacturing facility could be profitable if the cost of Earth to orbit transportation fell to about $1,200/kg. However, even if transportation costs from Earth reached that level, there would still be few products valuable enough to merit the transportation costs of the raw materials into space. The Materials Processing in Space Program has yielded very little after 30 years of effort. Industrialization of space appears to be a long way off.

(4) *Exploration and development of the solar system*—The exploration and development of the solar system is critically dependent on low-cost space transportation. If the cost of propellant delivered to particular points in space from the Moon is less than the cost of delivering that amount of propellant from Earth, propellant production on the Moon can be economically competitive, if the demand is large enough to amortize the cost of installing the production facilities. Prospects for refueling depots in space depend critically on reducing the tanker mass. This possibility is discussed extensively in Section 5.4.

(5) *The Moon as a planetary science laboratory*—The Moon provides a natural platform for its own study and for the study of planetary processes, particularly volcanism, crustal evolution, and impact.

(6) *Astronomical observatories on the Moon*—The Moon may provide a particularly useful platform for large astronomical instruments. Although the Moon appears to offer some advantages for astronomy, concepts that can be deployed in deep space may be superior.[89] Further examination of the tradeoffs are required between deep-space facilities (particularly at "L2") and facilities on the Moon.

In late 2004, prior to Dr. Michael Griffin coming aboard NASA, the Exploration Systems arm of NASA attempted to prepare a roadmap for exploration by appointing 26 teams to produce roadmaps for capabilities (13 teams) and strategic (13 teams) in 13 chosen areas. One of the capability teams (ISRU Capability Roadmap Team) was assigned the technology area of ISRU. Their Final Report, dated May 13, 2005, listed the applications of ISRU as shown in Table 5.1.

In situ manufacturing includes: "production of replacement parts, complex products, machines, and integrated systems from one or more processed resources."

---

[89] "Build astronomical observatories on the Moon?," *Scientific American*, p. 50, November 2006; "Does the Lunar Surface Still Offer Value as a Site for Astronomical Observatories?" Daniel F. Lester, Harold W. Yorke, and John C. Mather, *Space Policy*, Vol. 20, 99, 2004.

Table 5.1. Key ISRU capabilities according to ISRU Capability Roadmap Team.

Capability	Mission enabled	Key capabilities and status	
		Current state of practice	Need date
Lunar/Mars regolith excavation and transportation	All lunar ISRU and Mars water, mineral extraction, and construction ISRU	Apollo and Viking experience and Phoenix in 2007. Extensive terrestrial experience	2010 (demo) 2017 (pilot)
Lunar oxygen production from regolith	Sustained lunar presence and economical cis-lunar transportation	Earth laboratory concept experiments; TRL 2/3	2012 (demo) 2017 (pilot)
Lunar polar water/hydrogen extraction from regolith	Sustained lunar presence and economical cis-lunar transportation	Study and development just initiated in ICP/BAA	2010 (demo) 2017 (pilot)
Mars water extraction from regolith	Propellant and life support consumable production w/o Earth feedstock	Viking experience	2013 (demo) 2018 or 2022 (subscale)
Mars atmosphere collection and separation	Life support and mission consumable production	Earth laboratory and Mars environment simulation; TRL 4/5	2011 (demo) 2018 or 2022 (subscale)
Mars oxygen/propellant production	Small landers, hoppers, and fuel cell reactant generation on Mars	Earth laboratory and Mars environment simulation; TRL 4/5	2011 (demo) 2018 or 2022 (subscale)
Metal/silicon extraction from regolith	Large scale in situ manufacturing and in situ power systems	Byproduct of lunar oxygen experiments; TRL 2/3	2018 (demo) 2022 (pilot scale)
In situ surface manufacture and repair	Reduced logistics needs, low mission risk, and outpost growth	Terrestrial additive, subtractive, and formative techniques	2010 to 2014 (ISS demos) 2020 (pilot scale)
In situ surface power generation and storage	Lower mission risk, economical outpost growth, and space commercialization	Laboratory production of solar cells on lunar simulant at <5% efficiency	2013 (commercial demo) 2020 (pilot scale)

The notional architecture associated with this plan included:

- 9 robotic ISRU missions to the Moon prior to 2022.
- 4 human ISRU missions to the Moon prior to 2022.
- 5 robotic ISRU missions to Mars prior to 2022 including 3 prior to 2014.
- Polar lunar ISRU demonstrated in 2010 even without *in situ* prospecting!
- Lunar O_2 extraction demonstrated in 2011 with no defined process.
- Demonstration of *in situ* production of solar cells and utilization on the Moon by 2013.
- Water acquisition on Mars in 2013 with no *in situ* prospecting.
- ISRU robotic hopper demonstrated on Mars 2018.
- Full scale O_2 plant on Moon 2018.
- Fabrication and construction on Moon in 2020.
- ISRU sample return from Mars 2022.
- Phobos ISRU 2025.
- Metal/Si extraction on Mars 2025.
- Deep drill for water on Mars 2028.

The main problems with the scenarios presented by the ISRU Capability Team are:

(1) There is little distinction between elements that are first-generation technologies (oxygen and/or water production) and later-generation technologies (solar cell production, metal extraction, manufacturing on the Moon, . . .).
(2) The putative robotic and human missions prior to 2014 and 2022 will not occur, and even if they did they would not necessarily be primarily dedicated to ISRU.
(3) The emphasis on polar lunar ISRU and water acquisition without *in situ* prospecting is illogical. *In situ* prospecting is a critically important element of any sensible plan to exploit lunar polar resources.
(4) There is no way that the Mars Exploration Program will employ ISRU on a Mars Sample Return mission. JPL Mars Sample Return planners are adamantly opposed to inclusion of ISRU on the mission—for good reasons—it adds needless mass, complexity, and risk.
(5) The plans for implementing ISRU are excessive compared with any realistic rate of progress.

5.2.3   Lunar resources

There are basically four potential lunar resources:

- Silicates in regolith containing typically >40% oxygen.
- Regolith containing FeO for hydrogen reduction. FeO content may vary from 5% to 14% leading to recoverable oxygen content in the 1–3% range.
- Imbedded atoms in regolith from solar wind (typically, parts per million).
- Water ice in regolith pores in permanently shadowed craters near the poles (unknown percentage, but possibly a few percent in some locations).

5.2.4   Lunar ISRU processing

5.2.4.1   Oxygen from FeO in regolith

Hydrogen reduction of FeO in regolith is being developed by JSC as a means of extracting oxygen from regolith. Hydrogen reduction of regolith depends on the reaction of hydrogen with the FeO in the regolith to produce iron and oxygen. The remainder of the regolith does not enter into the reaction. The water (steam) produced in the reactor (at $\sim$1,300 K) is electrolyzed and the oxygen is saved while the hydrogen is recirculated. Some make-up hydrogen is needed, as the process is not 100% efficient. It is not clear how the regolith is fed into and withdrawn from the reactor. It is also not clear how one prevents "gunking up" within the reactor. Some heat recuperation can be accomplished by using heat from steam and perhaps spent regolith to pre-heat incoming regolith, but that will add considerable complexity and, for solid–solid heat transfer, introduce potential failure modes due to "gunking up."

The expected FeO content of two sources of regolith is summarized in Table 5.2.

The energy requirement to process X kg of regolith is the energy to heat the regolith from 200 K to 1,300 K. JSC hopes that the system can recuperate 50% of heat from spent regolith in solid–solid heat exchangers—heat losses were estimated at 10%. This would imply (if taken at face value) that the heat requirement is:

$$\text{Heat} = (X\text{ kg})(0.00023\text{ kWh/kg-K})(1{,}100\text{ K})(0.5 + 0.1) = (0.152X)\text{ kWh}$$

Power requirements to heat regolith to extract oxygen by hydrogen reduction based on the above formula are given in Tables 5.3 and 5.4, assuming that solar power is used and that the duty cycle for the process is 40% (3,500 hours of processing per year). These power requirements are only for the reactor and do not include power requirements for excavation, hauling, liquefaction, and cryogenics. If some sort of

Table 5.2. FeO content of two sources of regolith.

Location	% FeO in regolith	% recoverable O in regolith $= (16/72) * (\%\,FeO)$	mT of regolith needed to generate 1 mT of oxygen
Mare	14	3.1	32
Highlands	5	1.1	90

Table 5.3. Projected power requirement to extract oxygen from Mare regolith assuming 50% heat recovery.

Annual oxygen production rate (mT) $\Rightarrow$	1	10	50	100
Annual regolith rate (mT)	34	336	1,681	3,361
1,000s of kWh	5.1	51	255	510
Hours	3,500	3,500	3,500	3,500
kW to heat regolith	1.44	14.4	72	144

Table 5.4. Projected power requirement to extract oxygen from highlands regolith assuming 50% heat recovery.

Annual oxygen production rate (mT) $\Rightarrow$	1	10	50	100
Annual regolith rate (mT)	96	947	4737	9472
1,000s of kWh	14	143	719	1437
Hours	9864	9864	9864	9864
kW to heat regolith	4.1	40.6	202.9	405.8

magnetic or other pre-processing can be used to beneficiate the regolith, power requirements might be reduced.

Alternatively, if such heat recuperation is not feasible, the power requirement would roughly double.

The technical feasibility of this process has yet to be demonstrated.

The overall process includes systems for excavation of regolith, hauling regolith to the reactor, oxygen extraction in the reactor, and storage of the oxygen in a cryogenic storage system. The following aspects are not known:

- cost to develop and validate this technology;
- requirement for human oversight and control of the process;
- degree of autonomy that can be achieved.

5.2.4.2   Oxygen from regolith silicates

Lunar ISRU based on extraction of oxygen from regolith has two advantages:

(1) Regolith is typically >40% oxygen.
(2) Regolith is available everywhere and solar energy may be feasible for processing.

Unfortunately, the oxygen in regolith is tied up in silicate bonds that are amongst the strongest chemical bonds that are known, and breaking these bonds inevitably requires very high temperatures.

JSC is currently interested in the carbothermal process. This concept is based upon a high-temperature, direct energy processing technique to produce oxygen, silicon, iron, and ceramic materials from lunar regolith via carbonaceous high-temperature (carbothermal) reduction at $\sim$2,600 K. To prevent destruction of the container, they apply heat to a localized region of regolith and the surrounding regolith acts as an insulative barrier to protect the support structures.

The plan is to use a set of solar concentrators to beam direct heating of the regolith in the carbothermal reduction cells. Methane gas is injected into the reduction chamber. According to JSC:

> "The lunar regolith will absorb the solar energy and form a small region of molten regolith. A layer of unmelted regolith underneath the molten region will insulate

the processing tray from the solar energy. Methane gas in the reduction chamber will crack on the surface of the molten regolith producing carbon and hydrogen. The carbon will diffuse into the molten regolith and reduce the oxides in the melt while the hydrogen gas is released into the chamber. Some hydrogen may reduce the iron oxides in the regolith to form water, which will be recovered by the carbothermal system. A moveable solar concentrator will allow heating in the form of a concentrated beam on the regolith surface. A system of fiber optic cables will distribute the concentrated solar power to small cavities formed by reflector cups that concentrate and refocus any reflected energy. Solidified slag melts are removed from the regolith bed by a rake system. Slag waste and incoming fresh regolith are moved out or into the chamber through a double airlock system to minimize the loss of reactive gases."

This scheme would be difficult enough on Earth. On the Moon, it would be far worse. Preliminary testing has not produced any encouraging results. Despite the great challenges involved in extracting oxygen from regolith, documents indicate that JSC remains optimistic that they will succeed. It is difficult not to admire the tenacity of these stalwarts, for whom no engineering challenge is too great or any process too impractical. However, the probability that a practical process for autonomous lunar operation will come from any of this research appears to be very low.

In the extremely unlikely case that a high-temperature processor for oxygen from regolith on the Moon can be made into a practical unit, one would still be faced with the challenges (and costs) for development and demonstration of autonomous lunar ISRU systems for excavation of regolith, delivery of regolith to the high-temperature processor, operation of the high-temperature processor with free flow of regolith through it (without caking, agglomeration, and "gunking up" of regolith), and removal of spent regolith from the high-temperature processor to a waste dump.

5.2.4.3 *Extracting putative volatiles*

According to JSC documents, analysis of lunar rocks from the Apollo missions indicated that heating of the lunar rocks evolved a variety of volatile materials. Hydrogen and nitrogen were reported in the *Lunar Source Book* to be present at the concentration of 10–20 ppm. It was also claimed that "estimates of possible concentrations of H_2 and N_2 in lunar regolith at 50 to 150 ppm and 80 to 150 ppm, respectively," but there are no data that confirm these estimates.

Based on this, JSC is seriously considering the prospect of extracting hydrogen for use as a propellant, and nitrogen for use as an oxygen diluent in breathable air— assuming "a best case scenario" that these putative volatiles are available at the 150 ppm level. It is assumed that the volatiles will be released when the regolith is heated to $\sim$800 K. Assuming that the regolith starts at, say, 200 K, this involves raising the temperature of the regolith by 600°C.

JSC has developed several concepts for implementing this process. One process uses "a large inflatable dome that has a center-driven scraper-wand similar to an agricultural silo top-unloading device." The scraper moves in a circular sweep and the

Table 5.5. Requirements to heat regolith to drive off putative volatiles.

	Units	H_2 1130 kg/yr	N_2 4380 kg/yr
Regolith to be processed	10^3 m^3/yr	37.5	147
Regolith to be processed	10^6 kg/yr	56	219
Regolith processed @ 40% of time	m^3/hr	10.9	41.6
Regolith processed @ 40% of time	10^3 kg/hr	16	63

regolith is directed by a sort of Rube Goldberg arrangement to a ramp where it is heated by IR or microwave heaters. Evolved volatiles are collected by means of either a cryocooler (for N_2) or a hydride bed (for H_2). However, hydride beds are notorious for being easily poisoned by impurities requiring extremely pure H_2 to operate. The need for nitrogen for a crew of four would be about 4,380 kg/year assuming three parts nitrogen to one part oxygen in breathable air. The need for hydrogen for ascent propulsion would be about 1,130 kg/year to go along with 7,350 kg of oxygen.

Not being as optimistic as JSC, the author assumed volatile concentrations of 20 ppm, and solar availability at 50%, leading to a reactor duty cycle of ~40%. Based on this, Table 5.5 is derived.

For nitrogen, the need is to heat 63,000 kg/hr of regolith from 200 K to 800 K.

$$\text{Heat} = (63,000 \text{ kg})(0.00023 \text{ kWh/kg-K})(600 \text{ K})$$

$$= 8,700 \text{ kWh per hr or a steady rate of } 8,700 \text{ kW just for heating the regolith.}$$

Other power requirements are not included. Such a power rate would be prohibitive. Clearly, this process should be discarded immediately.

5.2.4.4   *Utilizing polar ice deposits*

Introduction

Another alternative is to hope for accessible ground ice in permanently shadowed areas near the poles. This approach has the great advantage that removal of water from regolith is a physical (rather than a chemical) process and requires far less energy and much lower temperatures. However, on the negative side, it will take a considerable investment to locate the best deposits of ground ice (if indeed they exist—which still remains to be proven conclusively—and if they are accessible); the percentage of water ice in the regolith is likely to be low, necessitating an extensive prospecting program to find the best, most accessible deposits, ultimately requiring processing a great deal of regolith; excavating ice-filled regolith may prove difficult; the logistics of autonomous regolith delivery, water extraction, and regolith removal from a reactor may prove difficult; and the water extraction process must be carried out in dark, permanently shadowed craters, necessitating the use of nuclear power.

JSC is considering a scheme in which regolith is excavated from a dark region of a crater, and processed in the dark to remove water (estimated at 1.5% water ice content) from the regolith. The extracted water is carried by a rover to a solar energy system located on the rim of the crater where the water is electrolyzed to hydrogen and oxygen.

It is difficult to be sure how much, if any water ice is present, and how deeply buried it is below a putative layer of desiccated regolith. The spent regolith is dumped ~100 m distant, and the extracted water is transported ~8 km to an electrolysis plant located at a rise on the crater rim where sunlight is available at a putative 70% of the time with a maximum of 100 hours maximum outage. Within the crater, all power for excavation, regolith transport, and water extraction is claimed to be nuclear, but there are no plans for installing a reactor, there couldn't possibly be enough radio-isotope thermal generators (RTGs) available to supply this power, and there certainly isn't enough plutonium available to enable such RTGs to be built. Driving water across a crater surface to an electrolysis plant appears to be a very inefficient process. The availability of solar energy on the crater rim will depend on the morphology of the surroundings. Whether 70% availability with a maximum of 100 hours maximum outage can be achieved is presently unknown, but it seems doubtful.

The Lunar Reconnaissance Orbiter (LRO) will use a neutron spectrometer (NS) to locate hydrogen signals at much higher spatial resolution than was possible with the Lunar Prospector. It is planned to map the H-signal to 5–10 km horizontal resolution. These "H pixels" can then be analyzed in the context of a 100 m grid of digital terrain along with thermal mapping that will produce 200 m pixels with 4–5 K sensitivity. So, the polar H will be mapped at a multi-kilometer scale in the context of 100–200 m support data on topology and temperatures. These are likely to encompass several subordinate craters within the south polar region, which should point the way to which of these extended areas are the best for further investigation with ground-based measurements.[90] Locating the best sites within such regions will require a series of *in situ* prospecting missions. Initially, long-range rovers equipped with neutron spectrometers (NSs) would be used to locate the best sites. At the best sites, follow-on missions would take subsurface samples to validate neutron spec-trometer indications, and make measurements of soil strength. This campaign to locate and validate accessible water ice resources is likely to require at least four and possibly as many as six *in situ* landed missions with long distance mobility, at a probable cost of over $1B each. If sorties with human crew are used for the final missions in this series, the cost will go up considerably. The NASA Robotic Lunar Exploration Program (RLEP) seems to have underestimated the requirements and cost of prospecting, the need for mobility on such precursor missions, the require-ments for taking subsurface samples with preservation of volatiles, and the extent and scope required for the overall campaign. To make matters much worse, a news release in March, 2007[91] indicates that "NASA has been rethinking its robotic lunar explora-tion strategy over the last 18 months and recently concluded that it has no immediate

---

[90] Information supplied by Jim Garvin, NASA.
[91] *NASA Plan Scales Back Lunar Robotic Program*, Brian Berger, Space Com, March 16, 2007.

need for any unmanned Moon missions beyond the heavily-instrumented Lunar Reconnaissance Orbiter (LRO) currently in development for a late 2008 or early 2009 launch. LRO's rocket will carry a secondary payload dubbed the Lunar Crater Observation and Sensing Satellite." NASA spokeswoman Beth Dickey was quoted as saying: "Near term funding is not going to be available for planning future lunar robotic programs. Those are going to be deferred until such a time as constellation requirements dictate a need for additional data beyond what will be provided by LRO ..." This strategy will be detrimental to implementation of lunar ISRU.

Development and demonstration of autonomous ISRU systems for excavation of regolith, delivery of regolith to a water extraction unit, operation of the water extraction unit with free flow of regolith through it (with no caking, agglomeration, and gunking up of regolith), and removal of spent regolith to a waste dump will require quite a few more billion. It is noteworthy that there is no evidence that NASA is planning to provide funds to develop the nuclear power systems needed for operation in the cold darkness of polar craters.

Overall, the required investment to do prospecting and validation of resources, and development and demonstration of regolith excavation and transport, and operation of a water extraction system, appears to be many billions of dollars. The benefit/cost ratio remains uncertain, but it may take many years to "break even" on the investment.

Any scenario that we develop for any step (whether that be prospecting or demonstration) should be elements of an overall campaign. A scenario for an individual step only has value as part of that campaign to the degree that it contributes to the campaign because the overall campaign produces the end result.

Power requirements

It seems likely that removal of ground ice from regolith would be implemented by heating to about 380 K to drive off steam at ~ 1 atm. This would also purify the water. If the regolith begins at say 40 K, then its temperature must be raised by 340 K. The power requirement to heat the regolith to drive off water is given in Table 5.6. This does not include the power required for excavation and hauling. This power is needed in the dark.

Table 5.6. Power requirement to heat the regolith to drive off water.

O_2 annual production (mT)	1	10	50	100
Annual regolith rate (mT)	80	800	4,000	8,000
Annual regolith rate (10^3 kg)	80	800	4,000	8,000
10^3 kWh	6.3	63	313	626
Hours	3,500	3,500	3,500	3,500
kW to heat regolith	1.8	17.9	89.4	178.7

Required campaign to utilize polar ice deposits

Unfortunately, NASA has not adequately defined the campaign for prospecting, demonstrating, and implementing lunar ISRU. [*Note:* In the present context "lunar ISRU" means oxygen (and possibly hydrogen) production, mainly for ascent propellants.] Although there are plans for manufacturing spare parts on the Moon, producing silicon solar cells on the Moon from regolith, beaming power back to Earth, and extracting parts per million of solar-wind deposited atoms, such work is (fortunately) not yet funded, even though it is included in JSC project plans.

Both JSC and ESAS appear to have simplistic notions about what it will take to prospect for polar ice resources and demonstrate ISRU systems, which will not hold up to any serious scrutiny. In addition, the RLEP Program—as of March, 2007— seems to have been eliminated altogether.[92]

A campaign is an end-to-end sequence of missions and programs to accomplish a goal.[93] Our view of the first five steps of the required campaign for developing lunar ISRU based on polar ice is as follows:

(1) The Lunar Reconnaissance Orbiter (LRO) will use a neutron spectrometer (NS) to locate hydrogen signals in horizontal spatial pixels of dimension $\sim$5–10 km. Despite the great improvement over Lunar Prospector, considerable uncertainty will remain as to which of these extended areas are the best for further investigation, and complete uncertainty as to the detailed distribution of hydrogen signal within each pixel.

(2) Even though neither JSC nor ESAS seem to have the slightest intention of doing this, what is required next is to send several long-distance rovers equipped with dynamic active NSs to several of these craters to cover a few tens of km in each, in order to determine at the outset: (a) to what extent the hydrogen signals and interpretations of them from LRO are substantiated by more reliable ground measurements, (b) how the hydrogen signal is distributed within each crater to $\sim$1 m pixel size (is the distribution fairly uniform or some kind of checkerboard?), (c) a much better estimate of the vertical distribution of hydrogen signal to a depth of perhaps 1 to 1.5 m and, in particular, the thickness of any desiccated upper layer covering the ice-containing layer.

(3) From the results of (2), a decision can be made as to which specific site (or sites) will be selected for more detailed measurement and verification. It is presumed that—if, for example, the LRO data indicate an average of 1% water-equivalent content across, say, $\sim$10 km—there are likely to be stretches of a km (or so) in extent with essentially constant water-equivalent content that are higher than average. Therefore, for purposes of very early planning we can guess that a several-km area has been located with perhaps 2% water-equivalent content. When actual data are available, this can be made more specific. [*Note:* The required areal extent of the ice field depends upon the water ice content and

[92] *NASA Plan Scales Back Lunar Robotic Program*, Brian Berger, Space Com, March 16, 2007.
[93] *Architecting Space Exploration Campaigns: A Decision-Analytic Approach*, Erin Baker, Elisabeth L. Morse, Andrew Gray, and Robert Easter, IEEEAC Paper #1176 (2006).

the cumulative need. If the outpost requirement is to produce $\sim$24 kg/day[94] of O_2, this requires 27 kg/day of water (with no losses) and perhaps 30 kg/day of water with losses. If we can roughly assume 2% water content in 70% of the top 1 m, then each square meter excavated yields about 1,500 kg $\times$ 0.7 $\times$ 0.02 $\sim$20 kg of water. Hence, the full-scale outpost ISRU system requires excavating about 1.5 m^2 (down to a depth of 1 m) per day, enabling processing of about 2,250 kg of regolith per day, and extracting about 30 kg of water per day. In one year, an area of about 1,100 m^2 is excavated. Over 10 years, an area of about 5,500 m^2 ($\sim$75 meters by $\sim$75 meters) is depleted.

(4) We would then send a short-range rover system to the selected site(s) to (a) map out the site with NS in great detail, (b) take subsurface samples to validate rover-mounted, dynamic, active NS measurements of water-equivalent content, (c) determine the actual form of hydrogen-containing compounds—which are almost surely dominated by water, (d) extract water from some samples and determine the water purity and the potential need for purification, and (e) determine the soil strength and requirements for excavation of the site. In some studies, this step would be implemented with support of a human crew who land in the Lunar Surface Access Module (LSAM). But, if step (4) can in fact be done robotically, why would anyone want to send a crew to do it?

(5) Develop a $\sim$1/10 scale lunar ISRU demonstration system for use at this site, deliver it with human oversight, get it started, and leave it to operate autonomously. In this task, several factors will be challenging:

(a) Even at 1/10 scale, there is a need to excavate 225 kg of regolith per day, transport it to the water extraction unit (WEU), heat the regolith to well over 300 K to drive off water vapor, remove spent regolith from the WEU, dispose of the spent regolith and any dry regolith layer that may lie atop the ice-containing layer, and deal with 3 kg/day of water produced. If the water is to be electrolyzed and the hydrogen and oxygen stored, that needs to be designed into the system. All of this takes place in the dark at very low temperatures.

(b) Definition of autonomous operations, including disposal of waste regolith, methods of excavation, and vehicles for transporting regolith to and from the WEU will require a great deal of study and analysis.

(c) Power is likely to be a major showstopper at every stage of this enterprise (see Table 5.6). If the demonstration must run autonomously after the crew leaves, how is it going to get sufficient power? It seems unlikely that enough RTGs will be available. Will NASA develop a nuclear reactor? There is no evidence that it will. Current planning for lunar ISRU by JSC is based almost entirely on solar energy.

It seems clear that neither JSC nor ESAS nor NASA have given adequate thought to the big picture of lunar ISRU, its requirements, and its benefit/cost ratio for the whole campaign. A sober assessment of the requirements for developing and imple-

---

[94] 24 kg/day is roughly 8 mT per year.

menting lunar ISRU, compared with the "value" of mass saved, creates significant doubt as to the net value of lunar ISRU.

Nevertheless, even if lunar ISRU is not a paying proposition, it still needs to be done effectively, and the approaches taken by JSC and ESAS do not seem to fit that requirement.

JSC campaign overview

In contrast to the campaign laid out above, the JSC/ESAS/NASA plan appears to provide only two lunar demonstrations prior to human sortie missions:

(A) "... a *lunar polar resource characterization mission* requiring hardware to be at TRL 6 by FY09 for a notional launch in FY12. In order to meet the requirement of polar volatile resource characterization, collection and separation an experiment to determine the form and concentration of the volatiles will be required. The RLEP 2 mission will carry this experiment and the lunar ISRU Project will dedicate a significant portion of its funding to design and develop this experiment package to TRL 6."

(B) "... a *lunar oxygen extraction demonstration* requiring hardware to be at TRL 6 by FY11 for a notional launch in FY14." The RLEP 3 mission would carry this experiment. While it is not explicitly stated as such, there is a strong implication that this would be an equatorial landing with oxygen extraction from regolith via a high-temperature process.

The JSC plan also says: "RLEP payload mass and power requirements are unknown at this time. However, notionally, payloads should be between 10 and 100 kg and not exceed 100 Watts of average power."

Note that neither NASA nor ESAS nor JSC have any plan or expectation to rove around the various craters that LRO identifies from space (via the neutron spectrometer [NS]) as containing hydrogen, to locate (a) the best local crater, and (b) the best site within the best local crater. The JSC/ESAS/NASA viewpoint has always been, and seems to remain, that they can just land anywhere in the region to "characterize, collect and separate volatiles."

The JSC/ESAS/NASA plan appears to be bottom-heavy with significant activity after sorties begin, and is woefully lacking in precursors to the sorties.

In addition to these two RLEP payloads, the JSC plan calls for developing "technologies to support human Sortie mission objectives of performing demonstrations and mission applications of ISRU subsystems and systems on the Moon." Requirements are stated as:

- "Demonstrations should be notionally 1/5th scale of early Outpost mission needs and no smaller than 1/10th scale in excavation/production rate."
- "Payloads should be ~250 to 500 kg in mass since total instrument payload is ~2,000 kg. Power requirements should be self-contained on ISRU demonstrations with recharging/refueling from the Lander as an option to be considered."

[*Note*: Preliminary estimates made by myself suggest that 250–500 kg will not be adequate for a 1/5-scale demonstration. Power remains a major question mark. While it is conceivable (though not necessarily likely) that the LSAM could recharge batteries on rovers and processors in a demonstration unit for a limited number of days while the crew is present, there does not seem to be enough RTG production or plutonium to power them, to provide "self-contained" power after the LSAM departs.]

- "Payloads should be autonomous with crew involvement only for setup, collection of raw material and product samples for return to Earth, and contingency recovery from failures."
- "Payloads should operate a minimum of 7 days. Provisions should be made to allow ISRU hardware to continue to operate after the crew left to obtain operation life, performance, and wear characteristic data on ISRU demonstration hardware." As mentioned above, power remains a major challenge for operation after the crew leaves.

5.2.5   Cost analysis for ESAS lunar ISRU

In this section, we compare the costs of an outpost with and without ISRU. The following basic assumptions are adopted:

- Costs to develop various lunar vehicles (CEV, LSAM, ...) are borne by NASA Project Constellation and do not enter the ISRU vs. non-ISRU comparison.
- The various lunar vehicles used for sorties (CEV, LSAM, ...) are also used for outpost deliveries and returns.
- The outpost is operated for 10 years with two exchanges of crew per year.
- A cargo delivery of 32 mT to the lunar surface is made once/year to deliver infrastructure at a launch cost of \$1.2B for launch and launch ops.
- LOX/methane propulsion is developed for ascent propulsion whether ISRU is used or not.
- Ascent from the Moon requires 4 mT of oxygen propellant.
- The "gear ratio" (mass in LEO/payload landed on the Moon) for cargo deliveries to the pole is ~4.
- The benefit of lunar ISRU is elimination of delivery of 4 mT of oxygen ascent propellant to the lunar surface, twice per year.

The net value of lunar ISRU is not transparent. Whether lunar ISRU is used or not, there will be two crew deliveries and two crew returns per year. The same vehicles are used whether ISRU is used or not. The only difference is that in the ISRU case the LSAM can be landed with empty oxygen ascent tanks. Simplistically, it would appear that in the lunar ISRU case an extra 4 mT of payload can be carried to the lunar surface with each crew delivery (8 mT per year). However, in order to keep the interior of the oxygen tank cold, it may be necessary to include some oxygen in the tank (probably a bit more than that which boils off in the non-ISRU case). The increased payload to the lunar surface would acquire value if we assume (as stated in the assumptions above) that periodic (maybe once a year) cargo deliveries are made to the Moon in which no CEV is used and no return is used. It is likely that such a cargo delivery system could deliver perhaps 32 mT to the lunar surface, and its IMLEO is about $4 \times 32 = 128$ mT. In that case, using ISRU with its annual increase of 8 mT of cargo to the lunar surface in crew deliveries, every fourth cargo delivery would be eliminated by supplying extra payload with each crew delivery. The net saving from use of ISRU is one cargo delivery every four years. If the cost of launching a cargo delivery mission is, say, \$1.2B, the annual saving from use of lunar ISRU is \$300M.

However, it is unlikely that the ISRU system would last for 10 years for a number of reasons. One is that, with each passing year, the local ice field adjacent to the processing unit tends to get depleted and the rovers transporting regolith to and from the processor must travel greater distances. Another is that these working rovers are constantly excavating and transporting regolith, and are likely to need periodic replacement. Other components are likely to need periodic replacement or upgrade. Hence, part of the assumed increase in cargo delivery capability with lunar ISRU would have to be used for lunar ISRU deliveries. Nevertheless, we will neglect this effect and optimistically assume that the benefit from use of lunar ISRU is \$300M/yr.

The cost of a lunar ISRU system includes the following items (not a complete list).

Development

- Development of processing technology. This includes a system that can receive regolith, heat it to drive off water vapor, condense and collect the water, and release spent regolith (with no caking, agglomeration, and gunking up of regolith).
- Development of excavation and regolith transport technology. This includes autonomous systems to excavate regolith, transfer it to a processor, and dispose of spent regolith.
- Simulation and test of systems on Earth. This may involve very large, cold, evacuated chambers where simulated field operations can be tested on Earth prior to test on the Moon.
- A nuclear reactor power system must be developed to provide power in the dark near the south pole.

The development cost for the lunar ISRU components is difficult to estimate. The JSC lunar ISRU Technology Development Plan says: "Limited funding, especially in first four (4) years of development will limit the scope and number of concepts that can be evaluated. Also, funding constraints may require early down-select before adequate characterization and mission studies have been performed." However, funding for the first five years of this program totals up to over \$60M, and it is limited to laboratory-type systems and preparation of payloads for RLEP-2 and RLEP-3. The cost of developing 1/5-scale or 1/10-scale demonstrations for sorties will be much higher. A rough guess is that the development cost for lunar ISRU components will be $800M. In addition, a wild guess for the nuclear reactor is $5B.

Prospecting

Prospecting will likely involve the following stages:

- LRO observation from orbit: $460M.
- Ground truth long-distance rovers equipped with NSs to locate sites (four missions at $800M each).
- Ground truth local mission to validate selected site with subsurface access (one mission at $1.2B)

Total cost for prospecting ~$5B.

In situ test and validation

Beginning with a 1/10-scale system, and extending this to a larger scale "dress rehearsal", two significant installations for autonomous lunar ISRU operations need to be developed, delivered, installed, debugged, and set to operating on the lunar surface. Since each of these involves sorties operated by human crews, the cost is roughly estimated to be $8B for the two demonstrations.

Total cost for lunar ISRU

The total cost to implement ISRU is estimated to be:

Development	$6B
Prospecting	$5B
In situ test and validation	$8B
Total	*$19B*

Saving $300M per year would require >60 years to break even, and it would be worse if account were taken of the fact that ISRU investment is upfront whereas return on investment is delayed many years.

5.2.6   A new paradigm for lunar ISRU

Introduction

The problems in using ISRU on the Moon within the current lunar architecture include:

- Lunar sorties must be fully capable of landing, ascending, and returning without utilizing ISRU.
- Lunar ISRU is tacked on as an afterthought to lunar missions well after outposts are set up.

- Required capabilities for landing, ascending, and returning that must be developed in the beginning without lunar ISRU are not mitigated by later use of lunar ISRU.
- Of all the many masses that must be sent to the Moon, ascent propellants (to lunar orbit) are only one moderate element (~4 metric tons).
- If lunar ISRU is used only to supply propellants for ascent to orbit, the mass benefits are modest—resulting in modest equivalent cost savings—at best.
- The investment needed for prospecting, validation of resources, validation of regolith excavation and handling, and validation of a lunar polar ISRU end-to-end system is large.
- If we ignore mass savings from lunar ISRU, and concentrate on return on investment in lunar ISRU, lunar polar ISRU does not pay back the initial investment in the current ESAS architecture.
- All of the above conclusions suggest that setting up an outpost near the pole has little justification based on ISRU, although the thermal environment is likely to be more benign than, say, near the equator.
- This, in turn, casts doubt on the entire basis for the enterprise of returning to the Moon.

Current NASA plans call for ascent propulsion based on space-storable NTO–MMH propellants.[95] If NASA does not replace this with an oxygen-based ascent propulsion system, then even the meager potential benefits from lunar ISRU in the current lunar ISRU architecture (providing ascent LOX) will disappear, and lunar ISRU would have essentially no value.

Given that lunar ISRU does not mesh with the current ESAS architecture, alternative architectures must be considered—either that or eliminate lunar ISRU entirely from the current architecture.

This leaves us with significant mass challenges. To justify retention of ISRU:

- Cryogenic propulsion utilizing LOX and possibly LH_2 must be used throughout descent and ascent from the Moon.
- A high-leverage user of lunar ISRU products must be found in addition to ascent propellants.
- Significant reductions in IMLEO must result from use of lunar ISRU.
- A very significant potential target is descent propellants—totaling some 25 to 30 mT of propellants.

In order for lunar ISRU to significantly impact the lunar exploration campaign, the following conditions must be fulfilled:

(a) Lunar ISRU must be built into the very fabric of the lunar campaign so that all space and Launch Vehicles are designed and sized to use lunar ISRU from the

[95] This plan was reaffirmed in 2007 by: "MoonHardware22Feb07_Connolly.pdf" on the *NASA Watch* website.

beginning. (This is as opposed to the ESAS approach of only using lunar ISRU rather late in the campaign as an add-on to a system that is sized without use of lunar ISRU.)

(b) In order for (a) to be possible, an extended robotic campaign must precede the human campaign, to establish a working lunar ISRU plant as a fundamental asset for the first human sorties. This would undoubtedly delay return of humans to the Moon by several years. However, NASA has apparently canceled plans for robotic precursors.[96]

(c) Lunar polar ice must be the lunar ISRU feedstock of choice because it is the only reasonable hope for a workable system.

(d) Oxygen must be retained as an ascent propellant. It would also be useful to use hydrogen for ascent as well.

(e) Utilization of lunar ISRU products must be expanded to include descent propellants as well as ascent propellants.

Although the benefit/cost ratio for this approach is still not favorable, it is far superior to that based on ISRU generation of only ascent propellants.

5.3   MARS ISRU

5.3.1   Introduction

The atmosphere of Mars is $\sim$95% CO_2, with the remainder made up mainly of Ar and N_2, and smaller amounts of CO and O_2. The CO_2 is the feedstock used to produce oxygen, and possibly hydrocarbon production if hydrogen is available. Atmospheric pressure varies with hour and season, but is typically a bit less than 1/100 of the atmospheric pressure on Earth. Water has been detected in the top $\sim$1 m of Mars subsurface by orbiting instruments, and is widely distributed on Mars, particularly at higher latitudes.

ISRU on Mars has significant advantages compared with lunar ISRU:

- The gear ratio for delivery of assets from LEO to the Mars surface is higher than for the lunar surface, thus requiring more IMLEO per unit mass delivered to the surface of Mars. This, in turn, makes mass replacement on Mars by ISRU more valuable than mass replacement by ISRU on the Moon.
- The Δv for ascent from the Mars surface to orbit is much greater than Δv for ascent in the lunar case, necessitating much greater propellant requirements for ascent. This, in turn, makes propellant production on Mars by ISRU more valuable than propellant production by ISRU on the Moon.
- By placing the Earth Return Vehicle (ERV) in an elongated elliptical orbit (in the Mars case) one can increase the required Δv for ascent (and thereby the amount of ascent propellants supplied by ISRU) while decreasing the Δv requirements

[96] *NASA Plan Scales Back Lunar Robotic Program*, Brian Berger, Space Com, March 16, 2007.

for orbit insertion of the ERV (propellants supplied from Earth) as well as for Earth return from Mars orbit. As it turns out, the mass savings by allowing the ERV to utilize an elliptical orbit are even greater than the mass saving due to reduced mass delivered to the Mars surface.

- The combination of the previous three points provides much greater mission impact (IMLEO reduction) for *in situ* production of ascent propellants on Mars than on the Moon.
- Because of the long round trip to Mars ($\sim$2.7 years), total consumption of life support consumables amounts to perhaps 200 mT. It is not clear whether an ECLSS system will have the longevity to provide fail-safe performance over that time period. Such an ECLSS system to provide air and water (if it is feasible) is likely to weigh >30 mT.
- Unlike the Moon, Mars has a ready supply of carbon and oxygen in its easily acquired atmosphere.
- Unlike the Moon, Mars has significant near-surface deposits of water (believed to be mainly in the form of ground ice)[97] widespread across much of the planet.
- The combination of atmospheric CO_2 and water from regolith provides feedstocks on Mars that enable proven, relatively simple Sabatier-electrolysis processing to produce methane and oxygen propellants, and water for life support.
- In conclusion, Mars ISRU is far more easily implemented and has far more mission impact than lunar ISRU.

The major unknowns regarding Mars ISRU are:

- What are the requirements for excavating water-bearing, near-surface regolith and extracting water?
- In the case of equatorial, water-bearing, near-surface regolith, is the water in the form of ground ice or mineral hydrates?

Unfortunately, neither ESAS nor the Mars Exploration (Science) Program has any specific plans to investigate these questions.

5.3.2 Timeline for ISRU on Mars

Launch opportunities to send vehicles to Mars are spaced at roughly 26-month intervals. The ISRU system would be launched $\sim$26 months prior to departure of the crew from LEO. The cargo delivery will take about 9 months to get to Mars and perhaps a month to set up operations on the surface. Therefore, ISRU operations could begin $\sim$10 months after launch. We would then have 16 months until the crew launches, and about 22 months until the crew arrives at Mars (assuming the crew transits via a fast $\sim$6-month trajectory). The full mission timeline is shown in Figure 5.1.

---

[97] In the equatorial areas it is not known whether it is ground ice or hydrated minerals.

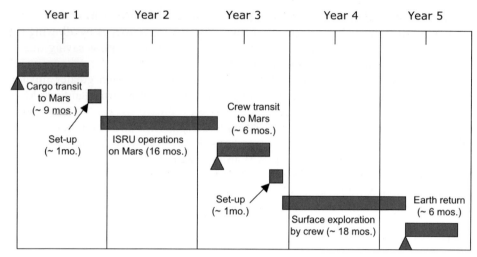

Figure 5.1. Hypothetical time-line for a Mars mission utilizing two launch periods and ISRU. Triangles represent departure dates.

The ISRU system could be sized to fill the MAV tanks in 16 months to assure that they are full prior to crew departure from Earth.

The situation for life support is less certain. The $\sim$100 mT of water needed on the surface has a volume of $\sim$100 m^3. It may be possible to store this amount of water in an inflatable tank, and let it freeze. Alternatively, it may be permissible to only extract (from regolith) and store some fraction of this during the 16 months prior to crew launch. The requirements for oxygen and buffer gas for a crew of six over 600 days are about 3.6 and 10.8 mT, respectively. One might not want to store all the needed buffer gas, and buffer gas might be recycled. As in the case of water, it will have to be decided whether to make all of the oxygen prior to human departure from Earth and then be faced with the problem of storing that large amount, or whether to be content with producing only part of the needed supply prior to crew arrival.

The decision regarding strategy for production of propellants and consumables requires further thought regarding mass, volumetric, and safety/risk considerations as well as power level. The most conservative approach would be to assume that the ISRU system would be sized so that all ascent propellants are produced in 16 months, so that the MAV tanks are full when the crew departs from Earth. The amount of life support consumables to be produced prior to crew departure from Earth remains open to question.

In regard to power, we need to first estimate power requirements for the $\sim$18 month period during which the humans are on the Mars surface, assuming that ISRU operations stop or are significantly reduced when humans arrive. This sets the minimum power level needed for the mission. If this same power level can supply the power needed for ISRU during its 16-month operational period, then it is fair to claim that the mass of the power system is attributable only to human support, and no mass (or cost) attribution for power is made to the ISRU system. If ISRU processing continues while humans are on the surface, the additional mass (and cost) required to scale up the power system beyond human support requirements must be attributed to ISRU.

A likely scenario (though by no means the only one) is that during the 16-month period of intense ISRU operations prior to crew departure from Earth, the ISRU system will produce the requisite amount of methane and oxygen for ascent, and store these as cryogenic liquids in the tanks of the ascent vehicle. There will be a gradual rate of boil-off of these propellants due to heat leaks into the tanks, unless zero boil-off techniques are used, with a consequent increase in mass, power, and complexity. Therefore, it will be desirable to operate the ISRU plant at a greatly reduced throughput while the crew is in transit to Mars (and after they land as well) to replenish boil-off. Alternatively, additional propellant tanks could be stored on Mars to hold extra propellants that could be piped to the Mars Ascent Vehicle prior to ascent to "top-off" the propellant tanks.

5.3.3 Mars ISRU products

A vital ISRU product of relevance to Mars missions is oxygen for use as a propellant in ascent from Mars. This oxygen would be stored as a cryogenic liquid in the Mars Ascent Vehicle. The amount of oxygen required will depend on several factors: (1) the mass of the capsule to hold the crew during ascent and rendezvous, (2) the number of crew-members, (3) the orbit in which rendezvous takes place, and (4) the fuel propellant used in conjunction with the oxygen. In a typical rocket using oxygen as the oxidizer (e.g., methane/oxygen rocket) the oxygen accounts for 75–80% of total propellant mass (depending on the actual mixture ratio used). Thus, if the methane (or hydrogen to produce methane from the CO_2 in the Mars atmosphere) is brought from Earth, and ISRU produces only the oxygen from Mars feedstocks, ISRU will nevertheless account for 75–80% of ascent propellant needs. Some forms of Mars ISRU produce not only oxygen propellant, but methane as well. In that case, 100% of ascent propellant needs are supplied by ISRU. The required mass of ascent propellants will be estimated in subsequent paragraphs for the case where both methane and oxygen are produced by ISRU.

Mars ISRU can also be used to produce life support consumables. The requirement for the surface phase of the mission was given in Table 4.2. The requirement is for about 100 mT of water and about 4 mT of oxygen. The mission benefit from using ISRU to supply these commodities depends on assumptions that are made regarding the efficiency and mass of recycling systems for these commodities. We have estimated the mass of the ECLSS systems for air and water in Table 4.3 based on JSC estimates (that may prove to be optimistic). According to this estimate, the air and water recycling system on the surface would weigh only about 15 mT. Whether such systems can be developed with sufficient longevity and reliability remains open to question.

In most conceptual human missions to Mars, some mass is delivered to Mars orbit (M_{MO}) and some mass is delivered to the Mars surface (M_{MS}). Multiplying each figure by its appropriate "gear ratio", the initial mass required in LEO (IMLEO) for the mission can be estimated. Thus, for assets transported to Mars orbit:

$$IMLEO_{MO} = M_{MO}G_{MO}$$

and for assets transported to the Mars surface:

$$IMLEO_{MS} = M_{MS}G_{MS}$$

and the total IMLEO is:

$$IMLEO = IMLEO_{MO} + IMLEO_{MS}$$

Gear ratios are tabulated in Chapter 3.

Of particular interest are components of these masses that depend upon which Mars orbit is utilized, and whether ISRU is used to produce ascent propellants. As a rough generalization, we use $G_{MS} \sim 9.3$ based on LOX/LH$_2$ for Earth departure and full aero-entry at Mars for landed assets. The gear ratios for propulsive Mars orbit insertion are 7.2 for circular orbits and 4.7 for elliptical orbits (see Table 3.13).[98]

Thus we set:

Mass injected into trans-Mars trajectory

    = Payload mass to Mars orbit

    + Mass of propulsion system for Mars orbit insertion

    + Mass of propellants for Mars orbit insertion

    + Mass of propulsion system for departure from Mars orbit

    + Mass of propellants for departure from Mars orbit

Mass injected into Mars orbit

    = Payload mass to Mars orbit

    + Mass of propulsion system for departure from Mars orbit

    + Mass of propellants for departure from Mars orbit

Mass delivered to Mars surface

    = Payload mass to Mars surface

    + Mass of propulsion system for ascent to orbit

    + Mass of propellants for ascent to orbit

---

[98] According to Table 3.16, the current best estimate for this gear ratio is around 11:1, but that estimate includes a 30% margin for entry system dry mass. With a 15% margin, the gear ratio drops to around 9.3, and we use this more optimistic figure here.

In mathematical terms,

$$M_{TM} = M_{PLO} + M_{PROI} + M_{POI} + M_{PROD} + M_{POD}$$

$$M_{MO} = M_{PLO} + M_{PROD} + M_{POD}$$

$$M_{MS} = M_{PLS} + M_{PRA} + M_{PA}$$

If the subscript begins PR this refers to the propulsion system, if it begins P that refers to propellants, and if it begins PL that refers to payload. The remainder of each subscript consists of OI = orbit insertion, OD = orbit departure, A = ascent, O = orbit, S = surface, and TM = trans-Mars.

In our calculations, we assume that the masses of the propulsion systems are proportional to the masses of the propellants:

$$M_{PROI} \sim 0.12 \times M_{POI}$$

$$M_{PROD} \sim 0.12 \times M_{POD}$$

$$M_{PRA} \sim 0.15 \times M_{PA}$$

Thus

$$M_{MO} = M_{PLO} + 1.12 \times M_{POI} + 1.12 \times M_{POD}$$

$$M_{MS} = M_{PLS} + 1.15 \times M_{PA}$$

For any propulsive step, the ratio of mass of propellant to mass of payload is given by

$$\frac{M_{PR}}{M_{PL}} = \left(\frac{q-1}{1 - K(q-1)} \right)$$

where K is the constant relating propulsion system mass to propellant mass (0.12 or 0.15) and q is the exponential in the rocket equation:

$$q = \exp\{(\Delta v)/(gI_{SP})\}$$

where Δv = change in velocity required for the transit;
         I_{SP} = specific impulse of the rocket (seconds);
         g = acceleration of Earth's gravity = 9.8 m/s^2.

Basically, we consider two potential orbits, (1) a circular orbit requiring a $\Delta v \sim 2.4$ km/s for orbit insertion, and (2) a 24-hr elliptical orbit requiring a $\Delta v \sim 1.2$ km/s for orbit insertion. We will also assume that these same values pertain to orbit departure as well. Ascent from the surface to a circular orbit at 300 km requires a Δv of about 4.3 km/s, whereas ascent to the elliptical orbit requires about 5.6 km/s. Thus, we see that for any Mars mission we can select a circular orbit in which case the Δv for orbit insertion and orbit departure is ~ 2.4 km/s, while the Δv for ascent is 4.3 km/s. Alternatively, we can select an elliptical orbit, in which case the

Table 5.7. Propellant mass/payload mass.

Transfer	Δv (m/s)	q	K	m_{PR}/m_{PL}
TMI to elliptical orbit	1,200	1.41	0.12	0.43
TMI to circular orbit	2,400	1.97	0.12	1.10
Surface to circular orbit	4,300	3.38	0.15	3.71
Surface to elliptical orbit	5,600	4.89	0.15	9.34

Δv for orbit insertion and orbit departure is ~1.2 km/s, while the Δv for ascent is 5.6 km/s.

We will assume that propulsion systems used for Mars orbit insertion, ascent, and Mars orbit departure are all CH_4/LOX with $I_{SP} \sim 350$ s. Then, we may calculate the data shown in Table 5.7.

Next, we compare the effect on IMLEO of using the elliptical orbit or the circular orbit, assuming that all propellants are brought from Earth.

The mass of the ascent capsule was estimated in several design reference missions. We assume here that for ascent, $M_{PL} \sim 5$ mT. Then, the mass of propellants required for ascent to orbit is $5 \times 3.71 = 18.5$ mT for ascent to circular orbit, and $5 \times 9.34 = 46.7$ mT for ascent to the elliptical orbit. With a gear ratio of 9.3 (based on aero-assisted EDL), the corresponding values of IMLEO for ascent propellants are $9.3 \times 18.5 = 172$ mT for the circular orbit and $9.3 \times 46.7 = 434$ mT for the elliptical orbit. The difference between the two values of IMLEO for the two orbits is $434 - 172 = 262$ mT. Thus, use of elliptical orbit increases IMLEO by 262 mT based on mass delivered to the surface (if ISRU is not used).

However, the propellant masses for Mars orbit insertion and departure are lower for the elliptical orbit case. It is difficult to predict the required mass for the payload to Mars orbit, but since it must house the crew in transit from Earth to Mars and back, and provide life support in case of abort to orbit, and it includes the Earth entry capsule, a rough guess for M_{PLO} (the payload to Mars orbit) is about 40 mT.

Using this figure, we find that the propellant mass needed for Mars orbit insertion is $1.1 \times 40 = 44$ mT for the circular orbit and $0.43 \times 40 = 17.2$ mT for the elliptical orbit. The gear ratio for transfer from LEO to Mars approach is about 3.

For Mars departure, the propellant masses are roughly the same as for Mars orbit injection (44 mT for the circular orbit and 17.2 mT for the elliptical orbit). However, the gear ratios from LEO to Mars orbit are ~7.2 for the circular orbit and ~4.7 for the elliptical orbit. Therefore, the corresponding values of IMLEO for orbit departure propellants are 317 mT for the circular orbit, and 81 mT for the elliptical orbit.

We may now compile Table 5.8. This table shows that the sum of propulsion system masses for orbit insertion, ascent, and orbit departure is somewhat higher when a circular orbit is used.

If ISRU is used to generate ascent propellants, we will assume that the mass of the ISRU plant is mainly determined by requirements for excavation, and therefore we

Table 5.8. Propellant masses and IMLEO associated with propulsion systems for orbit insertion, ascent, and orbit departure when ISRU *is not* employed.

Orbit	Payload (mT)	Δv (km/s)	Propellant/ payload ratio (from Table 5.7)	Propellant mass (mT)	Dry propulsion mass (mT)	Gear ratio	IMLEO (mT)
Ascent to orbit							
Elliptical	5	5.6	9.3	47	7	9.3	502
Circular	5	4.3	3.7	19	3	9.3	204
Mars orbit insertion							
Elliptical	40	1.2	0.43	17	2.0	3	57
Circular	40	2.4	1.1	44	5.3	3	148
Mars orbit departure							
Elliptical	40	1.2	0.43	17	2.0	4.7	89
Circular	40	2.4	1.1	44	5.3	7.2	355
Sum of three operations							
Elliptical							648
Circular							707

will simplistically assume that the mass of the ISRU plant is fixed at 4 mT regardless of the mass of ascent propellants. With this assumption, the reduction in the mass brought to the Mars surface is 15 mT for the circular orbit and 43 mT for the elliptical orbit when ISRU is used.

Thus, if ISRU is used to produce ascent propellants, Table 5.8 is replaced by Table 5.9.

Hence, using ISRU with the elliptical orbit, the value of IMLEO attributable to propulsion systems and propellants for ascent, orbit insertion, and orbit departure is reduced by about $(648 - 269) = 379$ mT if an elliptical orbit is used. If a circular orbit is used the IMLEO benefit of using ISRU drops to about $(707 - 568) = 139$ mT.

The mass saving due to ISRU propellant production is even greater if a mission architecture is used in which the Mars Ascent Vehicle (MAV) does not rendezvous with an Earth Return Vehicle (ERV) in Mars orbit, but instead the MAV goes directly from the Mars surface all the way back to Earth. The Mars Direct[99] and

---

[99] *Scientific American*, March, 2000; Addendum II: Mars Direct—A Practical Low-Cost Approach to Near-Term Piloted Mars Missions, *http://www.iaanet.org/p_papers/add2.html*; and *Mars Direct: A Simple, Robust, and Cost Effective Architecture for the Space Exploration Initiative*, Robert M. Zubrin, David A. Baker, and Owen Gwynne, AIAA-91-0328.

Table 5.9. Propellant masses and IMLEO associated with propulsion systems for orbit insertion, ascent, and orbit departure when ISRU is employed.

Orbit	Payload (mT)	Δv (km/s)	Propellant/ payload ratio	Propellant or ISRU mass (mT)	Dry propulsion mass (mT)	IMLEO (mT)
Ascent to orbit						
Elliptical	5	5.6	9.3	4	7	123
Circular	5	4.3	3.7	4	3	65
Mars orbit insertion						
Elliptical	40	1.2	0.43	17	2.0	57
Circular	40	2.4	1.1	44	5.3	148
Mars orbit departure						
Elliptical	40	1.2	0.43	17	2.0	89
Circular	40	2.4	1.1	44	5.3	355
Sum of three operations						
Elliptical						269
Circular						568

Mars Society[100] missions used this approach, and the MIT study[101] found significant benefits to this architecture. It appears from these studies that the amount of propellant (methane + oxygen) needed on the Mars surface to go all the way back to Earth is >100 mT, and if ISRU is not used this would imply the need for >900 mT initially in LEO with aero-assist, and >3,100 mT initially in LEO using propulsion for entry and descent. ISRU appears to be mandatory for this mission architecture.

5.3.4   Mars ISRU processes

5.3.4.1   Oxygen-only processes

Several schemes have been proposed for producing propellants from the Mars atmosphere. One approach utilizes only the CO_2 in the Mars atmosphere and

[100] *A New Plan for Sending Humans to Mars: The Mars Society Mission*, California Institute of Technology, 1999, Contributors: Christopher Hirata, Jane Greenham, Nathan Brown, and Derek Shannon, Advisor: James D. Burke, Jet Propulsion Laboratory.
[101] *From Value to Architecture: The Exploration System of Systems—Phase I*, P. Wooster, Presentation at JPL, August 23, 2005; *Paradigm Shift in Design for NASA's Space Exploration Initiative: Results from MIT's Spring 2004 Study*, Christine Taylor, David Broniatowski, Ryan Boas, Matt Silver, Edward Crawley, Olivier de Wec, and Jeffrey Hoffman, AIAA 2005-2766; "The Mars-Back Approach: Affordable and Sustainable Exploration of the Moon, Mars, and Beyond Using Common Systems," P. D. Wooster, W. K. Hofstetter, W. D. Nadir, and E. F. Crawley, *International Astronautical Congress, October 17–21, 2005*.

produces only O_2 via the reaction

$$2CO_2 \Rightarrow 2CO + O_2 \tag{5.1}$$

The two most developed concepts for utilizing Martian CO_2 are (1) the zirconia solid-oxide electrolysis (SOE) process, and (2) reverse water–gas shift (RWGS) developed at Pioneer Astronautics.

Solid-oxide electrolysis

Solid-oxide electrolysis (SOE) is based on the very unusual and unique electrical properties of some ceramics that conduct electrical current using oxygen ions (O^{2-}) as the charge carrier rather than electrons. Typically, a solid-state yttria-stabilized zirconia (YSZ) ion conductor is used. The doped crystal lattice contains "holes" allowing ions to move through the lattice when an electric field is applied across it. The electric field is generated by mounting porous metallic electrodes on each side of a zirconia wafer, and applying a difference in potential. In a zirconia cell, hot CO_2 is brought into contact with a catalyst on the cathode, thus causing some dissociation. Oxygen atoms in contact with the cathode pick up electrons to form O^{2-} that are transported through the zirconia to form pure oxygen on the other side at the anode (see Figure 5.2). YSZ has been under study for more than 20 years. The performance increases over the temperature range 800°C to 1,000°C, so the materials of all cell components are critical, and sealing the edges is difficult, particularly when the cell must be repeatedly thermally cycled through many cycles. Several investigators have built and tested single YSZ flat-disk designs, but these cannot provide the required YSZ surface area in a small volume. A "stack" of YSZ disks is needed to produce a significant flow rate of oxygen.

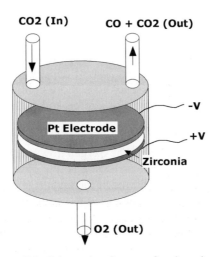

Figure 5.2. Schematic of one-wafer zirconia cell.

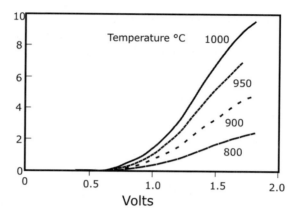

Figure 5.3. Ion current obtained by Crow and Ramohalli using "Minimox" at 60 sccm CO_2 flow rate. Vertical scale is in amps. [Based on data from *The MOXCE Project: New Cells for Producing Oxygen on Mars*, S. C. Crow, AIAA 97-2766, July, 1997.]

A basic quantity that relates the ion current to the oxygen gas flow rate is:

$$1 \text{ amp} = 3.79 \text{ std cc/min(sccm)} = 0.325 \text{ g/hr of } O_2$$

Therefore, the power (in watts) required to generate an oxygen production rate of 0.325 g/hr is simply the voltage across the cell.

For a system that can produce, say, 4 kg/hr of O_2, the required ion current is 12,300 amps. Typical current densities on YSZ disks range from 0.3 to 0.5 amps/cm^2. For a value of, say, 0.4 amps/cm^2, the required area of zirconia wafer is about 30,750 cm^2. If a zirconia disk is as large as, say, 5 cm × 5 cm square, its effective transport area is around 20 cm^2. This would imply that roughly 1,540 wafers of this size are needed for a full-scale unit. Thus, a full-scale system will require many zirconia wafers connected in series in "stacks".

The power requirement for the zirconia stack depends upon the voltage required to drive the ion current through the YSZ. It is found experimentally that, as the temperature and voltage are raised, the current density (amps/cm^2) increases. This allows use of less YSZ area, which leads to a more compact cell. However, as the temperature is increased, the problems of sealing and withstanding thermal cycling increase.

Figure 5.3 shows experimental data on a one-wafer YSZ cell.[102] Since the CO_2 flow rate was held constant at 60 sccm, the oxygen flow rate would be 30 sccm if there was 100% conversion. Since 30 sccm corresponds to a current of 7.9 amps, it can be seen that ~100% conversion was achieved at the highest temperatures and voltages— the approximate correction for end effects led to estimates higher than 100% at the far upper right.

[102] *The MOXCE Project: New Cells for Producing Oxygen on Mars*, S. C. Crow, AIAA 1997-2766, July.

Relatively little work has been reported on use of YSZ stacks, but it appears that sealing problems are very challenging. Whether a workable robust multi-wafer cell can ever be produced remains doubtful.

The reverse water–gas shift process

Robert Zubrin has made a number of innovations in ISRU technology.[103] One of these is the development of the reverse water–gas shift (RWGS) process.

The water–gas shift reaction is widely used by industry to convert relatively useless $CO + H_2O$ to much more useful hydrogen. However, if reaction conditions are adjusted to reverse the reaction, so it has the form:

$$CO_2 + H_2 \Rightarrow CO + H_2O \text{ (catalyst required)} \qquad (5.2)$$

then, CO_2 can be converted to water. Now, if that water is electrolyzed:

$$2H_2O + \text{electricity} \Rightarrow 2H_2 + O_2 \qquad (5.3)$$

the net effect of the two reactions is conversion of CO_2 to O_2—reaction (5.1). Ideally, all the hydrogen used in the first reaction is regenerated in the electrolysis reaction, so no net hydrogen is required. Actually, some hydrogen will probably be lost if the first reaction does not go to completion, although use of a hydrogen recovery membrane can minimize this loss. The above two reactions in concert represent what is referred to as the RWGS process.

Note that the reagents for the RWGS reaction are the same as for the S/E reaction (see Section 5.3.4.2). The main difference (aside from use of a different catalyst) is that the S/E process has a favorable equilibrium at lower temperatures (200–300°C), while the RWGS has a more favorable equilibrium at much higher temperatures (>600°C). If one considers the combined equilibria where catalysts are present which allow both reactions to take place, the S/E process will be dominant below about 400°C, and the reaction products will be mainly $CH_4 + 2H_2O$. At temperatures above about 650°C, methane production falls off to nil and the RWGS products ($CO + H_2O$) are dominant. Between about 400°C and 650°C, a transition zone exists, where both reactions take place. In this zone, CO production rapidly rises as the temperature increases from 400°C to 650°C, while methane production falls sharply over this temperature range. However, in the RWGS regime, no matter how high the temperature is raised, roughly half of the CO_2 and H_2 remain unreacted.

Figure 5.4 shows the product flow rates (assuming equilibrium is attained) for various chemical species when 44 mg/s of carbon dioxide and 2 mg/s of hydrogen (1 milli-mole of each) are introduced into a reactor at any temperature (note that the carbon dioxide flow rate is divided by 2 before plotting) and both the RWGS and S/E

---

[103] "Mars In-Situ Resource Utilization Based on the Reverse Water Gas Shift," R. Zubrin, B. Frankie, and T. Kito, AIAA-97-2767, *33rd AIAA/ASME Joint Propulsion Conference, Seattle, WA, July 6–9, 1997*; "Report on the Construction and Operation of a Mars in situ Propellant Production Unit Utilizing the Reverse Water Gas Shift," R. Zubrin, B. Frankie, and T. Kito, AIAA-98-3303, *34th AIAA/ASEE Joint Propulsion Conference, July 13–15, 1998, Cleveland, OH.*

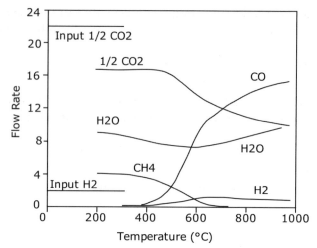

Figure 5.4. Product flow rates (assuming equilibrium is attained) for various chemical species when 44 mg/s of carbon dioxide and 2 mg/s of hydrogen are introduced into a reactor at any temperature. Note that the actual carbon dioxide flow rate is divided by 2 before plotting in order to reduce the range of the vertical scale.

reactions are catalyzed. It should be noted that this is the proper stoichiometric ratio for the RWGS reaction, but represents an excess of CO_2 of a factor of 4 for the S/E process. With this excess of CO_2, essentially all of the hydrogen will be reacted in those regimes where the S/E process takes place with high yield.

At lower temperatures (200–300°C), equilibrium would dictate that almost all the hydrogen is used up to produce methane and water, and the excess carbon dioxide is depleted by an equivalent amount. This is the S/E region. Almost no CO is formed. By contrast, at high temperatures (>650°C) CO and H_2O are the principal products, and very little methane is formed, but roughly half of the initial carbon dioxide and hydrogen remains unreacted in the product stream.

Operation at a temperature of >650°C is clearly a disadvantage. However, if one attempted to run the RWGS reactor at, say, 400°C, reaction (5.2) would only go about 25% to completion. Zubrin and co-workers have suggested several methods to force reaction (5.2) to the right, even at 400°C. These include:

(i) Water condensation and recirculation of $CO + CO_2$ (water produced by the RWGS reaction is condensed out downstream of the reactor and resultant gases are recirculated with continuous mixing of a smaller flow of feed gases).
(ii) Use of excess hydrogen (off-stoichiometric mixtures) to force the reaction to the right, with membrane recovery of unreacted hydrogen fed back into reactants.
(iii) Increasing the reactor pressure.

Using these techniques, Zubrin reported high conversion efficiency in a breadboard system. It remains to be seen how efficient and practical this system will be when further developed. NASA does not seem to have funded further development of this process after about 1995.

5.3.4.2   *The Sabatier/Electrolysis process*

Another approach utilizes both CO_2 in the Mars atmosphere and hydrogen (brought from Earth or made from Mars water deposits)—the Sabatier/Electrolysis (S/E) process. In the S/E process, hydrogen is reacted with compressed CO_2 in a heated chemical reactor:

$$CO_2 + 4H_2 \Rightarrow CH_4 + 2H_2O \text{ (catalyst)} \qquad (5.4)$$

The reactor is simply a tube filled with catalyst. Since the reaction is somewhat exothermic, no energy has to be supplied once the reaction is started.

The methane/water mixture is separated in a condenser, and the methane is dried, and stored for use as a propellant. The water is collected, deionized, and electrolyzed in an electrolysis cell:

$$2H_2O + \text{electricity} \Rightarrow 2H_2 + O_2 \qquad (5.5)$$

The oxygen is stored for use as a propellant and the hydrogen is recirculated to the chemical reactor. Note that only 1/2 as much hydrogen is produced by reaction (5.5) as is needed for reaction (5.4), showing that an external source of hydrogen is necessary for this process to work.

The equilibrium mixture of molecules in a mixture of $CO_2 + 4H_2$ is shown in Figure 5.5 as a function of temperature at a total pressure of 1 bar. As the temperature is raised, the equilibrium shifts away from the desired products of water + methane to $CO_2 + 4H_2$, but the rate of reaction increases. The challenge is then to operate the reactor at a temperature sufficiently high that the kinetics are fast enough to allow a small compact reactor, yet the temperature is not so high that the

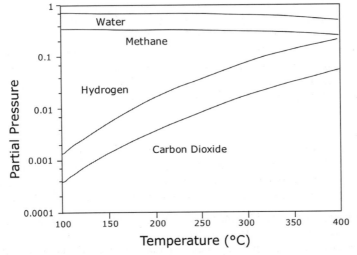

Figure 5.5. Equilibrium mixture at a pressure of 1 bar in a mixture of $CO_2 + 4H_2$.

equilibrium shifts to inadequate product yields. It has been found experimentally that at a reactor pressure of the order of $\sim$1 bar, if a mixture of $CO_2 + 4H_2$ enters a packed bed of catalyst, a temperature near $\sim$300°C is high enough to approach equilibrium in a small reactor, and the equilibrium is far enough to the right that yields of over 90% $CH_4 + 2H_2O$ are obtained. If the exit zone of the reactor is allowed to cool below 300°C, the yield can be >95%.[104]

In previous plans for use of the S/E process on Mars, it was conjectured that the hydrogen feedstock would be brought from Earth. There are two problems in application of the S/E process with hydrogen brought from Earth, which are closely coupled. The primary problem is the difficulty in bringing hydrogen from Earth and storing it on Mars. If the hydrogen is stored passively, boil-off could present significant problems. If boil-off is prevented by active cooling, the mass and power requirements increase.

The second problem is with the S/E process when used with a hydrogen feedstock—namely, that it produces one molecule of methane for each molecule of oxygen, while the ideal mixture ratio for propulsion is roughly one molecule of methane for each 1.75 molecules of oxygen. Thus, there is an excess of methane for the amount of oxygen produced. This in itself is not so fundamentally bad, except that it requires that we must transport extra hydrogen to Mars to create this wasted methane. Several schemes have been proposed to recover hydrogen from excess methane, and, in addition, other methods for reducing the required amount of hydrogen have been proposed by converting to higher hydrocarbons with higher C/H ratios than methane. These processes are undeveloped and are probably not needed since indigenous Mars water is likely to be used as the source of hydrogen. It has also been proposed that storage of cryogenic methane could be simplified by conversion to methyl alcohol,[105] but since cryogenic oxygen must be stored anyway, this appears to have only a minor benefit.

It is now evident that large amounts of H_2O are widespread in the near-surface regolith of Mars, and it may be possible to access and extract this resource for use as a feedstock in ISRU. There are three major virtues involved. One is that the need to bring hydrogen from Earth is eliminated. A second is that the products are closer to the desired mixture ratio of methane and oxygen. Finally, the availability of indigenous Mars water could eliminate the need to recycle water on Mars for $\sim$600 days. The process would then utilize equation (5.5) to produce hydrogen, followed by equation (5.4) to produce methane plus water, followed by a second use of equation (5.5) to produce oxygen and recover half of the initial hydrogen. The overall process would be:

$$CO_2 + 2H_2O \Rightarrow CH_4 + 2O_2 \tag{5.6}$$

This process produces a very slight excess of oxygen that can be used for life support.

[104] "In-Situ Propellant Production on Mars: A Sabatier/Electrolysis Demonstration Plant," D. L. Clark, *ISRU Interchange Meeting 1997.*
[105] *Report on the Construction and Operation of a Mars Methanol in situ Propellant Production Unit*, Robert Zubrin, Tomoko Kito, and Brian Frankie, Pioneer Astronautics report.

5.3.4.3   Timeline for ISRU

The Mars Ascent Vehicle (MAV) and the Nuclear Reactor Power System (NRPS) arrive at the same time as the ISRU system. The NRPS is deployed behind a berm at some distance from the main landing site and is connected by cables to the main site. The ISRU system is connected to the MAV via cryogenic connecting tubes. The ISRU system includes autonomous regolith-gathering rovers that dig up regolith, transport it to a processor that extracts and stores water, and discards the spent regolith in a slag heap (or for use as Habitat shielding). Some water is stored for life support and some is used as a feedstock for propellant production. Presumably, a robotic system has already visited the landing site, mapped out water deposits with a neutron spectrometer, and done spot checks (core drill and some excavation) here and there to validate neutron spectrometer data. The near-surface water source has been mapped out and designated by small flags.

5.3.4.4 CO_2 acquisition from the Mars atmosphere

All of the Mars ISRU systems described in the previous sections require a supply of relatively pure, pressurized CO_2 from the atmosphere. Since the atmospheric pressure on Mars is typically about 6 torr, it is desirable to compress this by at least a factor of $\sim$100 to obtain reasonable throughput in small vessels. These ISRU systems therefore implicitly utilize a subsystem that sucks in dust-free atmosphere, separates the CO_2 from other atmospheric constituents, and compresses the CO_2. (In this process, a limited amount of $Ar + N_2$ may be obtained as a by-product.)

One approach for pressurizing atmospheric CO_2 is a sorption compressor that contains virtually no moving parts and achieves its compression by alternately cooling and heating a sorbent material that absorbs low-pressure gas at low temperatures and drives off high-pressure gas at higher temperatures. By exposing the sorption compressor to the cold night environment of Mars (roughly 6 torr and 200 K at moderate latitudes), CO_2 is preferentially adsorbed from the Martian atmosphere by the sorbent material. During the day, when solar electrical power is available, the adsorbent is heated in a closed volume, thereby releasing almost pure CO_2 at significantly higher pressures for use as a feedstock in a reactor. A thermal switch isolates the sorbent bed from a radiator during the heating cycle. However, the energy required to heat up the sorbent is significant, and cooling down the sorbent overnight has been shown to be problematic. A large mass and volume of sorbent is needed.[106]

An alternate approach for compression and purification of CO_2 was developed by a team led by Larry Clark that appears to be superior in that it requires less energy, less mass, and less volume.[107] This approach is a cyclic batch process in which the first cycle is freezing out solid CO_2 (using a mechanical cryocooler) on a cold surface, while atmosphere is continuously blown over the surface. After a time, sufficient solid

---

[106] *Adsorption Pump for Acquisition and Compression of Atmsopheric CO_2 on Mars*, D. Rapp, P. Karlmann, D. L. Clark, and C. M. Carr, AIAA 97-2763, July, 1997.
[107] *Carbon Dioxide Collection and Purification System for Mars*, David L. Clark, Kevin S. Payne, and Joseph R. Trevathan, AIAA Paper 2001-4660.

CO_2 builds up, and the chamber is closed off from the atmosphere. The chamber is then allowed to warm up passively, which causes the CO_2 to sublime, producing a high gas pressure in the chamber. This high-pressure CO_2 can then be vented to a larger accumulation chamber in which successive inputs of CO_2 will gradually build up the pressure. Because N_2 and Ar remain as gases at CO_2 solidification temperatures, and therefore pass out through the exit of the chamber during acquisition, relatively high-purity CO_2 is produced in this process. A Lockheed-Martin prototype test produced very encouraging results. Unfortunately, NASA does not seem to have funded any further development of this process since 2001.

5.3.5   Mass and power requirements of a Mars ISRU system

The data in this section are based on a joint JPL/Lockheed-Martin Study in 2004–2005 (to be referred to as "JPLMS").[108] The system being modeled here utilizes the following elements:

- Water is obtained from near-surface regolith containing ~10% water by weight.
- The Sabatier/Electrolysis process is used.
- CO_2 is acquired by a cryogenic freezing process.
- ISRU is carried out in a 16-month cycle to produce 10 mT of CH_4, 35 mT of propellant oxygen, 5 mT of consumable oxygen, and 108 mT of water.
- Propellant tank masses are not included in the ISRU system mass because these are needed as part of the MAV, with or without ISRU. Requirements for maintaining CH_4 and O_2 as liquids in the MAV are also attributed to the MAV, not to ISRU. However, the one-time requirement to liquefy the gases produced by ISRU is attributed to the ISRU system.

The JPLMS estimated the mass and power requirements for various steps as shown in Table 5.10.

Acquisition of 1 kg of water requires excavating perhaps as much as 10 kg of regolith and heating it. The mass and power requirement for acquisition of water from regolith does not seem to have been estimated in any serious manner. In the absence of any reliable data on the subject, rough guesses for the mass and power requirements for excavation and extraction of water from regolith are:

$$\text{Mass} = 1,500 \text{ kg per kg/hr of water recovered}$$

$$\text{Power} = 12 \text{ kW per kg/hr of water recovered}$$

---

[108] *Preliminary System Analysis of Mars Isru Alternatives*, Donald Rapp, Jason Andringa, Robert Easter, Jeffrey H. Smith, and Thomas Wilson (Jet Propulsion Laboratory) and Larry Clark and Kevin Payne (Lockheed-Martin Space Systems), FY04 Report, JPL D-31341, October 25, 2004, revised January 25, 2005.

Table 5.10. Estimated mass and power requirements per unit feedstock processing rate for various Mars ISRU process steps. It will be assumed that for any feedstock rate the requirements scale linearly.

Process step	Unit feedstock rate	Mass of processor (kg)	Power required (kW)
CO_2 acquisition	1 kg/hr of CO_2	50	1.2
Sabatier conversion	1 kg/hr of CO_2	12.5	0.16
Water electrolysis	1 kg/hr of water	11.2	2.4
Liquefying O_2	1 kg/hr of O_2	35	1.1
Liquefying CH_4	1 kg/hr of CH_4	85	3.0
Regolith excavation and water extraction	1 kg/hr of water recovered	700	8.0

For the case where all ISRU is completed in 16 months, the total number of hours is $16 \times 30 \times 24 = 11,520$. The appropriate feedstock rates, masses, and powers to produce 10 mT of CH_4, 35 mT of propellant oxygen, 5 mT of consumable oxygen, and 108 mT of water in 16 months are given in Table 5.11.

Allowing for inefficiencies, we should probably increase the figures in Table 5.11 by at least 15–20%.

It should be noted that we have assumed that the ISRU system functions 24/7 for 16 months, and thus a relatively small system produces a huge amount of product. For example, regolith excavation is only about 20 kg per hour (or 10 liters of volume per hour) to produce 2 kg/hr of water. Note that the ISRU system mass is dominated by excavation systems.

Table 5.11. Estimated gross mass and power requirements for various Mars ISRU process steps.

Process step	Total feedstock used in 16 months	Feedstock rate (kg/hr)	Mass of unit (kg)	Power required (kW)
H_2O acquisition	22,500 kg H_2O	2.0	3,000	24
CO_2 acquisition	27,500 kg CO_2	2.4	120	2.9
Sabatier conversion	27,500 kg CO_2	2.4	30	0.4
Water electrolysis	33,750 kg H_2O^*	2.9	33	7.0
Liquefying O_2	35,000 kg O_2	3.0	105	3.3
Liquefying CH_4	10,000 kg CH_4	0.87	74	2.6
Total			*3,360*	*40.2*

* In addition to 22,500 kg of water feedstock, an additional 11,250 kg of recycled product water from the Sabatier process must be electrolyzed.

5.4   FUELING MARS-BOUND VEHICLES FROM LUNAR RESOURCES

5.4.1   Introduction

If Mars-bound vehicles could be fueled in LEO with H_2 and O_2 from the Moon, the required mass of Mars-bound vehicles to be delivered from Earth to LEO would be reduced to about 40% of the mass if propellants were brought from Earth. For example, a Mars-bound vehicle that weighs, say, 250 metric tons in LEO when propellants are brought from Earth, if fueled by hydrogen and oxygen from the Moon would weigh only about 100 metric tons. This would have a huge impact on the feasibility of launching large Mars-bound vehicles.

Using an extension of a model developed previously[109] the percentage of water mined on the Moon that can be transferred from the Moon to LEO for fueling Mars-bound vehicles can be estimated. It is implicitly assumed here that accessible water can be exploited on the Moon. If that is not the case, this entire concept becomes moot. Furthermore, the process may become untenable if the vehicle masses are too high.

We assume that a system is in place to extract water on the surface of the Moon. Part of the water mined on the Moon is electrolyzed to produce H_2 and O_2 propellants for transporting water to Lunar Lagrange point #1 (LL1). LL1 is an interesting place because it takes very little propellant to get there from lunar orbit or high Earth orbit (see Figure 5.6). It has sometimes been suggested that propellant depots be placed at LL1 and staging interplanetary spacecraft at this junction.

Two vehicles are used in this process. A *Lunar Water Tanker (LWT)* carries water from the lunar surface to LL1. At LL1, part of this water is electrolyzed to provide propellants to return the LWT to the Moon, and part is electrolyzed to provide propellants to send the *LL1-to-LEO Tanker (LLT)* to LEO with the remaining water. At LEO, the water is electrolyzed and part of the H_2 and O_2 is used to return the LLT to LL1. The remainder is used to fuel up a Mars-bound vehicle in LEO (see Figure 5.7). The figure of merit is the net percentage of water mined on the Moon that can be transported to LEO for use by Mars-bound vehicles.

The Δv values for various orbit changes provided by Blair *et al.* are listed in Table 5.12. The value for LL1–LEO requires aerocapture at LEO.

5.4.2   Value of lunar water in LEO

A major impediment to viable scenarios for human exploration of Mars is the need for heavy vehicles that must be landed on Mars. For each kg of payload landed on Mars, it may require about 9–11 kg in LEO, depending on systems used for entry, descent, and landing, assuming that hydrogen/oxygen propulsion is used for trans-

---

[109] *Space Resource Economic Analysis Toolkit: The Case for Commercial Lunar Ice Mining*, Brad R. Blair, Javier Diaz, Michael B. Duke, Elisabeth Lamassoure, Robert Easter, Mark Oderman, and Marc Vaucher, Final Report to the NASA Exploration Team, December 20, 2002.

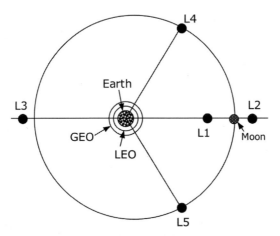

Figure 5.6. Earth–Moon Lagrange Points (not to scale). LEO and GEO are low Earth orbit and Geostationary Earth orbit.

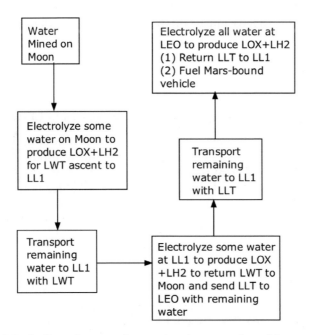

Figure 5.7. Outline of process for transporting water from Moon to LEO.

Mars injection (TMI). Thus, a 40 metric ton Mars lander would require perhaps 360 metric tons in LEO.

About 60% of the mass in LEO consists of H_2 and O_2 propellants for trans-Mars injection. If such Mars-bound vehicles could be fueled in LEO with H_2 and O_2

Table 5.12. Estimated Δv (m/s) for various orbit changes.

	Δv (m/s)
Earth–LEO	9,500
LEO–GEO	3,800
GEO–LEO with aerobraking	500
GEO–LL1 (assumption only)	800
LL1–LEO with aerocapture	500
LEO–LL1	3,150
LL1–LLO	900
LL1–Lunar surface	2,390
Lunar surface–LL1	2,500

from the Moon, the required mass of Mars-bound vehicles to be delivered from Earth to LEO would be reduced to about 40% of the mass that would be required if propellants were brought from Earth. For example, a Mars-bound vehicle that weighs, say, 250 metric tons in LEO when propellants are brought from Earth, if fueled by hydrogen and oxygen from the Moon would weigh only about 100 metric tons. This would have a huge impact on the feasibility of launching large Mars-bound vehicles.

5.4.3 Percentage of water mined on the Moon transferred to LEO

The percentage of the water mined on the Moon that can be transferred to LEO depends critically on the masses of the vehicles used for transport. While Blair *et al.* were concerned with a different issue—commercialization of orbit-raising communication satellites—the mass and propellant analyses are directly transferable to our concern: fueling Mars-bound vehicles in LEO with lunar-derived propellants.

5.4.3.1 Transfer via LL1

In the present analysis, the computations of Blair *et al.* are generalized by allowing the masses of the Moon–LL1 tanker and the LL1–LEO tanker to vary widely as parameters. Using the Δv estimates of Table 5.12, the efficiency of transfer of water mined on the Moon to LEO is estimated as a function of the tanker masses. The quantities that define each step are given in Table 5.13.

Electrolysis of water produces 8 kg of O_2 for every kg of H_2. Since the optimum mixture ratio for O_2/H_2 propulsion is assumed to be 6.5, 1.5 kg of excess O_2 will be produced per kg of H_2 that is produced. This O_2 would likely be vented, although O_2 in LEO might be useful for human life support. This indicates that, per kg of water electrolyzed, only $7.5/9 = 0.833$ kg of useful propellants are produced.

Table 5.13. Sample spreadsheet for calculating requirements to transfer 25 mT of water to LL1.

	C	D	E	F	G	H	I	J	K	L
	Total mass	Inert mass	Propellant mass	Water electrolyzed	Excess O_2	Water transferred	Water mined on the moon	Rocket equation factors		
3	M_t	M_i	M_p	M_{el}	M_{xs} O_2	M_w	M (mined)	R	$R-1$	K_1
4	$=H5/((1-L5-(K5/J5))$	$=L5*C5$	$=C5-D5-H5$	$=1.2*E5$	$=F5-E5$		$=H5+F5$		$=J5-1$	
5	53.50	5.35	23.15	27.78	4.63	25.00	52.78	1.763	0.763	0.10
6	58.51	8.19	25.32	30.38	5.06	25.00	55.38	1.763	0.763	0.14
7	64.55	11.62	27.93	33.52	5.59	25.00	58.52	1.763	0.763	0.18
8	71.99	15.84	31.15	37.38	6.23	25.00	62.38	1.763	0.763	0.22
9	81.36	21.15	35.20	42.25	7.04	25.00	67.25	1.763	0.763	0.26
10	93.53	28.06	40.47	48.57	8.09	25.00	73.57	1.763	0.763	0.30
11	109.99	37.40	47.60	57.12	9.52	25.00	82.12	1.763	0.763	0.34
12	133.49	50.72	57.76	69.31	11.55	25.00	94.31	1.763	0.763	0.38
13	169.74	71.29	73.45	88.14	14.69	25.00	113.14	1.763	0.763	0.42

The mass of either vehicle (LLT or LWT) is represented as a sum of three masses:

M_p = propellant mass

M_i = inert mass (including the structure, an aeroshell for the vehicle that goes to LEO, a landing system for the vehicle that goes to the lunar surface, the water tank, the propulsion stage, and the avionics)

M_w = water mass carried by the vehicle to the next destination

M_t = total mass
    = $M_p + M_i + M_w$

The inert masses of the LWT and LLT tankers are of critical importance in this scheme. We shall assume that the inert mass is some fraction of the total mass.

For each vehicle, we set

$$M_i = K(M_t)$$

where K is an adjustable parameter, and we define K_1 for the LWT and K_2 for the LLT independently.

We begin the calculation by assuming that we will extract enough water on the Moon to send 25 mT of water to LL1. We will then work backwards to estimate how much water would have to be extracted on the Moon in order to provide propellants to send 25 mT of water to LL1. The results can be scaled to any arbitrary amount of water transferred to LL1.

The rocket equation provides that

$$M_p/(M_i + M_w) = R - 1$$
$$M_t/(M_w + M_i) = (M_p + M_w + M_i)/(M_w + M_i) = R$$
$$M_t/M_p = R/(R - 1)$$

For transfer from the lunar surface to LL1, we have

$$R = \exp(\Delta v/(9.8 * I_{SP}) = \exp(2{,}500/(9.8 \times 450)) = 1.763$$

The total mass on the lunar surface is

$$M_t = M_p + M_i + M_w$$
$$M_t = M_t(R - 1)/R + KM_t + M_w$$
$$M_w = M_t[1 - (R - 1)/R - K]$$
$$M_t = M_w/[1 - (R - 1)/R - K]$$

Since we have specified M_w = 25 mT, M_t can be calculated. From this, all the other quantities can immediately be calculated. Table 5.13 shows the calculations for the transfer from the Moon to LL1.

The next step is returning the empty LLT from LL1 to the Moon using some of the 25 mT of water at LL1 to produce propellants. The spreadsheet for doing this is shown in Table 5.14. Negative values in Column H for water remaining at LL1 indicate that for sufficiently high values of K_1, no water can be transferred.

Table 5.14. Sample spreadsheet for calculating requirements to return the LLT from LL1 to the Moon.

	C	D	E	F	G	H	J	K	L
	Total mass	Inert mass	Propellant mass	Water electrolyzed	Excess O_2	Water remaining	Rocket equation factors		
	M_t	M_i	M_p	M_{el}	$M_{xs} O_2$	M_w	R	$R-1$	K_1
	=D17+E17	=D5	=D17*K17	=1.2*E17	=F17-E17	=25-F17		=J17-1	
17	9.20	5.35	3.85	4.62	0.77	20.38	1.719	0.719	0.10
18	14.08	8.19	5.89	7.07	1.18	17.93	1.719	0.719	0.14
19	19.98	11.62	8.36	10.03	1.67	14.97	1.719	0.719	0.18
20	27.23	15.84	11.39	13.67	2.28	11.33	1.719	0.719	0.22
21	36.37	21.15	15.22	18.26	3.04	6.74	1.719	0.719	0.26
22	48.25	28.06	20.19	24.22	4.04	0.78	1.719	0.719	0.30
23	64.30	37.40	26.90	32.28	5.38	-7.28	1.719	0.719	0.34
24	87.21	50.72	36.49	43.79	7.30	-18.79	1.719	0.719	0.38
25	122.57	71.29	51.28	61.54	10.26	-36.54	1.719	0.719	0.42

The next step is transfer of the remaining water from LL1 to LEO. Here, a trial-and-error procedure is used. We guess how much water can be transferred and the propellant requirements are calculated for this load, assuming some value of K_2. The amount of water that must be electrolyzed at LL1 is subtracted from the water remaining at LL1 (after sending the LLT back to the Moon) and the net amount of water is compared with the guessed value. The guessed value is varied until it agrees with the calculated value. A typical spreadsheet is shown in Table 5.15 for an assumed value of K_2. Each row corresponds to the K_1 values from Table 5.14. This process can be repeated for various values of K_2.

Finally, we estimate the requirements for sending the empty LWT back to LL1 from LEO as shown in Table 5.16.

The results of these calculations are summarized in Tables 5.17 and 5.18.

The crux of this calculation then comes down to estimates for K_1 for the LLT and K_2 for the LWT. For the LLT, the inert mass includes the landing structure, the spacecraft structure, the water tank, and the propulsion stage. The propulsion stage for H_2–O_2 propulsion is typically taken as roughly 12% of the propellant mass,[110] and since the propellant mass is likely to be about 42% of the total mass leaving the lunar surface (from calculations), the propulsion stage is perhaps 5% of the total mass leaving the lunar surface. The water mass is likely to be about 40% of the total mass leaving the lunar surface, and if the water tanks weighs, say, 10% of the water mass the tank mass would be about 4% of the total mass. The spacecraft and landing structures are difficult to estimate. A wild guess is 12% of the total mass. Thus, we crudely estimate the value of K_1 for the lunar tanker as $0.05 + 0.04 + 0.12 \sim 0.21$.

The LEO tanker does not require the landing system of the lunar tanker, so the spacecraft mass is estimated as 7% of the total mass of this vehicle. The water transported by the LLT is about 55% of the total mass, so the water tank is estimated as 5.5% of the total mass. In addition, an aeroshell is needed that is estimated at 30% of the mass injected into LEO, which is likely to be about 90% of the mass that departs from LL1 toward LEO, so this is roughly 27% of the total mass leaving LL1. Thus, K_2 for the LEO tanker is roughly estimated as 0.33. These are only rough "guesstimates".

If $K_1 \sim 0.21$ and $K_2 \sim 0.33$, only about 5% of the water extracted on the Moon is transferred to LEO. However, approximately 12% of the water lifted from the Moon is transferred to LEO.

5.4.3.2   Dependence on junction site

In this section, we briefly compare the delivery of water from the Moon to LEO with the junction site being either LL1 or low lunar orbit (LLO). The procedure is essentially the same as that given in the previous section, except that the Δv values of each step are those that either involve LL1 or LLO as the junction point. In doing this, we used the Δv values supplied by *Human Spaceflight: Mission Analysis and*

---

[110] However, the ESAS plan for ascent from the Moon indicates that this is more like 20%.

Table 5.15. Sample spreadsheet for calculating requirements to transfer water from LL1 to LEO. K_2 is constant at 0.1, and the values of K_1 correspond to those in Table 5.13.

	C	D	E	F	G	H	I	J	K	L
	Total mass	Inert mass	Propellant mass	Water electrolyzed	Excess O_2	Water transferred	Water transferred	Rocket equation factors		
	M_t	M_i	M_p	M_{el}	M_{xs} O_2	M_w (guess)	M_w (check)	R	$R-1$	K_2
	= H29/(1-L29-(K29/J29))	= L29*C29	= C29-D29-H29	= 1.2*E29	= F29-E29		= H17-F29		= J29-1	
29	22.12	2.21	2.37	2.85	0.47	17.54	17.54	1.12	0.12	0.1
30	19.46	1.95	2.09	2.50	0.42	15.43	15.43	1.12	0.12	0.1
31	15.26	1.53	1.64	1.96	0.33	12.10	13.01	1.12	0.12	0.1
32	12.29	1.23	1.32	1.58	0.26	9.74	9.75	1.12	0.12	0.1
33	7.30	0.73	0.78	0.94	0.16	5.79	5.80	1.12	0.12	0.1
34	0.83	0.08	0.09	0.11	0.02	0.66	0.67	1.12	0.12	0.1
35	0.13	0.01	0.01	0.02	0.00	0.10	-7.30	1.12	0.12	0.1
36	0.13	0.01	0.01	0.02	0.00	0.10	-18.80	1.12	0.12	0.1
37	0.13	0.01	0.01	0.02	0.00	0.10	-36.56	1.12	0.12	0.1

Table 5.16. Sample spreadsheet for calculating requirements to return the LWT from LEO to LL1. K_2 is constant at 0.1 and K_1 varies as shown in Table 5.13.

	C	D	E	F	G	H	J	K
	Total mass	Inert mass	Propellant mass	Water electrolyzed	Excess O_2	Water remaining	Rocket equation factors	
	M_t	M_i	M_p	M_{el}	M_{xs} O_2	M_w	R	$R-1$
	= D41+E41	= D29	= D41*K41	= 1.2*E41	= F41-E41	= 25-F41		= J41-1
41	4.52	2.21	2.31	2.77	0.46	14.77	2.043	1.043
42	3.98	1.95	2.03	2.44	0.41	12.99	2.043	1.043
43	3.12	1.53	1.59	1.91	0.32	11.10	2.043	1.043
44	2.51	1.23	1.28	1.54	0.26	8.21	2.043	1.043
45	1.49	0.73	0.76	0.91	0.15	4.89	2.043	1.043
46	0.17	0.08	0.09	0.10	0.02	0.57	2.043	1.043
47	0.03	0.01	0.01	0.02	0.00	-7.31	2.043	1.043
48	0.03	0.01	0.01	0.02	0.00	-18.82	2.043	1.043
49	0.03	0.01	0.01	0.02	0.00	-36.57	2.043	1.043

Table 5.17. Mass of water transferred from lunar surface to LEO vs. K_1 and K_2. The mass of water mined (which only depends on K_1) is also shown. The mass of water lifted from the lunar surface is 25 mT. All masses in mT.

$K_1 \Downarrow$ $K_2 \Rightarrow$	0.10	0.20	0.25	0.30	0.35	Mined
0.10	14.77	10.98	8.71	6.14	3.18	52.78
0.14	12.99	9.66	7.67	5.41	2.80	55.38
0.18	10.85	8.06	6.40	4.52	2.34	58.52
0.22	8.21	6.10	4.84	3.41	1.78	62.38
0.26	4.88	3.63	2.88	2.03	1.04	67.25
0.30	0.57	0.42	0.32	0.24	0.12	73.57
0.34						82.12
0.38						94.31
0.42						113.14

Table 5.18. Fraction of water mined that is transferred from lunar surface to LEO vs. K_1 and K_2. Blank cells are cases where no water can be transferred.

$K_1 \Downarrow$ $K_2 \Rightarrow$	0.10	0.20	0.25	0.30	0.35
0.10	0.28	0.21	0.17	0.12	0.06
0.14	0.23	0.17	0.14	0.10	0.05
0.18	0.19	0.14	0.11	0.08	0.04
0.22	0.13	0.10	0.08	0.05	0.03
0.26	0.07	0.05	0.04	0.03	0.02
0.30	0.01	0.01			

Table 5.19. Values of Δv used to compare transfer via LLO and via LL1.

Transfer step	Δv based on LL1	Δv based on LLO
Lunar surface to LLO or LL1	2,520	1,870
LLO or LL1 to lunar surface	2,520	1,870
LLO or LL1 to LEO	770	1,310
LEO to LLO or LL1	3,770	4,040

Design. For the LL1 case, these values are somewhat different than those used in the previous section. Table 5.19 lists these values.

Transfer via LL1 involves a higher Δv for transfers to and from the lunar surface, whereas transfer via LLO involves higher Δv for transfers to and from LEO. Therefore, transfer via LL1 is expected to be more sensitive to the value of K_1 and transfer via LLO is expected to be more sensitive to the value of K_2. This is the case, as illustrated by the results shown in Table 5.20.

Table 5.20. Fraction of mass of water mined on the Moon that is transferred to LEO as a function of K_1 and K_2.

	Based on transfer via LL1			
$K_1 \Downarrow$ $K_2 \Rightarrow$	0.10	0.20	0.25	0.30
0.10	0.23	0.14	0.09	0.03
0.14	0.19	0.12	0.07	0.02
0.18	0.15	0.09	0.06	0.02
0.22	0.10	0.06	0.04	0.01
0.26	0.05	0.03	0.02	0.01
0.30				
0.34				
0.38				
0.42				

	Based on transfer via LLO		
$K_1 \Downarrow$ $K_2 \Rightarrow$	0.10	0.20	0.25
0.10	0.25	0.11	0.03
0.14	0.22	0.10	0.02
0.18	0.20	0.09	0.02
0.22	0.17	0.07	0.02
0.26	0.14	0.06	0.02
0.30	0.10	0.05	0.01
0.34	0.09	0.04	0.01
0.38	0.03	0.02	
0.42			

5.5 LUNAR FERRY FOR DESCENT PROPELLANTS

The requirements to justify retention of ISRU in lunar mission plans include:

- Cryogenic propulsion utilizing LOX, and possibly LH_2 must be used throughout descent and ascent from the Moon.
- A high-leverage user of ISRU products must be found in addition to ascent propellants.
- Significant reductions in IMLEO must result from use of ISRU.
- A very significant potential ISRU target is descent propellants—totaling about 20 to 25 mT of propellants.

Furthermore, in order for ISRU to significantly impact the lunar exploration campaign, the following conditions must be fulfilled:

- ISRU must be built into the very fabric of the lunar campaign so that all space and Launch Vehicles are designed and sized to use ISRU from the beginning.

(This is opposed to the ESAS approach of only using ISRU rather late in the campaign as an add-on to a system that is sized without the benefit of ISRU.)
- In order for this to be possible, an extended robotic campaign must precede the human campaign, to prospect for resources and establish a working ISRU plant as a fundamental asset for the first human sorties. This would undoubtedly delay return of humans to the Moon by several years. However, NASA has apparently elected not to pursue robotic precursors.
- Lunar polar ice must be the ISRU feedstock of choice because it is the only reasonable hope for a workable system.
- Oxygen must be retained as an ascent propellant. It would also be useful to use hydrogen for ascent as well.
- Utilization of ISRU products must be expanded to include descent propellants as well as ascent propellants.
- The requirement for abort to orbit during descent must be eliminated.

Although the benefit/cost ratio for this approach is still not favorable, it is far superior to that based on ISRU generation of only ascent propellants.

A tanker ferry concept can be hypothesized for fueling incoming vehicles prior to landing on the Moon. In this concept:

- A permanent depot is established in lunar orbit with capability to
  - rendezvous with incoming vehicles;
  - electrolyze water delivered to it;
  - store LH_2 and LOX;
  - transfer LH_2 and LOX to tanks on attached vehicles.

- A long-term outpost is established in a lunar polar crater to
  - produce water continuously;
  - convert some water to LH_2 and LOX for ascent;
  - act as a launch pad and receiver for the tanker ferry.

- A tanker ferry vehicle shuttles back and forth between the outpost and the depot, delivering water for conversion to LOX and LH_2,
  - some of the water produced at the outpost is converted to $LOX + LH_2$ for ascent of the tanker ferry;
  - some of the water delivered to the depot is converted to $LOX + LH_2$ for the tanker ferry to return to the outpost.

- Incoming vehicles bound for the lunar surface (and ascent) are fueled at the depot prior to descent with propellants for both descent and ascent.

The efficiency with which one can transfer water from the lunar surface to lunar orbit can be derived from a table analogous to Table 5.14, but for lunar orbit instead of LL1. Only the Moon–lunar orbit transfer is considered. From this table we can derive Table 5.21.

Table 5.21. Percentage of water mined on the Moon that can be transferred to lunar orbit. K_1 is a parameter that determines the tanker mass (the inert mass is K_1 times the total mass).

K_1	Mass of water mined	Net mass of water transferred to lunar orbit	Net % water transferred
0.10	52.8	20.4	38.6
0.14	55.4	17.9	32.4
0.18	58.5	15.0	25.6
0.22	62.4	11.3	18.2
0.26	67.2	6.7	10.0
0.30	73.6	0.8	1.1

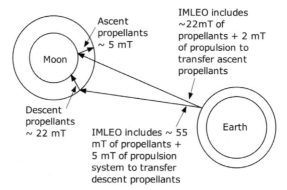

Figure 5.8. Potential saving in IMLEO using the lunar ferry to provide descent and ascent propellants to prospective lunar landers.

The potential mass saving in IMLEO is shown in Figure 5.8. The great stumbling block is: How to establish the outpost without requiring crew on the surface? The huge reductions in IMLEO only occur *after* this system is in place. If the crew must land (without ISRU) to establish the outpost, the full CEV/LSAM with heavy lift is needed to establish the outpost and benefits from using ISRU are greatly reduced.

5.6 STAGING, ASSEMBLY, AND REFUELING IN NEAR-EARTH SPACE

5.6.1 Introduction

One of the problems that seems to confuse discussions of future space missions is the difference in perspective of visionaries and realists. Visionaries tend to be far-sighted and look well beyond the current limitations and challenges to a conceptual realm where many things not presently possible can be imagined. Nevertheless, without a clear pathway leading from here to there, envisaging a bright future is not very useful. By contrast, realists (like myself) face up to the immediate difficulties, but the danger is that realists can easily get bogged down by the enormity of current limitations and challenges, ending up concluding that nothing is possible.

A subject that seems to keep "popping up" in various places is the concept of establishing bases in near-Earth space, typically at a Lagrange point, for assembly of space vehicles, storing propellants, and refueling.

In regard to propellant depots, the visionaries hold sway. For example, a Georgia Tech team[111] has proposed the use of LEO propellant depots and an innovative launch system for a futuristic space environment in which they plot performance metrics vs. number of flights per year, with the number of flights per year measured by hundreds up to 1,000.

5.6.2   Michael Griffin and orbiting fuel depots

In a presentation[112] given in November 2005, the NASA Administrator, Dr. Michael Griffin, said: "... our mission architecture hauls its own Earth-departure fuel up from the ground for each trip. But if there were a fuel depot available on orbit, one capable of being replenished at any time, the Earth departure stage could after refueling carry significantly more payload to the Moon, maximizing the utility of the inherently expensive Shuttle-derived heavy lift vehicle (SDHLV) for carrying high-value cargo." He then goes on to say that NASA cannot afford to develop such a fuel depot and "it is philosophically the wrong thing for the government to be doing ... It is exactly the type of enterprise which should be left to industry and to the marketplace."

Michael Griffin points out that in the future, when the lunar program is in operation, there will be two major missions to the lunar outpost per year, and each of these entails launching the crew on a ~25 mT launcher, and the Earth departure stage and additional payload on a ~125 mT SDHLV. About half of the mass sent to LEO consists of propellants for Earth departure. For two missions per year, the annual cost of sending the Earth departure propellants to LEO is estimated to be about $2.5B based on an estimate of $10,000/kg. He suggests that this potential $2.5B/year market would be attractive to private entrepreneurs.

He admits: "To maintain and operate the fuel depot, periodic human support may be needed. Living space in Earth orbit may be required ... Fuel and other consumables will not always be most needed where they are stored. Will orbital transfer and delivery services develop, with reusable 'space tugs' ferrying goods from centralized stockpiles to other locations? The fuel depot operator will need power for refrigeration and other support systems."

However, from his point of view, these challenges can become opportunities: the need for humans in space would "present yet another commercial opportunity for people like Bob Bigelow, who is already working on developing space habitats."

---

[111] *Tanker Argus: Re-supply for a LEO Cryogenic Propellant Depot*, Brad St. Germain, John R. Olds, Tim Kokan, Leland Marcus, Jeff Miller, Reuben Rohrschneider, and Eric Staton, IAC-02-V.P.10.
[112] "NASA and the Business of Space," Michael D. Griffin, NASA Administrator, *American Astronautical Society 52nd Annual Conference, 15 November 2005.*

The Boeing Company has endorsed this view and peered further into the distant future including a lunar water export business, water delivery to propellant stations, lunar hotels, daily scheduled lunar flights, and over 100 hotels and sport centers in Earth orbit to support an orbital population of 70,000.[113]

5.6.3 Propellants for Earth departure

For lunar and Mars missions, a significant fraction of the total mass that must be transported to LEO from Earth consists of propellants. Assuming use of LOX/LH_2 propulsion at a specific impulse of 450 s, about 50% of the mass in LEO is propellant for Earth departure toward the Moon, and about 60% of the mass in LEO is propellant for Earth departure toward Mars.

There are several conceptual approaches to providing propellants for Earth departure to space vehicles in LEO:

(1) Transport propellants from Earth in large Launch Vehicles along with Earth departure and other propulsion stages, Habitats, Ascent and Descent Vehicles, and other systems needed for human missions. In this scenario, Earth departure propellants are treated as simply one element of the total cargo inventory of a Heavy-Lift Launch Vehicle. This is the most straightforward approach and since NASA has adopted this approach for lunar transfers, it can be considered to be a baseline against which other approaches can be compared.

(2) A second approach would be to launch the lunar-bound or Mars-bound vehicles and systems with dry propellant tanks for Earth departure, and in the same time period send the Earth departure propellants to LEO with a separate dedicated tanker Launch Vehicle (LV) to rendezvous and fuel up the vehicle(s) bound for the Moon or Mars. Since Earth departure propellants amount to 50–60% of the mass in LEO,[114] this would replace one Heavy-Lift Launch Vehicle with two Launch Vehicles each with about half the capacity of the single Launch Vehicle. Presumably, the tanker LV would be less expensive than the cargo LV. However, the total cost to develop and implement two half-size LVs is likely to be greater than the cost of one full-size LV. NASA tends to reduce costs by reducing the number of distinct hardware systems. As an example, in going to the Moon they used a single propulsion system for both the final sub-orbital burn and Earth departure, rather than staging these with two smaller propulsion systems. Had they used staging, it would not have been necessary to carry the dead weight of the sub-orbital propulsion system during Earth departure, thus reducing the initial mass in LEO. Although staging would be mass-efficient, it is not cost-efficient.

(3) A third approach would be to launch Earth departure propellants separately in smaller, less reliable (and less costly) Launch Vehicles, and store them in space at

---

[113] *http://www.boeing.com/defense-space/space/constellation/references/presentations/2.6_Services_Intro.pdf*
[114] See Sections 3.9 to 3.12.

powered cryogenic fuel depots. When a human mission is launched from Earth, it would have empty Earth departure propellant tanks, and it would rendezvous in LEO with the fuel depot, fill its tanks, and then depart for its destination. This approach was advocated by Space Systems/Loral via their "Aquarius system".[115] The *Space Review* article says: "For this system to work, the consumables-only launch must be a lot cheaper than the launch of a high-value, possibly irreplaceable payload. Previously published studies[116] "show that allowing launch reliability to be reduced significantly, to between 0.67 and 0.8, can provide a way to cut launch cost by an order of magnitude. While a 0.67 delivery success rate might seem shockingly low from a traditional aerospace perspective, it is accepted routinely in terrestrial low-cost delivery systems. Aqueducts and high-tension power lines, for example, routinely lose one-third of their payloads en route, yet are highly successful." However, the argument that because some terrestrial systems operate at a delivery success rate of ~67%—therefore, it would be acceptable in space—can be construed to be equivalent to arguing that since all tables have four legs and all horses have four legs, therefore all tables are horses. When a terrestrial power line or aqueduct operates at 67% efficiency, the side effects are generation of heat and water vapor. When a space system fails, the result is either metal falling out of the sky or clogging up LEO with space debris. Furthermore, the difference between a 1/3 loss continuously and a 100% loss 1/3 of the time is dramatic.

(4) A fourth approach would be to mine putative water ice on the Moon and electrolyze some of it to produce LOX/LH_2 propellants for a round-trip tanker system to transport water from the Moon to LEO, where it would be electrolyzed to produce LOX/LH_2 for Earth departure. This approach would have virtue if the cost of developing and operating the lunar water ferry system is less than the cost of the baseline approach. However, analysis of the requirements for transporting water from the Moon to LEO suggests that this process has a very low efficiency and may be very expensive. In addition, the prospects for mining polar ice on the Moon are quite uncertain.[117] Since NASA does not seem inclined to develop a nuclear reactor, operations within the shaded area of a large crater would be several km distant from the solar energy source.

(5) A fifth approach would be similar to the fourth approach, except that the source of water ice would be an asteroid, rather than the Moon. The % of water in some asteroids is likely to exceed the indicated ~1% content in polar craters on the Moon, and the Δv for accessing the asteroid could be lower than for the Moon.[118]

[115] *http://www.thespacereview.com/article/544/1*
[116] I have not been able to locate these studies.
[117] See Section 5.2.4.4.
[118] "Profitably Exploiting Near-Earth Object Resources," Charles L. Gerlach, *2005 International Space Development Conference, National Space Society, Washington, D.C., May 19–22, 2005.*

(6) Within the scope of approaches (4) or (5) one could conceive of establishing fuel depots farther out in space than LEO to provide propellants for steps subsequent to Earth departure. Once a supply of propellants is available outside the influence of the Earth's gravity, many favorable possibilities exist for propellant depots. But, finding a source of extraterrestrial propellants outside a gravity field is a holy grail akin to the mice trying to figure out how to put a bell around the cat's neck.

Obviously, these approaches (other than the baseline) will become more competitive as the Earth departure traffic increases. However, for a few launches per year, it seems doubtful that any alternative can compete with the baseline.

5.6.4   On-orbit staging

A recent published paper[119] describes a concept called "on-orbit staging" (OOS) that the authors believe is critical to achieving the objectives of the human exploration initiative. The authors say: "We demonstrate with multiple cases of a fast (<245-day) round-trip to Mars, that using OOS combined with pre-positioned propulsive elements and supplies sent via fuel-optimal trajectories can reduce the propulsive mass required for the journey by an order-of-magnitude." The authors thereby imply strongly that such short round-trip missions to Mars can be made viable.

The basis for the claims made in this paper is well known. The more demanding a propulsion operation is, the more it benefits from staging the propulsion system into segments such that the amount of propellant used to accelerate propulsion stages (as opposed to payload) is minimized.

For moderately demanding propulsion steps, staging provides moderate benefits. For example, consider the case representing trans-Mars injection from Earth orbit, where the change in velocity (Δv) is $\sim$4,000 m/s, the specific impulse of the LH_2/LOX propulsion system is 460 s, and the ratio of stage mass to propellant mass is 0.1. These correspond to a typical long-stay Mars mission. If the total initial mass in LEO (sum of payload, stage, and propellant) is 1 mT, a single-stage propulsion system can send a payload of 0.338 mT on its way toward Mars. If this propulsion system is staged, the payloads that it can send toward Mars for various levels of staging are given in Table 5.22. The gains from staging are modest.

If we consider a round trip of a payload from LEO to the Mars surface and return to trans-Earth injection, using LOX/LH_2 propulsion for Earth departure and CH_4/LOX thereafter, with propulsion used for all steps (aero-assist is not used here), the values of Δv for a long stay (500–600 days at Mars) mission are estimated in Table 5.23. With these values, the effect of using staging for all steps is summarized in Table 5.24. Staging provides moderate benefits. However, staging beyond two stages produces diminishing returns.

[119] *Enabling Exploration Missions Now: Applications of On-Orbit Staging*, David C. Folta, Frank J. Vaughn, Jr., Paul A. Westmeyer, Gary S. Rawitscher, and Francesco Bordi, American Astronautical Society Paper AAS 05-273.

Table 5.22. Payload mass that can be sent toward Mars vs. level of staging.

	Payload/initial mass
1 stage	0.338
2 stage	0.354
3 stage	0.358
4 stage	0.360

Table 5.23. Estimated values of Δv for Mars long-stay round trip.

Step	Δv (m/s)
Earth departure	4000
Mars orbit insertion	2500
Mars descent	4000
Mars ascent	4300
Mars departure	2400

Table 5.24. Payload mass that can be sent on round trip to Mars surface vs. level of staging (long stay).

No. of stages	Round trip payload/ initial mass in LEO
1 stage	0.0097
2 stage	0.0116
3 stage	0.0122
4 stage	0.0124

Next, consider a short-stay mission with its much higher values of Δv. For a short-stay mission, the Δv values given by Folta *et al.* are summarized in Table 5.25.

With these higher values of Δv, staging has a greater effect. For the round trip mission, we find that the mission cannot even be implemented with a single stage. Multiple stages are necessary or the rocket equation "blows up". The results are given in Table 5.26.

Table 5.25. Estimated values of Δv for Mars short-stay round trip.

Step	Δv (m/s)
Earth departure	8320
Mars orbit insertion	8900
Mars descent	4000
Mars ascent	4300
Mars departure	9900

Table 5.26. Payload mass that can be sent on round trip to Mars surface vs. level of staging (short stay).

	Round-trip payload/ initial mass in LEO
1 stage	Cannot be done
2 stage	0.0000647
3 stage	0.0000893
4 stage	0.0001010

Several things should be noted:

- Staging has a much greater percentage effect in this case. Even the fourth stage produces a significant percentage improvement in payload mass. Folta *et al.* placed primary emphasis on this percentage gain.
- The absolute values of payload mass are far inferior to the payload masses for a long-stay mission. Even with four stages of propulsion, the payload mass fraction is about 1% of that for a long-stay mission.
- Even allowing for the fact that the short-stay mission requires far less in the way of life support and other resources, and therefore the vehicle mass for the short-stay mission is likely to be lower by perhaps a factor of 2, the short-stay mission nevertheless requires 50 times as many launches as the long-stay mission.

Thus, with or without staging, the short-stay mission is infeasible and unaffordable. Folta *et al.* reached the opposite conclusion. The best way to describe the paper by Folta *et al.* is by an analogy.

Suppose an incredibly grossly overweight person weighing 1,200 lb discovered a new innovative diet. By following this innovative diet, he can reduce his weight by 30% to 840 lb. He is very proud that he can produce such a large percentage weight reduction since a mildly overweight person weighing, say, 200 lb, could only expect to reduce his weight by say 10% to 180 lb using the same method. Now, the first person would be pleased with his performance on a percentage basis, and well he should. He could brag to the smaller person about his percentage gains. However, even after completion of his diet, he has a bloated weight. The smaller person, while undergoing a much smaller percentage reduction in weight, has an absolute weight that is much more appropriate.

5.7 TRANSPORTING HYDROGEN TO MARS

5.7.1 Terrestrial vs. space applications

Hydrogen is being widely touted by the U.S. Government as a terrestrial fuel of the future for use in fuel cell powered vehicles.[120] Government programs in fuel cell development and hydrogen storage have been implemented, although much of the apparent progress made so far has been in viewgraphs in PowerPoint rather than in hardware. There are a few overlaps between NASA needs and terrestrial transportation needs for hydrogen storage, although the requirements for H_2 storage are very different for terrestrial applications than those appropriate for space missions. The requirements for terrestrial applications include:

- More than 1,000 cycles.
- Rapid low-energy fill.
- Low energy fill. [*Note:* Liquefying H_2 requires about 1/3 of the heating value of H_2, making liquid storage less attractive for terrestrial use. For space applications involving a single fill at the launch pad, this is irrelevant.]
- Low-cost mass-produced units.
- Safety issues are different for vehicles than for space.
- Volume may be more important than mass for terrestrial use.
- The DOE seems to be interested in storage systems capable of containing $\gtrsim 6\%$ H_2 by weight.

In regard to terrestrial hydrogen storage for vehicles, if one starts with room temperature hydrogen at 1 bar, and stores this hydrogen cryogenically in a tank, roughly 30% of the ultimate heating value of the hydrogen stored is required to cool and liquefy the hydrogen. If the hydrogen thus stored is utilized in a fuel cell that is, say, 40% efficient, only $0.7 \times 0.4 = 28\%$ of the original heating value of the hydrogen is actually used in the final application. For vehicle use, this would be a very serious negative factor.[121] For one-time applications to space, it would not matter.

---

[120] The hydrogen economy concept has been shown to be a hoax by Robert Zubrin. "The Hydrogen Hoax," *The New Atlantis*, Winter, 2007.
[121] A recent report *Why a Hydrogen Economy Doesn't Make Sense* by Ulf Bossel elaborates on this point. See *http://technology.physorg.com/top_news/*

Whereas 6% H_2 by weight might be acceptable for terrestrial vehicles, it would be prohibitive for any imaginable space missions. In fact, one might as well transport water that contains 11.1% hydrogen. Space applications are driven by mass, volume, and power requirements. Furthermore, the energy required to cool and liquefy hydrogen is very high, and this makes storage as liquid hydrogen unsuitable for terrestrial applications where hydrogen tanks have to be repeatedly refilled. However, for space applications, typically involving a single fill of a tank for transport to a distant point, the energy required to initially liquefy the hydrogen is unimportant. Therefore, storage as a liquid has great advantages for space.

The DOE programs seem to be heavily immersed in go-go language: "Breakthrough", "Grand Challenge", and "Revolutionary". They have ambitious schedules and goals for future accomplishment. However, funding does not seem to be commensurate with the goals, and the use of many small parallel research activities, coupled with frequent turnover in management, casts doubt on whether these goals will be accomplished. The DOE-funded work appears to be heavily centered on solid-state sorbents for H_2 storage with a considerable amount of nano-technology (all of which appear to be under-funded and over-viewgraphed). It is a matter of concern that mass is not often mentioned in most publications and news releases, and year-end reports for FY02, 03, and 04 are similar, suggesting that progress has been slow.[122] Aside from the matter of technical progress, the whole concept of a hydrogen economy doesn't make much sense when net energy is compared with other alternatives.[123]

For our purposes, we are primarily interested in the requirements for (a) transfer of hydrogen from LEO to the Mars surface or lunar surface, and/or (b) long-term storage on the Moon or Mars. This hydrogen could be used as a feedstock for ISRU, or directly as a propellant, or for use in fuel cells for power generation.

Hydrogen is a potential propellant for space vehicles that has the advantages of high specific impulse (450 s in chemical propulsion with O_2; 900 s in a nuclear thermal rocket). It also has the logistic advantage that it can easily be produced from water, and therefore it is a natural propellant to use based on *in situ* resource utilization (ISRU) on the Moon or Mars, if accessible indigenous water supplies are available. If indigenous water is not available, transporting hydrogen to the Moon or Mars to be used as a feedstock in ISRU may also provide benefits. In addition, hydrogen is an appropriate fuel to use in fuel cells that would likely be elements of power systems involved in human exploration of the Moon and Mars.

Space applications for hydrogen include:

In present use:
- Use in Earth Launch Vehicles.
- Use as a propellant for chemical (H_2/O_2) propulsion for Earth departure from LEO toward the Moon or Mars or wherever.

[122] *http://www1.eere.energy.gov/hydrogenandfuelcells/*
[123] "Does a Hydrogen Economy Make Sense?", Ulf Bossel, *Proceedings of the IEEE*, Vol. 94, No. 10, October 2006.

Planned for mid-term use:
- Use as a propellant for lunar orbit insertion and descent to the surface of the Moon.
- Use as a feedstock for regenerative fuel cells.

Probably less likely to be used:
- Use as a propellant for ascent from the Moon (no foreseeable plans to do this— may ultimately depend on outcome of search for lunar polar ice).
- Transport from Earth to Moon for use as a feedstock for regolith-based ISRU.
- Transport from Earth to Mars for use as a feedstock for atmospheric-based ISRU.
- Use as a propellant for a nuclear thermal rocket for Earth departure toward the Moon or Mars (although it appears to be unlikely that the NTR will actually be developed and implemented).

However, the only systems that presently utilize hydrogen as a propellant are Earth Launch Vehicles and Earth departure propulsion systems, although it seems likely that the human exploration initiative will use hydrogen for propulsion as far out as descent to the lunar surface, within a week after launch. The reason that hydrogen is not used more widely in space missions is due to the perceived difficulty in storing and transporting hydrogen, particularly for extended periods of time, as well as problems meeting volumetric constraints. For Launch Vehicle applications, the hydrogen tanks can be topped off in the Launch Vehicle shortly before takeoff and it is burned up in rockets in a comparatively short time, before boil-off becomes a serious problem.

For space applications, the key figures of merit for hydrogen storage are:

$$F_I = \text{Initial mass fraction}$$

$$= (\text{initial mass of } H_2)/(\text{initial mass of } H_2 + \text{mass of storage system})$$

$$F_U = \text{Usable mass fraction}$$

$$= (\text{usable mass of } H_2)/(\text{initial mass of } H_2 + \text{mass of storage system})$$

The usable hydrogen mass is the initial mass of hydrogen, less the boil-off prior to use and any residuals left in the tank at the end of use. In the case of active systems, the mass of the storage system includes the active cryocooler system and its power system.

5.7.2   Storage of hydrogen in various physical and chemical states

5.7.2.1   *Storage as high-pressure gas at ~room temperature*

It would seem to be elementary that the first thing to do in studying hydrogen storage is to establish a baseline of characteristics of storage as a highly compressed gas at high pressure and room temperature. However, a search of the Internet for available information on the masses of such tanks reveals that, although there are many relevant sites, few of them deal in any detail with tank masses. In fact, it is truly

Table 5.27. Properties of historical high-pressure hydrogen tanks.

Pre-1980	Type I steel tanks	($\sim$1.5% H_2 by weight)
1980–1987	Type II hoop-wrapped tanks	($\sim$2.3% H_2 by weight)
1987–1993	Type III fully wrapped Al tanks	($\sim$3% H_2 by weight)
1993–1998	Type IV all-composite tanks	($\sim$4.5% H_2 by weight)
2000–2003	Advanced composite tanks	($\sim$7% H_2 by weight)

amazing how many websites discuss hydrogen storage at length, but rarely say anything useful or interesting, and almost never mention the word "mass".

Some information is available from the 2002, 2003, and 2004 U.S. DOE Progress Reports.[124] One DOE program of relevance is being carried out by QUANTUM Technologies, Inc. Specific goals of their project were to develop and demonstrate 5,000-psi tanks that can hold 7.5–8.5% hydrogen by weight, with % hydrogen based on the overall system (probably includes mountings, in-tank regulator, and plumbing) being $\sim$5.7% hydrogen. They also have a goal to develop 10,000-psi tanks that can hold 6.0–6.5% hydrogen, with overall system content of 4.5% hydrogen. These tanks are designed for use in terrestrial transportation and require capability for over 1,000 cycles, whereas in space applications such cycling is not necessary. It is worth noting that the advantage of the 10,000-psi tank is that it has a smaller volume for the same amount of hydrogen stored. However, the 5,000-psi tank is more mass-efficient and holds a higher ratio of hydrogen mass to tank mass. Furthermore, even though the 10,000-psi tank stores hydrogen at twice the pressure, the hydrogen density (at room temperature) is only 1.7 times as great as it is at 5,000 psi. The QUANTUM advanced composite tank technology incorporates three layers: (a) a seamless, one-piece, permeation-resistant, cross-linked, ultra-high molecular weight polymer liner, (b) a composite shell wrapped with multiple layers of carbon fiber/epoxy laminate, and (c) a proprietary external protective layer for impact resistance. QUANTUM's patented in-tank regulator confines high gas pressures to within the tank, and thus eliminates high-pressure fuel lines downstream of the fuel storage subsystem.

According to QUANTUM, the historical performance of tanks is given in Table 5.27.

QUANTUM also notes that the main cost driver for advanced tanks is the cost of carbon fibers.

A related DOE task by LANL says: "Recent theoretical results suggest that the best hydrogen containment solutions must store gas at pressures as high as 15,000 psi." They also say: "Preliminary results suggest volume improvements will be roughly twice as valuable as mass improvements (as ratios to existing technology)." LLNL built six 130-liter insulated storage tanks, but no mention was made of their mass.

---

[124] *Hydrogen, Fuel Cells, and Infrastructure Technologies 2002 Progress Report, Sec. III, Hydrogen Storage*, Department of Energy. *Ibid.*, 2003. *Ibid.*, 2004.

Table 5.28. Properties of the APL high-pressure
hydrogen tank.

Total empty weight	73 kg
Service pressure	340 bar
Total hydrogen capacity	4.2 Kg
Gas/container mass fraction	5.7%
External volume	266 liter
Internal gas volume	166 liter

   APL reported that they tested storage tanks at 5,000 psi and found a permeation
rate of 0.8 scc/hr/l for untreated plastic liner and 0.2 scc/hr/l for treated liners. The
tanks had the properties shown in Table 5.28.
   Realizing the inherent limitations to high-pressure storage, the DOE seemed to
be seeking revolutionary breakthroughs under its "grand challenge" solicitation in
2004.

5.7.2.2   *Storage as a cryogenic liquid*

Storage as a cryogenic liquid is almost surely the best approach for space applications.
   At a pressure of 1 bar and a temperature of 20.4 K, the density of liquid hydrogen
is 70.97 kg/m^3. If hydrogen liquid is stored at other pressures, the temperature and
density will be as shown in Table 5.29. There may be good reasons to store liquid
hydrogen at slightly elevated pressures if the system that utilizes the hydrogen

Table 5.29. Pressure–temperature–density data for saturated
liquid storage of hydrogen.

Temperature (K)	P (psi)	Density (g/cc)
16	3.17	0.075
17	4.83	0.074
18	7.07	0.073
19	9.99	0.072
20	13.71	0.071
21	18.33	0.069
22	23.98	0.068
23	30.77	0.067
24	38.82	0.065
25	48.24	0.064
26	59.17	0.062
27	71.72	0.060
28	86.03	0.058
29	102.24	0.056
30	120.52	0.053

operates at an elevated pressure. However, the density of liquid hydrogen decreases significantly at higher pressure (the rise in temperature more than compensates for the higher pressure), thus necessitating a larger tank.

Relatively little information seems to be available on the mass of hydrogen storage tanks for space applications. However, a few references dealt with mass.

As in the case of high-pressure storage, there are many Internet sites that deal in one way or another with storage of liquid hydrogen, including many DOE sites, but the word "mass" rarely appears in any of them.

Plachta and Kittel[125] provide an implicit estimate of tank masses for storing liquid hydrogen. For a spherical tank, they suggested using $5.4 \, \mathrm{kg/m^2}$ as the tank mass per unit area. For MLI insulation, they suggested using $0.02 \, \mathrm{kg/m^2}$ per layer, or roughly $0.9 \, \mathrm{kg/m^2}$ for 50 layers. They also suggested allowing 5% for ullage and residuals. For storage in a 3.3 m diameter spherical tank, this would lead to the following estimates:

- Area $= A = 34.21 \, \mathrm{m^2}$.
- Volume $= 18.82 \, \mathrm{m^3}$.
- Mass of hydrogen when full $= 1,320 \, \mathrm{kg}$ (at 1 bar).
- Usable mass of hydrogen $= 1,250 \, \mathrm{kg}$.
- Tank mass $= 185 \, \mathrm{kg}$.
- Insulation mass (50 layers of MLI) $= 35 \, \mathrm{kg}$.

Thus, prior to boil-off, the initial hydrogen loading amounts to $1,250/1,540 = 80\%$ ($F_I = 0.8$) of the total mass according to this model.

Other references[126] provide much less optimistic estimates of hydrogen tank mass requirements, but they do not provide enough data to check the details. Another reference[127] provides rough estimates of the tank mass to store liquid hydrogen for short periods appropriate for Earth orbit transfers. However, no details were provided on insulation. An aluminum tank to store 5 to 10 mT of hydrogen was estimated to weigh about 10% of the hydrogen mass, so the hydrogen would be about 90% of the total mass excluding insulation.

It appears from such approximate estimates that a rough rule of thumb is that one can store liquid hydrogen in a tank with a mass distribution of roughly 20% tank mass, 75% usable hydrogen mass, and 5% ullage and residual hydrogen ($F_I \sim 0.75$). This does not include other mass effects such as the need for shields, or implications

[125] *An Updated Zero Boil-Off Cryogenic Propellant Storage Analysis Applied to Upper Stages or Depots in an LEO Environment*, David Plachta and Peter Kittel, NASA/TM2003-211691, June 2003, AIAA20023589.

[126] "Simulations of Zero Boil-Off in a Cryogenic Storage Tank," Charles H. Panzarella and Mohammad Kassemi, *41st Aerospace Sciences Meeting and Exhibition, January 6–9, 2003, Reno, NV*, AIAA 2003-1159.

[127] "Evaluation of Supercritical Cryogen Storage and Transfer Systems for Future NASA Missions," Hugh Arif, John C. Aydelott, and David l. Chato, *28th AIAA Aerospace Sciences Meeting, Reno, NV, January 1990*, AIAA 90-0719.

for spacecraft structure induced by inclusion of cryogenic tanks. For short-term applications such as lunar transfers, these requirements might be minimal. For long-term applications such as Mars transfers, they could be very significant. For short-term applications, hydrogen storage would appear to be quite efficient mass-wise, and therefore the question of viability for various space applications of various durations will depend on the rate of heat leak into the tank, resulting in boil-off over the duration of the mission. Alternatively, one could possibly use active refrigeration to remove heat from the tank at the rate that heat leaks in, resulting in "zero boil-off" (ZBO). However, the mass and power requirements for ZBO, and issues related to reliability are complex. If ZBO is used, the rate of heat leak will determine the capacity of the required cryocooler system, which is likely to be large in many instances.

5.7.2.3   *Storage as a dense gas at reduced temperature*

The requirements for a cryocooler to remove heat from storage at 80–120 K are far less demanding than for heat removal at 20–30 K. Therefore, it is worth considering storage of hydrogen in this temperature range. Figure 5.9 shows the variation of density of gaseous hydrogen with pressure at various temperatures. Supercritical hydrogen can have about the same density as liquid hydrogen under various conditions of pressure and temperature.

For example, hydrogen gas at 90 K and 5,000 psi, or 120 K and 7,000 psi, has about the same density as liquid hydrogen at 100 psi. Storing hydrogen as a gas at such supercritical pressures has some attractive features because liquid–vapor separation is not needed and the hydrogen can be pressure-fed to the user with no

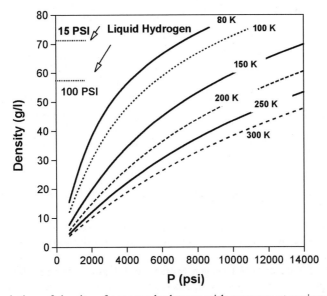

Figure 5.9. Variation of density of gaseous hydrogen with pressure at various temperatures.

need for pumps or compressors. However, the mass of such a tank would be determined mainly by its operating pressure, and the hydrogen content of the storage system at these pressures is expected to be 5 to 7%. This would be unattractive for transfers to the Moon or Mars.

A study[128] considered supercritical storage of hydrogen and oxygen as a supply for hydrogen/oxygen thrusters. They did not specify the storage temperature, but indicated that the hydrogen tank would start out at 250 psi with a density of about 34 kg/m^3. This corresponds to a temperature of roughly 35 K. As hydrogen is gradually removed from the tank, the pressure is maintained at 250 psi by adding heat. Two kinds of tanks were considered: (1) aluminum and (2) fiber-wound with a thin metallic liner. They estimated that the aluminum tank would have a mass about equal to the hydrogen mass, so it would store roughly 50% hydrogen by weight. The composite tank would store about 75% hydrogen by weight. It was also estimated that an aluminum tank to store liquid hydrogen at 1 bar would store about 92% hydrogen by weight.

5.7.2.4   *Storage as solid hydrogen*

The potential advantage of solid hydrogen over liquid hydrogen is that the heat of melting acts as a thermal buffer against heat leaks, increasing the longevity of hydrogen storage. However, solid-hydrogen storage presents several complications: (1) difficulties in supplying hydrogen to users at a high enough pressure and flow rate, (2) mass penalties associated with a full-volume internal metal foam, and (3) additional complexity of the ground loading system.

Delivery of hydrogen to users when stored as a sublimating solid will be difficult. A source of heat can be used to sublime some of the solid hydrogen, but hydrogen vapor must be constantly removed to keep the tank below the triple-point pressure $(7.0 \text{ kPa} = 0.07 \text{ bar} \sim 1.0 \text{ psi})$ and prevent the solid from melting. A compressor (basically, a vacuum pump) is needed that has an inlet (suction) pressure lower than the 1.0 psi triple-point pressure. Whether such a compressor is used to supply end-users directly or to charge accumulators, there are significant issues associated with the flow rate and/or the pressure ratio required, as well as issues of leakage past pistons, seals, etc. Power requirements could be significant. It may be possible to use metal hydrides to pressurize the hydrogen. Requirements do not seem to have been analyzed.

Regardless of the approach selected to add energy to sublime hydrogen for removal from the storage tank, some method is needed to hold the solid hydrogen in place. An aluminum foam with a relative density of 2% (2% of the density of a solid chunk of aluminum) has a density of about 56 kg/m^3. Since the density of solid hydrogen is 86.6 kg/m^3, the foam adds significantly to the mass of the tank.

---

[128] "Evaluation of Supercritical Cryogen Storage and Transfer Systems for Future NASA Missions," Hugh Arif, John C. Aydelott, and David I. Chato, *28th AIAA Aerospace Sciences Meeting, Reno, NV, January 1990*, AIAA 90-0719.

Table 5.30. Properties of liquid H_2, triple-point H_2, and 50% slush H_2.

Hydrogen state	T (K)	P (psi)	Density (kg/m³)	Heat sink (kJ/kg)
Liquid H_2 at normal boiling point	20.3	14.7	71.2	446
Triple-point H_2	13.8	1.1	77.0	497
50% slush H_2	13.8	1.1	81.8	526.3

The complexity of the ground support system and the cost of operations for solidifying hydrogen in large storage tanks appear to be formidable. The method used to fill the solid hydrogen is likely to involve circulation of liquid helium around the liquid hydrogen filled inner vessel to freeze the hydrogen. This is a very expensive method. The fact that the pressure within the hydrogen tank is only ~1 psi introduces concerns about the structural stability of the tank and air ingestion into the tank.

5.7.2.5 *Storage as solid–liquid slush*

The use of slush hydrogen (SLH) at the triple-point pressure (7.042 kPa or 1.02 psia) provides several advantages and several disadvantages compared with liquid hydrogen.[129] On the positive side, a 50% solid mass fraction in slush has 15% greater density and 18% greater heat capacity than normal boiling point liquid hydrogen. The increase in density allows a more compact tank, and the increased heat capacity reduces the amount of venting required for a given heat leak. Slush also has the advantage that the solid mass fraction can be allowed to vary to accommodate heat leaks without venting. During storage, heat leaks cause some of the solid in the slush to melt, reducing the solid mass fraction (some of the vapor condenses as well to maintain the tank at the triple-point pressure). However, it is not clear how to perform vapor extraction with vapor at 1 psi. A thermodynamic vent system, with its Joule–Thomson expansion device, might become clogged by the solid-hydrogen particles in the slush.

While the vapor extraction problem may possibly be surmountable, the low vapor pressure of slush hydrogen (the triple-point pressure) presents the same problems in compressor design as for the solid storage concept. In addition, the ground handling of slush is complicated by the fact that the volume of the slush increases as heat enters and the solid melts. As heat leaks into the slush, some of the solid will melt and the overall volume of the slush will increase. If left unchecked the volume of the

---

[129] "Long-term hydrogen storage and delivery for low-thrust space propulsion systems," Paul J. Mueller, Brian G. Williams, and J. C. Batty (all of Utah State University, Logan), *30th ASME, SAE, and ASEE Joint Propulsion Conference and Exhibition, Indianapolis, IN, June 27–29, 1994*, AIAA 1994-3025; *Benefits of Slush Hydrogen for Space Missions*, Alan Friedlander, Robert Zubrin, and Terry L. Hardy, NASA TM 104503, October 1991; "Technology Issues Associated with Using Densified Hydrogen for Space Vehicles," Terry L. Hardy and Margaret V. Whalen, *AIAA/SAE/ASME/ASEE 28th Joint Propulsion Conference. Nashville, TN, July 1992*, NASA TM 105642, AIAA 92-3079.

slush will exceed the volume of the tank and it will come out the vent. If the slush melted completely it would occupy the volume of the same mass of liquid. Making the tank large enough to accommodate this (or only filling it partially) negates the density/volume savings of using slush in the first place. Thus, slush conditioning will be required on the pad to maintain adequate solid fractions as heat leaks into the tank. Many of the ground-handling concerns were addressed in the National Aerospace Place (NASP) design effort, but developing the infrastructure for slush handling at launch facilities will be very expensive.

5.7.2.6   *Storage as hydrogen at its triple point*

Triple-point hydrogen (TPH)—liquid hydrogen at 1.02 psia and 13.8 K—offers increases in density (8%) and heat capacity (12%) in comparison to liquid H_2 at its normal boiling point. Although these benefits are not as large as those for slush, TPH does not have the added complication of solid particles, and thus may also be an option for future space vehicles.

5.7.2.7   *Storage as adsorbed hydrogen on a sorbent*

Introduction

Gas/sorbent pairs are characterized by whether sorption takes place by physical adherence due to polarization forces ("physisorption"), or whether the gas molecules chemically combine with the sorbent ("chemisorption"). The great advantage of chemisorption is that stronger gas surface forces are involved whereas physisorption involves weaker van der Waals forces. With these stronger forces, greater storage densities of gases can be achieved. However, it is difficult to find a suitable chemisorption sorbent material for a gas like hydrogen. Dissociative hydride alloys have been used, but they tend to be too massive and expensive for use in commercial storage systems. The advantage of physisorption sorbent materials is that they can be used for many gases, particularly hydrogen. All these choices are challenged by the demands of high gravimetric and volumetric density for any conceivable application.

Finely powdered ("activated") carbon is a widely used physisorption material used for absorbing and desorbing gases (see Figure 5.10). It derives its effectiveness from the very high surface area that can be produced if processed effectively. Typically, the adsorption energy (ε) to a sorbent is a few 10s of meV and, because of the weak forces, physisorption of hydrogen works best at cryogenic temperatures. Since adsorption scales with ε/kT, retaining good adsorption of hydrogen on carbon up to room temperature requires the adsorption energy ε to be increased to about 200 meV. Therefore, the "holy grail" in physisorption systems for effective storage of hydrogen at room temperatures is to find ways to structure the sorbent surfaces to increase the effective adsorption energy ε to about 200 meV.

Storage on nano-carbon

In the decade since their discovery, carbon nano-tubes have become a focal point of the nano-materials field, and many hopes have been pinned on them. Scientists

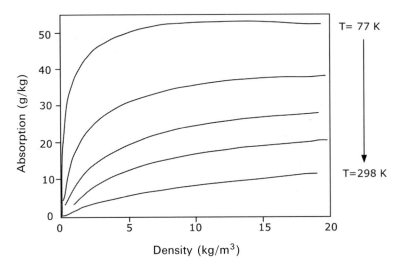

Figure 5.10. Isotherms for the temperature range of 77 K to 298 K for activated carbon. At 77 K, carbon can hold five times as much hydrogen as it can at room temperature.

have envisaged these molecular-scale graphitic tubes as the key to a variety of potentially revolutionary technologies ranging from super-strong composites to nano-electronics. It has been widely conjectured that carbon nano-tubes might provide the best medium for storing hydrogen.

When carbon nano-tubes first became available, they drew interest because these tubes are typically produced in bundles that are lightweight and have a high density of small, uniform, cylindrical pores (the individual nano-tubes). Under the right conditions, there is no reason nano-tubes would not allow hydrogen molecules into their interior space or into the channels between the tubes. But, the crucial question is this: Can nano-tubes store and release practical amounts of hydrogen under reasonable conditions of temperature and pressure?

An early paper by one group[130] reported that certain graphite nano-fibers can store hydrogen at levels exceeding 50 wt% at room temperature. The density was claimed to be higher than that of liquid hydrogen. Such results are incredible when one considers that methane has a hydrogen content of only 25 wt%, and it is hard to imagine packing more hydrogen around carbon atoms than in methane. Attempts at other labs to reproduce the Northeastern University findings were unsuccessful. A year later, another group[131] reported remarkable hydrogen uptake by alkali–metal doped, multi-walled nano-tubes formed by the catalytic decomposition of methane.

[130] "Hydrogen Storage in Graphite Nanofibers," A. Chambers, C. Park, T. K. Baker, and N. M. Rodriguez, *Journal of Physical Chemistry B*, Vol. 102, No. 22, 4253–4256, May 28, 1998.
[131] "Remarkable hydrogen uptake by alkali–metal," Jianyi Lin *et al.*, *Science*, Vol. 285, 91, 1999.

The H_2 uptake was claimed to be 20 wt% for lithium-doped nano-tubes at 380°C and 14 wt% for potassium-doped nano-tubes at room temperature. Subsequent studies in other laboratories cast doubt on these results, attributing them to the presence of water impurities. Both of these reports are now widely regarded as bogus aberrations due to experimental errors.

The era of exaggerated claims for nano-carbon storage of hydrogen is now hopefully past, but the question remains as to how much hydrogen such materials can possibly hold. The most optimistic view is that such materials may be able to store up to 8% by weight of hydrogen at room temperature and moderate pressure. Others claim that the hydrogen uptake could be as high as 6 wt% in nano-tubes—but only at 77 K, the temperature of liquid nitrogen. At room temperature, it is claimed that nano-tubes adsorb only minor amounts of hydrogen. The answer remains in doubt.

5.7.2.8   *Storage in metal hydrides*

Metal hydrides are metallic alloys that absorb hydrogen. These alloys can be used as a storage mechanism because of their ability to absorb and release hydrogen. The release of hydrogen is directly related to the temperature of the hydride. Typically, metal hydrides can hold hydrogen equal to approximately 1% to 2% of their weight. If the temperature is held constant the hydrogen is released at a constant pressure. A metal hydride tank can be used repeatedly to store and release hydrogen cyclically by alternately cooling and heating it. The limiting factor on its ability to store hydrogen is the accumulation of impurities within the tank.

The key tradeoff to utilizing a metal hydride storage system is whether there is sufficient heat available to extract the hydrogen from the hydride. The excess heat produced by a fuel cell (or other hydrogen-consuming component) must be greater than that required by the metal hydride to liberate the hydrogen at the desired flow rate and pressure. The heat available to the hydride must also take into account the inefficiencies associated with the heat transfer device used to transfer the heat from the source to the hydride. Another issue with hydride use is the time lag between initial heating and the release of hydrogen gas. Hydrides constructed of heavy metals such as vanadium, niobium, and iron–titanium will release hydrogen at ambient temperatures and avoid this problem. Other hydrides constructed of lighter materials will need to be heated from an auxiliary source until the temperature is sufficient to release hydrogen. Table 5.31 shows a list of potential metal hydride storage materials and the density of the hydrogen stored within the material. It should be noted that the density given in the table is for the hydrogen alone. This does not represent the hydride material or any other ancillary components needed for the storage system to operate.

Even though the storage density of hydrogen in a metal hydride is high, the total mass of the system is large due to the use of heavy metals as sorbents. Table 5.32 lists the specifications for commercially available, state-of-the-art, metal hydride storage tanks.

Table 5.31. Hydrogen densities on some metallic sorbents.

Metal hydride	Hydrogen density (kg/m^3)
Magnesium (MgH$_2$)	109
Lithium (LiH)	98.5
Titanium (TiH$_{1.97}$)	150.5
Aluminum (AlH$_3$)	151.2
Zirconium (ZrH$_2$)	122.2
Lanthanum (LaNi$_5$H$_6$)	89

Table 5.32. Percent hydrogen by weight on some metallic sorbents.

H$_2$ volume (m^3)	H$_2$ mass (kg)	Metal hydride tank mass (kg)	100 (H$_2$ mass)/ (Metal hydride tank mass)
0.042	0.0036	1	0.36%
0.068	0.0058	0.86	0.68%
0.327	0.0273	6.1	0.45%
0.906	0.0767	16.78	0.46%
1.274	0.1078	24	0.45%
2.547	0.214	36	0.59%

Complex metal hydrides such as alanate (AlH$_4$) materials have the potential for higher gravimetric hydrogen capacities than simple metal hydrides. Alanates can store and release hydrogen reversibly when catalyzed with titanium dopants.

Issues for complex metal hydrides include low hydrogen capacity, slow uptake and release kinetics, and cost. One of the major issues with complex metal hydride materials, due to the reaction enthalpies involved, is thermal management during refueling.

This type of hydrogen storage may some day be appropriate for vehicles, but it is very unlikely to be useful in space.

5.7.2.9   *Storage in glass microspheres*

Glass microspheres can store hydrogen in tiny hollow spheres of glass. If heated, the permeability of the microsphere walls to hydrogen will increase. This provides the ability to fill the spheres by placing the warmed spheres in a high-pressure hydrogen environment. Once cooled, the spheres lock the hydrogen inside. The hydrogen is released by subsequently increasing the temperature of the spheres. This method of storage provides a safe, contamination-resistant method for storing hydrogen. The fill rates of microspheres are related to the properties of the glass used to construct the spheres, the temperature at which the gas is absorbed (usually between 150°C and 40°C) and the pressure of the gas during absorption. Fill and purge rates are directly proportional to the permeability of the glass spheres to hydrogen, which increases with increasing temperature. At room temperature the fill/purge duration is on the order of 5,000 hours, at 225°C it is approximately 1 hour, and at 300°C it is approximately 15 minutes.[132] This dramatic increase in hydrogen permeability with temperature allows the microspheres to maintain low hydrogen losses at storage conditions, while providing sufficient hydrogen flow when needed. Engineered microspheres may allow up to perhaps 10% hydrogen by weight. There is a tradeoff between storage pressure and mass fraction of hydrogen stored. At lower storage pressures the mass fraction of stored hydrogen increases, but the overall hydrogen volumetric density decreases. This is caused by the increase in the glass sphere wall thickness needed to withstand the increase in storage pressure and maintain the same factor of safety.

Although the storage potential of glass microspheres looks impressive there are a number of drawbacks to their use from a system standpoint. The main issue is that to get the hydrogen into and out of the spheres, they must be heated. This heat transfer takes considerable energy and time to accomplish. The higher the heating, the quicker the hydrogen will purge from the spheres. But, from a system standpoint this heating must be accounted for.

5.7.3   Boil-off in space

5.7.3.1   *Rate of boil-off from MLI tanks*

The obvious choice for insulation in space is multi-layer insulation (MLI). However, MLI is ineffective in an atmosphere and it would be necessary to undercoat the MLI with an insulation that is effective in an atmosphere such as foam or evacuated microspheres for use during launch operations. This undercoat of insulation—which can function in an atmosphere—would be necessary for storage on Mars where there is an atmosphere, since the MLI will not provide good insulation capability on Mars. However, heat transfer through any insulation designed for an atmosphere will tend to be faster than it would be through evacuated insulation.

Boil-off on the launch pad is controlled by foam insulation. One estimate is 1.2% boil-off per hour.[133] That is why Launch Vehicles are "topped off" as near to launch time as possible.

Boil-off in space is controlled by MLI insulation. This same reference recommends use of the following equation for the heat flow (watts/m^2) through N layers of MLI, where T_H is the absolute external temperature and T_L is the absolute

---

[132] "Cryogenic Propellant Production, Liquefaction, and Storage for a Precursor to a Human Mars Mission," P. Mueller and T. C. Durrant, *ISRU III Technical Interchange Meeting 8015.pdf, Lockheed-Martin, Denver, CO, 1998,* also *Cryogenics,* Vol. 39, 1021–1028, 1999.
[133] *Cryogenic Propulsion with Zero Boil-Off Storage Applied to Outer Planetary Exploration,* Carl S. Guernsey, Raymond S. Baker, David Plachta (co-investigator), Peter Kittel, Robert J. Christie, and John Jurns, JPL D-31783, Final Report, April 8, 2005.

temperature of the liquid hydrogen in the tank:

$$Q_{MLI}/A = \{1.8/N\}\{1.022 \times 10^{-4} \times [(T_H + T_L)/2][T_H - T_L] + 1.67$$
$$\times 10^{-11}(T_H^{4.67} - T_L^{4.67})\}$$

For example, if $N = 40$, $T_H = 302$, and $T_L = 20$, we calculate

$$Q_{MLI}/A = 0.5 \text{ W/m}^2$$

The relationship between the heat leak rate Q_{MLI}/A (W/m^2) and the hydrogen boil-off rate can be derived by using the heat of vaporization (446 kJ/kg). The boil-off in kg/day per m^2 of tank surface is:

$$(Q_{MLI}/A) \times 24 \times 3,600/446,000 = 0.194(Q_{MLI}/A) \text{ kg/day per m}^2 \text{ of tank surface}$$

For a tank of, say, diameter 3.3 m with a surface area of 34 m^2, this would be

$$6.6(Q_{MLI}/A) \text{ kg/day for the entire tank}$$

This tank holds $(4/3)(\pi)(3.3/2)^3(70) = 1,917$ kg of hydrogen.
Therefore, the percentage loss of hydrogen is

$$6.6 \times 100(Q_{MLI}/A)/1,917 = 0.344(Q_{MLI}/A) \quad (\% \text{ per day})$$

or

$$10.3(Q_{MLI}/A) \quad (\% \text{ per month})$$

Using the previous estimate, $(Q_{MLI}/A) \sim 0.5$ W/m^2, implies a boil-off rate of ~0.17% per day or ~5% per month. This rate of boil-off could presumably be reduced by adding more layers of MLI.

This is an idealized rough estimate of the heat leak into a hydrogen tank surrounded by a vacuum environment at 302 K. In actuality, for any application, one should take into account the total environment of the hydrogen tanks, including their view to space, their view to other spacecraft elements, possible use of shades for shielding the tank from the Sun and planetary bodies, conduction through the support structure (struts), and obstructions that block the view to space, such as struts, thrusters, other tanks, and miscellaneous objects. A tank earmarked for landing on Mars would undoubtedly be mounted inside an aeroshell *en route* to Mars. Configurations that are planned for landing on the Moon or Mars would likely use a number of smaller tanks for packing density, rather than one large tank, and this would increase the ratio of area to volume compared with one large tank. A study of a Mars Sample Return–Earth Return Vehicle using cryogenic propellants[134] used an elaborate shield to reduce the temperature distribution surrounding a

---

[134] "Cryogenic Propulsion with Zero Boil-Off Storage Applied to Outer Planetary Exploration," Carl S. Guernsey, Raymond S. Baker, David Plachta, and Peter Kittel, *41st AIAA/ASME/SAE/ASEE Joint Propulsion Conference and Exhibition, July 10–13, 2005, Tucson, AZ*, AIAA 2005-3559; *Cryogenic Propulsion with Zero Boil-Off Storage Applied to Outer Planetary Exploration*, Carl S. Guernsey, Raymond S. Baker, David Plachta (co-Investigator), Peter Kittel, Robert J. Christie, and John Jurns, JPL D-31783, Final Report, April 8, 2005.

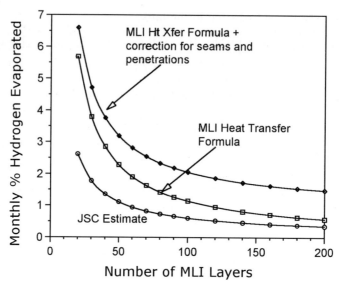

Figure 5.11. Comparison of predictions for boil-off rate. The heat transfer formula was provided by Guernsey *et al.* as $Q_{MLI}/A = \{1.8/N\}\{1.022 \times 10^{-4} \times [(T_H + T_L)/2][T_H - T_L] + 1.67 \times 10^{-11}(T_H^{4.67} - T_L^{4.67})\}$. The corrections for seams and penetrations were recommended by this same reference. The JSC estimate is given in a report published in 2004.[136]

hydrogen tank, and thereby reduced the heat leak by reducing the effective T_H in the equation for heat leak. This would not be practical for missions to Mars that require entry via an aeroshell.

In addition to the radiative environment, additional heat leaks from MLI degradation due to seams and penetrations and for conduction through supports and connecting pipes should be included. The MLI degradation factor is the ratio of heat leak due to seams and penetrations to the heat transfer rate through the MLI blanket. The conduction factor is the ratio of non-insulation heat transfer to heat transfer rates through the insulation system. One study[135] recommends use of an MLI factor 1.74 and a conduction factor of 0.14. Alternatively, Guernsey *et al.* discuss seams, penetrations, and conduction through supports, and recommend adding 20% to the calculated heat leak and then adding a 50% margin on top of that to cover uncertainties. These recommendations by Guernsey *et al.* are plotted in Figure 5.11. A JSC study[136] provided a much more optimistic estimate of boil-off (also in Figure 5.11); however, there appear to be contradictions in the JSC work.

[135] "Thermally optimized zero boil-off densified cryogen storage system for space," Mark S. Haberbusch, Robert J. Stochl, and Adam J. Culler, *Cryogenics*, Vol. 44, 485–491, 2004.
[136] *Lunar Architecture Focused Trade Study Final Report*, NASA Report ESMD-RQ-0005, October 22, 2004.

A more pessimistic view was expressed by another study.[137] This study says: "Actual thermal performance of standard multilayer insulation (MLI) is several times worse than laboratory performance and often 10 times worse than ideal performance." Handling and installation of MLI is tricky. If the MLI is compressed, it will "short out" layers and increase conductivity. Data are presented on heat leaks through MLI into cryogenic tanks. Their data are provided in terms of the equivalent effective thermal conductivity (k), which must be multiplied by the temperature differential, and divided by the thickness to obtain the heat flow per unit area. For 40-layer MLI in a high vacuum, they claim that k approaches about 0.0006 W/m-K. This MLI has a thickness of about 0.22 m. For a temperature difference of 280 K, the heat flow per unit area is $0.0006 \times 280/0.22 = 0.8$ W/m^2. This would indicate a boil-off rate of $\sim 8\%$ per month. This study would predict heat leak rates about double those of the upper curve in Figure 5.11. The difference between idealized and actual performance of MLI remains uncertain.

5.7.3.2 *Mass effect of boil-off in space*

Any system that stores liquid cryogen in space will inevitably have some heat leaks that cause vaporization of the cryogen, resulting in pressurization of the tank if excess vapor is not vented. Venting of vapor without loss of liquid cryogen is a tricky business in zero-g. Nevertheless, it can be done.

As we have seen, the viability of liquid-hydrogen storage for any space application is critically dependent on the rate of heat leak into the tank. If venting of boil-off is used, the rate of heat leak will determine how over-sized the initial tank must be to provide the required mass of hydrogen after the required storage duration. There are indications that hydrogen can be stored so that the initial amount of usable hydrogen accounts for 75% of the initial total mass, and based on this we can estimate the required initial total mass of storage if boil-off and venting amounts to $X\%$ per month for M months. In the following, subscripts "I" and "F" correspond to initial and final, and subscripts "H" and "T" refer to hydrogen and tank, respectively, and "tot" refers to total (hydrogen + tank). Thus:

$$M_{tot,I} = M_{H,I} + M_{T,I} = 1.33 M_{H,I}$$

$$M_{T,F} = M_{T,I}$$

$$M_{H,F} = M_{H,I}(1 - MX/100)$$

$$M_{tot,I} = 1.33 M_{H,F}/(1 - MX/100)$$

For example, if $M = 7$ months and $X = 7\%$, the initial total mass is 2.61 times the final delivered hydrogen mass, so the final hydrogen mass is 38% of the initial total mass. Clearly, when the product $[MX]$ approaches 100, the required initial mass blows up.

---

[137] "Cryogenic insulation systems," S. D. Augustynowicz, J. E. Fesmire, and J. P. Wikstrom, *20th International Refrigeration Congress, Sydney, 1999.*

Using the estimates of Plachta and Kittel,[138] we can derive the following inferences regarding what the initial mass must be in order to provide a given amount of hydrogen some time later, for any boil-off rate. For any tank of radius R initially filled with liquid hydrogen, the tank mass is estimated to be:

$$M_T = 6.3 \times A = 6.3 \times 4\pi R^2 \quad (\text{kg})$$

and the initial hydrogen mass is

$$M_{H,I} = 70V = 70 \times (4/3)(\pi)R^3 \quad (\text{kg})$$

The initial total mass is

$$M_{\text{tot},I} = 6.3 \times 4\pi R^2 + 70 \times (4/3)(\pi)R^3 \quad (\text{kg})$$

The rate of hydrogen boil-off in the tank is proportional to the tank surface area, $A = 4\pi R^2$. Hence, the final mass of hydrogen in the tank after storage for (N) days is

$$M_{H,F} = M_{H,I} - KAN$$

where K is a constant that depends on the rate of heat transfer into the tank and the heat of vaporization of hydrogen.

Plachta and Kittel estimated that to provide 1,250 kg of hydrogen after 62 days with a very well insulated tank, the required initial total mass of the tank plus initial charge of hydrogen is ~1,700 kg. Using the expression for $M_{\text{tot},I} = 1,700$, we find $R = 1.71$ m, $M_T = 232$ kg, $M_I = 1,468$ kg, and $A = 36.7$ m^2. Thus, we derive that according to this model:

$$K = (M_I - M_F)/(AN) = (1,465 - 1,250)/(36.7 \times 62) = 0.096 \quad (\text{kg/m}^2\text{-days})$$

Thus, the heat leak is estimated to be:

$$0.096 \times 446,000 \text{ J/kg}/(24 \times 3,600 \text{ s/day}) = 0.50 \text{ W per m}^2 \text{ of surface area.}$$

For a nine-month passage to Mars, $N = 270$ days. In this case,

$$M_{H,I} = M_{H,F} + K \times A \times 270$$

But, from the expression for $M_{H,I}$, we can convert A to $M_{H,I}$ using:

$$A = 4\pi R^2 = 4\pi [M_{H,I}/(70 \times (4/3)(\pi))]^{2/3}$$

Hence, we can solve the following equation for $M_{H,I}$ for any assumed values of $M_{H,F}$ and N by successive approximations:

$$M_{H,I} = M_{H,F} + K \times 4\pi [M_{H,I}/(70 \times (4/3)(\pi)]^{2/3} \times N$$

For the value $K = 0.096$ (kg/m^2-days) we find that, if $N = 270$ days, $M_{H,F}/M_{H,I} \sim 0.44$ and 56% of the hydrogen boils off during this transit.

---

[138] *An Updated Zero Boil-Off Cryogenic Propellant Storage Analysis Applied to Upper Stages or Depots in an LEO Environment*, David Plachta and Peter Kittel, NASA/TM2003-211691, June 2003, AIAA2002-3589.

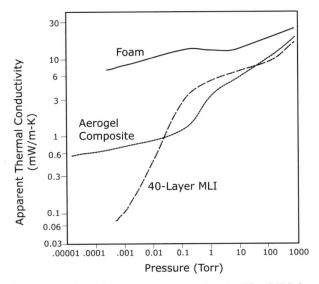

Figure 5.12. Performance of insulation vs. pressure (torr). The MLI is superior at high vacuums, whereas the aerogel is superior at Mars pressures (4–8 torr). [Based on data provided by D. Augustynowicz and J. E. Fesmire, "Cryogenic Insulation System for Soft Vacuums," *Montreal CEC, 1999.*]

In summary, a hydrogen tank insulated with MLI in a vacuum will acquire heat from surroundings at a rate that depends on the temperature distribution of its surroundings and the number of layers of MLI. There are various estimates of the heat leak rate and therefore the boil-off rate, and it is difficult to pinpoint the boil-off rate. For a tank surrounded by an environment at room temperature, with about 40–50 layers of MLI, the estimated heat leak is somewhere in the range 0.4 to 0.9 W/ m^2 of surface, and this will boil off at rates that might be in the range ~4 to 9% per month. In a nine-month transit to Mars it is estimated that, using 50-layer MLI, roughly half of the initial charge of hydrogen will boil off. Thicker blankets of MLI should reduce the boil-off, but penetrations and seams will still play their role.

5.7.3.3 Boil-off rate on Mars

The atmospheric pressure on Mars varies from about 4 to 8 torr. At these pressures, MLI is far less effective than it would be in a good vacuum. The performance of three types of insulation as a function of atmospheric pressure is shown in Figure 5.12.[139] It is noteworthy that all the insulations perform better (smaller k) at lower pressures.

The data presented in Figure 5.12 are in terms of equivalent effective thermal conductivity, which must be multiplied by the temperature differential, and divided

[139] "Cryogenic insulation systems," S. D. Augustynowicz, J. E. Fesmire, and J. P. Wikstrom, *20th International Refrigeration Congress, Sydney, 1999*; "Cryogenic Insulation System for Soft Vacuums," D. Augustynowicz and J. E. Fesmire, *Montreal CEC, 1999.*

by the thickness to obtain the heat flow per unit area. Using either foam or the best aerogel composite insulation for Mars pressures (4–8 torr) k is either 0.012 or 0.005 W/m-K with thicknesses of about 0.40 m and 0.32 m, respectively. For an average temperature difference (tank exterior − tank interior) of 200 K on Mars, the heat flow per unit area is $k \times 200/$(thickness). For foam and aerogel composites, the resultant heat leaks at Mars pressures are 6.0 W/m² for foam and 3.1 W/m² for aerogel composites. These heat leak rates will lead to boil-off rates of 61% and 32% per month, respectively. Furthermore, it is not clear how feasible it would be to use aerogel composites on very large tanks.

5.7.4   Zero boil-off systems

Basically, there are two approaches for cryogenic storage in space:

(1) Use a larger tank and allow boil-off, so that the desired amount of hydrogen is available at the end of storage duration.
(2) Use a tank that contains the ultimate amount of hydrogen needed from the start, and provide an active cooling system to remove an equivalent amount of heat that continually leaks in during the storage period, thus avoiding boil-off. This is called the "zero boil-off" (ZBO) approach.

The ZBO approach eliminates the need for a larger tank, but it requires a cryocooler, a controller, heat transfer equipment in the tank, and a power system to provide power. The greater complexity of this system implies more risk than for passive storage systems. Furthermore, ZBO technology is relatively immature and those who advocate it tend to be somewhat optimistic regarding heat leak rates.

Several references analyze ZBO systems.[140] The trade involved in deciding whether ZBO is appropriate depends on comparison of the larger tank loaded with hydrogen at the start for the passive storage case vs. the smaller tank with ZBO plus the cryocooler, radiator, controller, and power system required to implement ZBO. Plachta and Kittel performed such a comparison. For the ZBO case, they estimated the masses of the tank, insulation, cryocooler, solar array, and the radiator. As would be expected on general grounds, they found that for sufficiently long storage durations, ZBO has lower mass. For hydrogen, the cross-over duration in LEO where the estimated masses for ZBO and non-ZBO masses are equal varies from about

---

[140] "Simulations of Zero Boil-Off in a Cryogenic Storage Tank," Charles H. Panzarella and Mohammad Kassemi, *41st Aerospace Sciences Meeting and Exhibition, January 6–9, 2003, Reno, NV*, AIAA 2003-1159); *An Updated Zero Boil-Off Cryogenic Propellant Storage Analysis Applied to Upper Stages or Depots in an LEO Environment*, David Plachta and Peter Kittel, NASA/TM2003-211691, June 2003, AIAA2002-3589; "Propellant Preservation for Mars Missions," P. Kittel and D. Plachta, *Advances in Cryogenic Engineering*, Vol. 45, presented at the *Cryogenic Engineering Conference, Montreal, Canada, July 12–16, 1999*; "Cryocoolers for Human and Robotic Missions to Mars," P. Kittel, L. Salerno, and D. Plachta, *Cryocoolers*, Vol. 10, presented at the *10th International Cryocooler Conference, Monterey, CA, May 1998*.

90 days for a 2 m diameter tank to about 60 days for a 3.5 m diameter tank to about 50 days for a 5 m diameter tank. For the hydrogen tank described in Section 5.7.3.1, with a diameter of 3.3 m and a total mass of ~1,525 kg (including tank, insulation, and hydrogen), they estimated that the mass of the additional equipment to implement ZBO is roughly 175 kg, so the ZBO storage system has a mass of ~1,700 kg for all durations. In the non-ZBO case, the required mass increases with mission duration because a larger tank must be used for longer durations. In the ZBO case, the required mass is independent of duration. The required mass of the non-ZBO storage system is 1,525 kg for zero duration, increases to 1,700 kg for a duration of 62 days, and 2,210 kg for 270 days duration. Thus, for long-term storage, the ZBO approach appears to have significant advantages, based on the estimated heat leak of $0.5 \, \text{W/m}^2$. Plachta and Kittel therefore predict that hydrogen represents $\sim 1,250/1,700 = 73.5\%$ of the total mass using ZBO.

The surface area of this tank is $34.2 \, \text{m}^2$, so the estimated heat leak is likely to be somewhere around 17 W based on $0.5 \, \text{W/m}^2$. According to one estimate,[141] it would typically require 500 W of electrical power to remove 1 W of heat from a hydrogen tank at 20 K. Using ZBO, the electrical power requirement would appear to be $500 \times 17 = 8.5 \, \text{kW}$. Plachta and Kittel estimated that this could be provided within a mass of 175 kg. That would be an optimistic estimate for near-Earth applications. Furthermore, if we used a higher heat leak than $0.5 \, \text{W/m}^2$, the power requirement would rise and so would the mass estimate. These estimates appear to be idealized.

A study was carried out on long-term storage of cryogens for periods of 1–10 years.[142] They analyzed a spherical liquid hydrogen storage tank, with a passive multilayer insulation system, and an actively cooled shield system using a cryocooler. The study investigated the effects of tank size, fluid storage temperature (densification), number of actively cooled shields, and the insulation thickness on the total system mass, input power, and volume. The hydrogen temperature was 21 K and the ullage allowance was 2%. They considered a tank that could be cooled by two cryocoolers, with the inner shield cooled to either 16 K or 21 K, and the outer shield cooled to 80 K (see Figure 5.13), with a uniform external temperature of 294 K. The analysis was carried out for three cases:

(1) 250 or 4,000 kg of liquid hydrogen stored without an outer shield.
(2) 250 kg of liquid-hydrogen storage with an 80 K outer shield.
(3) 4,000 kg of liquid-hydrogen storage with an 80 K outer shield.

Blankets were constructed of double-aluminized Mylar with Dacron net spacing.

---

[141] "Cryogenic Propulsion with Zero Boil-Off Storage Applied to Outer Planetary Exploration," Carl S. Guernsey, Raymond S. Baker, David Plachta, and Peter Kittel, *41st AIAA/ASME/SAE/ASEE Joint Propulsion Conference and Exhibition, July 10–13, 2005, Tucson, AZ,* AIAA 2005-3559.

[142] "Thermally optimized zero boil-off densified cryogen storage system for space," Mark S. Haberbusch, Robert J. Stochl, and Adam J. Culler, *Cryogenics,* Vol. 44, 485–491, 2004.

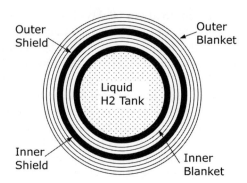

Figure 5.13. Tank arrangements used in study.[143] The inner shield could be cooled to either 16 K or 21 K, and the outer shield could be cooled to 80 K, using active cryocoolers.

It was assumed that the effective conductivity model for MLI can be extrapolated to 300 layers since it is fairly linear. Experimentally derived factors to model additional heat leak from MLI degradation due to seams and penetrations and for conduction through supports and pipes resulted from testing. The MLI factor as defined in the model was the ratio of heat leak due to seams and penetrations to the heat transfer rate through the blanket. The conduction factor was the ratio of non-insulation heat transfer to heat transfer rates through the insulation system. The MLI factor used was 1.74 and the conduction factor used was 0.14. The MLI factor is a strong function of thickness and seam type, but not temperature. Therefore, the assumption was made that the same MLI factor can be used regardless of the boundary temperatures used in the analyses for MLI with up to 300 layers.

The study found significant benefits from cooling the inner shield, and even greater benefits from cooling the inner and outer shields.

In each model, the power and mass requirements were estimated for a ZBO system to match the heat loads, although it was not specified what the heat loads were.

One series of runs was made for the case where only the inner shield was actively cooled and the outer shield was allowed to "float" without active cooling. It was found that when only the inner shield was cooled, the power and mass requirements for zero boil-off were relatively high with a moderate thickness of MLI, but decreased significantly with increasing blanket thickness. For a tank holding 4,000 kg of hydrogen (3 m diameter by 9 m length), use of blankets with 100 layers of MLI resulted in a total storage system mass of 2,000 kg, a power requirement of 10 kW, and a volume requirement of about 60 m^3 (only slightly greater than the actual hydrogen volume of about 57 m^3). The surface area of the tank was about 86 m^2. Using the rule-of-thumb that it takes 500 W of power to remove 1 W of heat from a liquid-hydrogen tank, the apparent heat leak into the tank is estimated at $10,000/500 = 20$ W, or 0.23 W/m^2.

---

[143] "Thermally optimized zero boil-off densified cryogen storage system for space," Mark S. Haberbusch, Robert J. Stochl, and Adam J. Culler, *Cryogenics*, Vol. 44, 485–491, 2004.

However, the use of models previously described would predict that for 100 layers of MLI, the heat leak would be somewhat higher (about $0.36\,\text{W/m}^2$). Using cooling of only the inner shield, hydrogen represents $4,000/6,000 = 67\%$ of the total mass.

Two additional series of runs were made—for 250 kg and 4,000 kg hydrogen tanks—when both the inner shield and the outer shield were actively cooled. For the 4,000 kg tank, the storage system mass was minimized with about 75 layers of MLI in the outer blanket and decreases as the inner blanket is made thicker. For an inner blanket of $\sim$150 layers and an outer blanket of $\sim$75 layers, the estimated storage total mass is 1,750 kg, the power level is $\sim$2 kW, and the volume is $\sim$64 m^3. Hydrogen would then represent about $4,000/5,750 = 70\%$ of the total mass. These are very attractive figures if they can be achieved.

5.7.5   Potential space applications for hydrogen

Space applications for hydrogen were discussed in Section 5.7.1.

5.7.5.1   Shuttle H₂ tanks

The large External Tank used for Shuttle lift-off stores liquid oxygen in the front and liquid hydrogen in the rear portion of the tank. The liquid hydrogen is stored at $\sim$32 to 34 psi. The hydrogen tank weighs 13,166 kg and has a volume of 1,515 cubic meters. It stores 102,700 kg of hydrogen at a density of 68 g/l. The tank mass is 11.4% of the total mass, and the hydrogen mass is 88.7% of the total mass.

The hydrogen tank is filled hours before launch and it separates 9 minutes after blast-off. It is covered with a foam insulation that has become a source of controversy because pieces of foam can fall off during launch, causing damage to Shuttle tiles. A thorough search of the Internet does not provide much information on the rate of boil-off from this tank. However, one estimate is that there is a loss of 1.2% per hour from a hydrogen tank on the launch pad with foam insulation.[144]

A study[145] was carried out to determine whether the Shuttle Orbiter could be utilized, along with a refueled External Tank (ET), to deliver payloads to lunar orbit. To address concerns about boil-off of the cryogenic propellants from the ET during a 14-day lunar mission, it was noted that the foam insulation currently used on the exterior of the ET would be ineffective as a cryogen insulator for an extended duration. Therefore, it was assumed that part of the ET's foam insulation would be replaced with multi-layer insulation (MLI). However, some of the foam insulation would have to be retained as the MLI would have to be protected from aerodynamic loads during ascent to orbit. However, the changes in storage mass due to this change in insulation were not assessed. They assumed that enough MLI was added so that

---

[144] *Cryogenic Propulsion with Zero Boil-Off Storage Applied to Outer Planetary Exploration*, Carl S. Guernsey, Raymond S. Baker, David Plachta (co-investigator), Peter Kittel, Robert J. Christie, and John Jurns, JPL D-31783, Final Report, April 8, 2005.

[145] *Feasibility Analysis of Cis-lunar Flight Using the Shuttle Orbiter*, Davy A. Haynes, NASA Technical Memorandum 104084, June 1991.

propellant boil-off was limited to 4% per month. Finally, they decided that the Shuttle is far too heavy to be a practical transfer vehicle to the Moon.

A related study[146] examined strategies for propellant supply and storage options for a hypothetical fleet of reusable orbital transfer vehicles. The system included liquid-hydrogen and liquid-oxygen storage tanks in low Earth orbit, refilled periodically by propellant brought up from the Earth. They considered a tank for on-orbit storage of propellants for a reusable orbital transfer vehicle containing a total of 59,400 kg of liquid oxygen and hydrogen, at a mixture ratio of 6 to 1. This equates to 8,486 kg hydrogen, and 50,914 kg oxygen, before boil-off. They used multilayer insulation consisting of 50 layers of double-aluminized Kapton (0.15 mm thick), separated by Dacron net. Heat leakage included that through the insulation, as well as through tank/shell struts and fill, feed, and vent lines. The heat leakage causes a boil-off of 0.45 kg/hour of liquid hydrogen, and 0.36 kg/hr of liquid oxygen. Given the loading of each propellant, this translates into these percentage losses:

- Hydrogen boil-off rate = 0.13% per day = 3.8% per month.
- Oxygen boil-off rate = 0.016% per day = 0.49% per month.

If only 10 rather than 50 layers of insulation were used, the hydrogen boil-off rate would increase to about 2.2 kg/hr rather than 0.45, and the oxygen boil-off rate would increase to about 1.6 kg/hr rather than 0.36. The boil-off of 0.45 kg of hydrogen per hour corresponds to a heat leak of

$$[0.45 \ (\text{kg/hr})/3{,}600 \ (\text{s/hr})] \times 446{,}000 \ \text{J/kg} = 55.8 \ \text{W}$$

A cylindrical tank to hold 8,486 kg of hydrogen would have a volume of 120 m^3. A possible layout for this tank might have the dimensions (length ~12.3 m and radius ~1.76 m), so the surface area is about 136 m^2. Hence, the heat leak rate implied for this system is 55.8/136 = 0.4 W/m^2 of surface.

It should be noted that boil-off is governed by heat leakage, and the rate in kilograms per hour does not depend on the amount of propellant in the tanks. With partly filled tanks, the percentage loss per day or month would be higher. Also, note that heat leakage is driven by surface area, while the original mass of propellant in the tanks is governed by volume. Thus, the smaller the tank, the faster that liquid will boil off (square/cube law). Boil-off rates in this study were estimated for tanks in LEO. Lower boil-off rates would be expected for tanks at a further distance from the Sun. Lower boil-off rates can also be expected if tanks can be maintained at an angle that minimizes their cross-sectional area exposed to the Sun, or if the tanks are protected by a sunshade. Taking into account the enthalpy of vaporization, the oxygen tank heat flux across the walls is 0.36 × 213.1 = 76.7 kJ per hour. For the larger and colder hydrogen tank, the figure is 0.45 × 445 = 200.25 kJ per hour. The hydrogen tank vents 0.45 kg/hour of very cold vapor.

[146] *Future Orbital Transfer Vehicle Technology Study, Volume II—Technical Report*, Eldon E. Davis (study manager), NASA Contractor Report 3536, Boeing, 1982.

5.7.5.2   *Hydrogen storage trade study for a solar electric orbital transfer vehicle*

An analysis[147] was made of alternatives for storage of hydrogen for a solar electric orbital transfer vehicle employing hydrogen arcjets. Although this reference does not state how long the storage must last, it seems likely that the transfer might take several months. However, these results are not directly applicable to our study because, in the SEOTV application, "it is assumed that the heat leak into all cryogenic systems is balanced, or 'tuned' to match the required hydrogen propellant outflow for the arc-jets. This heat load is a combination of the system parasitics and additional tuning heaters attached to the cryogen tank walls." Thus, the tanks used in this study were not conceived for low heat leak; on the contrary, a sizable heat leak was desirable to produce a constant flow of hydrogen from the tank. Nevertheless, it is worth mentioning that this reference estimates that the tank mass would be about 15% of the hydrogen mass for storing liquid hydrogen.

5.7.5.3   *Estimates of boil-off by a lunar mission focused study*

A JSC study[148] estimated rates of hydrogen boil-off in lunar missions. Their calculations were for a 5.75 m diameter tank with surroundings at 268 K. This tank holds 8,500 kg of hydrogen, and a simple calculation shows that it has an effective length of 4.7 m.

They do not seem to specify the tank mass, but it is likely to be of the order of 2,200 kg. As the number of MLI layers is increased from 20 to 200, the MLI mass will increase from about 50 kg to about 420 kg. The heat leak through the MLI was assumed to be linear in duration of storage, and it falls off approximately as $1/N$ where N is the number of layers. The total heat leak will include leaks through supports, connections, and penetrations, which may depend very weakly on N. Therefore, the heat leak is expected to have a form

$$H = C_1 + C_2/N$$

where C_1 presumably represents seams and penetrations and C_2 represents heat transfer through the MLI. A best fit to their data indicates that the boil-off is predicted as

$$\text{Boil-off (\% per day)} = 0.0033 + 1.68/N$$

These estimates are about a factor of 3 lower than estimates made by others, as shown previously in Figure 5.11.

If these results were correct, it would appear feasible to store hydrogen on the Moon for seven months with a loss of only a few percent due to boil-off. This seems very unlikely based on other data.

[147] *Hydrogen Storage Trade Study for a Solar Electric Orbital Transfer Vehicle (SEOTV)*, R. S. Bell, AIAA 1993-2398.
[148] *Lunar Architecture Focused Trade Study Final Report*, October 22, 2004, NASA Report ESMD-RQ-0005.

5.7.5.4   *Transporting hydrogen to the lunar surface*

Transporting hydrogen (and oxygen) to the lunar surface and storage on the surface should be achievable with hydrogen constituting roughly 75% of initial total storage mass. The rate of boil-off and venting would probably be in the range 3–7% loss per month, depending on the number of layers of MLI used.

If hydrogen boil-off on the lunar surface can be minimized to, say, 3–5% per month, it may be feasible to use hydrogen as the fuel for ascent from the Moon (along with oxygen). This, of course, will be highly dependent on the latitude of the site. The advantage in specific impulse over methane and oxygen is considerable (450 vs. 360 s) and even if, say, as much as 10–15% of the hydrogen is lost to boil-off, hydrogen would still out-perform hypergolics. Furthermore, if lunar polar ice is accessible, electrolysis of water is a relatively simple process, and the products could be used directly for ascent.

The problem right now is that we have conflicting estimates of the rate of boil-off, and it is difficult to precisely estimate the performance of hydrogen storage systems under realistic conditions.

5.7.5.5   *Transporting hydrogen to Mars for ISRU*

Four of the previous major design reference missions (DRMs) for sending humans to Mars required transportation of hydrogen to the surface of Mars for use in production of propellants and life support consumables from the atmosphere, so that these commodities need not be brought from Earth. This produces a sizable reduction in IMLEO provided that the hydrogen storage mass is not too large.

JSC's DRM-3 required 39 mT of methane/oxygen for ascent propellants, so that 29.3 mT of oxygen is needed. An ISRU system that reacts hydrogen with carbon dioxide produces $CH_4 + O_2$. Hence, to produce 29.3 mT of oxygen plus excess methane requires 3.66 mT of hydrogen. The net benefit of *in situ* resource utilization (ISRU) is then the difference between the 39 mT of propellants produced and the mass of the system that does the production. This latter mass includes the hydrogen, the hydrogen tank, the ISRU process plant, the connecting system from ISRU to the Mars Ascent Vehicle, and the power system. If, as is likely, the power comes for "free" because the same power system subsequently to be used by humans is used for ISRU prior to Earth departure by the crew, then the power system can be neglected in this comparison. Obviously, a high-pressure gas tank that provides 6% hydrogen would weigh $3.66/0.06 = 61$ mT, and this would make this form of ISRU utterly useless. In fact, one could simply bring water to Mars as a source of hydrogen. Water contains 11.1% hydrogen, so this would be better than a high-pressure gas tank, but still not adequate to justify use of ISRU, since 29.3 mT of water would have to be brought to Mars and most of the leverage of ISRU would disappear.

Assuming that the hydrogen tank would be embedded in an aeroshell, it is possible that a tank storing liquid hydrogen brought to Mars in a 9-month transit with 100 layers of MLI would lose about 3–4% of the hydrogen per month due to boil-off, and would have to be over-sized by about 30–35% to account for this boil-off. The net usable hydrogen approaching Mars would then be roughly 50% of the initial storage mass. That is not a bad figure. Unfortunately, storage on Mars would be very problematic due to rapid boil-off on the Mars surface. A one-year ISRU process with storage of hydrogen on Mars is unthinkable, even though all of the existing Mars design reference missions assumed that this could be done. To avoid a lengthy storage time for hydrogen on Mars, Robert Zubrin has suggested an ISRU process that uses up the hydrogen rapidly via the Sabatier reaction (which is slightly exothermic) and storing methane and water until the water can slowly be converted to oxygen via electrolysis. However, Zubrin's scheme requires acquisition of a huge amount of CO_2 in a short time which requires significant power, and other volumetric and logistic challenges in which hydrogen, oxygen, and methane would be moved around from tank to tank. It is not clear how practical this scheme is.

5.7.6   Summary and conclusions

1   Although the current NASA exploration plans call for use of hypergolics as propellants for ascent from the Moon, it may be possible to substitute hydrogen and oxygen and obtain the benefit of higher specific impulse (450 s vs. 320 s). The potential IMLEO benefits are significant, but they hinge upon the achievable boil-off rate for hydrogen on the Moon. This may be very different for equatorial and polar locations.

2   A number of papers on hydrogen storage were reviewed, and there is some diversity of opinion as to what level of boil-off may be achievable with a passive storage system. One can (at least, in principle) keep adding layers of MLI, but there is bound to be a region of diminishing returns where penetrations, connections, seams, and handling factors lead to an asymptotic plateau for high numbers of layers. It appears that such a plateau may be in the range 2–7% per month boil-off in the real world, but it is difficult to fix the boil-off range any more precisely.

3   If boil-off rates as low as 2–3% per month can be achieved with a passive storage system, use of hydrogen as an ascent propellant on the Moon should be feasible based on hydrogen brought from Earth, even for outposts. On the other hand, it if it is near the top of the range, at 6–7%, use of hydrogen brought from Earth would be more marginal. Nevertheless, if polar ice can be obtained on the Moon, all the propellants for ascent can be produced on the Moon, regardless of boil-off, and this would produce savings in IMLEO. Clearly, the two things needed in this regard are (1) upgrading of our ability to predict boil-off rates on the Moon, and (2) prospecting for lunar polar ice.

4   Because of the longer trip time to get to Mars, and the fact that the hydrogen tank would be imbedded within an entry aeroshell, boil-off in transit to Mars is a much more serious problem than for the Moon. Nevertheless, it may be possible to transport hydrogen to Mars, arriving with partly empty tanks. For a 9-month cargo transit, clearly there is a large difference between 2 or 3% per month and 7% per month boil-off rates for passive storage systems. Alternatively, a zero

boil-off (ZBO) system might be effective, but this would add complexity and risk. Depending on the rate of heat leak, the power requirement might be excessive.

5   But, aside from the problem of transporting hydrogen to Mars, an even bigger problem is storing it on the surface of Mars where MLI is ineffective and other insulation systems are much less effective than MLI would be in a high vacuum.

6

Mars mission analysis

From 1988 to 2006, NASA conducted a number of paper studies of requirements and approaches for human missions to the Moon and Mars.

6.1 PREVIOUS MARS MISSION STUDIES

6.1.1 Office of Exploration case studies (1988)

The NASA Office of Exploration (OExP) conducted a series of studies of human and robotic exploration beyond LEO during the 1987–1988 time frame. Four focused case studies were examined: Human Expeditions to Phobos, Human Expeditions to Mars, Lunar Observatory, and Lunar Outpost to Early Mars Evolution. Unfortunately, it is not clear how to obtain copies of these studies. They seem to be unavailable to the public at large.

The ESAS Report says:[149] "The case studies were deliberately set at the boundaries of various conditions in order to elicit first principles and trends toward the refinement of future options, as well as to define and refine prerequisites." Recommendations resulting from the 1988 case studies included the following key points:

- A space station is the key to developing the capability to live and work in space; *that is a mantra that was often chanted by NASA management, but it now seems to have fallen into oblivion.*[150]
- Continued emphasis on research and technology (R&T) will enable a broad spectrum of space missions and strengthen the technology base of the U.S. civilian space program; *a nice generalization, but will NASA do what is needed,*

---

[149] *ESAS Report*, p. 79. The ESAS Report is discussed in Sections 6.1.10 and 7.4.
[150] For good reasons.

despite the constituencies?[151] *And is R&T being squeezed out by the need to maintain the Shuttle and the ISS? R&T does not seem to have a high priority in current ESAS plans.*

- A vigorous life science research-based program must be sustained; *if it leads to useful products.*
- A heavy-lift transportation system must be pursued with a capability targeted to transport large quantities of mass to LEO; *provided we know how much we need to lift. If Mars missions end up requiring a good deal more mass than current plans allow for, this could be a problem in the future.*
- Obtaining data via robotic precursor missions is an essential element of future human exploration efforts; *however, NASA has most recently (February, 2007) canceled the whole program.*
- An artificial gravity research program must be initiated in parallel with the zero-gravity countermeasure program if the U.S. is to maintain its ability to begin exploration in the first decade of the next [21st] century; *this is not needed for the Moon, and ESAS does not seem to be interested in Mars technology, at least prior to 2020.*

6.1.2   Office of Exploration case studies (1989)

The ESAS Report says that,[152] following the 1988 studies, the OExP continued to lead the NASA-wide effort to provide recommendations and alternatives for a national decision on a focused program of human exploration of the solar system. Three case studies were formulated during 1989 for detailed development and analysis: Lunar Evolution, Mars Evolution, and Mars Expedition. Results from the 1989 OExP studies were published in the FY 1989 OExP Annual Report. Key conclusions from the 1989 studies included the following:

- Mars trajectories: OExP point out that human missions to Mars are character-ized by the surface stay-time required—short-stay referring to opposition-class missions and long-stay pertaining to conjunction-class Mars missions. However, the OExP studies actually concluded that:[153] "Because of the large number of Earth launches, the high level of in-space assembly and fueling operations and the resulting cost, and the required short-stay times at Mars, this short-stay case was not considered practical within the 1988 development ground rules." *What is interesting here is that, back in 1988, they realized that Mars short-stay missions were not practical, and yet JSC kept the short-stay torch burning for many years afterward, even in 2006 (see Zubrin quote at end of Section 3.2.2).*
- In-space propulsion: all-chemical-propulsive transportation results in prohibitive total mission mass for Mars missions (1,500–2,000 mT per mission). On the other hand, aero-assist at Mars can provide significant mass savings (50%) as com-

---

[151] "Getting Space Exploration Right," Robert Zubrin, *The New Atlantis*, Spring 2005.
[152] *ESAS Report*, p. 80.
[153] *OExP 1989 Annual Report*, p. 5.

pared with all-chemical-propulsive transportation. Incorporation of advanced propulsion, such as nuclear thermal rockets or nuclear electric propulsion, can result in mission masses comparable with chemical/aerobraking missions; *yes, but NASA has not adequately invested in either NTR or large-scale aerocapture, and does not seem to be on a path for either. Furthermore, if the NTR cannot be fired up in LEO, it loses a good deal of its value* (see Section 3.4.3).

- Reusable spacecraft: employment of reusable spacecraft is predominantly driven by economic considerations; however, reusing spacecraft requires in-space facilities to store, maintain, and refurbish the vehicles, or the vehicles must be designed to be space-based with little or no maintenance; *later studies have apparently concluded that reusable spacecraft are not a good idea.*
- *In situ* resources: the use of *in situ* resources reduces the logistical demands on Earth of maintaining a lunar outpost and helps to develop outpost operational autonomy from Earth; *yet, the odd thing here is that NASA is neglecting Mars ISRU which is very practical and has significant mission benefits and is putting all its emphasis on lunar ISRU which is unlikely to be practical and has dubious mission benefits.*
- Space power: as the power demands at the lunar outpost increase above the 100 kWe level, nuclear power offers improved specific power; *yes, but is NASA going to develop nuclear reactors? Apparently not.*

6.1.3   NASA 90-Day Study (1989)

On July 20, 1989, the President announced a major new space exploration vision ... To support this endeavor, the NASA Administrator created a task force to conduct a 90-day study of the main elements of a human exploration program. Five reference approaches were developed. Regardless of the reference architecture, the study team concluded that Heavy-Lift Launch Vehicles (HLLVs), space-based transportation systems, Surface Vehicles, Habitats, and support systems for living and working in deep space are required. Thus, the reference architectures made extensive use of the Space Station (Freedom) for assembly and checkout operations of reusable transportation vehicles, ISRU (oxygen from the lunar regolith), and chemical/aerobrake propulsion. *But, reusable vehicles are now widely considered to be a bad idea, trying to extract oxygen from lunar regolith is likely to be impractical, and no significant funds have gone into (or are going into) large-scale aerocapture.*

6.1.4   America at the threshold—"The Synthesis Group" (1991)

In addition to the internal NASA assessment of the Space Exploration Initiative (SEI) conducted during the NASA 90-Day Study, an independent team called the Synthesis Group examined potential paths for implementation of the exploration initiative. This group examined a wide range of mission architectures and technology options. In addition, the group performed a far-reaching search for innovative ideas and concepts that could be applied to implementing the initiative.

The Synthesis Group's four candidate architectures were Mars Exploration, Science Emphasis for the Moon and Mars, The Moon to Stay and Mars Exploration, and Space Resource Utilization. Supporting technologies were identified as key to future exploration.

Their list of technologies was a good one. But, what has NASA done about most of these technologies beyond the PowerPoint level? The Synthesis Group also conducted an extensive outreach program with nationwide solicitation for innovative ideas. *A great many ideas were submitted, but was any really good new idea generated?*

In addition, the Synthesis Group provided specific recommendations for the "effective implementation of the Space Exploration Initiative," including:

- Recommendation 1: establish within NASA a long-range strategic plan for the nation's civil space program, with the SEI as its centerpiece; *NASA plans seldom last longer than one year before they are discarded. In fact, this writer never saw a NASA plan last more than a year.*
- Recommendation 4: establish a new aggressive acquisition strategy for the SEI; *first figure out what we need and why.*
- Recommendation 5: incorporate SEI requirements into the joint NASA-DoD Heavy-Lift Program; *bad idea.*
- Recommendation 6: initiate a nuclear thermal rocket technology development program; *Why? On what basis? Do we have to go to >1,000 km altitude before turning it on? How much will it really cost? What are the risks? Why not put the money into better H_2 storage and enable long-term cryogenic propulsion?*
- Recommendation 7: initiate a space nuclear power technology development program based on the SEI requirements; *we've had SP-100 and Prometheus, and it seems clear that NASA will not follow it through to completion and build it.*
- Recommendation 8: conduct focused life sciences experiments; *the object is not to conduct experiments. The object is to produce results and products.*
- Recommendation 9: establish education as a principal theme of the SEI; *it is not clear to this writer why education should be a principal theme of the SEI.*
- Recommendation 10: continue and expand the Outreach Program; *while the NASA outreach programs do some good things, they have generated a new bureaucracy that sometimes seems very contrived.*

6.1.5   First lunar outpost (1993)

Following the Synthesis Group's recommendations, NASA began planning the implementation of the first steps of the SEI after completion of the Space Station—namely, "back to the Moon, back to the future, and this time, back to stay." This activity was termed the First Lunar Outpost, an Agency-wide effort aimed at understanding the technical, programmatic, schedule, and budgetary implications of restoring U.S. lunar exploration capability. Emphasis was placed on minimizing integration of elements and complex operations on the lunar surface and high reliance on proven systems in anticipation of lowering hardware development costs. Key features of First Lunar Outpost activity included:

- Graceful incorporation of advanced operational and technology concepts into downstream missions; *the problem is that NASA tends to persist with proven operational approaches and technologies that are inefficient and outdated but are known to work. That may have been adequate for the Moon, but it will not work for Mars.*
- A Heavy-Lift Launch Vehicle with a 200 mT delivery capability; *the current HLLV has a capability of 125 mT to LEO. The ESAS Report assures us that this is sufficient for Mars missions, but this seems very doubtful* (see Section 6.6).
- A mission strategy of a direct descent to the lunar surface and direct return to Earth; *this strategy, though strongly supported by the MIT-Draper Study,[154] and an independent study by Robert Zubrin, was rejected in the 2005 ESAS Plan.*

6.1.6   Human lunar return (1996)

In September 1995, the NASA Administrator challenged engineers at JSC to develop a human lunar mission approach, the Human Lunar Return (HLR) study, which would cost significantly less (by one to two orders of magnitude) than previous human exploration estimates. *It is not clear what these estimates were.* Key objectives of the HLR activity were to demonstrate and gain experience on the Moon with those technologies required for Mars exploration, initiate a low-cost approach for human exploration beyond LEO [*there is no "low-cost approach for human exploration beyond LEO"*], establish and demonstrate technologies required for human development of lunar resources, and investigate the economic feasibility of commercial development and utilization of those resources. The HLR study represents the minimum (but completely unworkable) mission approach for a return-to-the-Moon capability.

6.1.7   Mars Exploration Design Reference Missions (1994–1999)

According to the *ESAS Report*: "From 1994 to 1999, the NASA exploration community conducted a series of studies focused on the human and robotic exploration of Mars. Key studies included Mars Design Reference Mission (DRM) 1.0, Mars DRM 3.0, Mars Combo Lander, and Dual Landers. Each subsequent design approach provided greater fidelity and insight into the many competing needs and technology options for the exploration of Mars."[155] *Actually, it appears as if each subsequent design approach represented further attempts to reduce apparent mass and cost, even if they required impractical technologies and mission designs.* These JSC DRMs are discussed in Section 6.2.

---

[154] *From Value to Architecture: The Exploration System of Systems—Phase I, Final Presentation at JPL*, P. Wooster, MIT-Draper Study, August 23, 2005; also *Paradigm Shift in Design for NASA's Space Exploration Initiative: Results from MIT's Spring 2004 Study*, Christine Taylor, David Broniatowski, Ryan Boas, Matt Silver, Edward Crawley, Olivier de Wec, and Jeffrey Hoffman, AIAA 2005-2766).

[155] *ESAS Report*, p. 83.

6.1.8   Decadal Planning Team/NASA Exploration Team (2000–2002)

In June 1999, the NASA Administrator chartered an internal NASA task force, termed the Decadal Planning Team (DPT), to create a new integrated vision and strategy for space exploration. The efforts of the DPT evolved into the Agency-wide team known as the NASA Exploration Team (NExT). Early emphasis was placed on revolutionary exploration concepts. Several architectures were analyzed during the DPT and NExT study cycles, including missions to the Earth–Sun Libration Point (L2), the Earth–Moon Gateway and L1, the lunar surface, Mars (short and long stays), near-Earth asteroids, and a 1-year round trip to Mars.[156] *A 1-year round trip to Mars would require an immense propulsion system.*

6.1.9   Integrated Space Plan (2002–2003)

During the summer of 2002, the NASA Deputy Administrator chartered an internal NASA planning group to develop the rationale for exploration beyond LEO. This *Team*, termed the "Exploration Blueprint Team", performed architecture analyses to develop roadmaps for accomplishing the first steps beyond LEO through the human exploration of Mars. Results from this study produced a long-term strategy for exploration with near-term implementation plans, program recommendations, and technology investments. Recommendations from the Exploration Blueprint Team included the endorsement of the Nuclear Systems Initiative, augmentation of bio-astronautics research, a focused space transportation program including heavy-lift launch and a common exploration vehicle design for ISS and exploration missions, and an integrated human and robotic exploration strategy for Mars.

6.1.10   Exploration Systems Mission Directorate (2004–2005)

On January 14, 2004, the President announced a new "Vision for Space Exploration", NASA's Exploration Systems Mission Directorate (ESMD) was created in January 2004 to begin implementation of the President's Vision. ESMD's Requirements Division conducted a formal requirements formulation process in 2004 to better understand the governing requirements and systems necessary for implementing the Vision. Included in the process were analyses of requirements definition, exploration architectures, system development, technology roadmaps, and risk assessments for advancing the Vision. The analyses were intended to provide an understanding of requirements for human space exploration beyond LEO. In addition, these analyses attempted to identify system "drivers" (i.e., significant sources of cost, performance, risk, and schedule variation, along with areas needing technology development).[157]

---

[156] *ESAS Report*, p. 84.

[157] It is remarkable how similar the 2004 initiative was to the 1989 initiative described at the beginning of Section 6.1.3.

The initial ESMD response to this (prior to Michael Griffin taking the helm at NASA) included (1) lunar mission studies by NASA Centers (the so-called "Focused" and "Broad" Studies),[158] (2) an advanced technology program that was hurriedly implemented based on a short-turnaround proposal solicitation, and (3) a NASA-wide roadmap planning exercise.

The mission studies were carried out by competent teams, but the significant divergences between their results indicated the sensitivity to assumptions inherent in such studies, showing that "the devil is in the details." Unfortunately, it was typically difficult to extract detailed data from the reports because of the way they were written, particularly on the states and steps involved in the various transfers involved in the end-to-end missions (see Sections 7.3.3 and 7.3.4).

The advanced technology program though intriguing did not lead directly to support for missions. The gulf between advanced technology and mission implementation increased further.

The roadmap activity was a disaster of the first magnitude. Planning was divided into strategic planning and capabilities planning. Approximately 26 independent teams were created to plan for aspects of strategy and needed capabilities. Each of the teams operated fairly independently with almost no guidelines or constraints, and very little communication between teams. This unwieldy arrangement diverged with time into a gangling mess of fragments, although some of the fragments may have included valuable content. Nevertheless, this exercise failed to produce a cohesive plan for development of needed capabilities, validation of capabilities *in situ*, and evolving implementation of human missions.

Although the activities of ESMD planning in the 2004–2005 time frame did not converge to any viable framework for the future, it did convey a sobering impression that lunar missions were difficult enough to occupy us for some time, and the "human exploration of Mars" indicated by the President as our ultimate goal would likely have to be pushed to the back burner.

When Michael Griffin took the helm of NASA in late spring of 2005, he quickly came to the same conclusions as those listed above. Initial steps included (a) establishment of a JSC-led NASA Study Team to define upcoming lunar missions and vehicles using a minimum of new technology, (b) cancelation of most of the advanced technology contracts issued by his predecessors, and (c) the 26 roadmap teams were disbanded, and their reports were put in limbo.

The NASA Administrator realized that, in order to meet desired time-goals for return to the Moon, he needed a substantial budget to implement development of new vehicles and capabilities. At the same time, he realized that the Space Shuttle (SS) and the International Space Station (ISS) were usurping a majority of NASA's budget with little or no payoff. After completion of the Apollo Program, NASA elected to develop a generic access to space in the form of the Space Shuttle. But, the Shuttle has reached the point where recent flights seemed to have one major mission objective,

---

[158] Lunar Architecture Broad Trade Study (ESMD-RQ-0006, "Broad Study") and Lunar Architecture Focused Trade Study (ESMD-RQ-0005, "Focused Study"). Neither of these study reports is available to the public.

and that was to do enough repairs in space to assure a safe landing on return! As unproductive as the Shuttle Program was, the ISS produced less. It provides no products and little assistance to anyone getting anywhere in the solar system. It is a poor place to mount telescopes and has almost no utility other than as a sink for funds. Therefore, an attempt was made to phase out the SS and ISS to create budget room for the new Exploration Program. This met with a storm of resistance from certain Congressmen who will not stand for any interruption in "America's access to space"—as if that in itself has any innate value. Some Congressmen would prefer to prolong operation of the Shuttle so it can service the International Space Station (ISS) (apparently, merely to follow through with our international partners) when there is no longer a need for the Shuttle, and there never was a need for ISS. Such prolongation can accomplish very little and will significantly delay the onset of the new Program. Good sense would dictate that NASA should kill both the Shuttle and the ISS and get on with the Exploration Program, but that seems to be fraught with political difficulties.

Faced with strong resistance to the phasing out of the SS and ISS, it was decided to tap other programs within NASA for funds. For example, in spite of JPL's exemplary work over many years on planetary probes, particularly Mars rovers, it has faced severe cutbacks in funding for future robotic missions and technology development. But, this is typical of NASA's on-again, off-again management style that plans, replans, starts, cancels, and rarely sticks with a plan for any significant time (e.g., Jupiter Icy Moons Orbiter).

The NASA Administrator assembled a set of technical and system experts to work together as a cohesive team to plan out sequential missions and technology needs for the Human Exploration Initiative. This "60-day study" took a bit more than 60 days to prepare, and even longer to release (via *NASA Watch*)—probably because they found this Human Exploration Initiative to be far more complex, difficult, multi-faceted, and costly than any of the original forecasts suggested. Their report is known as the "Exploration Systems Architecture Study—ESAS Report."[159]

Despite the fact that the details of human exploration planning activities were shrouded in secrecy, and internal contradictions seem to be widespread, several conclusions seem to be emerging from all of this:

(A) The requirements for a new program of lunar exploration are much more demanding than George W. Bush realized, and almost all of the attention, focus, and detailed planning has been aimed at lunar missions. Mars planning has been relegated to the back burner. This seems to be contradictory to the fundamental view that the Moon is a stepping-stone to Mars, since establishment of a Moon–Mars connection requires definition of the Mars program.

(B) The emphasis in lunar planning seems to be focused on the so-called Crew Exploration Vehicle (CEV). The plan defines the CEV first, and then builds missions around it. This is counter to more logical planning methodologies

[159] The *ESAS Report* can be downloaded from *http://www.nasa.gov/mission_pages/exploration/news/ESAS_report.html*

in which missions are planned first, and vehicles are designed to meet mission needs.

(C) The lunar planning activities are shrouded in secrecy. It is difficult for independent observers to determine whether the plans (i) have been reviewed by independent review boards, (ii) include adequate margins, and (iii) are technically viable. The fact that contractors will bid competitively for the CEV has been used as an excuse for such secrecy, but this makes no sense since surely these contractors must have access to the NASA studies anyway.

(D) The requirements for the lunar cargo LV are stated in many places to be the ability to deliver 125 metric tons (mT) to LEO. However, independent estimates of mission requirements suggest that this is inadequate to launch the vehicles defined by NASA. It appears that, as of 2006, NASA has come to a similar conclusion and is working on redesigns.

(E) In a number of places in the ESAS Report, NASA indicated that the 125 mT LV for the Moon would also suffice for Mars. Considering the early stage of Mars planning, and the fact that earlier Mars design reference missions in the 1990s made optimistic assumptions, there is no credible assurance that 125 mT to LEO is adequate for Mars missions. Indeed, it is not clear that human missions to Mars are feasible and affordable in any architecture (see Section 6.6).

(F) The treatment of *in situ* resource utilization (ISRU) in NASA exploration plans is often misguided. On the one hand, the JSC mission planners seem to push ISRU into the future with phrases such as "support a long-term strategy for human exploration" implying that ISRU would only be employed after initial missions are carried out without use of ISRU. Mission planning for the Moon does not utilize ISRU, except as a potential embellishment rather late in the sequence of lunar exploration. It is difficult to be certain exactly what NASA has in mind for Mars, but the latest releases imply that, similarly, ISRU would not be utilized on initial Mars missions. Furthermore, the mission design in the ESAS Report does not allow for ISRU because the Mars Ascent Vehicle is not landed 26 months prior to crew arrival. On the other hand, JSC ISRU enthusiasts continue to advocate ISRU processes on the Moon that are very problematic and have unlikely feasibility, such as oxygen recovery from regolith and extraction of solar wind deposited hydrogen/methane volatiles from regolith. ISRU is conceived as going far beyond the immediate and obvious applications of producing ascent propellants and life-support consumables, to include far-fetched ideas such as developing industrial and electronic fabrication facilities on extraterrestrial bodies, and beaming energy down to Earth from the Moon. Meanwhile, water is by far the best feedstock for ISRU on the Moon and Mars, and there is a good deal of water on Mars in the near subsurface. Yet NASA has no plans to investigate this resource and determine its accessibility on Mars.

(G) In order to make Mars missions more palatable to those not in the know, NASA continues to imply in places that a human mission to Mars can be cut short if, on arrival, things are not going right. This is simply not true. The crew must stay in the vicinity of Mars (either on the surface or in orbit) for roughly 18 months until the positions of Mars and the Earth are situated for a return with affordable propulsion requirements.

6.1.11   Constellation (2006–?)

Having completed the ESAS study in late 2005, implementation of the plan was assigned to a new NASA organization called Project Constellation. Officially, Constellation appears to be assigned the task of designing the transportation systems for the lunar missions defined in the ESAS Report. However, it became apparent that there were inconsistencies and serious problems in the mission designs contained in the ESAS Report, and Constellation therefore took on a broader role, including refinement of lunar mission architecture as well as vehicle and transportation system design. Constellation also became involved in planning future Mars missions.

It is particularly noteworthy that the approach taken by NASA is unusual because of its heavy emphasis on designing the so-called Crew Exploration Vehicle (CEV) even though end-to-end lunar missions are not clearly defined. Whereas JPL always designs missions first and then derives requirements for vehicles and transportation systems from the end-to-end mission, NASA seems to design vehicles and transportation systems before detailing end-to-end mission designs. It would seem that they must end up dealing with the question of what can be accomplished mission-wise, given a set of vehicles and transportation systems. That seems to be a strange way of doing business.

The documents prepared by Project Constellation are embargoed and not generally available. Unless one is a member of the Constellation Team, it is impossible to be certain exactly what is being pursued within Project Constellation. However, based on some interim and draft reports that became available to me a few comments can be made.

The Constellation Team has to a great extent adopted many conclusions and viewpoints from the ESAS Report. However, some of these seem highly questionable. Hopefully, the Constellation Team will re-think some aspects of mission design from the ESAS Report. For example, an early draft of a Constellation Requirements Document said: "The Constellation Architecture shall utilize an Earth Orbit Rendezvous (EOR)–Lunar Orbit Rendezvous (LOR) mission approach." The rationale for choosing this approach was given as: "This architecture requirement is a balance between numerous factors including minimizing IMLEO and overall mission costs while maximizing crew safety and the probability of mission success. EOR/LOR also provides extensibility of the lunar architecture elements to later crew missions to Mars where an EOR/Mars Orbit Rendezvous architecture is the current baseline. The EOR/LOR architecture successfully addresses each of these areas. The EOR requirement was, in part, driven by the requirement for separate launches of the crew in the Crew Exploration Vehicle (CEV) using the Crew Launch Vehicle (CLV) and the Lunar Surface Access Module/Earth Departure System (LSAM/EDS) on the Cargo Launch Vehicle (CaLV). This launch strategy requires the Earth-rendezvous of the CEV and LSAM/EDS prior to trans-lunar injection using the EDS. In addition, the LOR requirement was driven by the decision to leave the CEV in lunar orbit while the

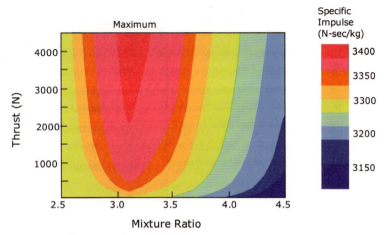

Figure 3.2. Specific impulse of methane/oxygen rocket as a function of mixture ratio and thrust. [Adapted from "A 2007 Mars Sample Return Mission Utilizing In-Situ Propellant Production," Daniel P. Thunnissen, Donald Rapp, Christopher J. Voorhees, Stephen F. Dawson, and Carl S. Guernsey, AIAA 99-0851; by permission of the American Institute of Aeronautics and Astronautics.]

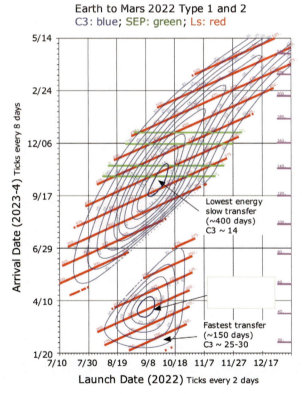

Figure 3.14. Typical "pork-chop" plot for 2022. Contours of constant C3 are blue. Transit time to Mars (days) is in red. [Adapted from a JPL report.]

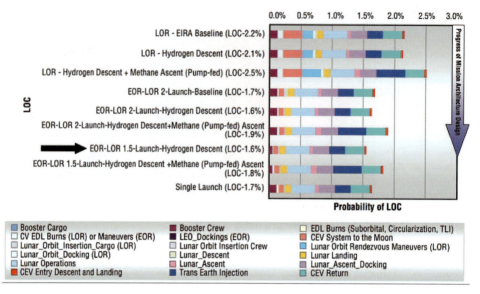

Figure 7.7. ESAS Report estimates of probability of loss of crew.

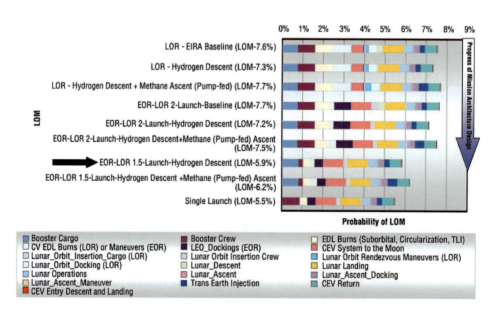

Figure 7.8. ESAS Report estimates of probability of loss of mission.

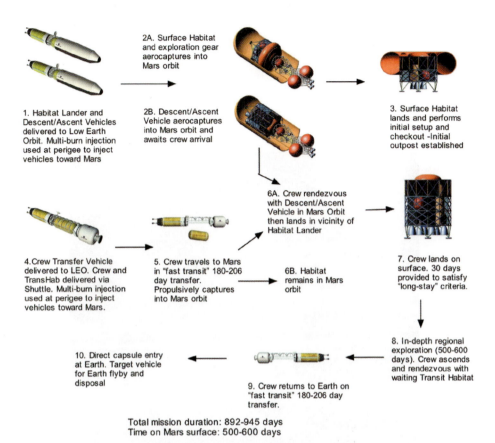

1. Habitat Lander and Descent/Ascent Vehicles delivered to Low Earth Orbit. Multi-burn injection used at perigee to inject vehicles toward Mars

2A. Surface Habitat and exploration gear aerocaptures into Mars orbit

2B. Descent/Ascent Vehicle aerocaptures into Mars orbit and awaits crew arrival

3. Surface Habitat lands and performs initial setup and checkout -Initial outpost established

4. Crew Transfer Vehicle delivered to LEO. Crew and TransHab delivered via Shuttle. Multi-burn injection used at perigee to inject vehicles toward Mars.

5. Crew travels to Mars in "fast transit" 180-206 day transfer. Propulsively captures into Mars orbit

6A. Crew rendezvous with Descent/Ascent Vehicle in Mars Orbit then lands in vicinity of Habitat Lander

6B. Habitat remains in Mars orbit

7. Crew lands on surface. 30 days provided to satisfy "long-stay" criteria.

8. In-depth regional exploration (500-600 days). Crew ascends and rendezvous with waiting Transit Habitat

9. Crew returns to Earth on "fast transit" 180-206 day transfer.

10. Direct capsule entry at Earth. Target vehicle for Earth flyby and disposal

Total mission duration: 892-945 days
Time on Mars surface: 500-600 days

Figure 8.2. NASA Mars DRM (c. 2006). [Adapted from 2006 NASA presentation.]

Descent / Ascent Vehicle
(Two Heavy-Lift Launches)
Payload   72mt
Stage Mass 30
Propellant   61
Total Mass **163**

Surface Habitat
(Two Heavy-Lift Launches)
Payload   49 mt
Stage Mass 30
Propellant   52
Total Mass 131

Crew Transit Vehicle
(Two Heavy-Lift Launches)
Payload   27mt
Stage Mass 39
Propellant   86
Total Mass **152**

Figure 8.3. Description of Mars DRM. Masses are in mT in LEO. Total IMLEO for the entire mission is predicted to be 446 mT. [Adapted from 2006 NASA presentation.]

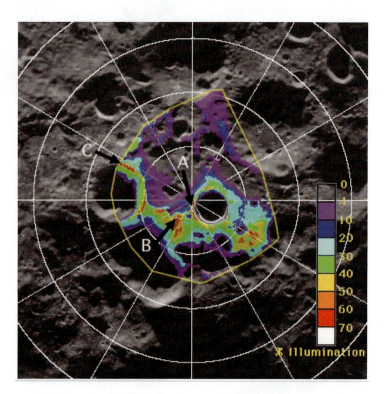

Figure A.22. The illumination map for the lunar south pole showing the percentage of the time that a point on the surface is illuminated during an Earth year. The three most illuminated regions are labeled A, B, and C. A and B are only 10 km apart and collectively (i.e., one or the other) are illuminated for more than 98% of the time. [Reproduced from "Permanent Sunlight at the Lunar North Pole," D. B. J. Bussey, M. S. Robinson, K Fristad, and P. D. Spudis, *Nature*, Vol. 434, 842, April 14, 2005, doi:10.1038/434842a; by permission of Nature Publishing Group.]

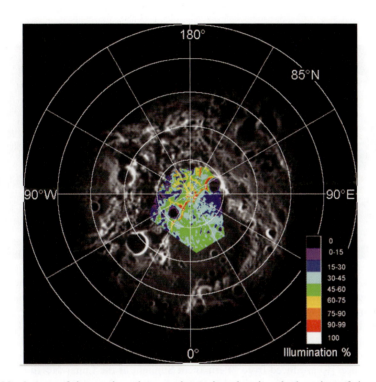

Figure A.23. A map of the northern lunar polar region showing the location of simple craters that contain permanent shadow. The total area of permanent shadow within 12° latitude of the pole is 7,500 km^2. [Reproduced from "Permanent Sunlight at the Lunar North Pole," D. B. J. Bussey, M. S. Robinson, K Fristad, and P. D. Spudis, *Nature*, Vol. 434, 842, April 14 , 2005, doi:10.1038/434842a; by permission of Nature Publishing Group.]

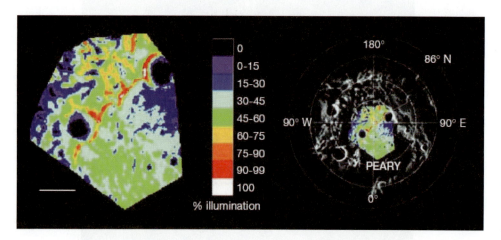

Figure A.24. A quantitative illumination map of the Moon's north pole. The color scale indicates the percentage of time that a point on the surface is illuminated during a lunar day in summer. On the left, several areas on the rim of Peary crater (78 km in diameter) can be identified (white) that were continuously illuminated in summer. The spatial extent of the map is within about 11.5° of the pole. The scale bar at the left is 15 km. At the right, the color illumination map is superimposed on a grey-scale composite Clementine mosaic, for spatial reference. [Reproduced from "Permanent Sunlight at the Lunar North Pole," D. B. J. Bussey, M. S. Robinson, K Fristad, and P. D. Spudis, *Nature*, Vol. 434, 842, April 14, 2005; by permission of Macmillan.]

Figure B.9. Photograph of the Sun near sunset by Mars Pathfinder.

Figure B.10. Photograph of the Sun near sunset by Mars Pathfinder.

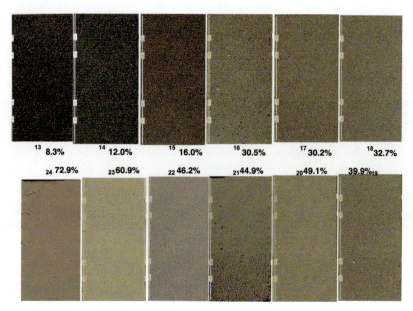

Figure B.46A. Photograph of 12 solar cells coated with various amounts of simulated Mars dust ("JSC 1"). Small numbers are sample numbers, and large numbers are percent obscuration for each sample. Without dust, cells would be black. [Adapted from *Solar Energy on Mars*, Donald Rapp, JPL Report D-31342.]

Figure B.46B. Photograph of 12 solar cells coated with various amounts of simulated Mars dust ("Carbondale Clay"). Small numbers are sample numbers, and numbers in parentheses are percent obscuration for each sample. Without dust, cells would be black. [Adapted from *Solar Energy on Mars*, Donald Rapp, JPL Report D-31342.]

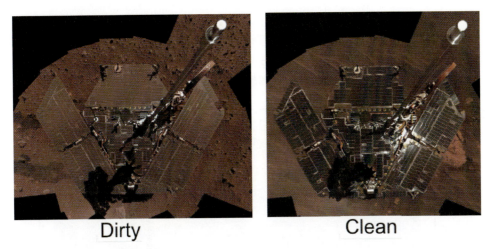

Dirty   Clean

Figure B.44. Example of self-cleaning of Opportunity solar array. [Adapted from a figure at *http://www.space.com/imageoftheday/image_of_day_050222.html*]

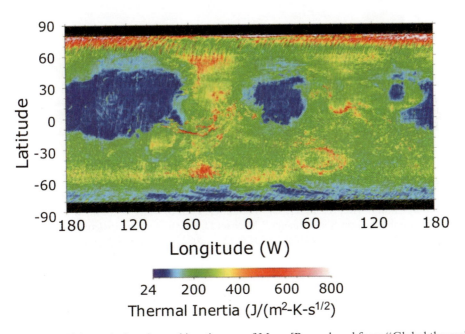

Figure C.1. High-resolution thermal inertia map of Mars. [Reproduced from "Global thermal inertia and surface properties of Mars from the MGS mapping mission," Nathaniel E. Putzig, Michael T. Mellon, Katherine A. Kretke, and Raymond E. Arvidson, *Icarus*, Vol. 173, 325–341, 2005; by permission of Elsevier.]

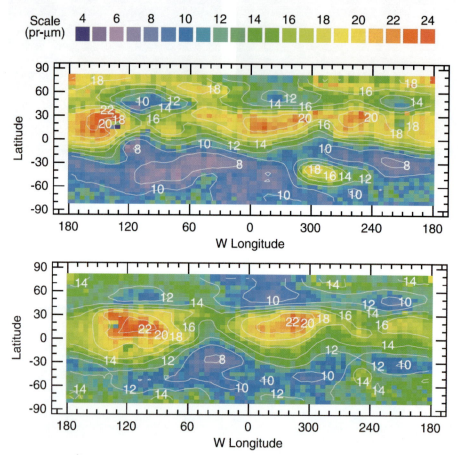

Figure C.5B. The column abundance of water vapor as a function of L_S and latitude: (top) as observed by TES.[*] Contours show a smoothed representation of the results, and (bottom) as observed by Viking.[†] [Reproduced from "The annual cycle of water vapor on Mars as observed by the Thermal Emission Spectrometer," Michael D. Smith, *Journal of Geophysical Research*, Vol. 107, No. E11, 5115, doi:10.1029/2001JE001522, 2002; as permitted by *http://www.agu.org/pubs/copyright.html*]

[*] Michael Smith, *loc. cit.*
[†] "The seasonal and global behaviour of water vapor in the Mars atmosphere: Complete global results of the Viking Atmospheric Water Detector Experiment," B. M. Jakosky and C. B. Farmer, *J. Geophys. Res.*, Vol. 87, 2999–3019, 1982.

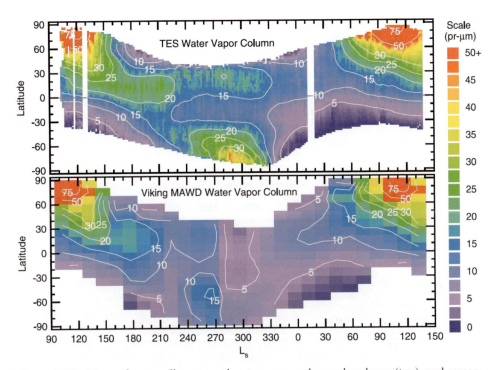

Figure C.5C. Maps of seasonally averaged water vapor column abundance (top), and season-ally averaged water vapor column abundance divided by ($p_{surf}/6.1$) to remove the effect of topography. Contours show a smoothed representation of the maps.[*] [Reproduced from "The annual cycle of water vapor on Mars as observed by the Thermal Emission Spectrometer," Michael D. Smith, *Journal of Geophysical Research*, Vol. 107, No. E11, 5115, doi:10.1029/2001JE001522, 2002; as permitted by *http://www.agu.org/pubs/copyright.html*]

[*] M. Smith, *loc. cit.*

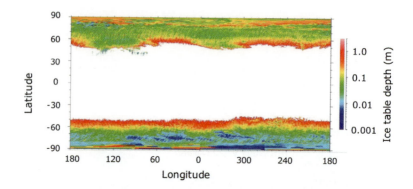

Figure C.11. Ice table depth according to Mellon *et al.* (2003) based on a global annual average water vapor pressure of 10 pr µm. [Reproduced from "The presence and stability of ground ice in the southern hemisphere of Mars," J. T. Mellon, W. C. Feldman, and T. H. Prettyman, *Icarus*, Vol. 169, 324–340, 2003; by permission of Elsevier.]

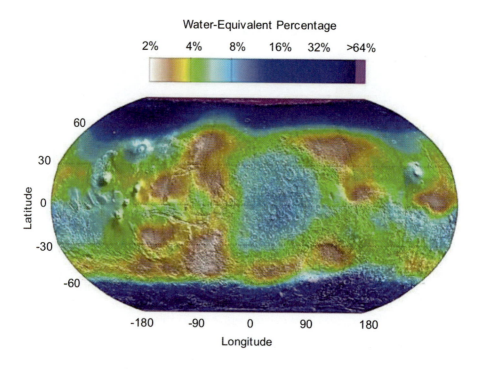

Figure C.15. Global variation of water content in upper ∼1 m of Mars based on a uniform regolith model (no layers) using epithermal neutron data. [Reproduced from "Global distribution of near-surface hydrogen on Mars," W. C. Feldman, T. H. Prettyman, S. Maurice, J. J. Plaut, D. L. Bish, D. T. Vaniman, M. T. Mellon, A. E. Metzger, S. W. Squyres, S. Karunatillake *et al.*, *J. Geophys. Res.*, Vol. 109, E09006, doi:10.1029/ 2003JE002160, 2004; as permitted by *http://www.agu.org/pubs/copyright.html*]

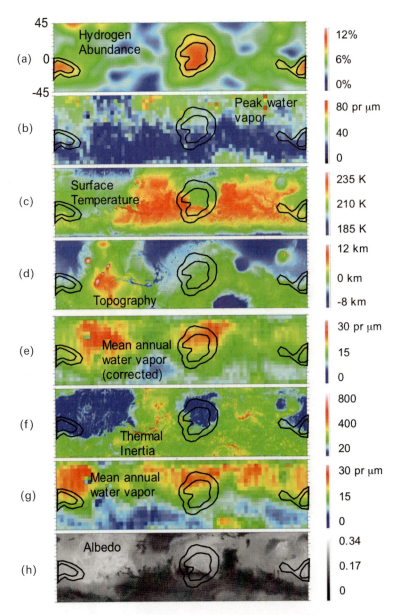

Figure C.18. Maps showing: (a) abundance of water as measured by neutron spectrometer; (b) annual peak abundance of water vapor; (c) mean annual surface temperature; (d) topography; (e) mean annual water vapor corrected for topography; (f) thermal inertia; (g) mean annual water vapor uncorrected for topography; and (h) albedo. [Adapted from "Mars low-latitude neutron distribution: Possible remnant near-surface water ice and a mechanism for its recent emplacement," Bruce M. Jakosky, Michael T. Mellon, E. Stacy Varnes, William C. Feldman, William V. Boynton, and Robert M. Haberle, *Icarus*, Vol. 175, 58–67, 2005; by permission of Elsevier.]

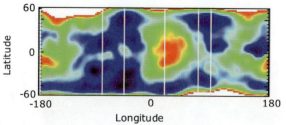

Figure C.20. Distribution of water-equivalent hydrogen (red ~10% water, deep purple ~2% water). The six vertical white lines are the traverses along which WEH is compared with elevation. [Reproduced from "Topographic Control of Hydrogen Deposits at Mid- to Low Latitudes of Mars," W. C. Feldman, T. H. Prettyman, S. Maurice, R. Elphic, H. O. Funsten, O. Gasnault, D. J. Lawrence, J. R. Murphy, S. Nelli, R. L. Tokar, and D. T. Vaniman, *Journal of Geophysical Research*, Vol. 110, E11009, doi:10.1029/2005JE002452, 2005; as permitted by *http://www.agu.org/pubs/copyright.html*]

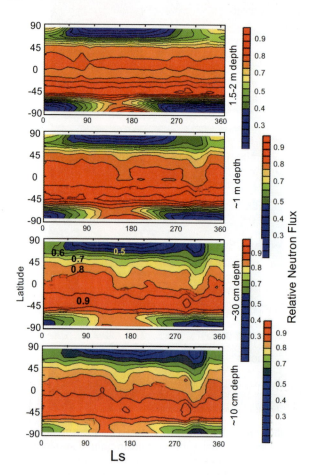

Figure C.22. Dependence of neutron flux on season (L_S) and latitude: uppermost = very slow neutrons (depth = 1.5–2 m); second down = slow neutrons (~1 m); third down = fast (20–30 cm); lowermost = very fast (~10 cm). High neutron fluxes indicate low water concentrations, and *vice versa*. [Adapted from "Seasonal Redistribution of Water in the Surficial Martian Regolith: Results of the HEND Data Analysis," R. O. Kuzmin, E. V. Zabalueva, I. G. Mitrofanov, M. L. Litvak, A. V. Parshukov, V. Yu. Grin'kov, W. Boynton, and R. S. Saunders, *Lunar and Planetary Science*, Vol. XXXVI, Paper 1634, 2005.]

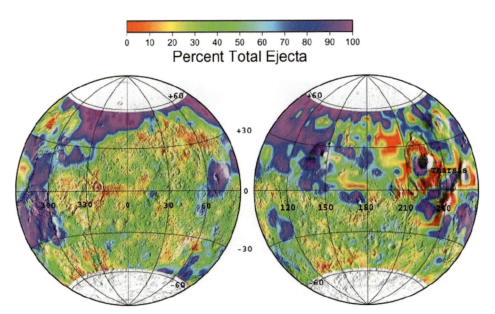

Figure C.33. Percentage of all craters that display ejecta attributable to volatiles. [Reproduced from "Martian impact crater ejecta morphologies as indicators of the distribution of subsurface volatiles," Nadine G. Barlow and Carola B. Perez, *Journal of Geophysical Research*, Vol. 108, No. E8, 5085, doi:10.1029/2002JE002036, 2003; as permitted by *http://www.agu.org/pubs/copyright.html*]

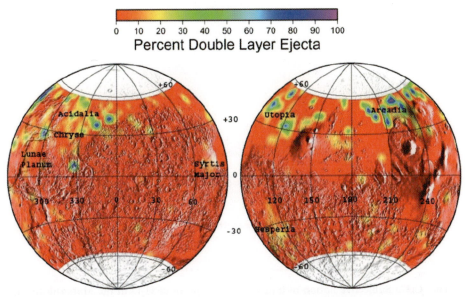

Figure C.34. Percentage of all ejecta craters that have DLE morphology. [Reproduced from "Martian impact crater ejecta morphologies as indicators of the distribution of subsurface volatiles," Nadine G. Barlow and Carola B. Perez, *Journal of Geophysical Research*, Vol. 108, No. E8, 5085, doi:10.1029/2002JE002036, 2003; as permitted by *http://www.agu.org/pubs/copyright.html*]

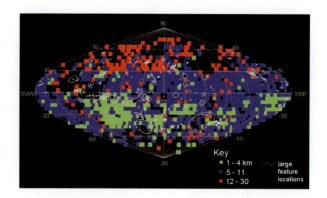

Figure C.35. Distribution of onset diameters for rampart formation. [Adapted from"Global Distribution of On-Set Diameters of Rampart Ejecta Craters on Mars: Their Implication to the History of Martian Water," Joseph M. Boyce, David J. Roddy, Lawrence A. Soderblom and Trent Hare, *Lunar and Planetary Science*, Vol. XXXI, Paper 1167, 2000.]

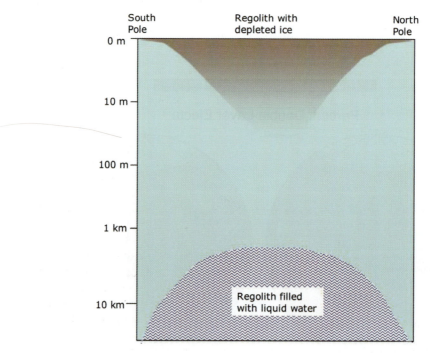

Figure C.37. Subjective notion of a possible distribution of H_2O in the Mars subsurface.

crew takes the LSAM to the surface instead of taking the CEV to the surface and returning to Earth directly at the conclusion of the surface stay period." However, this architecture requires the greatest number of maneuvers, connections, and transfers, and therefore would appear to have the highest risk. The MIT-Draper Study and an independent study by Bob Zubrin show that a direct return from the Moon has significant advantages and benefits the most from use of ISRU. The chosen architecture minimizes ISRU benefits.

Project Constellation and, more generally, NASA appear to have inconsistent views regarding use of ISRU in lunar and Mars missions. Whereas some decisions appear to favor eliminating oxygen as an ascent propellant from the Moon, others insist on using oxygen for ascent, because ISRU would more or less be eliminated without it. The requirement for anytime abort capability even during descent from lunar orbit would require bringing ascent propellants from Earth and would also seem to eliminate ISRU, although it is not clear that Project Constellation has taken this into account. At the same time, Project Constellation has insisted that: "The Constellation Architecture shall establish a lunar outpost located within 5 degrees latitude of the lunar South Pole." The rationale for this was given as: "Polar regions of the Moon present unique opportunities for lunar resource utilization, scientific investigations, advantages for transportation system flexibility, efficiency. Specific outpost site selection criteria will be developed and documented in a separate HQ controlled document as was done during Apollo." One of the problems in this whole lunar endeavor is that certain mantras get repeated and repeated and repeated until everyone believes them. There are transportation advantages to a polar site. However, the main rationale for a polar site is implied to be the hope for acquisition of ground ice as a resource. Yet, we have only preliminary evidence of its existence and no evidence of its vertical and horizontal distribution, and not a hint of its accessibility. Furthermore, if ground ice exists, it is in deeply shaded areas where solar energy is not available. Power becomes a major problem. Some have suggested solar power on top of polar peaks or crater rims, but that is not where the ice is, and there are other problems. Then, to add to the confusion, no one has done an analysis to show that providing the equipment, power systems, and human effort to recover lunar ice is more efficient than bringing water from Earth. Furthermore, if NASA is determined to establish an outpost at the pole, it needs to land precursors to map out the ice with rovers. Instead, NASA is spending millions on technology to extract oxygen from molten regolith at 2,000°C.

In regard to Mars missions, Project Constellation requires that the "Architecture shall provide the capability to conduct human Mars missions at any injection opportunity in the Earth–Mars synodic cycle for conjunction-class missions." The rationale for this is stated as: "The propulsive energy requirements for round-trip conjunction-class Mars missions can vary from opportunity to opportunity. This requirement insures that the transportation systems are not limited to infrequent mission opportunities. This will also envelope return mission velocities." However, by making this demand, we are assured that the design must accommodate the worst-case scenario and this implies an even more demanding propulsion system for Mars missions where IMLEO is already a major challenge.

It appears that Project Constellation is adopting the Mars architecture described in Section 6.3.2, and if so that will prove to be unfortunate.

6.2   MARS DESIGN REFERENCE MISSIONS (DRMs)

6.2.1   DRM-1

6.2.1.1   Introduction

The JSC Mars mission study known as "DRM-1"[160] was carried out in the mid-1990s and, unlike later JSC DRMs that appear to be sales documents intent on proving against all logic that Mars missions are feasible, DRM-1 appears to be a sincere effort to identify requirements and challenges for human missions to Mars.

DRM-1 is well documented and provides an intelligent credible self-appraisal:

"The Reference Mission is not implementable in its present form. It involves assumptions and projections, and it cannot be accomplished without further research, development, and technology demonstrations. It is also not developed in the detail necessary for implementation, which would require a systematic development of requirements through the system engineering process.

One principal use of the Reference Mission is to lay the basis for comparing different approaches and criteria in order to select better ones.

The primary purpose of the Reference Mission is to stimulate further thought and development of alternative approaches which can improve effectiveness, reduce risks, and reduce cost."

A number of ground rules were established for DRM-1. Two of these are:

- Mars surface stay is 500–600 sols.
- Utilize three sequential human missions to Mars. Each mission will return to the site of the initial mission, thus permitting evolutionary establishment of capabilities on the Mars surface.

DRM-1 assumed that crews will explore out to a radius of a few hundred km from the outpost, implying use of a highly capable pressurized rover.

DRM-1 utilized ISRU to produce methane and oxygen propellants for the Mars Ascent Vehicle (MAV), and brought hydrogen from Earth to facilitate this process. However, in an apparent oversight, no mention was made of hydrogen storage requirements (volume, power, ...).

[160] *Human Exploration of Mars: The Reference Mission of the NASA Mars Exploration Study Team*, Stephen J. Hoffman and David I. Kaplan (eds.), Lyndon B. Johnson Space Center, Houston, TX, July 1997, NASA Special Publication 6107.

6.2.1.2   DRM-1 vehicles

The following vehicles were defined:

ERV Cargo Vehicle—the Earth Return Vehicle (ERV) is transported to Mars orbit where it awaits the return of astronauts from Mars surface. The ERV consists of a Habitat for six astronauts plus a propulsion system fully loaded with propellants for return to Earth, and an entry system for landing on Earth.

MAV/Infrastructure Cargo Vehicle—delivers a payload consisting of (a) the Mars Ascent Vehicle (MAV) with empty methane and oxygen tanks, (b) an ISRU propellant production plant including a supply of hydrogen,[161] (c) a 160 kW nuclear electric power plant, and (d) ~40 mT of additional infrastructure to the surface.

Habitat Landing Vehicle—delivers Habitat #1 (includes a laboratory) to the surface of Mars in the vicinity of the MAV/Infrastructure Cargo Vehicle. This also includes a redundant 160 kW nuclear electric power plant.

Crew Lander—delivers Habitat #2 (similar to Habitat/Laboratory #1) to the surface of Mars in the vicinity of the Habitat Landing Vehicle and the MAV/Infrastructure Cargo Vehicle. This provides a redundant Habitat that is "will be connected to" Habitat #1.

From a series of volume, mass, and mission analyses, a common Habitat structural cylinder, 7.5 meters in diameter, bi-level, and vertically oriented, was derived for DRM-1. The three habitation element types identified for DRM-1 (the ·surface laboratory, the transit/surface habitation element, and the Earth return habitation element) would contain substantially identical primary and secondary structures, windows, hatches, docking mechanisms, power distribution systems, life support, environmental control, safety features, stowage, waste management, communications, airlock function, and crew egress routes.

The following are brief descriptions of the unique aspects of the three primary habitation elements developed for Reference Mission analysis.

- The *Mars Surface Laboratory* will operate only in 3/8 gravity. It contains crew support elements on one level and the primary science and research lab on the second level.
- The *Mars Transit/Surface Habitats* contain the required consumables for the Mars transit and surface duration of approximately 800 days (180 days in transit and 600 days on the surface) as well as all the required equipment for the crew during the 180-day transfer trip. This is the critical element that must effectively operate in both zero and partial gravity. Once on the surface of Mars, this

---

[161] In an apparent oversight, no mass seemed to be assigned to these tanks, and no boil-off seems to have been assumed.

element will be physically connected with the previously landed Surface Lab, thereby doubling the pressurized volume for the crew.

- The *Earth-Return Habitat*, functioning only in zero-*g* and requiring the least amount of volume for consumables, will be volume-rich but must be mass-constrained to meet the limitations of the trans-Earth injection (TEI) stage. Since little activity (other than conditioning for the one-*g* environment on Earth and training for the Earth-return maneuvers) is projected for the crew during this phase of the mission, mass and radiation protection were the key concerns[162] in the internal architecture concepts created. However, no mass was allocated for radiation shielding.

6.2.1.3   *Mission sequence*

An important consideration was whether to land all crews at the same location or land each crew at a different location. The principal tradeoff is between the additional exploration that might be accomplished by exploring three distant sites vs. the benefits of building up the capability to test settlement technologies (such as closed life support systems) and the reduced risk provided by accumulating surface assets at one site. As the range of exploration provided in a single-location Mars outpost was proposed to be high (hundreds of kilometers with a pressurized rover), the advantages of exploring several landing sites were considered of lower priority for DRM-1. Therefore, the first three missions would land at the same site and build infrastructure.

The strategy chosen for the DRM-1, generally known as a "split mission" strategy, breaks mission elements into pieces that can be launched directly from Earth with very large Launch Vehicles, without rendezvous or assembly in low Earth orbit (LEO). The strategy has these pieces rendezvous on the surface of Mars, which will require both accurate landing and mobility of major elements on the surface to allow them to be connected or to be moved into close proximity. Another attribute of the split mission strategy is that it allows cargo to be sent to Mars without a crew at one or more opportunities prior to crew departure. This allows cargo to be transferred on low-energy, longer transit time trajectories and the crew to be sent on a higher energy, shorter transit time trajectory. Dividing each mission into two launch windows allows much of the infrastructure to be emplaced and checked out before committing a crew to the mission, and also provides a robust capability, with duplicate launches on subsequent missions providing either back-up for the earlier launches or growth of initial capability.

In DRM-1 four vehicles would be launched from Earth to Mars in the first mission sequence and, thereafter, three vehicles would be launched with each subsequent mission sequence. The first three launches will not involve a crew but will

---

[162] Radiation may have been a "concern" but no specific mitigation plans seem to have been included.

send infrastructure elements to low Mars orbit and then onto the surface of Mars for later use. Each of the subsequent mission sequences will send one crew and two cargo missions to Mars. These cargo missions will consist of an Earth Return Vehicle (ERV) on one flight and a lander carrying a Mars Ascent Vehicle (MAV) and additional supplies on the second. This sequence gradually builds up assets on the Martian surface so that at the end of the third crew's tour of duty, the basic infrastructure could be in place to support a permanent presence on Mars. For the nominal mission sequence, Launches 1 through 4 are required to support the first crew. The DRM says that Launches 5 and 6 provide back-up systems for the first crew and, if not used, are available for the second crew. However, while the ERV in Launch 5 provides back-up for the ERV in Launch 1, it is not clear that Launch 6 provides back-up for anything because it arrives with empty tanks.

The mission sequence involves three sequential deliveries of astronauts to the surface of Mars at roughly 26-month intervals as shown in Figure 6.1. For the first astronaut delivery, four vehicles are involved: the ERV Cargo Vehicle, the MAV/ Infrastructure Cargo Vehicle, the Habitat Landing Vehicle, and the Crew Lander. The three cargo vehicles depart Earth about 26 months before the astronauts, on minimum energy trajectories direct to Mars (i.e., without assembly or fueling in LEO). After the descent stage lands on the surface, the nuclear reactor autonomously deploys itself several hundred meters from the Ascent Vehicle. The MAV tanks are gradually filled with propellant prior to astronaut departure from Earth. The Crew Lander provides a second redundant Habitat.

The first crew of six astronauts departs for Mars (Launch 4) in the second opportunity. They leave Earth after the two cargo missions have been launched, but, because they are sent on a fast transfer trajectory of only 180 days, they will arrive in Mars orbit approximately 2 months before the cargo missions (Launches 5 and 6). The crew lands on Mars in a Surface Habitat substantially identical to the Habitat/Laboratory previously deployed on the Martian surface. After capturing into a highly elliptic Mars orbit, the crew descends in the Transit Habitat to rendezvous on the surface with the other elements of the surface outpost. The DRM says: "The crew carries sufficient provisions for the entire surface stay in the unlikely event that they are unable to rendezvous on the surface with the assets previously deployed." But, rendezvous on the surface with the MAV is still required in order to lift off.

The second and third astronaut deliveries eliminate the separate Habitat Landing Vehicle and utilize only the ERV Cargo Vehicle, the MAV/Infrastructure Cargo Vehicle, and the Crew Lander.

DRM-1 assumed that the capability of very precise landing on Mars can be developed technically, and that all assets for each flight can be integrated on Earth and simply joined on Mars. These capabilities can be demonstrated on precursor robotic missions, although NASA has not demonstrated a proclivity to implement precursor missions in the past.

After their stay on Mars, each crew will use the previously landed ISRU-fueled Ascent Vehicle to return to orbit where they will rendezvous with the waiting ERV. The crew will return to Earth in a Habitat on the ERV similar to the one used for the

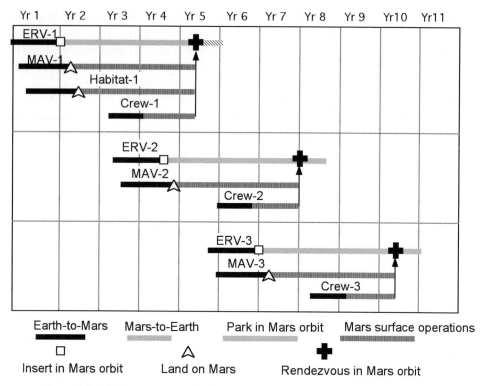

Figure 6.1. Mission sequence for three astronaut deliveries to Mars in DRM-1.

outbound transit leg. This Habitat, which is part of the ERV deployed in a previous opportunity by one of the cargo flights, typically will have been in an untended mode in Mars orbit for nearly 4 years prior to the crew arrival.

6.2.1.4   *Mars Ascent Vehicle (MAV)*

The ISRU system fills the MAV tanks in about one year prior to departure of astronauts from Earth. The MAV is about 9 m high and 4 m in diameter. The ascent propulsion system requires 5.6 km/s of velocity change. Normally, one expects the Δv to be more like 4.4 km/s to a Mars circular orbit, but DRM-1 (and DRM-3) placed the ERV in a highly elliptical orbit that required an addition to the impulse requirement for a circular orbit, bringing the Δv up to about 5.6 km/s.[163]

The amount of propellant required for ascent is proportional to the mass that must be lifted from the surface to orbit. This mass includes the capsule to house the crew of six, the dry propulsion system (tanks, thrusters, plumbing, ...) and the control system for implementing rendezvous in orbit. DRM-1 assumed a very

[163] See Section 5.3 for a more complete discussion of Mars ISRU.

optimistic mass for these systems,[164] and they concluded that only 26 mT of methane and oxygen propellants are needed to accomplish the nearly 5.6 km/s of velocity change required for a single-stage ascent to orbit and rendezvous with the previously deployed ERV. DRM-3 later increased this to 39 mT, but it is probable that even this figure may be optimistic. Both DRM-1 and DRM-3 assumed that the specific impulse of the methane/oxygen rocket is 379 s, whereas it appears likely that 360 s might be more realistic.

6.2.1.5 Surface power

The DRM-1 plan was to include 160 kW nuclear reactor electric power plants on both the Cargo Lander and the Habitat Lander. The first crew will thereby have access to 320 kW of nuclear electric power, or alternatively 160 kW with full back-up. Each subsequent crew will add another 160 kW. No explanation is provided as to why 160 kW is needed or how it would be used. It is possible that a lower power level might suffice. The surface power systems would have 15+ year lifetimes to allow them to serve the three mission opportunities with good safety margins. Surface transportation power systems would have 6+ year lifetimes to minimize the need for replacement over the program lifetime.

Additionally, each Habitat would retain the solar arrays used during transit, and they can also be operated on the Martian surface. Each solar power system can produce approximately 30% of the power generated in space. For emergency situations, it was claimed that the pressurized rover's Dynamic Isotope Power System can supply 10 kW of continuous power. (It remains to be seen how much this DIPS system will weigh, whether sufficient isotopes would be available, and whether it can be made sufficiently reliable.)

6.2.1.6 Interplanetary transportation system

The DRM-1 interplanetary transportation system consists of a trans-Mars injection (TMI) stage, a biconic aeroshell for Mars orbit capture and Mars entry, a descent stage for surface delivery, an ascent stage for crew return to Mars orbit, an Earth return stage for departure from the Mars system, and a crew capsule for Earth entry and landing.

The transportation strategy adopted in DRM-1 eliminates the need for assembly or rendezvous of vehicle elements in LEO, but it does require a rendezvous in Mars orbit for the crew leaving Mars.

DRM-1 employed a trans-Mars injection (TMI) stage (used to propel a payload from LEO onto a trans-Mars trajectory) based on nuclear thermal propulsion (NTP). However, as discussed in Section 3.4.3, the advantages of NTP are minimized if it is required to transport it to a higher Earth orbit (>1,000 km) prior to start-up for safety reasons. DRM-1 (and DRM-3) made the unlikely assumption that the NTP

---

[164] They assumed that the capsule and propulsion stage were 2.8 and 2.6 mT, respectively.

can be fired up in LEO. The TMI stage uses four 15,000 lb thrust NERVA (Nuclear Engine for Rocket Vehicle Application) derivative reactor engines ($I_{SP} = 900$ seconds). This stage is assumed to have a maximum diameter of 10 meters and an overall length of 25.3 meters. Much of this volume is taken up by stored hydrogen. Each TMI stage utilizes 86 mT of hydrogen. If the hydrogen is stored as liquid at 1 bar the required volume is about 1,300 m^3. For a 10 m diameter, the required length of hydrogen storage is 16.5 m. The dry mass of the TMI stage was estimated to be about 30 mT, or 35% of the propellant mass. That may prove to be optimistic.

After completion of its role, the TMI stage is inserted into a trajectory that will not reencounter Earth or Mars over the course of one million years. The TMI stage used for the crew incorporates a shadow shield between the engine assembly and the LH$_2$ tank to protect the crew from radiation that builds up in the engines during TMI burns. The same type of TMI stage is used in all cargo missions, which allows the transportation system to deliver up to approximately 65 mT of useful cargo to the surface of Mars or nearly 115 mT to Mars orbit on a single launch from Earth. As stated previously, this is based on the assumption that NTP can be fired up in LEO, coupled with an optimistic estimate of the propulsion system dry mass.

Mars orbit capture and the majority of the Mars descent maneuver is performed using a single biconic aeroshell. The decision to perform the Mars orbit capture maneuver was based on the facts that (1) an aeroshell will be required to perform the Mars descent maneuver no matter what method is used to capture into Mars orbit, (2) the additional demands on a descent aeroshell to meet the Mars capture requirements were claimed to be modest, and (3) a single aeroshell eliminated one staging event, and thus one more potential failure mode, prior to landing on the surface. However, aerocapture technology for very large vehicles will require very challenging and expensive development, as discussed in Section 4.6, and this was not acknowledged by DRM-1 (or DRM-3). The JSC DRMs made very optimistic projections for aero-entry system masses that are much lower than those estimated by Braun *et al.* (see Section 4.6).

The crew is transported to Mars in a Habitat that is fundamentally identical to the Surface Habitat deployed robotically on a previous cargo mission. By designing the Habitat so that it can be used during transit and on the surface, a number of advantages to the overall mission are obtained:

- Two Habitats provide redundancy on the surface during the longest phase of the mission.
- By landing in a fully functional Habitat, the crew does not need to transfer from a "space-only" Habitat to the Surface Habitat immediately after landing, which allows the crew to readapt to a gravity environment at their own pace.
- The program is required to develop only one Habitat system. The Habitat design is based on its requirement for surface utilization. Modifications needed to adapt it to a zero-*g* environment must be minimized.

A common descent stage was assumed for the delivery of the Transit/Surface Habitats, the Ascent Vehicle, and other surface cargo. The Descent Vehicle is capable of landing up to approximately 65 mT of cargo on the Mars surface. The Landing Vehicle was somewhat oversized to deliver the crew; however, design of a scaled-down lander and the additional associated costs were avoided. To perform the post-aerocapture circularization burn and the final approximately 500 m/s of descent prior to landing on the Mars surface, the common descent stage employed four RL6-class engines modified to burn LOX/CH_4. The use of parachutes was assumed to reduce the Descent Vehicle's speed after the aeroshell has ceased to be effective and prior to the final propulsive maneuver. The selection of LOX/CH_4 allows a common engine to be developed for use by both the descent stage and the ascent stage, the latter of which is constrained by the propellant that is manufactured on the surface using ISRU. The Ascent Vehicle is delivered to the Mars surface atop a cargo descent stage. It is composed of an ascent stage and an ascent crew capsule. The ascent stage is delivered to Mars with its propellant tanks empty. However, the descent stage delivering the MAV included several tanks (about 5 mT with >70 m^3 volume) of seed hydrogen for use in producing 26 mT of LOX/CH_4 propellant by ISRU for ascent to orbit and rendezvous with the ERV. The Ascent Vehicle used two RL6-class engines modified to burn LOX/CH_4. However, it is not clear what allowance was made for the storage and cooling of this large volume of hydrogen. No mass seems to be allocated to the hydrogen storage system. The requirement for only 26 mT of ascent propellants is based on extremely optimistic estimates of the masses of the crew capsule and the ascent propulsion stage. It seems likely that a propellant mass >40 mT might actually be required, and indeed DRM-3 increased ascent propellants to 39 mT.

The ERV is composed of the trans-Earth injection (TEI) stage, the Earth return Transit Habitat, and a capsule that the crew will use to reenter the Earth's atmosphere. The TEI stage is delivered to Mars orbit fully fueled, where it waits for nearly 4 years before the crew uses it to return to Earth. It uses two RL6-class engines modified to burn LOX/CH_4. These are the same engines developed for the ascent and descent stages, thereby reducing engine development costs and improving maintainability. The return Habitat is a duplicate of the outbound transit/surface. No mention was made of the requirements or methodologies for cryogenic storage of propellants.

6.2.1.7 Launch vehicles

The scale of the required Earth-to-orbit launch capability was determined by the mass of the largest payload intended for the Martian surface. The nominal design mass for individual packages to be landed on Mars in DRM-1 was 50 mT for a crew Habitat sized for six people that is transferred on a high-energy orbit. This was estimated to require the capability for a single-launch vehicle to be about 240 mT from Earth to a 220-mile circular orbit (LEO). Because 200-ton-class Launch Vehicles raise development cost issues, consideration was given to the option of launching subunits to LEO using smaller vehicles and assembling (attaching) them in space prior to launching them to Mars. A smaller Launch Vehicle (110 to 120 mT) would have the advantage of lower development costs. However, the smaller Launch Vehicle introduced several potential difficulties to the DRM-1 scenario. The simplest, most desirable implementation using this smaller Launch Vehicle is to simply dock the two elements in Earth orbit and immediately depart for Mars. To avoid the boil-off loss of cryogenic propellants in the departure stages, all elements must be launched from Earth in quick succession. This places a strain on a single-launch facility and its ground operations crews or requires the close coordination of two or more launch facilities. Assembling the Mars vehicles in LEO and loading them with propellants from an orbiting depot just prior to departure may alleviate the strain on the launch facilities, but the best Earth orbit for a Mars mission is different for each launch opportunity.

A 240-ton payload-class Launch Vehicle was assumed by DRM-1. However, such a vehicle is beyond the experience base of any space-faring nation. While such a vehicle might be possible, it would require a significant development effort for the Launch Vehicle, launch facilities, and ground processing facilities; and its cost would represent a considerable fraction of the total mission cost. The choice of a Launch Vehicle remained an unresolved issue for any Mars mission at the completion of the DRM-1 study.

6.2.1.8   In situ *resource utilization*

In DRM-1, ISRU was planned to provide two basic resources: (1) propellants for the MAV and (2) cached reserves for the Life Support Systems (LSSs).[165] ISRU production includes two virtually redundant ISRU plants, the first delivered with the first Ascent Vehicle and the second delivered with the second Ascent Vehicle. Each ISRU plant can produce propellants for at least two MAV missions. However, only the first plant is required to produce life support caches. For each MAV ascent, DRM-1 projected that the ISRU plant must produce 20.2 mT of oxygen and 5.8 mT of methane propellants at a 3.5 to 1 mixture ratio.[166] Further, the first ISRU system is required to produce 23.2 mT of water, 4.5 mT of breathable oxygen, and 3.9 mT of nitrogen/argon inert buffer gases as a back-up cache for use by any of the three Mars crews. The system liquefies and stores all of these materials as redundant life support reserves or for later use by the MAV. The approach to ISRU production uses the Martian atmosphere for feedstock and imports hydrogen from Earth. The main processes used are common to both ISRU plants. The significant difference between

---

[165] There is a rather strange logic working here. On the one hand, the crew depends entirely on ISRU for propellants to return home, but ISRU is only used as a back-up for consumables. It is not clear why ISRU is not used as a primary source for both, because both are necessary for survival.

[166] However, as stated previously, these figures appear to be optimistically low. DRM-3 revised the requirement upward from 26 mT to 39 mT, and it seems likely that the actual requirement would be > 40 mT.

the two is that the second plant is smaller and excludes equipment for buffer gas extraction. DRM-1 indicates that, should sources of indigenous and readily available water be found, this system could be simplified.[167]

DRM-1 required that hydrogen must be imported from Earth. However, no discussion was provided on requirements and feasibility for transporting hydrogen to Mars and storing hydrogen on Mars.[168] The DRM-1 ISRU system used Sabatier, water electrolysis, carbon dioxide electrolysis, and buffer gas absorption processes to achieve these ends.

The Sabatier process results in a water-to-methane mass ratio of 2.25:1 and requires 0.5 mT of hydrogen for each mT of methane produced. The resultant methane is stored cryogenically as fuel. The water can either be used directly as cached life support reserves or can be electrolyzed to oxygen for storage and hydrogen to be recycled.

Oxygen production can be accomplished with two different processes. DRM-1 uses water electrolysis to produce oxygen from water produced by the Sabatier process, as well as carbon dioxide electrolysis to directly convert the Mars atmosphere to oxygen. Water electrolysis is a mature technology. The combined Sabatier and electrolysis processes generate oxygen and methane for use as propellants at a mass ratio of 2:1. In this combined process case, the hydrogen is recycled into the Sabatier process so that 0.25 mT of hydrogen are needed for each mT of methane. The engines selected for DRM-1 use oxygen and methane at a mass ratio of 3.5 to 1. Therefore, an additional source of oxygen is needed to avoid overproduction of methane. The carbon dioxide electrolysis process is used to provide the needed additional oxygen. The process converts the atmospheric carbon dioxide directly into oxygen and carbon monoxide using zirconia cells at high temperature.[169] This process eliminates the over-production of methane during propellant production except during the first mission when Sabatier-produced water is also needed.

The buffer gas extraction process was not examined in detail during this study. It was suggested that it would most likely be a nitrogen and argon absorption process in which compressed atmosphere is passed over a bed of material which preferentially absorbs the nitrogen and argon. The gases are then released by heating the bed and the products are passed on to the cooling and storage system. Parallel chambers are used so that one bed is absorbing in the presence of atmosphere while the other is releasing its captured gases.[170]

Ancillary systems for atmosphere intake, product liquefaction, and product storage and transfer will be needed. These systems were not detailed for DRM-1.

---

[167] Note that DRM-1 was prepared prior to the discovery by neutron spectrometers of widespread near-surface water on Mars (see Appendix C).

[168] See Section 5.7 regarding storing cryogens in space.

[169] However, carbon dioxide electrolysis technology is immature, and may not even be feasible (see Section 5.3.4.1).

[170] However, the advent of a newer cryogenic atmosphere acquisition process could produce buffer gases as a side-product.

The first ISRU plant is delivered to Mars over a year prior to the first departure of humans from Earth, and during that year the plant produces all the propellants and life support caches that are needed. Thus, humans do not even leave Earth until reserves and return propellants are available on Mars. This plant also produces propellants for the MAV mission of the third crew in the overall DRM-1 scenario. The plant integrates all the processes needed for both propellant and life support products.

The size of the ISRU plant was only estimated roughly. These estimates were based on some previous work[171] on the options for ISRU and on the rates needed to produce requisite materials over a 15-month period. The mass and power requirements were tabulated, but it is difficult to rationalize these figures.

6.2.1.9   *Life support consumables*

Consumables are needed in the form of water and breathable air (consisting of some mixture of inert gases and oxygen). Unfortunately, the topic of consumables is not covered in any comprehensive way in DRM-1.[172] A search on the word "consumable" reveals more than a dozen places where the word occurs in the document, but most of these are tangential, peripheral, or lacking content. The Habitat mass budget provides 7.5 mT for consumables under "crew accommodations", and 3 additional mT of consumables are attributed to plant growth and life support. It would have been valuable to discuss options for breathable air composition and potential for recycling, but this was not done. Another thing that is lacking is a table showing water and breathable air requirements for each leg of the trip (trans-Mars, descent, surface, ascent, and trans-Earth). It appears that DRM-1 was counting on ≫90% recycling.

6.2.2   DRM-3

DRM-3 was prepared by JSC in 1997 as a refinement of DRM-1. DRM-3 reaffirmed that the primary roles of the Reference Mission (as stated in DRM-1) are to (1) form a template by which subsequent exploration strategies may be evaluated for the human exploration of Mars, and (2) stimulate additional thought and development in the exploration community and beyond.

A major concern regarding DRM-1 centered on its launch system; specifically, a large, yet-to-be developed Launch Vehicle was required to place the mission elements into low Earth orbit (LEO). A ~240 mT Launch Vehicle would be required to achieve a human mission in four launches. It was recognized in DRM-1 that development of the large 200+ mT Launch Vehicle posed a significant technology and development challenge to the mission strategy. Design of the large launcher raises several cost

---

[171] This work is unpublished and unavailable outside JSC.
[172] Life support consumables, which could potentially be a major mass requirement as well as a critical safety issue, seem to be barely treated.

issues (development, new launch facilities, etc.), and the physical size of the Launch Vehicle is itself a potential limitation to implementing DRM-1. The requirement of a heavy lift booster was driven primarily by the initial mass to low Earth orbit (IMLEO); therefore, an effort was initiated to reduce the required mass and volume of each launch. These efforts were undertaken while balancing the need to minimize the number of launches to reduce ground launch costs and limit added operational complexity due to low Earth orbit (LEO) rendezvous and docking. In order to reduce the size of the Launch Vehicle, a critical examination of the payloads was conducted in terms of their physical size and mass. The goal of this modification was to reduce the mass of each payload element to ~160 mT, and divide it onto two smaller (~80 mT) Launch Vehicles rather than the single large vehicle.

In addition, significant system repackaging to reduce the physical size of the launch elements was important from many aspects of the Launch Vehicle design, including reducing the mass of the systems and reducing the aerodynamic loads on the payload shroud. The geometry of the large (10 m diameter) aeroshell used for the Mars landers in DRM-1 revealed significant unused volume between the lander and the aeroshell. In DRM-3 the aeroshell was redesigned so that the Habitat structure was integrated with the Mars entry aeroshell and launch shroud. In addition to reducing the structural mass of the element, the integrated design served several additional functions: (1) the integrated Habitat/pressure hull with a thermal protection system (TPS) served as both an Earth ascent shroud and Mars entry aeroshell, (2) the need for on-orbit assembly/verification of the aeroshell was eliminated,[173] (3) it allowed for stowage in an 80-metric-ton-class Launch Vehicle.

During the outbound and return interplanetary journeys, DRM-1 allowed for 90 m^3 of pressurized volume per crew-member. This was retained in DRM-3.

Another step in changing the launch strategy focused on reducing the vehicle masses. The payload masses were critically examined, and any duplications were eliminated. In addition, studies were undertaken to "scrub" the system masses to achieve the required weight savings. The goal of this work was to reduce each payload delivery flight to accommodate the approximate volume and weight limitations of two 80-metric-ton launchers per vehicle delivered to Mars. This design, delivering the interplanetary propulsion system and cargo into Earth orbit separately, would require one rendezvous and docking operation prior to each outbound journey to Mars. While doubling the number of launches, this strategy eliminates the high costs of developing the large >200 mT Launch Vehicle of DRM-1.[174]

While reviewing the original mission strategy, the initial Habitat lander was identified as a launch component that could potentially be eliminated. Instead, a lightweight inflatable Habitat was included with the MAV/Infrastructure Lander, and thus one major vehicle was eliminated. The mass of the inflatable module

---

[173] Note there didn't appear to be any indication in DRM-1 that this was necessary.
[174] However, the process of "scrubbing" masses was carried out in a manner that it is difficult to trace how it was done or how reliable the estimates may be.

(estimated at 3.1 mT without crew accommodations or life support) could be sub-stituted for the mass of the pressurized rover (stated to be $5\,mT$)[175] originally manifested on the Cargo-1 flight. The pressurized rover, deferred to the second Cargo delivery flight, would arrive a few months after the crew and would still be available for the majority of the mission. In essence, the redundancy of the pressurized rover (for the first Mars crew) was traded for the elimination of an entire Mars-bound Habitat flight. DRM-3 claims that, although elimination of Habitat-1 would reduce system redundancy, there is sufficient redundancy already built into DRM-1 even without Habitat-1. DRM-3 claims that:

1. *In situ* resource utilization processes would generate enough water and oxygen for the entire surface mission to run "open loop".
2. The Ascent Vehicle/ISRU plant on Cargo-2 of the subsequent mission, arriving at the surface a few months after the crew, could be used to supply life support rather than for propellant production.
3. If necessary, the surface could be abandoned for the orbiting Earth Return Vehicle, which has a sufficient food cache to last until the next trans-Earth injection window.[176]
4. The Earth Return Vehicle (ERV-2) of the subsequent mission, arriving a few months after the first crew, would provide an additional refuge for the crew if necessary.

All of the above statements are dependent on assumptions that are highly dubious and do not stand up to scrutiny. It is implicit in all of these statements that recycling systems exist that have recycling efficiencies near 98–99% and unlimited lifetime. Use of ISRU while the crew is on Mars, rather than prior to their departure from Earth, is a risky business. Transporting large amounts of hydrogen to Mars and storing it there raises questions regarding mass, volume, and thermal requirements, and solid oxide electrolysis is an unproven technology.

In regard to "abort-to-orbit", human factors are likely to be problematic with the crew locked into a small vehicle orbiting Mars for ~600 days. Jack Stuster has investigated human factors in historical human exploration expeditions and used these results as analogs for predicting human issues in isolated surroundings as may be encountered in space missions. His findings are reported in a paper available on the Internet.[177] The dimensions used to describe the analog conditions were: duration of tour, amount of free time, size of group, physical isolation, psychological

---

[175] However, DRM-1 had estimated the mass of a pressurized rover to be $15.5\,mT$, and it seems unlikely that it could be reduced to $5\,mT$.

[176] However, the crew would then be exposed to higher radiation levels and zero-g for an additional 500+ days, and the ERV would have to provide life support for an additional 500+ days. In addition, the psychological effects of being encapsulated in an orbiting Habitat for that length of time could be significant.

[177] *http://www.ingentaconnect.com/content/asma/asem/2005/00000076/A00106s1/art00012*

isolation, personal motivation, composition of group, social organization, hostility of environment, perceived risk, types of tasks, preparedness for mission, quality of life support conditions, and physical quality of the Habitat. By studying the diaries of various expeditions, Stutser was able to determine which factors led to the greatest number of negative diary entries. He found that group interaction was the most prevalent source of discord. It is recommended that anyone contemplating plans for a human mission to Mars should take a careful look at Stuster's reports.

The DRM-3 report says:

"The combination of repackaging the mission elements into smaller launch vehicles along with elimination of the initial habitat lander has allowed significant reduction in launch vehicle size, from 200-metric tons down to 80-metric tons, while only introducing two additional flights to the overall launch manifest."

This does not appear to be possible. Figure 6.1 shows that even though Habitat-1 is eliminated in DRM-3, all the other launches remain and are split in two. Therefore, it appears that whereas DRM-1 involved 10 launches, DRM-3 involves 18 smaller launches with 9 assemblies in LEO.

No discussion was provided in DRM-3 of how each major vehicle (ERV, Cargo Lander, Crew Lander) would be split into two pieces for launch and assembled in Earth orbit.

DRM-3 adopted a common descent/ascent propulsion system approach. The ascent stage propulsion system shares common engines and propellant feed systems with the descent stage. This eliminates the need for a separate ascent propulsion system, reducing the overall mass and subsequent cost. These common engines are the same RL6-class engines modified to burn LOX/CH$_4$ as the descent stage. However, it was estimated that the ascent propulsion system will require approximately 39 metric tons of propellant whereas DRM-1 required 26 mT of ascent propellants. This was due to an increase in assigned mass to the crew capsule and ascent propulsion system.

6.2.3 Mass comparisons: DRM-3 and DRM-1

It is difficult to make sense of the mass tables provided by DRM-1 and DRM-3, because they are not complete, they do not show the masses at each step, and in some cases they do not seem to satisfy the rocket equation. However, using some of the data in their tables, it is possible to make some plausible interpretations, although the numbers will differ somewhat from those provided by JSC. In all of these calculations, the optimistic JSC assumptions regarding use of the nuclear thermal rocket in LEO, high-efficiency long-life recycling, mass-less transport of hydrogen to Mars, and ISRU production of ascent propellants were adopted.

Table 6.1. Mass estimates for the Earth Return Vehicle.

ERV	DRM-1	DRM-3	Comments
Payload	45	29	Payloads adjusted to make IMLEO = JSC estimates
TEI propulsion	8	4	Assumed to be 15% of propellant mass
TEI propellant	51	25	Calculated from rocket eq. using $\Delta v = 2.4$ km/s and $I_{SP} = 360$ s. Assumes that 7.4 mT of food is jettisoned prior to Mars departure.
Total in Mars orbit	104	57	
Aero-entry system	16	9	Uses optimistic JSC estimate of 15% of injected mass. 60% is probably more realistic—see Section 4.6
Total in TMI	120	66	
TMI propulsion	29	23	Figures from JSC reports
TMI propellants	96	58	Calculated from rocket eq. using $\Delta v = 4.4$ km/s and $I_{SP} = 900$ s. Figures in the table are higher than JSC figures (86 and 50)
IMLEO	246	147	JSC estimates

Based on figures provided by JSC, we can derive Tables 6.1, 6.2, and 6.3. If the nuclear thermal rocket is replaced by a hydrogen/oxygen chemical rocket for Earth departure, the values of IMLEO tend to rise by about 58%.

The mass reductions in going from DRM-1 to DRM-3 are difficult to trace to specific implementations.

6.2.4   The Dual Landers Reference Mission

The Dual Landers Reference Mission was an outgrowth of its predecessors, DRM-1 and DRM-3. In DRM-1 the emphasis was on supplying full capabilities using conservative engineering estimates for vehicle masses. This led to relatively heavy vehicles, and to avoid assembly in LEO an extremely large Launch Vehicle (LV) was postulated. In the DRM-3 modification to DRM-1, masses of all vehicles and systems were significantly reduced, and—instead of postulating a huge LV—a ~80–85 mT LV was utilized, although this necessitated doubling the number of launches with assembly of two sub-units in LEO for each vehicle. Both DRM-1 and DRM-3 utilized a nuclear thermal rocket (NTR) for trans-Mars injection (TMI) in order to reduce the required mass in LEO needed to land any given mass on Mars. Without NTR TMI, the masses in LEO would have been roughly 58% higher. The Dual Landers DRM was never documented in a report, although "dog-eared" versions of a viewgraph presentation seem to be in existence in various places.

Table 6.2. Mass estimates for MAV/Cargo Lander.

MAV/CL	DRM-1	DRM-3	Comments
Mass to orbit	4	6	= Ascent Capsule
Ascent propellants	0	0	Propellants are produced by ISRU and need not be brought from Earth
Ascent propulsion stage	3	5	Adjusted to make ascent propellants 26 and 39 mT for DRM-1 and DRM-3
Ascent Capsule	4	6	
Cargo	57	32	Adjusted to make IMLEO correspond to JSC
Mass to surface	64	42	
Descent system	32	17	Data were not available. 50% of landed mass used for DRM-1 and 40% for DRM-3. Note that Braun (Section 4.6) estimates the mass of the orbit insertion system plus the descent system to be 2.3 times the payload delivered to the surface
Total in Mars orbit	96	59	Aeroshell not discarded in Mars orbit
Total in TMI	96	59	
TMI propulsion	29	23	Figures from JSC reports
TMI propellants	81	53	Calculated from rocket eq. using $\Delta v = 4.4$ km/s and $I_{SP} = 900$ s. Results are different than JSC figures (86 and 45)
IMLEO	205	135	JSC estimates

Table 6.3. Mass estimates for Crew Lander.

CL	DRM-1	DRM-3	Comments
Mass to surface	61	40	Adjusted to make IMLEO correspond to JSC
Descent system	33	16	Data were not available. 54% of landed mass used for DRM-1 and 40% for DRM-3. Note that Braun (Section 4.6) estimates the mass of the orbit insertion system plus the descent system to be 2.3 times the payload delivered to the surface.
Total in Mars orbit	94	57	
Total in TMI	94	57	
TMI propulsion	32	27	Figures from JSC reports
TMI propellants	82	54	Calculated from rocket eq. using $\Delta v = 4.4$ km/s and $I_{SP} = 900$ s. Results are different than JSC figures (86 and 50)
IMLEO	208	137	

The goal of the Dual Landers study seems to have been to make the human mission to Mars more practical and more politically acceptable by:

- Avoiding the use of an NTR for TMI.
- Eliminating the need for on-orbit assembly of vehicles.
- Utilizing a "Magnum" LV capable of placing 100 mT in LEO. Utilizing the same propulsion system for descent and ascent from Mars (although this had already been adopted by DRM-3).
- Delivering MAV propellants from Earth on the first cycle, but using ISRU-compatible propellants for ascent from Mars. Use of ISRU is delayed until second or third human crew arrivals.[178]

Because the mission is designed initially to avoid use of ISRU, the requirement to transport MAV propellants to Mars as part of the cargo mission increases IMLEO substantially. To avoid this problem, the Dual Landers mission cut the masses of all systems and vehicles very severely.[179] Nevertheless, even with massive cuts the requirement for IMLEO per vehicle would be as high as ~180 mT per vehicle if chemical propulsion is used for TMI from LEO. To avoid this problem, the Dual Lander mission adopted use of a reusable "tug" powered by solar electric propulsion to lift the vehicles from LEO to an elongated elliptical Earth orbit, because TMI with chemical propulsion requires far less propellant from this orbit than it would from a circular orbit in LEO. This scheme was under study at the end of the DRM-3 study. By carrying out such a solar-powered orbit raising, the gear ratio for delivering payload from LEO to Mars is reduced by almost 50%. The electric propulsion tug concept is illustrated in Figure 6.2.

Documentation on the Dual Landers mission is sparse, and it is difficult to appraise the feasibility of the masses used in this mission. The slow spiraling out of the SEP tug creates time delays and operational difficulties. Because the SEP tug drags the vehicles slowly through the radiation belts, the crew would have to wait until the Trans-Habitat Vehicle reached HEO before using yet another vehicle, a "crew taxi" to rendezvous with the Trans-Habitat Vehicle. The viability of the SEP tug depends critically on use of a hypothetical, high-efficiency, lightweight, thin-film solar array that is likely to be difficult to develop, and using current array technology all mass benefits from using SEP disappear. Furthermore, it is not clear that the required amount of xenon propellant could be obtained, and if obtainable whether the cost would be affordable. The difficulties and impediments to implementing such a system are discussed in Section 3.4.4.

---

[178] This has become the *modus operandi* of all JSC lunar and Mars mission plans. ISRU is never incorporated into the main mission architecture, but, rather, is tacked on late in the campaign as an afterthought. As a result, all vehicles and the Launch Vehicle are sized without the benefit of ISRU, and later application of ISRU does not change any of these space systems, although it might improve the payload percentage in late departures.

[179] The cuts were so deep that their credibility must be questioned.

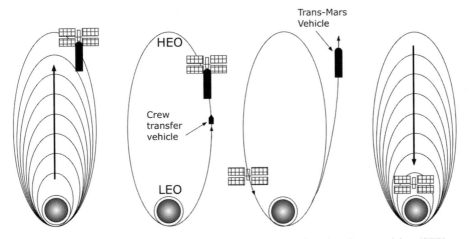

Figure 6.2. Dual Landers scheme for orbit raising using a solar electric propulsion (SEP) tug.

As in DRM-3, two landers are planned for the Dual Landers mission, but the Earth Return Vehicle is now designated the Trans-Habitat Vehicle. In Dual Landers, the landed vehicle is the Habitat Lander and the other is the piloted Descent/Ascent Vehicle. The Habitat Lander descends, lands, and performs an autonomous checkout. After the Habitat Lander is verified to be operational, the crew lands in the Piloted Descent/Ascent Lander. The Descent/Ascent Lander provides accommodation for the crew for the first 30 days of the surface mission. After this 30-day acclimation period, the crew transfers to the Habitat Lander for the remaining 18 months of the mission. The Descent/Ascent Lander is available at any time during the surface mission for an "abort-to-orbit". It is unclear whether an abort-to-orbit within the first 30 days of the surface mission would allow an immediate return to Earth in the TransHab vehicle, and thus turn the mission into a short-stay mission. It is probable that this is not a possibility due to greatly increased propellant requirements for the return trip in a short-stay architecture. So, an abort-to-orbit at any time during the mission would require the crew to remain in Mars orbit until the next window opens for a return trajectory. This would require life support capability aboard the TransHab vehicle for an extra 600 days, and it would expose the crew to zero-g and higher radiation levels for that period. In addition, the psychological effects of encapsulation in an orbiting habitat are likely to be serious (see discussion in Section 6.2.2).

A slender-body aeroshell using the Ellipsled configuration with a diameter of 8.5 m and a length of 21.3 m is used for aerocapture and descent to the surface for both the Habitat Lander and Descent/Ascent Lander. The aeroshell can also function as the Earth Launch Vehicle shroud.

The baseline Dual Lander architecture relies on solar power for both the Habitat Lander and the Descent/Ascent Lander. The assumptions for this architecture include a near-equatorial landing site between 15°N and 15°S and the development of high-efficiency (18%), large-area, integrated, thin-film photovoltaic cells. Regenerative fuel cells are used for initial checkout and initial operation on both the Descent/Ascent Lander and the Habitat Lander. Solar arrays are deployed within the first 30 days of the surface mission. The solar array design relies on flat roll-out arrays and a "phased" array deployment strategy. The idea is to deploy "fair weather/nominal ops" arrays (25–50% of total) initially and only deploy fresh "storm" arrays (remaining 50–75%) when needed during a dust storm. It is assumed that there will be a few days' warning prior to dust storm arrival. Lithium batteries are added for night energy storage.[180] A nuclear option was also considered as an alternative. This architecture uses regenerative fuel cells for initial power and then switches to nuclear power for the nominal surface mission. The power system for the Dual Landers mission does not seem to have been developed beyond initial conjecturing at the viewgraph level of detail.

The "Dual Landers" Reference Mission does not rely on ISRU up front, but utilizes ISRU-compatible propellants to make eventual inclusion of ISPP a possibility. However, a token ISRU oxygen production system is included to provide breathable oxygen to support EVA. It is also expected to provide back-up oxygen production for the Habitat's environmental control life support system (ECLSS). The system includes a sorption pump for atmosphere acquisition, a zirconia stack for oxygen production, and compression and storage hardware. However, zirconia stack technology is unlikely to prove to be practical. It is not clear that any significant benefits will ensue from late introduction of ISRU as a tacked-on afterthought late in the campaign.

The overall plan for the Dual Landers mission is shown in Figure 6.3. It does not seem worthwhile to attempt to outline the vehicle masses in the Dual Landers DRM, because the whole concept seems to be based on impractical systems, such as the SEP orbit-raising tug, use of solar power, use of inflatable Habitats, and use of an extremely low-mass ascent system.

It would be useful if tabular comparisons of vehicles from DRM-1, DRM-3, and Dual Landers could be prepared to make "apples-to-apples" comparisons. However, after attempting to do this in several ways, it was found to be difficult because of incompatible definitions and mass breakdowns. Indeed, none of the DRMs provided a glossary with definition of terms such as "crew accommodation" which is not always easy to interpret in terms of consumables and installations.

6.2.5 Common characteristics of 1990s' JSC Mars DRMs

Amongst the optimistic assumptions used in JSC DRMs are:

● Use of *nuclear thermal propulsion* for departure from LEO (but not for Dual Landers). However, it is noteworthy that the ESAS Report indicates starting

---

[180] However, the mass and volume of batteries was not specified and these are likely to be huge. One must be skeptical that a fully solar-powered mission is feasible. It would add a great deal of risk and uncertainty and make use of ISRU even more difficult because ISRU processing works much better without start-up and shut-down.

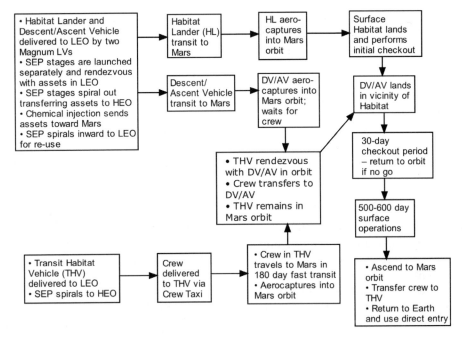

Figure 6.3. Overall plan for Dual Landers mission.

the NTP at 800–1,200 km altitude,[181] which significantly reduces the advantage of using NTP for Earth departure (see Section 3.4.3).

- The ESAS Report suggests that "The ESAS architecture does not address the Mars phase in detail, but it is recognized that traditional chemical propulsion cannot lead to sustainable Mars exploration with humans. Nuclear Thermal Propulsion (NTP) is a technology that addresses the propulsion gap for the human Mars era."[182] When the ESAS Report uses the word "sustainable" it seems to be a way of saying that we will eventually need it but perhaps not on the early missions. Yet the ESAS Report appears to intend to use this propulsion system on early missions,[183] so it is not clear what they mean. Furthermore, there is no indication that anyone understands the requirements (fiscal, logistic, health, and political) to develop, test, validate, and implement NTP, and whether the whole concept is feasible and affordable at all.

- *Low-mass aero-entry system for large systems at Mars.* Aerocapture entry systems are assessed at around 15% of the mass inserted into orbit, whereas the work reported in Section 4.6 suggests that this estimate is low by perhaps a factor of 5.

- *Transport of hydrogen* to Mars for ISRU via weightless tanks with zero boil-off.

---

[181] "The MTV loiters in a circular orbit of 800- to 1,200-km altitude," *ESAS Report*, Section 1.4.2.7, p. 35.
[182] *ESAS Report*, Section 9.3.3, p. 631.
[183] *ESAS Report*, Section 2.4.7, p. 69.

- Reliable, long-life, ~98% efficient *consumable recycling system* for life support.
- No provision for *radiation protection*.
- Use of *nuclear reactors* for power including placement up to a km distant from the main outpost with connecting cables (except the Dual Landers DRM).
- Use of *inflatable Habitats*.
- Use of large-scale *solar electric propulsion* for cargo orbit-raising prior to Earth departure (Dual Landers only).

6.3   MARS DRMS IN THE 2005–2006 ERA

6.3.1   The ESAS Report Mars plans

In late 2005 the Exploration Systems Architecture Study (ESAS) released a lengthy report defining plans for sending humans to the Moon in some detail, with a rather sketchy outline of how that would be followed by human missions to Mars.[184] The ESAS version of a proposed human mission to Mars is illustrated in Figure 6.4.

As can be seen from Figure 6.4, each human mission to Mars is comprised of three vehicle sets; two cargo vehicles and one round-trip piloted vehicle. The surface exploration capability is implemented through a split mission concept in which cargo is transported in manageable units to the surface, or Mars orbit, and checked out in advance of committing the crews to their mission. The split mission approach also allows the crew to be transported on faster, more energetic trajectories, minimizing their exposure to the deep-space environment, while the vast majority of the material sent to Mars is sent on minimum energy trajectories. The first phase of the mission architecture begins with the pre-deployment of the first two cargo elements, the Descent/Ascent Vehicle (DAV) and the Surface Habitat (SH). These two vehicle sets are launched, assembled, and checked out in low Earth orbit. After all systems have been verified and are operational, the vehicles are injected into minimum energy transfers from Earth orbit to Mars. Upon arrival at Mars, the vehicles are captured into a high Mars orbit. The SH then performs entry, descent, and landing on the surface of Mars at the desired landing site. After landing, the vehicle is remotely deployed, checked out, and all systems verified to be operational. Periodic vehicle checks and remote maintenance are performed in order to place the vehicles in proper configuration prior to crew arrival. The DAV remains in Mars orbit in a quiescent mode, waiting for arrival of the crew two years later. A key feature is the deployment of significant portions of the surface infrastructure (but not ISRU) before the human crew arrives. Pre-deployed and operated surface elements include the Surface Habitat, power system, thermal control system, communications system, robotic vehicles, and navigation infrastructure. This strategy includes the capability for these infrastructure elements to be unloaded, moved significant distances, connected to each other, and operated for significant periods of time without humans present.

---

[184] The *ESAS Report* can be downloaded from *http://www.nasa.gov/mission_pages/exploration/news/ESAS_report.html*

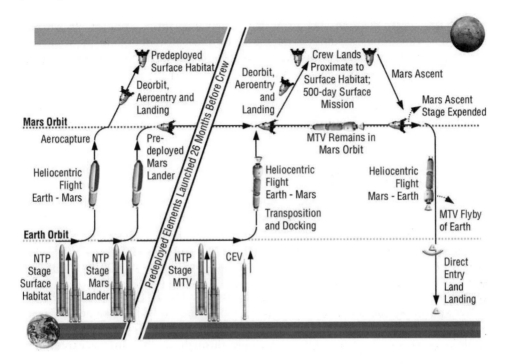

Figure 6.4. ESAS version of Mars DRM. This diagram is based on the belief that a Mars mission can be accomplished with six 125 mT cargo launches.

In fact, the successful completion of these various activities will be part of the decision criteria for launch of the first crew from Earth.

The second phase of this architecture begins during the next injection opportunity with the launch, assembly, and checkout of the Mars Transfer Vehicle (MTV) in Earth orbit. The MTV serves as the interplanetary support vehicle for the crew for a round-trip mission to Mars orbit and back to Earth. Prior to departure of the flight crew, a separate assembly and checkout crew is delivered to the MTV to perform vital systems verification and any necessary repairs prior to departure of the flight crew. The lunar CEV is used for crew delivery to and from the MTV. After all vehicles and systems—including the DAV, SH, and the MTV—are verified operational, the flight crew is injected on the appropriate fast-transit trajectory towards Mars. The length of this outbound transfer to Mars is nominally 180 days. Upon arrival at Mars, the crew performs a rendezvous with the DAV, which serves as their transportation leg to and from the Mars surface.

The DAV serves as the primary transportation element for the crew in the vicinity of Mars. The vehicle is designed to transport the mission crew from a high Mars orbit to the surface of Mars, support the crew for the initial post-landing acclimation period (up to 7 days), the 30-day start-up of the SH, and return the crew from the surface to the high Mars orbit,[185] whereby it performs a rendezvous with the MTV. The functional capabilities of the DAV must accommodate the ability to operate in a fully automated mode, since it is anticipated that the crew will not be capable of performing complicated tasks due to the long exposure to micro-gravity while in transit from Earth. Vehicle terminal phase targeting/control, post-landing check-out, initial flight-to-surface transition, and appendage deployments must occur without crew exertion. Thus, the vehicle must provide adequate time for the crew to re-adapt to $0.38g$ on Mars.[186] During this period, no strenuous activities (e.g., EVA) will be scheduled for any crew-members and the focus of the operations will be on developing adequate crew mobility and maintaining systems operability. NASA believes that "current human health and support data indicates that it may take the crew up to one week to acclimate to the partial gravity of Mars."[187] After the crew has acclimated, the focus of the initial surface activities is on transitioning from the DAV to the SH. This includes performing all remaining setup, checkout, and maintenance that could not be performed remotely from Earth. The crew has up to 30 days after landing to perform all necessary start-up activities of the SH. During this period local science is also conducted to insure that the initial science objectives can be met if early ascent from the surface is required. Lastly, the DAV is connected to the SH power system and placed in a semi-dormant mode, since it will not be needed again until ascent from the surface is required. Although the DAV is in a semi-dormant mode, emergency abort-to-orbit is available throughout the surface exploration phase of the mission.[188]

The concept for the DAV is very similar to that planned for lunar exploration. Unlike previous DRMs, the Ascent Vehicle does not descend to the surface at the previous opportunity prior to the crew departing from Earth, and this greatly reduces the prospects for use of ISRU.

It is stated: "Emphasis is placed on ensuring that the space transportation systems are designed to be flown in any Mars injection opportunity. This is vital in order to minimize the programmatic risks associated with funding profiles, technology development, and system design and verification programs." However, there are large variations (including an occasional really bad year) in required Δv from launch opportunity to launch opportunity. By demanding that the system is capable of going to Mars at any opportunity, the design is constrained by the worst case. Since there can be no "rescue missions" or aborts from Mars, this may be a poor

---

[185] Returning to a high orbit requires a lot more propellant than return to a low circular orbit. ISRU could produce that propellant. But, since ISRU is not used, adoption of a high orbit does not make much sense.

[186] Evidently, no artificial gravity is planned.

[187] See Chapter 6 for a more pessimistic view of the effects of lengthy exposure to zero-g. In fact, it is not known what deleterious effects might result from 500–600 days exposure to $0.38g$.

[188] As stated before, abort-to-orbit introduces large mass increases on the MTV for consumables, great demands on longevity of ECLSS, radiation doses that are way over the limit, and lengthy exposure to zero-g.

decision. A design that allows travel to Mars at most, but not necessarily all, opportunities might be far more affordable.

Although the ESAS Report says: "the ESAS team chose Mars DRM-3 as the baseline Mars mission,"[189] the DRM described in the ESAS Report, while similar in some ways to DRM-3, includes some significant differences.

The elements that are similar include:

- Pre-deployed elements 26 months prior to crew arrival.
- Three major vehicles: Surface Habitat, Mars Lander, and Mars Transit Vehicle (previously known as the Earth Return Vehicle).
- General timing of mission: 500–600 day surface stay, 180+ day crew transfer.
- Launch Vehicle with ~100–125 mT capability to LEO. (The ESAS Report reassures us many times that such a Launch Vehicle is adequate for Mars missions. However, this is not backed up by credible supporting analysis.)

The elements that are different include:

- In DRM-3 the Mars Ascent Vehicle and its accompanying ISRU plant land 26 months early and fill the Ascent Vehicle tanks prior to crew arrival.
- There is no provision for ISRU in the ESAS DRM. Indeed, ISRU is impractical unless the Ascent Vehicle lands 26 months early. The ESAS technology program calls for development of lunar ISRU (but not Mars ISRU). The ironic thing is that lunar ISRU is far more difficult to implement and has a much lesser mission impact than Mars ISRU.

The following elements are unclear:

- Surface power system: nuclear or solar? The ESAS technology plan does not call for development of a reactor.
- Elliptical or circular orbit for MTV? (This changes the propulsion requirement balance between the DAV and the MTV.)

The ESAS Report says: "... it is recognized that traditional chemical propulsion cannot lead to sustainable Mars exploration with humans. Nuclear Thermal Propulsion (NTP) is a technology that addresses the propulsion gap for the human Mars era. NTP's high acceleration and high specific impulse together enable fast transit times with reasonable initial mass in LEO." But, the ESAS Report does not mention the costs, environmental impacts, and safety considerations required to develop and

---

[189] "In order to view its lunar mission design work in the larger context of a future human mission, the ESAS Team chose Mars DRM 3.0 as the baseline Mars mission. This allowed the ESAS Team to choose technologies, spacecraft designs, Launch Vehicles, and operational demonstrations that were extensible to future Mars missions."

demonstrate an NTP, nor is the issue of minimum altitude for initiation of the NTP considered.[190]

The treatment of the Mars DRM by the ESAS Report is sketchy, sparse, and in many ways not credible. It represents a distinct step backwards from previous efforts by NASA.

6.3.2   2006 Mars DRM

During 2006, NASA continued to refine the Mars DRM concept originated in the ESAS Report in the "Constellation Program". However, the reporting on this work remains very sketchy and lacking in detail, making it impossible to check and evaluate the claims. This evolving 2006 Mars DRM has only been documented in a few PowerPoint slides, and the presentation in which these slides are included has not been released to the public.

One of the surprising things about the 2006 Mars DRM is that there are actually two DRMs under consideration: a "short stay" and a "long stay". The "short stay" mission is discussed at the end of Section 3.2.2. It is clearly infeasible, yet NASA continues to retain it as an option.

An NTR is used for Earth departure (presumably from LEO) for all three vehicles. In the case of the Crew Transit Vehicle (CTV) in the long-stay mission, the NTR is used for Earth departure, Mars orbit insertion, and Earth return. This necessitates that liquid hydrogen be stored for the 6+ month transit to Mars, during Mars entry into orbit, and for about a year and a half in Mars orbit. It seems unlikely that this could be accomplished passively (see Section 5.7) and the technical feasibility and requirements for using an active cooling system for zero boil-off remains uncertain.

According to Section 5.7, a reasonable expectation for the heat leak into a very well insulated hydrogen tank is about $0.5 \, \text{W/m}^2$. For a tank that holds, say, 50 mT of hydrogen at 100 psi, the volume is about $830 \, \text{m}^3$. If this is a cylinder of diameter ~ 7.5 m, the length would have to be about 19 m, the surface area $535 \, \text{m}^2$, and the total heat leak about 270 W. Since the cryocooler power requirement for storing hydrogen is roughly 100 times the heat leak, the power for the cryocoolers might be roughly 27 kW. At Mars, this would require a solar array of area greater than $200 \, \text{m}^2$.

The masses claimed by NASA for the vehicles in the short- and long-stay missions are summarized in Table 6.4. The "payloads" represent masses (other than a nuclear thermal rocket) sent into trans-Mars injection on its way toward Mars. The origin of these masses is impossible to trace. However, the rocket equation dictates some basic constraints and these can be explored.

One can (to some extent) reverse-engineer the data in Table 6.4 to determine what the data imply regarding masses delivered to destinations. For the DAV and SH, the results are shown in Table 6.5. These results suggest that the figures in Table 6.4 for

---

[190] Section 3.4.3 shows that if the NTP must be taken up to $\sim 1{,}250$ km before initiating, almost all of its advantage disappears.

Table 6.4. NASA estimates of masses (mT) for 2006 Mars vehicles (DAV = Descent/Ascent Vehicle, ERV = Earth Return Vehicle, CTV = Crew Transit Vehicle, SH = Surface Habitat).

	Short stay			Long stay		
	DAV	ERV	CTV	DAV	SH	CTV
Payload (in TMI)	72	0	27	72	49	27
Earth departure stage mass	30	54	58	30	30	39
Earth departure propellant mass	59	158	114	61	52	86
Vehicle mass (IMLEO)	161	212	199	163	131	152
Total mass (mission IMLEO)		*582*			*446*	

the DAV and the SH are compatible with the landed mass being 49 mT for the DAV and 39 mT for the SH. For the DAV it was assumed that ISRU is not used and rendezvous takes place after ascent with methane and LOX propellants, in a low circular orbit.

When reverse-engineering is applied to the CTV, however, problems arise. The results are shown in Table 6.6. In Table 6.6, the column marked "no staging" implies that the NTR propulsion system (mass = 39 mT) is carried through all steps: trans-Mars injection, Mars orbit insertion, and Earth return. The column marked "staging" implies that, with each successive burn, part of the dry propulsion system from the previous burn is jettisoned. In either case, the value of IMLEO exceeds the value claimed by NASA in Table 6.4 (152 mT) by about a factor of 2.

Table 6.5. Reverse-engineered mass estimates for DAV and SH in Mars 2006 DRM.

Element	DAV	SH
Mass to orbit	5	
Ascent propellants	25	
Ascent propulsion stage	5	
Ascent capsule	5	
Cargo	15	39
Mass to surface	49	39
Descent system	*25	*20
Total in Mars orbit	74	59
Total in TMI	74	59
TMI propulsion	29	29
TMI propellants	61	52
IMLEO	163	139

* Note that Braun (Section 4.6) estimates the mass of the orbit insertion system plus the descent system to be 2.3 times the payload delivered to the surface. In the above table, this mass is taken as 0.5 times the payload delivered to the surface, apparently a significant underestimate.

Table 6.6. Reverse-engineered mass estimates for CTV in Mars 2006 DRM based on optimistic JSC estimates for entry system mass.

	No staging of NTR	Staging of NTR
Payload	30	30
TEI propulsion	39	27
TEI propellant	23	19
Total in Mars orbit	92	76
MOI propulsion	39	29
MOI propellant	43	34
Total in TMI	173	139
TMI propulsion	39	39
TMI propellant	141	118
IMLEO	354	296

Aside from the matter of assuring that mass estimates satisfy the rocket equation, there remains the question of whether the vehicle masses assumed in these calculations are realistic. It is too early to be sure about this.

6.4 NON-NASA DESIGN REFERENCE MISSIONS

The two non-NASA DRMs that stand out are "Mars Direct" developed by Robert Zubrin and co-workers, and the Mars Society Mission (MSM) developed by James Burke of JPL along with several Caltech students.

6.4.1 Mars Direct

During the 1990s, Robert Zubrin, founder and President of the Mars Society, advocated an approach to sending humans to Mars that he claimed was significantly less complicated and less costly. This was designated "Mars Direct". Zubrin claimed: "all the technologies needed for sending humans to Mars are available today." The plan relied on relatively small spacecraft launched directly to Mars with booster rockets similar to the Saturn V for the Apollo missions.

The Mars Direct plan proposed that astronauts use the strategy of earlier explorers: "travel lightly and live off the land." The Mars Direct plan travels lightly by requiring only two types of landed assets: an unmanned Earth Return Vehicle (ERV) and a manned Habitat carrying four astronauts (a crew of six would have driven the masses too high to meet the self-imposed mission limits). Unlike JSC missions where the ERV remains in Mars orbit, the MAV ascends, rendezvous with the ERV, and transfers the crew to the ERV, Mars Direct combined the MAV and the ERV into a single vehicle that takes the crew home directly from the surface of Mars. The MAV "lives off the land" by using propellants produced *in situ*. Because

Table 6.7. Mass allocations for Mars Direct components at arrival on the surface of Mars.

ERV components	mT	Habitat components	mT
ERV cabin structure	3.0	Habitat structure	5.0
Life support system	1.0	Life support system	3.0
Consumables	3.4	Consumables	7.0
Solar arrays (5 kW)	1.0	Solar arrays (5 kW)	1.0
Reaction control system	0.5	Reaction control system	0.5
Communications and information management	0.1	Communications and information management	0.2
Furniture and interior	0.5	Furniture and interior	1.0
Space suits (4)	0.4	Space suits (4)	0.4
Spares and margin (16%)	1.6	Spares and margin (16%)	3.5
Aeroshell (for Earth return)	1.8	Pressurized rover	1.4
Rover	0.5	Open rovers (2)	0.8
Hydrogen feedstock	6.3	Lab equipment	0.5
ERV propulsion stages	4.5	Field science equipment	0.5
Propellant production plant	0.5	Crew	0.4
Nuclear reactor (100 kW)	3.5		
ERV total mass	*28.6*	*Habitat total mass*	*25.2*

the MAV must go all the way back to Earth, its propellant requirement is far greater than it is in JSC missions; however, these propellants would be produced by ISRU in the Mars Direct DRM. Consumables would be carried from Earth in sufficient quantity to support the mission, but *in situ* resources may also be used to augment the consumable supply. A very high recycling efficiency was assumed.

The unmanned ERV is launched from Earth in "year 1" of the mission. It has a mass of 45 mT in LEO. The ERV arrives at Mars, aerocaptures, and uses a parachute and retrorockets for landing. It carries 6 tons of liquid hydrogen as a feedstock for ISRU, an ISRU plant, a 100 kW nuclear reactor mounted on the back of a large rover powered by methane and oxygen, and other equipment as indicated in Table 6.7. Note that 6 mT of liquid hydrogen requires a volume of ~86 m^3. When insulation is added, the volume will increase considerably. For example, if the storage tanks had 0.2 m thick insulation, the volume would increase to over 110 m^3. The power requirements to maintain the hydrogen did not seem to be clearly specified.

Some simple reverse-engineering can be attempted on the ERV to check the consistency of these numbers. With 45 mT of payload in LEO and 28.6 mT landed on the surface of Mars, one possible set of requirements[191] would be as shown in Table 6.8.

According to Mars Direct, after landing, the nuclear reactor is autonomously deployed several hundred meters from the ERV. Then, an ISRU chemical-processing

[191] Reverse-engineering is not unique, so this is only one of many possible interpretations. But, it is believed that this is a reasonable interpretation.

Table 6.8. Reverse-engineering guesses for Mars Direct ERV masses (mT).

Mass to surface	29
Aero-assist system @ 57% of landed mass*	16
Total in Mars orbit (retain aeroshell)	45
Total in TMI	45
TMI propulsion @ 14% of TMI propellants	12
TMI propellants	88
IMLEO	145

* Note, as before, that Braun would have estimated the aero-assist mass as 230% of the landed payload mass whereas Mars Direct uses 57%. Using Braun's estimate, the entire Mars Direct mission becomes inoperative.

plant uses Sabatier/Electrolysis to provide 48 mT of oxygen and 24 mT of methane. An additional 36 mT of oxygen is generated by dissociation of atmospheric CO_2 for a total of 108 mT of methane/oxygen propellant at a 3.5:1 mixture ratio. This process takes 10 months and is accomplished before the first crew departs for Mars. Of this total, 96 mT of propellant is allocated to the ERV (that includes the MAV) for ascent and return to Earth, leaving 12 mT for operation of rovers. Additional stockpiles of oxygen and water can be produced (in principle) for consumables, but more hydrogen would have to be brought from Earth.

On the return trip to Earth from Mars, Mars Direct would presumably use those elements in Table 6.7 needed for return, totaling ~16.5 mT. Using a Δv of 6.8 km/s for return to Earth from Mars, and a specific impulse of 360 s, we calculate a requirement of ~97 mT of $CH_4 + O_2$ propellants produced by ISRU on Mars.

Note that the propulsion requirements for the ERV are much higher in Mars Direct than they are in the JSC DRMs because in Mars Direct the ERV acts as both the MAV and the ERV of the JSC DRMs, and must carry propellant for both to attain Mars orbit from the surface, and propellant for return to Earth. The advantage of Mars Direct is that the JSC DRM requirement of placing a separate ERV in Mars orbit is eliminated, and the propellant used to return to Earth is produced on Mars, instead of being brought from Earth. However, this propellant must be lifted from the surface to Mars orbit, thus increasing the propellant load for ascent to orbit. Returning directly to Earth eliminates the need for a separate ERV in Mars orbit, but it significantly increases the size of the ascent propulsion system as well as the entire gamut of ISRU plant size and propellant storage capabilities. The initial two launches are followed by additional pairs of launches timed to the Mars mission launch opportunity spacing of 26 months.

The data in Table 6.7 represent optimistic estimates of required masses. For example:

- The estimated mass of the cabin structure and interior furnishings for a crew of four for ~6 months is only 3.5 mT, whereas it seems likely that a Habitat for a 6-month duration might be >20 mT.

- The life support system on the ERV that supports the crew for both the outbound and inbound legs of the trip is estimated to have a mass of only 1.0 mT and consumables are listed as only 3.4 mT, for a total mass of 4.4 mT. However, according to Table 4.3, even assuming 99% water recovery the need for ECLSS and consumables on the two legs of the trip require a mass of about 9 mT (for a crew of four), and if JSC is unable to meet the 99% water recovery goal it could be much higher.
- No mass is allocated for hydrogen storage, nor is the method of storage stated.
- The ERV propulsion stages that utilize 97 mT of cryogenic propellants are estimated to weigh 4.5 mT or 5%, whereas conventional rules of thumb would suggest ~15% of the propellant mass.
- The mass of the nuclear reactor is estimated to be 3.5 mT, whereas a figure of 8 to 10 mT seems more credible.

Therefore, the mass of the ERV is likely to be much greater than the allocations of 45 mT in TMI and 29 mT on Mars. Using the rocket equation for return to Earth from Mars with a Δv for transport from the Mars surface to Earth of about 6.8 km/s, we find that for 97 mT of cryogenic propellants (specific impulse = 360 s) the mass transported (including the dry mass of the propulsion system) is limited to 16.5 mT. This tallies with an estimate of the ERV return mass from Table 6.7 if only those elements of the ERV needed for the return leg are included. However, as previously stated, the Mars Direct estimates of the masses of the elements of the ERV were quite optimistic. In this regard, of greatest importance is the mass of the dry propulsion system. Mars Direct estimated this mass as 4.5 mT, whereas standard rules of thumb suggest a mass of perhaps 15% of the propellant mass. If the mass of the ERV (leaving out the dry propulsion system) is increased from 12 mT (as indicated by Mars Direct) to a more realistic value of 35 mT, and if the mass of the dry propulsion system is assumed to be 15% of the propellant mass, the rocket equation requires that the dry propulsion system weighs 259 mT and 2,017 mT of propellants are needed. Even if we assume that the mass of the dry propulsion system is only 10% of the propellant mass, the rocket equation requires that the dry propulsion system weighs 50 mT and 583 mT of propellants are needed. Acceleration through a Δv for transport from the Mars surface to Earth of 6.8 km/s creates a very high leverage condition for the propellant and propulsion system masses. Only by reducing the ERV mass estimate for the return leg to a very low value did Mars Direct succeed in reducing the propellant requirement to just under 100 mT.

All of the above leaves out the issue of aero-entry system masses. As discussed in Table 6.8, it is likely that these masses are a factor of 4 too low.

Mars Direct attempts to deal with low gravity *en route* to Mars by having the manned Habitat employ artificial "Mars-g" gravity using a tether between the Habitat and the burned-out booster rocket's upper stage and spinning at approximately 1 rpm. A small storm shelter is also available in case of a radiation storm.

Table 6.9. Vehicles involved in the MSM.

Vehicle	Launch date	Content	Pathway
Earth Return Vehicle (ERV)	1st	Provide life support and Habitat for return from Mars (a CRV plus propulsion)	Mars orbit; remain until ascent; transport crew back to Earth
Mars Ascent Vehicle (MAV)	1st	Provide MAV and ISRU	Land on Mars; transport crew to ERV in Mars orbit; accompany ERV back to Earth
Cargo Lander (CL)	1st	Provide power, hydrogen, mobility, science, etc.	Land on Mars. Fill tanks of MAV
Habitat (Hab-1)	2nd	Provide Habitat and life support for transit to Mars and surface stay	Transport crew to Mars and provide habitat on Mars
Crew Return Vehicle (CRV)	2nd	Capsule and life support as back-up to Hab-1 on outward leg	Transfer crew to Hab-1 in LEO. Accompany Hab-1 to Mars but return to Earth on free return trajectory

6.4.2   The Mars Society Mission

In 1998 Jim Burke of JPL led a team of four Caltech students in a study called the Mars Society Mission (MSM). This was a paper design of a human mission to Mars combining various aspects of Mars Direct and JSC DRM-3, with the intent of improving upon the safety, cost, and political viability of DRM-3, with significant additions of vehicle redundancy. The MSM is documented better than most DRMs. Unfortunately, the website that provided a pdf download no longer exists.[192]

In the MSM, three launches take place at the first launch opportunity. The vehicles are an Earth Return Vehicle (ERV), a Mars Ascent Vehicle (MAV), and a Cargo Lander (CL). At the second launch opportunity, two more vehicles are launched, a Habitat (Hab-1) for transfer of the crew from LEO to the surface of Mars, and a Crew Return Vehicle (CRV) that is a small crew capsule that tags along with the Habitat on the outbound trip to Mars to provide the crew with a back-up spacecraft that can keep them alive in the event a critical system on the Habitat fails. The MAV returns to Earth alongside the ERV, providing redundancy for the homeward leg. The vehicles are summarized in Table 6.9.

The mass budgets for the vehicles launched at the first opportunity are given in Table 6.10. The mass budget for Habitat-1 is given in Table 6.11.

---

[192] However, it is now available at *http://www.mars-lunar.net/References/Mars.Soc.Mission.pdf*

Table 6.10. MSM vehicle mass budget.

Cargo Lander	Mass (mT)	Explanation
Nuclear reactor (160 kW)	9.3	DRM-3. Uses lander mobility for deployment
Hydrogen	11.8	Stoichiometry
H_2 tank	4.7	40% of liquid-hydrogen mass
Power line from reactor	0.8	DRM-3 mass scrub
Science and exploration	4.7	Based on remaining launch to Mars surface capability
Fuel cell	0.3	5 kWe power (DRM-1)
Cargo Lander mobility	5.5	Assumed 15% of total landed mass
Transit power 5 kWe solar	0.5	DRM-1
Interplanetary RCS	0.8	Provides 45 m/s Δv
Landed mass	*38.5*	Sum of above
Descent propulsion	0.6	Four RL-10M engines
Descent propellant	7.2	For 632 m/s Δv
Propellant tanks	0.6	9% of propellant
Parachutes	0.7	DRM-3
Aeroshell	8.2	18% of payload
Total injected mass	*54.9*	Sum of above

MAV Lander	Mass (mT)	Explanation
MAV	15.0	
ISRU	9.0	
1st stage	12.4	9% of propellant mass + 14 RL-10M engines
2nd stage	2.4	9% of propellant mass + 2 RL-10M engines
Fuel cell	0.3	DRM-1
Interplanetary RCS	0.8	Provides 45 m/s Δv
Transit solar power	0.5	DRM-1
Landed mass	*40.4*	
Landing propellant	3.7	Enough to provide $\Delta v = 324$ m/s
Parachutes	0.7	DRM-3
Aeroshell	8.4	18% of payload
Total injected mass	*53.2*	Sum of above figures

ERV	Mass (mT)	Explanation
CRV	15.5	
TEI stage structure	2.4	9% of propellant mass + 2 RL-10M engines
TEI stage propellant	23.0	Propellant needed to return crew to Earth
Power supply	0.8	NASA Reference Mission 1.0
Aerobrake	7.5	18% of payload mass
Interplanetary RCS	0.8	Provides 45 m/s Δv
Total injected mass to TMI	*50.0*	Sum of above figures

Table 6.11. Mass budget for Habitat-1 in MSM compared with DRM-3 and Mars Direct.

	Mars Direct	DRM-3	MSM	Explanation for MSM figures
Habitat Module structure	5.0	5.5	4.8	Scaled from DRM-3
Life support system	3.0	4.7	3.8	NASA model for crew of six
Consumables	7.0		3.2	98% closed H_2O/O_2 + food
				= 0.000630 mT per person per day
				900 days total
Descent fuel cell	1.0	3.0	1.3	
Reaction control system	0.5		0.5	Mars Direct
Comm/info	0.2	0.3	0.3	DRM-3
Science	1.0			
Crew	0.4	0.5	0.4	
EVA suits	0.4	1.0	1.0	DRM-3
Furniture and interior	1.0		1.5	
Open rovers	0.8	0.5		Mass budgeted with surface power
Pressurized rover	1.4			Not included in Hab payload
Hydrogen and Hab ISRU	0.4			
Spares and margin	3.5	0.0		Included in individual listings
Health care	1.3			
Thermal		0.6	0.5	DRM-3 scaled
Crew accommodation		11.5		
Surface power (reactor)		1.7	5.0	At least 25 kWe needed
EVA consumables		2.3		Produced by ISRU on MAV and Hab
Power distribution		0.3	0.3	DRM-3 scaled
Total landed	*25.2*	*31.8*	*24.2*	Total of above
Terminal propulsion + propellant			5.3	
Parachutes		0.7	0.5	
Orbital power (solar)			1.7	
Aeroshell structure and TPS			9.5	30% of (24.2 + 5.3 + 0.5 + 1.7)
Artificial gravity (125 m)			1.4	
Transit power (solar)			1.7	
Reaction control propellant	Above		1.7	
Total injected to TMI			*46.0*	

Table 6.12. ISRU products in DRM-3 compared with MSM.

	DRM-3	MSM
O_2	30.33	106.38
CH_4	8.67	30.40
Consumables	23.00	23.00
Total	*62.00*	*159.78*

The ERV is designed to return the crew from Mars orbit. It consists of a CRV-based crew capsule and a trans-Earth injection stage that is identical to the MAV's second stage, the only difference being that the methane/oxygen bipropellant of the ERV will come from Earth.

The presence of both an ERV and MAV capable of returning the crew provides for a number of contingencies. If the ERV's trans-Earth injection stage fails, the crew abandons the ERV and continues on to Earth in the MAV alone. If the ERV's life support, communications system, or other critical system is disabled and the ERV is rendered unable to support the crew before or after TEI, the ERV will still accompany the crew. This is because (1) a faster trajectory is possible using both ERV and MAV stages, regardless of life support capabilities and (2) the ERV could still provide spare parts to the MAV or be repaired after aerobraking into Earth orbit.

Because the Ascent Vehicle must travel from the surface of Mars all the way back to Earth, it requires 137 mT of propellants, produced by ISRU. This requires bringing 11.8 mT of hydrogen to Mars. The MSM allocated 4.7 mT for hydrogen storage. Neither Mars Direct nor DRM-3 appears to have allocated mass for hydrogen storage.

An artificial gravity system was deemed necessary for the MSM's outbound Hab flight to (1) minimize bone loss and other effects of freefall; (2) reduce the shock of deceleration during Mars aerobraking; and (3) have optimal crew capabilities immediately upon Mars landing. To save mass, the MSM used an artificial gravity system with the Habitat counterbalanced by a burned-out Launch Vehicle stage, as in Mars Direct. If 3 rpm is taken as the maximum rotational rate that humans may be subjected to for long-duration missions, and Mars gravity (which is easier to provide than Earth gravity, but will of course condition the crew for the gravitational environment of their destination) is desired, the distance between the spacecraft and its burned-out upper stage is 125 meters. However, we don't know the physiological effects of prolonged exposure to $0.38g$. An aluminum truss between the Hab and burnt-out upper stage was chosen over a tether, because the truss had (1) a much lower risk of failure when impacted by a micrometeorite; (2) no risk of snag; (3) less energy is stored in the tension of the connecting structure that could be potentially damaging if released. An artificial acceleration due to gravity of $3.7\,\text{m/s}^2$ was chosen as a compromise between desired fitness of the crew and mass budget concerns stemming from a larger truss. The final mass budgeted to the artificial gravity system was enough for a truss system capable of bearing six times the expected load.

Neither the ESAS Report, nor any JSC publications that this writer has seen, make any reference to or even acknowledge the existence of either Mars Direct or the MSM.

6.4.3 TeamVision approach to space exploration

An alternative approach to space exploration was proposed by the TeamVision corporation. The TeamVision concept involves five eras of space exploration:

- Manned Exploration Transition (2004–2016).
- Lunar Return Missions (2012–2020).

- Manned Lunar Surface Missions (2016–2020).
- Lunar Resource Development Utilizing Mars Class Hardware (2020–2030).
- Manned Mars Missions Utilizing Lunar Resources (2024–2030).

While the timing of these eras appears to be overly optimistic and unattainable, particularly for Mars, nevertheless, the sequence makes a good deal of sense. The first era is focused on replacing the Space Shuttle's International Space Station mission role. This is accomplished through the use of existing medium-class Expendable Launch Vehicles (ELVs) using a lunar-class Crew Module and lower mass lunar precursor Service Module. In the Lunar Return Mission era they introduce a new family of hybrid Heavy Lift Vehicles (HLVs) based on ELV and Shuttle-derived launch systems that enable the performance of lunar precursor missions. These lunar precursor spacecraft are then combined and launched via a growth version of the hybrid HLV family to perform two crew direct ascent/return manned lunar surface missions. These manned lunar surface missions work in close concert with the previous era's remote-controlled lunar surface robots. The higher lunar surface payload delivery capabilities of a direct return architecture are then expanded into a lunar surface rendezvous (LSR) architecture. Utilizing Mars mission precursor equipment, a significant expansion of lunar resource development is implemented. It is claimed that these lunar resources and facilities are then used to significantly lower all future launch expenses enabling a significant expansion in both lunar and Mars mission scopes. Whether this is feasible remains to be seen. Mars precursor missions to the asteroids and Mars vicinity are then performed utilizing the lunar-tested Mars-class hardware forming the final foundation for manned Mars surface missions.

TeamVision[193] adopted a philosophy of mixing robotic and human endeavors— whereas NASA seems determined to maximize humans and minimize robotics. TeamVision's plan for gradually increasing Launch Vehicle capability by adding to a core system is a very good one (this approach was used in a much lesser form by the Mars Society DRM). Also, their plan to go beyond NASA's limit of 125 mT to LEO is very credible (and necessary).

6.4.4 The MIT Study

There are many conceivable architectures for carrying out human missions to Mars. The MIT Team has analyzed a wide selection of conceptual architectures.[194]

---

[193] *An Alternate Approach towards Achieving the New Vision for Space Exploration*, Stephen Metschan, AIAA 2006-7517.

[194] *From Value to Architecture: The Exploration System of Systems—Phase I*, P. Wooster, Presentation at JPL, August 23, 2005; *Paradigm Shift in Design for NASA's Space Exploration Initiative: Results from MIT's Spring 2004 Study*, Christine Taylor, David Broniatowski, Ryan Boas, Matt Silver, Edward Crawley, Olivier de Wec, and Jeffrey Hoffman, AIAA 2005-2766; "The Mars-Back Approach: Affordable and Sustainable Exploration of the Moon, Mars, and Beyond Using Common Systems," P. D. Wooster, W. K. Hofstetter, W. D. Nadir, and E. F. Crawley, *International Astronautical Congress, October 17–21, 2005.*

During 2005, MIT carried out an extensive analysis of 1,162 variants of Mars architectures and compared them on the basis of relative values of IMLEO as well as risk and cost. The values of IMLEO varied from about 750 mT up to more than 10,000 mT, but there were about 200 architectural variations with close to the minimum IMLEO.

Vehicle definitions:

ERV—Earth Return Vehicle (used to return the crew from Mars orbit, and in some cases to deliver the crew to Mars surface and return).

ITV—Interplanetary Transfer Vehicle (used to transfer crew from LEO to Mars orbit, and return).

LSH—Landing and Surface Habitat (used to transfer the crew to the Mars surface from Mars orbit).

MAV—Mars Ascent Vehicle (used to transfer the crew from the Mars surface to Mars orbit).

TSH—Transfer and Surface Habitat (used to transfer the crew to the Mars surface from LEO).

MIT investigated the question of how to stage human missions to Mars. Three basic types of architectures were emphasized as shown in Figure 6.5:

(1) Direct Return: an ERV and a TSH are landed on Mars. The ERV returns the crew to Earth directly from the Mars surface.
(2) Mars Orbit Rendezvous (on return): a MAV and a TSH are landed on Mars and an ERV is placed into Mars orbit. The MAV performs a rendezvous in Mars orbit and transfers the crew to the ERV for return to Earth.
(3) Mars Orbit Rendezvous (both ways): a MAV lands on Mars. An LSH goes to Mars orbit. An ITV takes the crew to Mars orbit where the crew transfers to the LSH for descent. On return, the MAV transfers the crew to the ITV, which returns the crew to Earth.

They then systematically examined the effects of changing propulsion systems, adding or removing ISRU, various power options, etc. on IMLEO.

MIT evaluated IMLEO for each of the architectures in Figure 6.5 using various propulsion systems (chemical, electric, and nuclear thermal). It is difficult to trace the specific assumptions used by MIT for vehicles and propulsion systems. In general, their estimates for IMLEO seem to be rather on the low side. One important result was that the direct return architecture had a very high IMLEO if ISRU is not used, because of the huge amount of propellant that must be delivered to the Mars surface. However, when ISRU is used in the mission, the mass of the direct return option is comparable with the other options. It also seems likely that MIT used rather optimistic performance estimates (based on JSC DRMs) of advanced propulsion and consumable recycling efficiency.

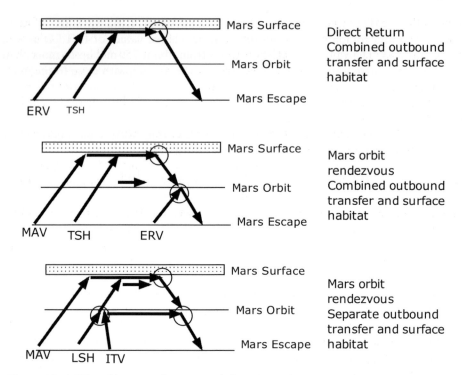

Figure 6.5. MIT architectures for Mars missions. [Based on MIT papers previously cited.]

6.4.5   ESA Concurrent Design Facility Study (2003)

The European Space Agency (ESA) carried out a rather extensive study of human missions to Mars that was reported in early 2004.[195] This study utilized a number of ground rules. The following are excerpted from these:

- Crew of six—with three landing on Mars and three remaining in orbit. (How the three in orbit would cope with the ill-effects of zero-*g* is not stated.)
- Development of a new "on-purpose" launcher of the 100 mT class to LEO is excluded.
- No previous cargo mission with surface infrastructure or consumables.
- No nuclear power either for cruise and Martian surface.
- No nuclear thermal propulsion or electric propulsion.
- No ISRU, either for propellant or for life support.
- No food production (e.g., greenhouse).
- No inflatable structure technology for the Habitation Module.

[195] *Human Missions to Mars—Overall Architecture Assessment*, Executive Summary, CDF Study, CDF-20(C), February 2004.

Some of these ground rules are very unfortunate, because they lead to very inefficient mission designs. The launch scenario using a variety of existing Launch Vehicles with maximum throw to LEO of 80 mT leads to a very unwieldy assembly sequence. The lack of a previous cargo mission adds risk and severe mass requirements to the system that lands the crew. The non-use of nuclear power on the Mars surface makes the mission power system problematic, risky, unwieldy, and probably infeasible. The lack of ISRU adds to the required launch mass.

The ESA Study examined three possible mission scenarios: (1) a conventional conjunction-class mission with a total mission duration of 963 days and 533 days on the surface, (2) an opposition-class mission with 30 days on the surface and a mission duration of only 376 days, and (3) a short-stay mission with 28 days on the surface utilizing a Venus swing-by on return for a total mission duration of 579 days. The total mission Δv for concept (1) was fairly independent of launch date at around 8.5 to 9 km/s. The total mission Δv for concept (2) varied widely with launch date, varying from 17 km/s to 23 km/s over the period 2025 to 2038. Using the Venus flyby on return lowers this to the range 10–15 km/s. For a reference year of 2033, they found the data shown in Table 6.13.

The radiation figures are difficult to comprehend, considering that while on Mars the crew is protected by the planet below and the atmosphere above. The Venus swing-by with 551 days in space should have the highest radiation exposure (not to mention the greatest ill-effects of microgravity). The estimates of consumables seem to be enormously low, and must perforce be based on an assumption that there is an ECLSS with extremely high efficiency. However, this was not discussed in the Executive Summary.

Fortunately, the ESA Study discarded the short missions and devoted the bulk of their analysis to the conjunction mission.

To ensure that the astronauts do not receive a 30-day dose in excess of the limits, it was planned to include a storm shelter with at least 25 g/cm^2 of shielding to protect against solar proton events. It was also planned to provide a minimum of 9 g/cm^2 of shielding throughout the Habitation Module "to ensure the yearly limit and career dose limits are not exceeded by the galactic cosmic ray radiation." However, it is doubtful that 9 g/cm^2 will do the job.

Table 6.13. ESA estimates for 2033 missions.

	Conjunction	Opposition	Venus swing-by
Total mission duration (days)	963	376	579
Possible surface duration (days)	533	30	28
Δv (m/s)	8,368	15,120	10,230
Radiation dose (GCR, Sv, BFO)	1.087	0.496	0.756
Consumables (mT)	10.2	4.2	6.4
Mass to LEO (tonnes)	1,336	45,938	2,481

The main Launch Vehicle for the mission was chosen to be the Russian Energia capable of delivering 80 mT to a 200 km × 200 km orbit and then raised to a 400 km orbit for Earth departure. Several smaller launchers (~20 mT to LEO) are also employed in the designed assembly strategy that uses a total of 28 launches. In this ESA concept, assembly in LEO is nothing short of a nightmare. At least five service platforms are needed and it takes 4 to 5 years to assemble the whole thing in LEO. The total mass launched (including service platforms) is estimated to be ~1,500 mT. The decision not to develop a Heavy-Lift Launch Vehicle was based on expediency, but in a gigantic enterprise, like sending humans to Mars, expediency is not appropriate.

Continuing with the philosophy of expediency, the ESA mission utilizes space-storable propulsion throughout, thus avoiding the need to develop aero-assisted Mars entry.

The assembled spacecraft consists of a Propulsion Module, an Earth Re-entry Capsule, a Transfer Habitat Module for transfer from Earth to Mars orbit and back, and a so-called Mars Excursion Vehicle that hosts the crew from orbit to the surface, on the surface, and during ascent. This vehicle is partitioned into three parts: Descent Module, Surface Habitat, and Ascent Vehicle.

The Transfer Habitat Module has an estimated mass of 67 mT including about 10 mT of consumables.

The description of the Mars Excursion Vehicle (MEV) is difficult to assimilate. One problem is that the Surface Habitat is described as being designed for a 30-day stay on the surface, whereas the actual stay is >500 days. The use of space-storable propellants adds mass. Only 0.3 mT of consumables are allotted to the MEV, whose total mass is estimated to be 46.5 mT, of which 20.5 mT is propellants. Power is provided by H_2/O_2 fuel cells that might be feasible for 30 days, but certainly not for >500 days. None of the masses are very credible.

The ESA Study was predicated on maximum use of existing capabilities that led to an unwieldy mission design that contains internal contradictions and varies from extreme conservatism in propulsion, to gross optimism for life support consumables, to incredible concepts for power systems. This Study would best be filed away and ignored.

6.5   IMLEO FOR HUMAN MISSIONS TO MARS

We have emphasized the importance of IMLEO as a measure of the cost of a human mission to Mars, and we have explored the mass gear ratios (before and after) for various steps involved in an overall Mars mission. In the present section, we will integrate previous results into simplified models for IMLEO in several variations of a representational Mars mission. The following assumptions are made:

- A habitat and life support system for a 600-day stay on Mars will be transferred from LEO to the Mars surface. This system is assumed to have a mass of 30 mT.

This does not include any radiation protection, and it assumes a very high recycling efficiency.

- Miscellaneous cargo will be transferred to the surface of Mars including power, mobility, EVA capability, science lab, etc. This system is assumed to have a mass of 30 mT.
- An ascent system will be transferred to the surface of Mars that includes an ascent capsule, a dry-propulsion system, and either an ISRU system or methane/oxygen propellants ($I_{SP} = 360$ s). The capsule has a mass of 5 mT and the propulsion system has a mass that is 15% of the ascent propellant mass. Propellants for ascent can either be taken from Earth or produced *in situ* by ISRU. If ISRU is used, the ISRU system is assumed to have a mass of 6 mT.
- A Mars Transit Vehicle (MTV) will transfer the crew from LEO to Mars orbit, transfer the crew to a Descent Vehicle, wait in Mars orbit, and eventually transfer the crew back to Earth. In case of an abort-to-orbit, the MTV would have to provide an extra ~600 days of life support. This system is assumed to have a mass of 40 mT. This does not include any radiation protection, and it assumes a very high recycling efficiency. Of this 40 mT, 8 mT are used as the Earth entry capsule, leaving 32 mT for the Habitat, life support, and spacecraft functions. An elliptical Mars orbit is used, and Mars orbit insertion is accomplished with $CH_4 + O_2$ propulsion.
- A Mars Descent Vehicle will be transported to Mars orbit, from where it will descend to the surface with the crew after a rendezvous with the MTV (mass ~8 mT).
- The mass of an aerocapture system for Mars orbit insertion is 67% of the mass inserted into Mars orbit. The mass of an EDL system for descending to the surface from Mars orbit is 100% of the payload delivered to the surface from Mars orbit.

These masses are undoubtedly optimistic, because they assume very high recycling efficiencies for lengthy periods, no radiation protection, no low-gravity mitigation, weightless hydrogen storage, zero boil-off of all cryogens, and a number of other mission elements that may prove to be more massive than were assumed. We shall assume use of LOX/LH_2 propulsion for Earth departure, LOX/methane propulsion thereafter, and full aero-assist at Mars.

Step 1: Transfer 30 mT of habitat from LEO to the surface of Mars. It requires about 9.3 mT in LEO to land 1 mT of payload on Mars with full aero-assist. Therefore, this step requires $9.3 \times 30 = 279$ mT in IMLEO.

Step 2: Transfer 30 mT of cargo to the surface of Mars. This step also requires $30 \times 9.3 = 279$ mT in IMLEO with full aero-assist.

Step 3: Transfer the ascent system to the surface of Mars. The payload mass transferred to orbit is the capsule (5 mT). As shown in Section 5.3.3, the ascent propellant mass is 47 mT and the ascent propulsion system weighs 7 mT.

Table 6.14. Summary of rough IMLEO estimates for human mission to Mars.

IMLEO (mT)	Earth departure propulsion	Mars descent and landing	ISRU
730	NTR from LEO	Aero-assist	Yes
1,000	NTR from LEO	Aero-assist	No
990	Chemical (LH$_2$/LOX)	Aero-assist	Yes
1,390	Chemical (LH$_2$/LOX)	Aero-assist	No
1,960	NTR from LEO	Chemical (CH$_4$/LOX)	Yes
2,880	NTR from LEO	Chemical (CH$_4$/LOX)	No
2,660	Chemical (LH$_2$/LOX)	Chemical (CH$_4$/LOX)	Yes
4,000	Chemical (LH$_2$/LOX)	Chemical (CH$_4$/LOX)	No

    Hence, the mass on the Mars surface prior to liftoff is 59 mT. The IMLEO required to deliver 59 mT to the Mars surface is $9.3 \times 59 = 549$ mT (if ISRU is not used). If ISRU is used, the mass delivered to the Mars surface is the sum of the capsule, propulsion, and ISRU masses $= 5 + 7 + 4 = 16$ mT and the corresponding IMLEO is $9.3 \times 16 = 149$ mT.

Step 4: Transfer Mars Transit Vehicle (MTV) from LEO to Mars orbit and transfer from Mars orbit back to Earth. It requires about 1.5 mT in Mars elliptical orbit to transfer 1 mT on a transfer route toward Earth. For a 40 mT MTV, the required mass in Mars orbit is $1.5 \times 40 = 60$ mT, of which 40 mT is payload and 20 mT is Earth return propulsion and propellants. The gear ratio for transfer to Mars elliptical orbit from LEO is 4.7 using propulsion for MOI. Thus, the required IMLEO to transfer 60 mT from LEO to Mars orbit is $4.7 \times 60 = 282$ mT.

    If ISRU is not used, the cumulative IMLEO for all four steps is $279 + 279 + 549 + 282 = 1,390$ mT, or approximately 10 launches of a Heavy-Lift Launch Vehicle. Had we used ISRU, IMLEO would have been $279 + 279 + 149 + 282 = 992$ mT for a mass saving of about 400 mT.

    Had we not used aero-assist for entry, descent, and landing, but instead used LOX/methane propulsion, the gear ratio for transfer from LEO to the Mars surface in Steps 1, 2, and 3 would have increased from 9.3 to 31.2. Thus, the IMLEO would have been $936 + 936 + 1841 + 282 = 4,000$ mT if ISRU is not used, and $936 + 936 + 499 + 282 = 2,657$ mT.

    If a nuclear thermal rocket can be used for Earth departure from LEO, with a dry mass/propellant ratio[196] of 0.5, the required values of IMLEO are multiplied by a factor of 0.72. A summary of the foregoing estimates is given in Table 6.14.

    The absolute estimates of IMLEO in Table 6.14 are very crude and approximate, but the general trends from case to case are likely to be real.

---

[196] This ratio was defined as K in Section 3.2.2.

We may therefore draw these conclusions:

1   By any yardstick, the requirements for IMLEO in a human mission to Mars are formidable.

2   Full aero-assist is a critically needed technology for Mars human missions to reduce IMLEO by a significant amount. Development of aerocapture and aero-assisted descent for full-size vehicles is an urgent need.[197]

3   Using aero-assist at Mars, and chemical propulsion for Earth departure, ISRU eliminates approximately two heavy-lift launches.

4   Use of an NTR for Earth departure would provide a significant reduction of IMLEO if departure occurs from LEO. However, if the NTR had to be lifted to, say, 1,250 km prior to ignition, essentially all of the benefits would disappear.

---

[197] "Mars exploration entry, descent and landing challenges," R. D. Braun and R. M. Manning, *Journal of Spacecraft and Rockets*, Vol. 44, 310–323, 2007.

7

How NASA is dealing with return to the Moon

7.1 INTRODUCTION

Some of the important questions that can be asked in regard to NASA's plan to return to the Moon include:

- What size Launch Vehicle is optimum, and how should it be configured?
- How critical is returning to the Moon at the earliest opportunity?
- How should the mission architecture be configured, with particular emphasis on the need for rendezvous in space, and which assets should be delivered to the surface?
- Is the best rendezvous point low lunar orbit or a Moon–Earth libration point?
- Should the establishment of an outpost be preceded by a series of short sortie missions to validate systems?
- Should sortie missions be preceded by a series of robotic landers to validate technologies and prospect for resources?
- What role should ISRU play?
- How can the Moon–Mars connection be strengthened?

7.1.1 Need for a Heavy-Lift Launch Vehicle

In his 1999 lectures at the University of Wisconsin entitled: "Heavy Lift Launch for Lunar Exploration," Dr. Michael Griffin concluded that: "A heavy-lift launch vehicle is, if not strictly mandatory, highly desirable for lunar operations." By any yardstick, the amount of mass that must be sent to LEO to enable a round trip to the Moon is clearly going to be in the range 150 mT to 200 mT, and use of a Heavy-Lift Launch Vehicle provides economies of scale as well as elimination (or minimization) of on-orbit assembly. Robert Zubrin[198] has provided evidence that the cost per kg to LEO decreases with Launch Vehicle size, all other factors being equal.

NASA appears to be committed to development of a Heavy-Lift Launch Vehicle. However, it is still not clear what lifting capability or configuration of stages is optimum. The ESAS Report seems to regard 125 mT to LEO as a fundamental limit to Launch Vehicle capability, and this figure is used 32 times in the ESAS Report. For example, it says in several places that "a Cargo Launch Vehicle (CaLV) capable of delivering at least 125 mT to Low Earth Orbit (LEO)" will be developed for lunar exploration.[199] In the section of the ESAS Report dealing with Launch Vehicles, it says: "The ESAS LV heavy-lift CaLV is recommended to provide the lift capability for lunar missions. It is approximately 357.5 ft tall and is configured as a stage-and-a-half vehicle composed of two five-segment RSRMs and a large central LOX/LH_2-powered core vehicle utilizing five RS25 SSMEs. It has a gross liftoff mass of approximately 6.4M lbm and is capable of delivering 54.6 mT to TLI, or 124.6 mT to 30×160 nmi orbit inclined 28.5 deg."[200]

For Mars, the ESAS Report says in at least two places that 100 to 125 mT to LEO[201] is the required capability, although in many other places it uses only the figure 125 mT.

The possibility that perhaps 125 mT to LEO may not be adequate for the Moon or Mars does not seem to be considered in the ESAS Report. The requirements for IMLEO depend on a number of things, but particularly the vehicle masses and the values of Δv for various transfers involved in a lunar mission. These are discussed in Section 7.3.

The expected value of IMLEO for human missions to Mars is somewhere in the range 800–4,000 mT, depending on the assumptions made, with 1,400 mT being a likely value. At 125 mT per launch, that requires 12 launches for a Mars mission. It seems likely that a larger Launch Vehicle would lead to a less expensive and safer mission. But, NASA seems to be locked onto the 125 mT figure, perhaps because NASA believes that a human mission to Mars can be done with 446 mT in LEO (Table 6.4 or Figure 8.3, color section).

Another aspect of the Heavy-Lift Launch Vehicle is the staging. For lunar missions (as described in the ESAS Report) a significant burn of the second stage is required to place the lunar spacecraft into LEO. This so-called sub-orbital burn must supply a Δv of about 3.0 km/s. The subsequent Earth departure burn for lunar missions must supply a Δv of about 3.1 km/s. There would be a mass advantage in using two separate propulsion systems for these two burns, in which case the mass of the dry propulsion system for the sub-orbital burn would be jettisoned in orbit, and it would not have to be accelerated during the Earth departure burn. However, it would be more expensive to develop two propulsion stages than a single larger stage, so ESAS plans to use a single propulsion stage for both the sub-orbital burn and the Earth departure burn. Evidently, ESAS placed higher emphasis on cost control than

---

[198] *Review of NASA Lunar Architectures*, Robert Zubrin, Informal Report, Pioneer Astronautics, January 10, 2005.
[199] *ESAS Report*, pp. 13, 72,
[200] *ESAS Report*, p. 374.
[201] *ESAS Report*, pp. 51, 76, 385.

mass efficiency. Because the mass in LEO includes extra mass for the sub-orbital stage, the mass delivered to LEO is actually greater than 125 mT, but the mass in excess of 125 mT is assigned to the propulsion stage and is not available for payload use. A rough estimate is that this mass is about 13 mT, so ~138 mT is delivered to LEO including about 13 mT of propulsion stage that theoretically could have been jettisoned.

As we have seen, ESAS seems to have stuck rigidly to a 125 mT IMLEO capability despite the fact that it is inadequate for lunar missions and it is grossly inadequate for Mars missions. If they staged the sub-orbital and Earth departure burns, they could have improved the mass situation somewhat.

7.1.2   Urgency in returning to the Moon at the earliest possible opportunity

It seems evident from the ESAS Report and other NASA plans, that NASA is determined to return humans to the Moon at the earliest possible opportunity. Almost all of NASA's efforts are aimed at designing vehicles and propulsion systems based on current capabilities, while development of advanced capabilities has been placed onto a side-track that might possibly contribute late in the campaign. It is particularly noteworthy that, after making a big point that use of LOX as an ascent propellant has the benefits of enabling the most near-term form of ISRU and creating a direct connection to Mars mission technology, NASA subsequently replaced the LOX/methane propulsion system with one that uses space-storable propellants, because of the cost and time required for development of the LOX/methane propulsion system. However, as noted in the previous section, with mass already a major problem for lunar missions, use of a less efficient space-storable propulsion system is likely to create all sorts of problems for lunar mission designers.

Aside from LOX/methane propulsion systems, it is more generally evident that Michael Griffin's plan is to assemble workable systems using currently available technology to put humans on the Moon at the earliest possible date. Plans for use of robotic precursors to prospect for resources, and validate capabilities, have been minimal. Finally, during early 2007 the entire robotic precursor program was scrapped.

7.2   MOON–MARS CONNECTION

7.2.1   The Moon as a means of risk reduction for Mars

W. W. Mendell discussed the Moon–Mars connection in regard to the 2004 President's Vision for Space Exploration:[202] "The President asks NASA to conduct human expeditions to Mars, after gathering adequate knowledge about the planet and after successfully demonstrating sustained human exploration missions to the Moon ... A central theme to the declaration is the execution of lunar missions and surface activities for the purpose of learning how to successfully carry out human

---

[202] "Meditations on the new space vision: The Moon as a stepping stone to Mars," *Acta Astronautica*, Vol. 57, No. 2-8, 676–83, July–October, 2005.

expeditions on Mars." Mendell claims that "Just how the Moon will be 'used to go to Mars' and why it will be 'used to go to Mars' has been the subject of a great deal of misunderstanding and has been misrepresented to bolster the view that the Vision for Space Exploration is unrealistic. Particularly problematic is a sentence in the President's speech: 'Spacecraft assembled and provisioned on the Moon could escape its far lower gravity using far less energy and thus far less cost.' While it is true that energy required for a lunar launch is much less than that for a terrestrial launch, elementary systems analysis shows that building a launch complex on the Moon solely for human expeditions to Mars is not practical." Mendell suggested: "this statement came from a misunderstanding by the speechwriter and managed to slip through proofreading."

Mendell dealt in particular with two directives in the Vision: "Undertake lunar exploration activities to enable sustained human and robotic exploration of Mars and more distant destinations in the solar system;" and "Use lunar exploration activities to further science, and to develop and test new approaches, technologies, and systems, including use of lunar and other space resources, to support sustained human space exploration to Mars and other destinations." He also noted another relevant directive from the Vision for Space Exploration: "Develop and demonstrate power generation, propulsion, life support, and other key capabilities required to support more distant, more capable, and/or longer duration human and robotic exploration of Mars and other destinations."

Mendell discussed the first President Bush's 1989 Space Exploration Initiative (SEI). Mendell discussed the question of whether to use the Moon as a stepping-stone to Mars or whether to go directly to Mars. He points out that "many voices in the space community called for bypassing the Moon to concentrate on human expeditions to Mars." This was partly because they "characterized the Moon as boring when compared to Mars," and the fear that the space program "would 'get stuck' on the Moon. In other words, NASA and its aerospace industry clients would create such an investment in lunar missions and facilities that they would find excuses to postpone Mars missions indefinitely."[203] Mendell dismissed these concerns because of the "dramatic disparities in the technical resources that must be brought to bear to conduct the two programs" as well as other major differences such as restricted launch windows and minimum mission duration. However, this would seem to beg the question of how Moon expeditions are supposed to enable Mars expeditions if they are so different?

Mendell then proceeded to discuss the Moon-vs.-Mars debate in terms of four risk issues:

- Uncertainty in assuring the physiological, medical, and psychological health and performance of the crew.
- Lack of experience with mission operations of the scale and scope of a human expedition to Mars.

---

[203] It is perhaps ironic that the new space initiative by the second President Bush seems to be bogging down in 2006 on the Moon just as predicted in the first space initiative.

- Reliability and maintainability of the hardware and software systems for a Mars mission scenario of 1,000-day duration, including a 500-day stay on Mars, lacking an abort-to-Earth capability.
- Political viability—public disenchantment with long periods of apparent inactivity with lack of demonstrable tangible accomplishments in the interim.

Mendell claimed: "A lunar program of human missions can provide a venue for mitigating all these risk categories." He argued that there will not be sufficient time and funding to adequately test Mars systems at Mars, and therefore, even though testing at the Moon may not be a perfect simulation for Mars, it is the only affordable approach for reducing risks involved in complex space systems for long durations. He also claimed that lunar missions can be implemented with far less delay, thus mitigating the political risk associated with a long-drawn-out Mars development program. He then argued that the principal goal of the lunar program should be to mitigate risk "at all levels but particularly with respect to human performance, mission operations, and system reliability. All three of these risk categories are driven by the extreme duration of the Mars voyage, the lack of abort-to-Earth options, and the absence of logistical support. Therefore, the ultimate objective of the lunar program is … a mission scenario [in which] design teams, operation teams, management teams, and technology levels deal robustly with those issues. Such a scenario is a physical facility where a crew of at least six spends at least a year out of sight of the Earth, i.e., on the lunar far side. The lunar far side location is critical to mimic the psychological isolation that will face the Mars explorers."

He accepted that lunar sorties should precede establishment of an outpost that would test and validate a variety of technologies and systems prior to establishment of the far side base.

Finally, he discussed a lunar exit strategy, acknowledging "the concern of many people that a lunar program will be self-perpetuating and human expeditions to Mars will be postponed indefinitely." He insisted that: "The human lunar program must incorporate an exit strategy at its initiation. Decisions must be made as to the fate of any habitats, rovers, power stations, or resource extraction plants once the exploration program emphasis shifts away from the Moon."

Dr. Mendell's arguments make a good deal of sense at a high level. However, the devil is in the details, and the question of how much Mars risk is actually mitigated by lunar precursors remains quite murky. Furthermore, there are some technologies such as aero-assisted entry, descent, and landing at Mars that are critical to mission success, that cannot be tested at the Moon, and which will require two decades of time and many billions of dollars to implement. And, finally, there is no indication that Dr. Mendell's proposition of a long-term base on the far side of the Moon will be implemented.

7.2.2 ISRU as a stepping-stone from the Moon to Mars

The 2005 NASA Vision for Space Exploration clearly views the human return to the Moon as a stepping-stone to Mars. For example, one of the five elements of NASA's vision statement is:

"Extend human presence across the solar system, starting with a human return to the Moon by the year 2020, in preparation for human exploration of Mars and other destinations."

One of NASA's main strategic objectives includes the statement:

"As a stepping-stone to Mars and beyond, NASA's first destination for robotic and human exploration is the Moon."

The Moon–Mars Science Linkage Science Steering Group (MMSSG) was formed within the oversight of the Mars Exploration Program Analysis Group (MEPAG) in response to a request from NASA HQ to develop an analysis of the science-based activities on the lunar surface that would benefit the scientific exploration of Mars. Their report[204] says:

"[Within] the new solar system exploration initiative proposed by the President of the United States, the scientific linkages between Moon and Mars have become increasingly important. Owing to the proximity of the Moon to Earth, there are important technological and scientific concepts that could be developed on the Moon that will provide valuable insights into both the origin and evolution of the terrestrial planets and be fed-forward to the scientific exploration of Mars ... The committee also identified important technological demonstrations that would be instrumental to the scientific exploration of the Moon and Mars, and important for eventual manned occupancy of both planetary bodies."

The MMSSG emphasized that "Water is an important resource for human exploration missions and can be used both for life support and for fuel." The MMSSG goes on to say:

"The lunar and Martian regoliths contain potential components of rocket propellants that could be produced in-situ, avoiding the need to transport them from Earth. Water and carbon compounds, if present at the lunar poles, solar wind hydrogen and carbon in the regolith, and oxygen produced by reduction of ilmenite or pyroclastic glass are the principal potential sources on the Moon ... The extraction of propellants from regolith minerals would use similar approaches for both Moon and Mars, including excavation, thermal extraction, water electrolysis, and oxygen, hydrogen or methane liquefaction. A lunar propellant production demonstration could influence the design of human mis-

[204] "Findings of the Moon–Mars Science Linkage Science Steering Group," C. Shearer, D. W. Beaty, A. D. Anbar, B. Banerdt, D. Bogard, B. A. Campbell, M. Duke, L. Gaddis, B. Jolliff, R. C. F. Lentz *et al.*, Unpublished White Paper, 29 pp., posted October 2004 by the Mars Exploration Program Analysis Group (MEPAG) at *http://mepag/reports/index.html*

sions to Mars as well as demonstrate the validity of the concepts for propellant production from the regolith on Mars."

However, the similarities between lunar and Martian resources have been grossly exaggerated. Concentrations of solar wind deposited hydrogen and carbon are undoubtedly extremely low,[205] requiring processing of huge amounts of regolith to acquire significant amounts of H and C. Production of oxygen on the Moon by reduction of ilmenite or pyroclastic glass requires very high temperature processing of large quantities of regolith. Water ice may exist and might be accessible in shadowed craters near the lunar poles, but we have no firm evidence that this is a viable resource. Power supply is an apparent showstopper. By contrast, water deposits on Mars are present in much higher concentrations, and may exist as accessible near-surface ground ice in many locations. Carbon is readily available in the atmosphere of Mars, for which there is no analog on the Moon. The Sabatier/Electrolysis process provides a ready means to produce methane and LOX. In actuality, lunar ISRU will provide relatively modest support for Mars ISRU, and furthermore there are doubts that lunar ISRU will be implemented in any serious way. The following observations are made:

1. The availability of the atmosphere on Mars provides a ready supply of C and O. No such resource exists on the Moon.
2. Water is potentially an extremely valuable resource on the Moon and Mars.
3. Water is widely available on Mars in concentrations in the upper $\sim$1 m or regolith ranging from $\sim$8–10% by weight in the best equatorial areas to $\sim$100% in polar areas. At higher latitudes, the water exists as ground ice and is easy to process. At equatorial regions, the water might exist as ground ice or mineral hydration—we do not know at this time. Possible existence of liquid water at multi-kilometer depths on Mars remains a matter of speculation.
4. On the Moon, water may exist in local deposits of ground ice in polar shadowed areas. Finding, assessing, acquiring, and processing such resources appears to be more challenging than on Mars. However, there may be similarities in the acquisition and processing steps. There is no evidence that NASA will carry out the *in situ* prospecting missions needed to locate these deposits, and JSC has repeatedly implied [illogically] that such prospecting is not needed. Power remains a serious problem for exploiting these shadowed areas.
5. Acquisition of oxygen by heating silicate rocks on the Moon is very energy-intensive, and requires complex handling and processing of solids and gases in very high temperature reactors. Such a process, which has been widely suggested for lunar ISRU, appears to be far more challenging than other ISRU approaches. It is not clear how regolith would be inserted and removed from such reactors.
6. Acquisition of resources deposited by the solar wind on the Moon is difficult to appraise due to lack of data. However, concentrations appear to be very low,

---

[205] Measured in 10s of parts per million.

which requires processing extremely large amounts of regolith, making the process extremely inefficient with very high power requirements.

Based on these observations, the following conclusions are drawn:

- Lunar ISRU is far more challenging than Mars ISRU, and Mars ISRU provides much greater mission benefits.
- One fundamental technology overlap between lunar and Mars ISRU is the need to liquefy and store cryogenic propellants. However, the thermal environments are quite different.
- If lunar polar ice can be excavated for ISRU, there may be overlaps between the Moon and Mars in acquisition and processing technologies of ground ice.
- The less tangible benefit of experience in implementing and operating a remote ISRU installation might prove to be the greatest benefit of lunar ISRU for Mars ISRU.

Despite these possibilities, it appears likely that the development and demonstration of lunar ISRU will not provide the critically needed validations required to assure that Mars ISRU is feasible and desirable. There are benefits to Mars ISRU from carrying out lunar ISRU, but these benefits are not great enough to conclude that lunar ISRU is a necessary or even important precursor to Mars ISRU.

7.2.3   The ESAS viewpoint on the Moon–Mars connection

The ESAS Report uses the word "Mars" 297 times. However, relatively little attention was addressed to the Moon–Mars connection. "Mars forward technology demonstrations" were mentioned in a few places, and briefly discussed. The ESAS Report admits that:

> "The short duration of lunar sortie crew missions will limit operational similar-ities to Mars surface exploration. However, valuable information will be gained in the areas of geologic field techniques, tele-operation of robotic systems and possibly crew mobility systems, and dust mitigation and planetary protection strategies. The small number of surface systems required for sortie missions also limits the demonstration of technologies linked to Mars exploration. Technologies that can be demonstrated on lunar sortie crew missions include: oxygen–methane rocket propulsion, EVA suits and portable life support, surface mobility such as un-pressurized rovers, long-lived scientific monitoring packages, ISRU technologies, a small long-lived power supply, and sealed sample containers.

There is no doubt that Mars missions will benefit from preceding lunar missions. The question is how much, and is it enough? Merely mentioning that there are some benefits does not answer this question.

The ESAS Report also says in regard to "Mars-forward testing" that: "although the Moon and Mars are two very different planetary environments, the operational techniques and exploration systems needed to work and live on both surfaces will have similar strategies and functions. While it is not likely that exact 'copies' of Mars-bound systems will be operated or tested on the Moon, it is very likely that components and technologies within those systems ultimately destined for the surface of Mars will undoubtedly find their heritage based on lunar surface operations. Therefore, the philosophy is to do what is proper and required for the lunar environment and use the knowledge, experience, and confidence gained from lunar operations as the foundation for the design of surface systems for Mars and other destinations in the solar system."

The ESAS Report says: "Two important operational techniques that should be developed on the Moon are crew-centered control of surface activities and tele-operation of robotic explorers from a central planetary outpost. As Mars is very distant from the Earth–Moon system, and due to the speed of light, one-way communications with Mars can take up to approximately 20 minutes. Communications with a human crew on Mars will be hindered by this time delay, and the crew will need to be able to operate in an autonomous mode without constant supervision from Earth. As this is different from the way human space missions have been conducted to-date, the Moon will provide the opportunity to transition from Earth-centered to crew-centered control of daily operations."

The ESAS Report also says: "Regardless on which planetary surface human crews are living and working, similar supporting infrastructure will be necessary. Habitation, power generation, surface mobility (i.e., space suits and roving vehicles), surface communication and navigation, dust mitigation, and planetary protection systems are all common features that will first be provided on the Moon. Repetitive and long-term use of these systems at a lunar outpost will allow the design of these systems and their components to be refined and improved in reliability and maintainability—system traits that will be essential for the exploration of destinations much more distant from Earth. The presence of an atmosphere and a stronger gravitational field on Mars, however, will require modifications to some components of the planetary surface systems used on the Moon, particularly those systems that directly contact the Martian surface or atmosphere. Lunar surface systems that are internal to a pressurized environment, such as a habitat's regenerative life support system, are more likely to be directly applicable to Mars with fewer or no modifications."

The above claims made by the ESAS Report for a Moon–Mars connection seem reasonable as far as they go. The problem is that they do not go far enough. They do not deal with the question of whether the risk reduction for Mars missions by virtue of these lunar activities is sufficiently great to justify the investment in the lunar program. Clearly, the lunar program can provide benefits to the Mars program. But is the benefit worth the investment?

7.2.4   A European viewpoint

A paper by a European group provides some insights.[206] Their viewpoint is that: "A main goal of the lunar missions is the development, testing and updating of procedures and operations that will be necessary for a human Mars mission ...

---

[206] *Proposed Lunar Surface Operations in Preparation for a Human Mars Mission,* Tina Büchner, Melinda M. Gallo, Robert Gresch, Inka Hublitz, and Michael M. Schiffner, AIAA 2005-2699.

Examples include ground operations, extraterrestrial construction, housekeeping, planetary science, running a lunar mission on a Martian day cycle, introducing an artificial time delay into radio communication with Earth, or implementing planetary protection and decontamination procedures." They suggest that the Moon is a good place to test out medical procedures, emergency and rescue procedures, and communications, but it is not clear why this has to be done on the Moon rather than on Earth. Operating a Habitat on the Moon with life support systems would be a valuable Mars-forward experience. Extra-vehicular activities on the Moon would also have likely important value for subsequent Mars missions.

It is suggested that: "the Moon also offers the best analog for researching and developing radiation management. Data regarding the effectiveness of shielding materials in reducing biological dose during lunar missions may be applicable for radiation protection during a human Mars mission. Other requirements such as establishing acceptable dose limits, dose monitoring, and treatment of acute radiation exposure will also be developed for long-stay missions. Another profitable outcome could be the establishment of a procedure to build a safe haven on the surface itself."

They point out that lunar missions provide a chance to study physiological deterioration and psycho-social problems that arise in confined Habitats. They also say:

> "Lunar surface-stay studies monitoring crew member condition (including muscle atrophy, endurance and bone density) as well as other biological experiments can be used to improve countermeasures and physical training programs ... and may allow better understanding of the degree of physiological deterioration and determination of the best countermeasure regime for a surface stay on Mars ... The Moon provides a better analog than LEO or terrestrial facilities [for psycho-social studies] because it provides the two crucial stressors of complete isolation and reduced gravity."

All of these points seem reasonable, but, again, does it justify the lunar program?

7.3 LUNAR ARCHITECTURE OPTIONS

7.3.1 Zubrin's analysis

In an informal report, Robert Zubrin[207] carried out a comparison of various mission options for delivering a crew to the Moon. His description appears to pertain to a lunar sortie, and concomitant deliveries of cargo would have to be coupled with crew deliveries to create an outpost.

---

[207] *Review of NASA Lunar Program Architectures,* Robert Zubrin, Pioneer Astronautics, *Zubrin@aol.com* January 10, 2005.

The following vehicles are utilized:

CEV = Crew Exploration Vehicle—used to transport crew from Earth to LEO or LLO, depending on the mission.

EDS = Earth Departure System—propulsion module for transferring payloads in LEO to more distant points.

LSAM = Lunar Surface Access Module—used to transfer crew to lunar surface and ascend from the lunar surface.

Zubrin conjectured four architectures for transportation to and from the Moon. The first architecture is the so-called "point of departure" (POD) mode that seems to have been initially selected by NASA mission planners as a baseline. It involved four launches and four rendezvous operations, and has the virtue of requiring the smallest Launch Vehicle (LV), but requires the most maneuvering in space, and therefore is the most risky. The other modes successively reduce the number of launches and rendezvous operations, but require more capable LVs.

The following paragraphs provide short descriptions of the four architectures (also see Figure 7.1).

(i) *POD*: Four launches are carried out in parallel to LEO: (1) LSAM (15 mT), (2) EDS for LSAM (27 mT), (3) CEV (12.3 mT), and (4) EDS for CEV (33 mT). The LSAM and its EDS are joined in LEO, and the CEV and its EDS are joined in LEO. The EDS vehicles transport the LSAM and the CEV to LLO, where the crew transfers to the LSAM to land on the Moon. On return, the crew transfers from the LSAM to the CEV waiting in lunar orbit. The largest single item launched is 33 mT.

(ii) *Two-launch, two rendezvous*: Two launches are implemented in parallel to LEO: (1) LSAM and its EDS (39.3 mT), and (2) the CEV and its EDS (48 mT). The LSAM, CEV, and two EDSs are joined in LEO, and the combined EDS transports the LSAM and the CEV to LLO, where the CEV remains in lunar orbit while the LSAM lands the crew on the Moon. On return, the crew transfers from the LSAM to the CEV waiting in lunar orbit. The largest single item launched is 48 mT.

(iii) *One-launch, one rendezvous*: A single launch puts a larger EDS, LSAM, and CEV combination in LEO (87.3 mT). The EDS sends the CEV and LSAM to LLO, where the CEV separates and the LSAM descends to the surface. On return, the crew transfers from the LSAM to the CEV waiting in lunar orbit.

(iv) *One-launch, no rendezvous*: A single launch puts the largest EDS, LSAM, and CEV combination in LEO (119.3 mT). The EDS sends the CEV and LSAM to LLO, where the CEV and the LSAM descend to the surface. On return, the crew occupies the CEV that is lifted to LLO by the LSAM, where the CEV then departs for Earth under its own power.

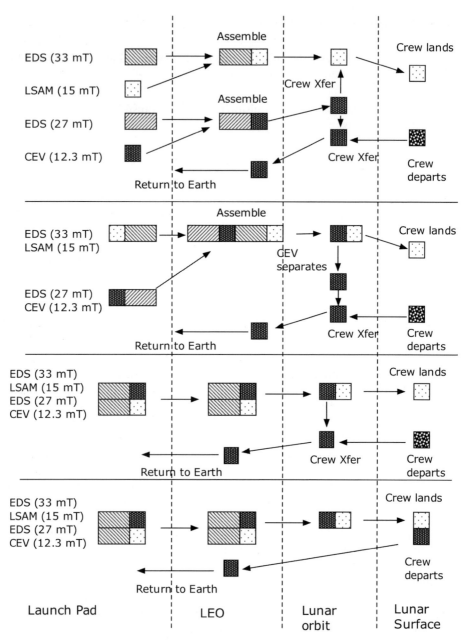

Figure 7.1. Four competing architectures for lunar sortie missions. The uppermost scheme involves four launches with two rendezvous in LEO and another rendezvous in lunar orbit. The second scheme involves two launches. The second scheme involves two launches with rendezvous in LEO. The third scheme involves one launch with the CEV remaining in lunar orbit. The fourth scheme involves one launch and sends the CEV to the lunar surface. [Based on *Review of NASA Lunar Program Architectures*, Robert Zubrin, informal report.]

These scenarios are illustrated in Figure 7.1. The mass estimates provided in Figure 7.1 were made by Zubrin. These illustrations are without the use of ISRU. If ISRU is used for ascent propellants, the fourth architecture benefits the most because it requires the most ascent propellants. According to Zubrin's estimates, without ISRU the value of IMLEO is the same for the first three architectures at 87.3 mT, whereas for architecture 4 it is 120 mT. However, when ISRU is employed he estimates that IMLEO for the first three architectures drops merely to 72.1 mT, whereas for the fourth architecture it drops to 69.8 mT.

The mass estimates made by Zubrin were made prior to the ESAS Study and Zubrin did not have access to the ESAS Report at that time. Hence, his mass estimates are rather optimistic and the actual required values of IMLEO will be higher. Of particular note is the fact that for the LSAM Descent and Ascent Stages, he used an inert mass of 5,000 kg, and LOX/CH_4 propulsion with $I_{SP} = 370$ s and a stage dry-mass fraction of 0.1. His total wet mass of the LSAM was 15 mT. As Table 3.4 shows, the wet mass of the LSAM is actually 35 mT, more than double Zubrin's estimate, and the stage dry-mass fractions for descent and ascent are 0.13 and 0.22. Nevertheless, Zubrin's trends are undoubtedly realistic.

Clearly, the fourth scheme in Figure 7.1 has the virtue of no rendezvous and transfers in space, and benefits the most from use of ISRU. However, it requires the largest Launch Vehicle. When ESAS considered this scheme with their higher mass estimates for the LSAM, they concluded that the required Launch Vehicle would have to deliver something of the order of 200 or more mT to LEO, and that was rejected by ESAS.

ESAS decided early in lunar mission planning that the crew would launch separately from the cargo and rendezvous the two in LEO. It is not exactly clear why this decision was made, but it seems likely that the rationale is that by doing this the very high safety and abort capabilities required for launching humans into space can be relegated to a smaller Crew Launch Vehicle (CLV) while a much larger Cargo Launch Vehicle (CaLV) need not utilize all of the "bells and whistles" to make it man-rated. However, additional risk is introduced by requiring the crew to rendezvous and link to the Earth departure stage sent to LEO by the CaLV. We therefore take it as a fundamental axiom that a minimum of two launches will be used, one for the crew, and the other(s) for cargo. Zubrin's analysis was prepared before this decision was made by ESAS.

7.3.2   MIT analysis

An MIT Team[208] has independently investigated a large number of options for lunar architecture.

Each of these architectures was evaluated using parametric models for a set of technology options and screened using a series of proximate metrics related to cost and risk. They also employed a "Mars-back" approach of first looking at the requirements for Mars exploration systems and then projecting their capabilities towards lunar exploration systems. The goal was to develop a common set of systems to enable affordable and sustainable exploration of both locations. Using this

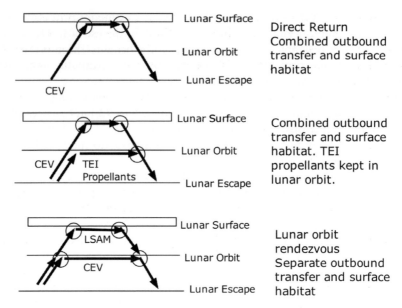

Figure 7.2. Lunar architectures of interest from the MIT Study. Crew transfers between vehicles are shown by circles.

approach, they identified three lunar architectures that appear to be of particular interest as potential candidates for the lunar crew transportation system for short-duration lunar sortie missions. The three architectures are described in Figure 7.2. In terms of destination, two of them involve the CEV descending to the lunar surface and one has the CEV remaining in a staging location in the lunar vicinity.

In the lunar direct return architecture, the crew transfers to the surface of the Moon in the CEV, makes use of the CEV for surface operations (potentially including a transfer to a surface Habitat for long-duration missions), and then returns directly to the Earth from the surface of the Moon in the CEV. Propulsion stages are used in series to accomplish the required maneuvers and remain with the CEV until expended. The second architecture also has the CEV proceeding to the surface of the Moon.

However, in this case the propulsion system for trans-Earth injection (TEI) remains in lunar orbit, so that it does not have to be transported down and up.

[208] *Crew Exploration Vehicle Destination for Human Lunar Exploration: The Lunar Surface,* Paul D. Wooster, Wilfried K. Hofstetter, and Edward F. Crawley, AIAA 2005-6626; also see *From Value to Architecture: The Exploration System of Systems—Phase I,* P. Wooster, Final Presentation at JPL, August 23, 2005; *Paradigm Shift in Design for NASA's Space Exploration Initiative: Results from MIT's Spring 2004 Study,* Christine Taylor, David Broniatowski, Ryan Boas, Matt Silver, Edward Crawley, Olivier de Wec, and Jeffrey Hoffman, AIAA 2005-2766; "The Mars-Back Approach: Affordable and Sustainable Exploration of the Moon, Mars, and Beyond Using Common Systems," P. D. Wooster, W. K. Hofstetter, W. D. Nadir, and E. F. Crawley, *International Astronautical Congress, October 17–21, 2005.*

After ascending from the surface, the CEV performs a rendezvous in lunar orbit to acquire the TEI propulsion system that sends the CEV on its way toward Earth. The third architecture is a lunar orbit rendezvous with separate Lunar Surface Access Module. In this architecture, "the CEV proceeds only as far as lunar orbit. At that point the crew transfers to a separate Lunar Surface Access Module, descends to the lunar surface, and is supported by the LSAM on the surface for short periods of time or until the crew transfers to a long-duration surface habitat. Upon concluding the surface mission, the crew ascends to lunar orbit in the LSAM, rendezvous with and transfers to the CEV, and subsequently returns to Earth." As it turns out, ESAS has selected this architecture.

The MIT Team notes that there are two primary options: (1) whether the architecture has the CEV go to the surface or whether a separate LSAM is used and (2) whether lunar orbit rendezvous is used or a direct return to Earth is performed. While sending the CEV to the surface is traditionally associated with the direct return architecture, sending the CEV to the surface is also compatible with orbital rendezvous as in the second architecture above (see Figure 7.2).

The MIT Team tested the widely held belief that architectures in which the CEV goes to the lunar surface—in particular, direct return architectures—incur a substantial mass penalty when compared with Lunar Orbit Rendezvous architectures.

In determining the IMLEO of human space exploration architectures, the mass of the crew compartments involved are of utmost importance as they are the primary payload and thus determine the size of the propulsion stages. The MIT Study assumed that the LSAM Habitat mass was 6,800 kg, the orbit CEV was 9,150 kg, and the landed CEV was 10,050 kg. They were based on a Draper/MIT analysis and study. The surface mission duration was 7 days and a crew size of four was assumed, Propulsion stages were sized based upon these crew compartments to determine the IMLEO of each of the three architectures for a series of mission and technology options. In all cases, Earth departure was based on LOX/LH$_2$ with an $I_{SP} = 460$ s.

The MIT Study found that, just as in the Apollo case, the lunar orbit rendezvous (LOR) architecture has a lower IMLEO than direct return for equatorial access if space-storable propulsion ($I_{SP} \sim 316$ s) is used beyond Earth departure, and ISRU is not used. However, if global access is demanded, then TEI propulsion rendezvous has a lower IMLEO than LOR. But, MIT points out that when methane/oxygen propulsion is introduced, and particularly so with a LOX/LH$_2$ lander, the IMLEO of the lunar direct return architecture is only slightly higher than that of the LOR architecture. The TEI propulsion rendezvous IMLEO is lower than the LOR IMLEO with cryogenic propellants. The landed CEV architectures benefit greatly from use of ISRU because of their higher ascent propellant requirements. However, no allowance was made for the mass of the ISRU system that would offset some of the benefit of ISRU relative to non-ISRU missions. While ISRU provides a significant benefit, MIT felt that it was unlikely that this technology would be used on initial lunar missions (and the ESAS Report shows that MIT was right). Nevertheless, MIT concluded there are significant benefits associated with use of cryogenic propellants downstream of Earth departure.

Table 7.1. MIT-estimated values of IMLEO for three architectures and variable parameters.

Parameters				IMLEO (mT) for each architecture		
I_{SP} (s) for descent	I_{SP} (s) elsewhere	Lunar access	ISRU used?	Direct return	Landed CEV w/ TEI propulsion rendezvous	Lunar orbit rendezvous
316	316	Equatorial	No	213	161	144
316	316	Global	No	213	173	178
362	362	Global	No	163	140	149
430	362	Global	No	150	131	143
430	362	Global	Yes	89	97	119

Having discussed IMLEO variations with architecture, MIT then went on to include factors in this analysis such as mission risk, crew safety, and overall development and operational cost.

Mission risk and crew safety

The three architectures discussed previously have notable differences regarding their operational sequences and the number and type of mission-critical events and crew safety hazards. These can be translated into differences in risk and crew safety. Each of the following features was analyzed by MIT:

- Crew safety for an Apollo 13 style emergency.
- Rendezvous in lunar orbit.
- Docking in lunar orbit.
- Hardware accessibility in the lunar vicinity.

MIT concluded that if adequate power, ECLSS, and propulsion subsystem redundancy is provided, the direct return architecture is the safest, with the LOR architecture being second in safety, and the TEI propulsion rendezvous being least safe.

MIT then reviewed development and production costs.

MIT discussed the merits of separate CEV and LSAM vehicles vs. a CEV that goes to the surface with no LSAM. It was concluded that the development cost of a single vehicle would be lower than that for two vehicles.

While the primary analysis presented by MIT focused on the crew transportation solution for lunar exploration, MIT pointed out that, to establish an outpost, long-duration Habitats and other surface systems will need to be emplaced. Cargo delivery to the lunar surface on the order of 25 to 30 mT may be required to meet such needs. There would be an advantage if the system to deliver payloads of this magnitude will be required for the crew transportation system, in which case the same system could be used for cargo delivery. Table 7.2 presents the payload capacity of the Descent

Table 7.2. MIT-estimated values of payload delivered to the surface for three architectures and variable parameters.

Parameters				Payload delivery to the surface (mT) for each architecture		
I_{SP} (s) for descent	I_{SP} (s) elsewhere	Lunar access	ISRU used?	Direct return	Landed CEV w/ TEI propulsion rendezvous	Lunar orbit rendezvous
316	316	Equatorial	No	33	23	16
316	316	Global	No	33	23	16
362	362	Global	No	28	21	15
430	362	Global	No	28	21	15
430	362	Global	Yes	17	14	10

Stage employed in each of the three architectures for the mission and technology options analyzed previously. The lander capacity for the direct return architecture is consistently the highest and, other than in the ISRU case, should be able to deliver large cargos to the lunar surface with the same system.

MIT also looked at the applicability of the architectural elements to future Mars missions and concluded that the direct return vehicles would have great direct applicability to a direct return mission from Mars. The other architectures were not examined in as much detail.

The MIT study concludes: "architectures in which the CEV travels to the lunar surface are comparable to Lunar Orbit Rendezvous architectures ... With the introduction of ISRU, CEV to surface architectures would offer significant mass savings relative to LOR. From a risk perspective, the Lunar Direct Return architecture offers significant advantages as all mission assets are accessible by the crew for inspection and maintenance, no rendezvous or docking are required for return to Earth, and no assets must be operated autonomously in lunar orbit. These advantages will become particularly relevant in the long-duration lunar missions envisioned to prepare for expeditions to Mars. From a cost perspective, having the CEV travel to the lunar surface offers significant savings by eliminating the development of a second crew compartment with all of its life-cycle cost implications in development and operations, including design, testing, launch processing, flight control, software maintenance, and logistics. The Lunar Direct Return architecture offers the additional benefits of eliminating one propulsion stage and allowing the use of a common lander for both the crew transportation system and the emplacement of large surface assets, again with significant life-cycle benefits ... Given the numerous benefits of the Lunar Direct Return architecture, we recommend selecting it as the baseline architecture for human lunar exploration. Whether or not the Direct Return architecture is initially the baseline, we recommend developing the ISS CEV such that it is extensible to lunar surface missions should such architectures be selected in the future."

7.3.3   The Broad Study

In the summer of 2004, the Requirements Division (RQ) of NASA's Exploration Systems Mission Directorate (ESMD) commissioned two major lunar mission trade studies to evaluate the relative benefits of alternative manned lunar mission architectures. The two efforts, the Lunar Architecture Broad Trade Study[209] and the Lunar Architecture Focused Trade Study[210] examined a variety of mission options and provided estimates of the initial launch mass to low Earth orbit (IMLEO) necessary to meet short- and long-term mission objectives involving the transport of humans to the surface of the Moon and their return to Earth. The studies used different assumptions and were based on different tools, but because the Broad Study included a subset of the options considered in the Focused Study—particularly, using chemical propulsion throughout—there was significant overlap in their studies, even though the results showed some differences. These studies provided a starting point for the ESAS study of 2005. However, both reports are very voluminous and, although executive summaries are provided, it is very difficult to summarize their extensive work. Neither report was released to the public. Similarly, the ESAS Report is 758 pages long with no table of contents. Finding things in this report is difficult.

One thing that immediately stands out from reading the various reports on lunar architecture and comparing them with the ESAS Report[211] is that trajectories from Earth to the Moon are complicated, particularly in regard to the all-important dependence of Δv for various steps on parameters such as (1) the range of access to the lunar surface (global, equatorial only, poles, . . .), (2) the flexibility of departure and return dates and times (anytime return vs. required stay times), and (3) the need for Δv-costly plane changes both in going to the Moon and returning, and how this depends on the latitude of the landing site. Since the values of Δv for various steps determine the mass of propellants needed, this is very important in determining IMLEO.

The Broad Study provides an extensive discussion of Δv requirements for various steps involved in lunar transfers. All estimates are based on a 7-day stay with global access to all sites on the Moon using a simplified patched conic model.

Earth departure to lunar orbit (and back)

For Earth departure from LEO and transfer to lunar orbit, the Broad Study defines two values of Δv:

Δv_1 for Earth departure
Δv_2 for lunar orbit insertion

Depending on Δv_1 for Earth departure, Δv_2 for lunar orbit insertion will vary as shown in Figure 7.3. Clearly, one can obtain a minimum in both Δv values for

---

[209] *Broad Study*, ESMD-RQ-0006, 245 pp., September 1, 2004.
[210] *Focused Study*, ESMD-RQ-0005, 822 pp., October 22, 2004.
[211] *ESAS Report*: *http://www.spaceref.com/news/viewsr.html?pid=19094*

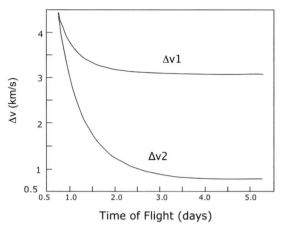

Figure 7.3. Dependence of Δv_2 for lunar orbit insertion on Δv_1 for Earth departure. [NASA Broad Study, *loc cit.*]

sufficiently long transit times ($>\sim$ 3.5 days). The minima are Δv_1 for Earth departure $\sim$3.07 km/s, and Δv_2 for lunar orbit insertion $\sim$0.78 km/s. However, the Broad Study recommends Δv_2 for Earth departure $\sim$3.15 km/s, and Δv_2 for lunar orbit insertion $\sim$1.07 km/s with a trip time of about 2.35 days. They also point out that departure from lunar orbit to Earth requires the same Δv values, but in the reverse order.

Earth departure direct to lunar surface

For Earth departure from LEO and direct transfer to the lunar surface, the Broad Study defines three values of Δv:

  Δv_1 for Earth departure
  Δv_2 for lunar direct descent and landing
  Δv_3 for maneuvering, hovering, and gravity losses during descent

They take Δv_1 for Earth departure to be, as before, 3.15 km/s. They then calculate Δv_2 for lunar direct descent and landing to be 2.76 km/s and they roughly "guesstimate" Δv_3 for maneuvering, hovering, and gravity losses during descent at 0.35 km/s.

Low lunar orbit to lunar surface

For descent from low lunar orbit ($\sim$100 km) to the lunar surface, the Broad Study defines three values of Δv:

  Δv_1 for de-orbiting
  Δv_2 for lunar descent and landing
  Δv_3 for maneuvering, hovering, and gravity losses during descent

They take Δv_1 for de-orbiting to be ~0.03 km/s. They calculate Δv_2 for lunar descent and landing to be 1.70 km/s and they roughly "guesstimate" Δv_3 for maneuvering, hovering, and gravity losses during descent at 0.35 km/s. The sum is 2.08 km/s.

Low lunar orbit to LEO

For transfer from LLO to LEO, the process is the reverse of LEO to LLO. Depending on the duration of the transfer, a range of possible Δv is possible. The Broad Study recommends:

Δv_1 for trans-Earth injection = 1.07 km/s
Δv_2 for Earth orbit insertion = 3.15 km/s

If aerocapture is used for Earth orbit insertion, a small (but unknown) Δv is required during the aerocapture process for orbit adjustment.

Lunar surface to LEO

For lunar surface departure transfer to LEO, the Broad Study says that the process is almost the reverse of Earth departure and direct transfer to the lunar surface. The only difference is that there is no hovering and maneuvering, but there is a gravity loss of roughly 0.2 km/s. Thus, they define three values of Δv:

Δv_1 for lunar surface departure
Δv_2 for trans-Earth injection
Δv_3 for gravity losses during ascent

They take Δv_1 for lunar surface departure and trans-Earth injection to be 2.76 km/s and Δv_2 for Earth orbit insertion to be 3.15 km/s and they roughly "guesstimate" Δv_3 for gravity losses during ascent at 0.20 km/s. If aerocapture is used for Earth orbit insertion, a small (but unknown) Δv is required during the aerocapture process for orbit adjustment.

Lunar surface to low lunar orbit

For lunar surface departure transfer to LLO (~100 km) the Broad Study says that the process is almost the reverse of lunar orbit to the lunar surface. The only difference is that there is no hovering and maneuvering, but there is a gravity loss of roughly 0.2 km/s. Thus they define three values of Δv:

Δv_1 for lunar surface departure
Δv_2 for lunar orbit insertion
Δv_3 for gravity losses during ascent

They take Δv_1 for lunar surface departure to be 1.70 km/s, Δv_2 for lunar orbit insertion to be 0.03 km/s, and they roughly "guesstimate" Δv_3 for gravity losses during ascent at 0.20 km/s. If—instead of a 100 km orbit—a 10,000 km orbit is chosen, then Δv_1 for lunar surface departure increases from 1.93 km/s to be 2.22 km/s.

Ascent from lunar surface and rendezvous in lunar orbit

Because the Moon rotates once per month with respect to inertial space, the landing site on the lunar surface will rotate out of the orbital plane of the orbiting spacecraft, which is essentially inertially fixed without propulsive maneuvers. Also, due to the lunar orbit around the Earth, the orientation of the lunar parking orbit will change with respect to the Earth–Moon line. Both of these effects are considered when determining the Δv required to transfer from the lunar surface (LS) to low lunar orbit (LLO), to account for the necessary plane change for the rendezvous between the Ascent Vehicle and the Orbiting Vehicle, and to ensure proper orientation of the departure asymptote from the Moon for the return to Earth. The exact calculation of the required plane change in any application depends on the landing site latitude and longitude and the length of stay on the surface. The worst case is a 90° plane rotation. If this rotation is performed at 100 km altitude, the required $\Delta v = 2.31$ km/s. This is a very large perturbation.

  Although large savings in Δv are possible by performing the plane change at high altitude, it may not be the best place for the lunar Ascent Vehicle and Orbiter to rendezvous. From a circular parking orbit of 100 km, the Orbiter requires $\Delta v_1 = 0.51$ km/s to inject into a 100 km × 10,000 km elliptical transfer orbit, $\Delta v_2 = 0.48$ km/s to perform a 90° plane change at 10,000 km altitude (the apoapse of the transfer orbit), and $\Delta v_3 = 0.51$ km/s to circularize back in a 100 km circular orbit where the rendezvous takes place. The total Δv needed to perform the maneuver is 1.50 km/s (vs. 2.31 km/s if the rotation is performed at 100 km altitude) and the total time required is 14 hours. A TEI burn of 1.07 km/s is applied to leave the circular lunar orbit on the way to LEO or direct Earth entry.

  A summary of Broad Study recommended values of Δv for various transfers is provided in Table 7.3.

  The Broad Study did not seem to overtly deal with the number of allowable launch days per month, and their requirement was stated as one launch per year. It is not clear, but it appears that launch is possible any day in this model; however, there is a price to be paid in the Δv required for plane change during rendezvous in lunar orbit.

7.3.4   The Focused Study

The Focused Study provides an extensive discussion of Δv requirements for various steps involved in lunar transfers with global access to all points on the Moon. In the Focused Study a number of variables were considered with particular emphasis on

Table 7.3. Lunar mission segment Δv (km/s) and time of flight (days). [NASA Broad Study, *loc. cit.*]

To ⇒		Earth surface (ES)		Low Earth orbit (LEO)		Earth–Moon L1		Low lunar orbit (LLO)		Lunar surface (LS)	
From ⇓		Δv	Δt	Δv	Δt	Δv	Δt	Δv	Δt	Δv	Δt
ES	Δv_1			9.27(a)	0.03						
	Δv_2			0.12(b)							
	Total			9.39							
LEO	Δv_1	0.10				3.07	3.92	3.15	2.32	3.15	2.36
	Δv_2	—				0.64		1.07		2.76	
	Loss	—				—		—		0.35(d)	
	Total	0.10				3.71		4.22(c)		6.26	
L1	Δv_1	0.64	3.92	0.64	3.92			0.08	2.66	0.08	2.66
	Δv_2	—		3.07				0.64		2.34	
	Loss	—		—						0.35(d)	
	Total	0.64		3.71				0.72		2.77	
LLO	Δv_1	1.07(e)	2.32	1.07	2.32	0.64	2.66			0.03	0.04
	Δv_2	—		3.15(g)		0.08				1.70	
	Loss	—		—		—				0.35	
	Total	1.07		4.22		0.72				2.08	
	Δv_1	0.56(f)									
	Δv_2	—									
	Total	0.56									
LS	Δv_1	2.76	2.36	2.76	2.36	2.34	2.66	1.70	0.04		
	Δv_2	—		3.15(g)		0.08		0.03			
	Loss	0.20(h)		0.20(h)		0.20(h)		0.20(h)			
	Total	2.96		6.11		2.62		1.93(i)			
	Δv_1							2.22	0.29		
	Δv_2							0.32			
	Loss							0.20(h)			
	Total							2.74			
	Δv_3							2.31(k)	0.58		
	Δv_3							1.50(l)			
	Δv_3							0.91(m)			
	Δv_3							0.99(n)			

(a) Elliptical orbit, (b) circularization at 400 km, (c) same for equatorial and polar orbit, (d) powered descent, (e) return from circular lunar orbit with direct aero-entry at Earth, (f) return from elongated elliptical lunar orbit with direct aero-entry at Earth, (g) propulsive Earth orbit insertion, (h) powered ascent, (i) ascent to 100 km (no plane change), (j) ascent to 10,000 km (no plane change), (k) 90° (worst case) plane change at 100 km and rendezvous, (l) 90° (worst case) plane change at 10,000 km and rendezvous at 100 km, (m) 90° (worst case) plane change at 10,000 km and rendezvous at 10,000 km, (n) rendezvous at 100 km, plane change at 10,000 km, remain in 100 × 10,000 km orbit (typically for polar landings).

Table 7.4. Focused Study estimates of Δv requirements for lunar missions (m/s). [NASA Focused Study, *loc cit.*]

Transfer or operation	1	2	3	4	5	6	7
	"Loitering" used to minimize Δv for Earth–Moon–Earth transfers			Plane change requires higher Δv for Earth–Moon–Earth transfers			Anytime departure and return
Δv requirements (m/s):							
Trans-lunar injection	3,074	3,074	3,074	3,074	3,074	3,074	3,074
Lunar orbit insertion	978	978	978	1,416	1,416	1,416	879
Descent to lunar surface	1,881	1,881	1,881	1,881	1,881	1,881	3,510
Ascent to lunar orbit	1,868	2,025	2,344	1,868	2,025	2,344	3,463
Lunar orbit rendezvous	100	100	100	100	100	100	100
Trans-Earth injection	865	865	865	1,410	1,410	1,410	864
Overall total Δv	*8,766*	*8,923*	*9,242*	*9,749*	*9,906*	*10,225*	*11,890*
Days/month:							
Launch opportunities	3	7	11	3	7	11	28
Lunar orbit departure	1	1	1	3	7	11	28
Total days lunar vicinity	23	23	30	5	9	13	28

options for cutting the mission short if something goes awry. The parameters that were included were:

- Length of surface stay (3, 7, or 11 days).
- Available days/month for launch from Earth (up to 28).
- Available days/month for lunar orbit departure (up to 28).
- Total days/month spent in lunar vicinity (up to 30).

As with the Broad Study, while acknowledging that the Focused Study addressed an admittedly complex and difficult subject (lunar trajectories and mission design), their descriptions, like many other NASA and JSC reports and publications, are almost impossible to follow. Table 7.4 provides a summary of some of their results. There is a uniform progression in overall total Δv requirements from column 1 to column 7. Columns 1–3 and columns 4–6 allow progressively increasing flexibility in launch and return dates. In addition, columns 1–3 utilize loitering in lunar orbit to minimize the required plane changes involved in transfer from LEO to LLO and back from LLO to LEO. The Focused Study does not elaborate on how this loitering step is carried out, but it seems likely that this involves (1) upon arrival, insertion into a lunar orbit with minimum plane change from the Earth departure trajectory, and (2) after rendezvous, delaying return to Earth until the orientation of departure plane and the ideal trajectory back to Earth are in consonance.

According to Table 7.4, the Focused Study used 1.42 km/s for lunar orbit insertion and 1.41 km/s for Earth departure from lunar orbit if there is no "loitering" after rendezvous. However, if "loitering" is employed, they were apparently able to avoid plane changes in Earth–Moon–Earth transfers, and thereby reduce requirements to 0.98 km/s for lunar orbit insertion and 0.87 km/s for Earth departure from lunar orbit. Thus, it appears that, depending on the allowable delay after rendezvous prior to Earth return, the required Δv for lunar orbit insertion and Earth departure can vary considerably. If it is required that launch from Earth or return from the Moon is possible at any time, then the requirements become those shown in column 7 of Table 7.4.

7.3.5   Comparison of Broad and Focused Study estimates of Δv

It is difficult to compare the estimates of Δv in the Broad and Focused studies because they utilize different scenarios.

The Broad Study provides specific estimates for all transfer steps except for the required plane change involved in rendezvous of the Ascent Vehicle with the Earth Return Vehicle in lunar orbit. The required Δv for the plane change depends on many factors including the duration of stay, the latitude and longitude of the landing site, and whether rendezvous takes place in low or high lunar orbit, but it can range up to a maximum of 2.31 km/s, a very high value.

The Focused Study apparently deals with the plane change requirement by building it into the required Δv values for ascent to lunar orbit and trans-Earth injection.

A summary of Focused Study and Broad Study estimates for Δv is provided in Table 7.5.

7.3.6   ESAS Report Δv estimates

A step-by-step review is provided next.

(a) Earth departure: trans-lunar injection

The Broad Study estimated $\Delta v \sim 3{,}150$ m/s based on a simplified trajectory model. The Focused Study estimated 3,074 m/s based on a more detailed model. The ESAS Report estimated 3,120 m/s on p. 443 and 3,150 m/s on p. 151. There seems to be good agreement on a value of $\sim$3.1 km/s.

(b) Lunar orbit insertion

The Broad Study estimated that $\Delta v \sim 1{,}070$ m/s for global access based on a simplified trajectory model. Using a more detailed model, the Focused Study estimated that Δv is the sum of two terms due to basic orbit insertion and a plane change in orbit. The basic orbit insertion step was estimated to require 978 m/s. The Δv required for the plane change depends on the insertion orbit and the desired inclination of the final orbit. In the worst case (90°) the plane change adds another

Table 7.5. Comparison of Focused Study and Broad Study estimates of Δv based on 7-day stay on surface (km/s).

Step or process	Broad Study	Focused Study (with loitering and 7 launch days/month)	Focused Study (without loitering and 7 launch days/month)
Trans-lunar injection from LEO	3.15	3.07	3.07
Lunar orbit insertion	1.07	0.98	1.42
Descent to surface from LLO	2.08	1.88	1.88
Ascent to LLO from surface	1.93	2.03	2.03
Plane change	Variable		
Trans-Earth injection	1.07	0.87	1.41
Direct return to LEO from lunar surface	6.11		
Direct transfer from LEO to lunar surface	6.26		

438 m/s for a total of 1,416 m/s. However, if the spacecraft loiters in lunar orbit for up to 14 days (depending on the plane change required) the Moon will rotate under the spacecraft whose orbit is fixed in inertial space, and the added Δv can be eliminated. In other words, one can trade loitering time for Δv. The Focused Study also estimated that lunar orbit insertion within an inclination range of $\pm 30°$ requires 1,143 m/s, implying that this limited plane change range requires up to $1,143 - 978 = 165$ m/s. The ESAS Report estimated Δv for lunar orbit insertion at 890 m/s plus a plane change of 510 m/s for a total of 1,400 m/s.[212]

In general, one desires to place the Lunar Orbiter into an inclined orbit so that it can release a Descent Vehicle that can make an in-plane descent to the lunar surface to reach a chosen landing site. Therefore, the ultimate landing site determines the lunar orbit from which the Descent Vehicle is released, and this in turn determines how much plane change may be required, which in turn determines Δv for orbit insertion. On p. 151 of the ESAS Report, calculated values of Δv are given for orbit insertion relevant to a variety of suggested landing sites. These range from 826 m/s to 1,078 m/s. However, the ESAS Report also says: "The maximum Δv is 1,078 m/s for a mission to the far side South Pole–Aitken Basin floor. Vehicle sizing for LOI in all design cycles has included 1,390 m/s to protect for a worst-case 90-deg plane change at arrival." This does not seem to agree with any of the other estimates. But. then the ESAS Report goes on to present global maps of Δv for orbit insertion depending on the latitude and longitude of the landing site to which the Descent Vehicle will be released from that orbit. One map is without loitering and the other is for loitering up to 3 days.[213] These are reproduced here as Figures 7.4 and 7.5.

---

[212] *ESAS Report*, p. 443.
[213] Three days of loitering is not enough to eliminate the plane change, but it can reduce it.

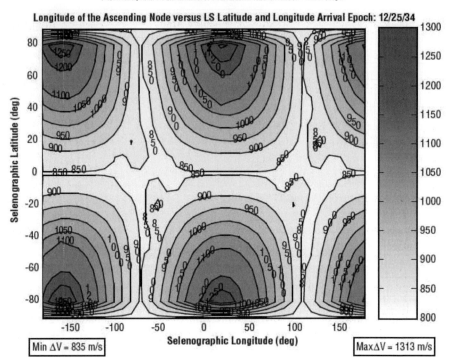

Figure 7.4. Global map of estimated Δv for orbit insertion depending on the latitude and longitude of the landing site to which the Descent Vehicle will be released from that orbit with no loitering involved.

The ESAS Report says that the global minimum Δv is 835 m/s for polar or near-equatorial sites, while local Δv maxima are found near 75°N or S latitudes and 25°E/160°W longitudes. The global maximum Δv without loitering was 1,313 m/s. It is claimed that, by loitering up to 3 days, the net reduction in maximum LOI Δv is 212 m/s, from 1,313 m/s with no loiter time to 1,101 m/s with up to 3 days' loiter time.

I have shown these figures to a variety of experts on trajectory analysis and none of them can make any sense out of them. There doesn't seem to be any obvious physical basis for these diagrams. Incidentally, these figures were prepared for an ostensible arrival on 12/25/2034, which ESAS claims is a "worst-case" arrival date.[214] It is not clear why this is a "worst case" or why it is necessary to be able to arrive at the Moon on Christmas in 2034. There seems to be a plethora of estimates of Δv for orbit insertion, and none of them agrees with any other.

---

[214] The claim is made that "the Moon is at perigee, at its minimum inclination in the 18.6-year metonic cycle (18.3 deg), and at maximum declination (18.3 deg)."

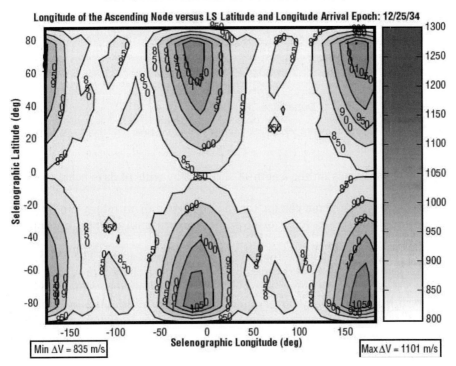

Figure 7.5. Global map of estimated Δv for orbit insertion depending on the latitude and longitude of the landing site to which the Descent Vehicle will be released from that orbit with up to 3 days loitering utilized.

(c) Coplanar descent

The Focused Study indicates a $\Delta v = 1,881\,\text{m/s}$ and the ESAS Report claims $1,900\,\text{m/s}$, which is good agreement compared with other mission steps.

(d) Ascent, rendezvous, and trans-Earth injection

The main issue here is that it is desired to provide the capability for "anytime return"[215] as an abort option. According to the MIT study, the Orbiter is initially positioned for coplanar descent to desired landing site, and the landing site rotates 13 degrees/day away from the Orbiter plane. It is claimed that the Ascender can always launch into optimal TEI plane. However, the Orbiter needs to make a plane change to rendezvous with the Ascender. The worst case is 90° (it would seem

---

[215] "Anytime return" may actually require a delay of a few hours, so that the orientation of the ascent orbit is proper for rendezvous. This point is not clear in the various documents.

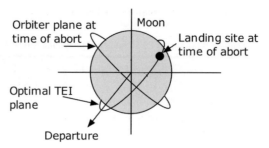

Figure 7.6. MIT figure showing 90° plane change for worst-case rendezvous of Orbiter with Ascender.

likely that this depends on the length of surface stay with 14 days being worst—see Figure 7.6).

Alternatively, the plane change for the Orbiter prior to return can be reduced or avoided by loitering in lunar orbit prior to return until the next coplanar TEI opportunity is reached.

All of the above pertains to the general case of global access. The ESAS Report mentions the word "ascent" 453 times, but none of these clarifies the ascent process. The ESAS Report says: "Outpost missions will also have anytime return capability if the outpost is located at a polar or equatorial site. For other sites, loitering on the surface at the outpost may be required to enable ascent to the orbiting CEV." It appears that for equatorial or polar landing sites one can put the Orbiter into an equatorial or polar orbit, and there would then be no need for a plane change with anytime return. The geometry of this is difficult to comprehend.

Ascent is the most confusing of all the steps. The Focused Study asserts that use of a "retrograde parking orbit" with inclination in the range 90° to 180° provides favorable abort performance. It goes on to say that the required plane changes on ascent (and associated plane change Δv) are 1.2° (34 m/s), 6.7° (191 m/s), and 17.9° (510 m/s) for surface stay times of 3, 7, and 11 days, respectively. It is difficult to understand why these plane changes are so small and the Δv are so large. The Δv for ascent and rendezvous is claimed by the Focused Study to be the sum of 1,834 m/s plus plane rotation (varies from 1,868 m/s to 2,344 m/s for surface stays of 3 to 11 days). However, it is difficult to make sense of this because, apparently, only the Orbiter (not the Ascender) is required to make a plane change. Note that the ESAS Report says that Δv for ascent is 1,888 m/s, which seems to imply no plane rotation is needed.

The Focused Study indicates that the Δv for Earth return is 865 m/s plus a plane change of up to 445 m/s for 90° (or loiter up to 14 days before Earth return). The ESAS Report indicates Δv for Earth return is 1,450 m/s (if no loiter prior to return).

The values of Δv for various lunar trajectory steps, and their dependence on latitude and longitude of landing site, arrival date, capability for anytime return, and effect of loitering (both on arrival and departure) remain quite murky with significant disagreements between different studies. Yet, these figures are critical in determining how much propellant is required for lunar missions.

7.3.7   Lunar orbit rendezvous vs. L1 rendezvous or direct return

The Earth–Moon libration points are shown in Figure 5.6. Both the Broad and Focused studies examined the pros and cons of using L1 (vs. lunar orbit rendezvous [LOR]) as a staging point. They also considered the direct return alternative.

L1 rendezvous can be viewed as a special case of LOR in which the L1 libration point represents a very high lunar orbit. In terms of inherent mission flexibility for lunar landing site access and Earth return opportunities, L1 rendezvous resembles direct return, but with an interim L1 return target rather than Earth.

7.3.7.1   Broad Study

In regard to mission duration, transfer from LEO to lunar orbit requires about 3 days whereas transfer to L1 requires about 4 days. Transfer from lunar orbit to the lunar surface takes a few hours, but transfer from L1 to the lunar surface requires about 2 days. Taking into account going and returning, the use of L1 adds about six days to the mission, so that a 7-day surface sortie requires a minimum of about 15 days via lunar orbit (with no loitering) and about 21 days via L1. For a 90-day surface stay at an outpost, the total mission duration would be about 98 days via lunar orbit and about 104 days via L1.

The Broad Study estimated IMLEO for comparable missions using either L1 or lunar orbit (LO) as the staging point, with LOX/LH_2 propulsion used for all steps (TLI, ascent and descent, and TEI). For sortie missions, a plane change of up to $90°$ had to be taken into account. For long-term missions to an outpost, this was not included. They estimated that, for sortie missions, IMLEO via LLO is typically about 70% of IMLEO via L1, and for outpost missions (with lower Δv for the LLO case) the ratio is ∼50%. With other propulsion systems having lower specific impulse, the disparity would increase.

The Broad Study concluded that staging from lunar orbit is preferable to staging from Earth–Moon L1 based on: (a) IMLEO, (b) mission duration, (c) abort options, (d) time required to return crew to Earth in the event of a contingency, and (e) mission risks and system hazards.

7.3.7.2   Focused Study

The Focused Study considered three architectures: (1) lunar orbit rendezvous, (2) L1 rendezvous, and (3) direct return. The L1 rendezvous and direct return architectures inherently provide a wide range of mission flexibility in terms of landing site access and surface stay times. Both of these architectures support short-duration expeditions as well as longer duration missions anywhere on the lunar surface. Both architectures also inherently offer the capability to initiate a return to Earth at any time during the lunar surface stay. However, the direct return architecture incurs the mass penalty of transporting the propellant required for Earth return ($\Delta v \sim 850 \, \text{m/s}$) to the lunar surface and back to lunar orbit. The L1 rendezvous approach has higher Δv requirements than the lunar orbit rendezvous approach. In addition, the L1 rendezvous approach adds approximately 2.5 days to the one-way transit time between the Earth and the Moon. The Focused Study is not always self-consistent in its comparison of the L1 rendezvous approach with the lunar orbit rendezvous approach. While Δv requirements are considerably higher for the L1 rendezvous approach, this is not an "apples-to-apples" comparison because the lunar orbit rendezvous approach in its simple form has constraints on landing site access, mission duration, and return flexibility. By including substantial plane change capability on the CEV and the lander elements, the lunar orbit rendezvous architecture can provide the same mission flexibility in terms of global lunar access and variable mission duration with anytime Earth return. However, according to the Focused Study, the additional Δv required for these plane changes increases the IMLEO of the lunar orbit rendezvous architecture when sized for maximum mission flexibility, so that it is roughly comparable with the IMLEO of the L1 rendezvous approach. But, the Focused Study also says that "... it is possible to provide a lower mass solution using the lunar orbit rendezvous architecture by taking advantage of the limited duration of the lunar surface mission ... The lowest mass solution uses an optimized lunar parking orbit that minimizes the lunar descent and ascent plane change required for anytime Earth return over the seven-day surface stay. Using this approach, a lunar orbit rendezvous mission results in an IMLEO estimate of 199 mT in comparison to the L1 rendezvous value of 230 mT—a reduction of 31 mT (13.5%)." However, these findings are in disagreement with the results of the Broad Study that found a greater disparity between the IMLEO values for the L1 rendezvous approach and the lunar orbit rendezvous approach. The Focused Study report is 822 pages long. Obviously, it is difficult to summarize their findings.

7.3.8   ESAS report architecture comparison

The ESAS Report immediately eliminated L1 rendezvous based on: "Recent studies performed by NASA mission designers[216] concluded that equivalent landing site access and 'anytime abort' conditions could be met by rendezvous missions in LLO with less propulsive delta-V and lower overall Initial Mass in Low Earth Orbit (IMLEO)." With L1 rendezvous eliminated, the ESAS Report considered four alternative architectures: (1) Earth orbit rendezvous–lunar orbit rendezvous (EOR–LOR), (2) lunar orbit rendezvous (no EOR), (3) EOR–direct return from Moon, and (4) direct–direct (no rendezvous in LEO or lunar orbit). Architecture (4) was summarily dismissed because it would require the largest Launch Vehicle (without ISRU). The three remaining mission modes (LOR, EOR–LOR, and EOR–direct return) were analyzed in significant detail.

---

[216] Presumably, this refers to the Broad and Focused Studies. But, since these were not released to the public they could not be mentioned by the ESAS Report. For practical purposes, they do not exist outside of JSC.

The EOR–direct return mission mode was eliminated from further consideration primarily because in this mode the CEV must operate in, and transition among, $1g$ pre-launch and post-landing, hyper-gravity launch, zero-gravity orbital and cruise, powered planetary landing and ascent, and $\frac{1}{6}g$ lunar surface environments. It was claimed that this added significant complexity to a vehicle that must already perform a diverse set of functions in a diverse number of acceleration environments. Another reason given was reduced commonality between the ISS and the lunar CEV. These arguments do not necessarily make sense to this writer. Yes, the CEV is more complex in this mode, but the overall mission, shorn of lunar orbit rendezvous, would appear to be simpler and safer.

The baseline approach chosen by the ESAS Team was a 2-launch LOR split mission termed the ESAS Initial Reference Architecture (EIRA). For the final analysis cycle, the EIRA mission architecture was compared with a 2-launch EOR–LOR approach, which used two launches of a 100 mT LEO payload vehicle, and a "1.5-launch" EOR–LOR approach, which used a launch of a 125 mT LEO Cargo Launch Vehicle (CaLV)[217] and a smaller Crew Launch Vehicle (CLV). They were also compared for three different levels of propulsion technology. The baseline option used pressure-fed LOX/methane engines on the CEV Service Module (SM) and the lander Ascent and Descent Stages to maximize commonality. A second option substituted a pump-fed LOX/hydrogen system on the lander Descent Stage to improve performance. The third option also used LOX/hydrogen for the lander Descent Stage and substituted a pump-fed LOX/methane system for the Ascent Stage propulsion system.

The assumed mission mode for the EIRA is a 2-launch "split" architecture with LOR, wherein the LSAM is pre-deployed in a single launch to low lunar orbit (LLO) and a second Launch Vehicle of the same type delivers the CEV and crew to lunar orbit, where the two vehicles initially rendezvous and dock. The entire crew then transfers to the LSAM, undocks from the CEV, and performs descent to the surface. The CEV Crew Module (CM) and SM are left unoccupied in LLO. After a lunar stay of up to 7 days, the LSAM returns the crew to lunar orbit and docks with the CEV, and the crew transfers back to the CEV. The CEV then returns the crew to Earth with a direct-entry-and-land touchdown, while the LSAM is disposed of on the lunar surface.

The EOR–LOR architecture is functionally similar to the EIRA, with the primary difference being that the initial CEV–LSAM docking occurs in LEO rather than LLO. Whereas the EIRA incorporated two smaller Earth departure systems (EDSs) in two parallel launches to deliver the CEV and LSAM to the Moon, the EOR–LOR architecture divides its launches into one heavy launch for a single, large EDS and the second launch for the CEV, crew, and LSAM. The combined CEV and LSAM dock with the EDS in Earth orbit, and the EDS performs TLI. Another difference between the EIRA and EOR–LOR architectures is that, for the baseline pressure-fed LOX/methane propulsion system, the EDS performs LOI for the EIRA. Due to launch performance limitations of the single EDS with EOR–LOR, LOI is

[217] Note that according to my calculations, 125 mT to LEO is not adequate for the CaLV.

instead executed by the CEV for optimum performance in the EOR–LOR architecture. Once the CEV and LSAM reach LLO, this mission mode is identical to the EIRA.

The ESAS study introduced another variant of the EOR–LOR architecture, known as 1.5-launch EOR–LOR, so named due to the large difference in size and capability of the LVs used in the architecture. Whereas the previous EOR–LOR architecture used one HLLV to launch the EDS and another HLLV to launch the CEV, crew, and LSAM, this architecture divides its launches between one large cargo (CaLV) and one relatively small crew (CLV). The 1.5-launch EOR–LOR mission is an EOR–LOR architecture with the LSAM and EDS pre-deployed in a single launch to LEO with the heavy-lift CaLV. A second launch of a 25 mT-class CLV delivers the CEV and crew to orbit, where the two vehicles initially rendezvous and dock. The EDS then performs the TLI burn for both the LSAM and CEV and is then discarded. Upon reaching the Moon, the LSAM performs the LOI for the two mated elements, and the entire crew transfers to the LSAM, undocks from the CEV, and performs descent to the surface. The CEV is left unoccupied in LLO. After a lunar stay of up to 7 days, the LSAM returns the crew to lunar orbit, where the LSAM and CEV dock and the crew transfers back to the CEV. The CEV then returns the crew to Earth with a direct or skip-entry-and-land touchdown, while the LSAM is disposed of via impact on the lunar surface.

The ESAS team generated mission performance analysis for each option (IMLEO, number of launches required, and launch margins), integrated program costs through 2025, safety and reliability estimates such as probability of loss of crew—P(LOC), probability of loss of mission—P(LOM), and other discriminating FOMs.

Three mission modes were analyzed, with three different propulsion technologies applied. In addition to the LOR, EOR–LOR "2-launch", and EOR–LOR "1.5-launch" modes, analysis was also performed on a single-launch mission that launched both the CEV and lander atop a single heavy-lift CaLV (the same used for the 1.5-launch solution), much like the Apollo/Saturn V configuration. However, the limited lift capability provided by this approach restricted its landing site capabilities to the same equatorial band explored by Apollo, in addition to the lunar poles. For each of the mission modes, end-to-end single-mission probabilities of LOC and LOM were estimated for (1) a baseline propulsive case using all pressure-fed LOX/methane engines, (2) a case where a LOX/hydrogen pump-fed engine was substituted on the lander Descent Stage, and (3) a third case where the lander Ascent Stage engine was changed to pump-fed LOX/methane. Estimates were made of P(LOC) and P(LOM) for each of the 18 steps in each architecture. Figures 7.7 and 7.8 (both in color section) show the results.

According to the ESAS Report, P(LOC) was dominated by propulsive events and vehicle operating lifetimes. The LOR options had added risk (compared with EOR–LOR options) due to the LSAM being sent to the lunar orbit separately from the CEV, and thus not having a back-up crew volume during transit to handle "Apollo 13" like contingencies. The LOR mission also required the CEV SM to perform an LOI maneuver. Generally, each time a pump-fed engine technology was introduced to replace a pressure-fed system risk increased. When all the mission event probabilities were summed, all mission options fell within a relatively narrow range (1.6 to 2.5%), but the difference between the highest risk and lowest risk options approached a factor of 2. Missions using the LOR mode were estimated to be the highest risk options, while EOR–LOR "1.5-launch" options were the lowest. The lowest P(LOC) option was the 1.5-launch EOR–LOR mission using a pump-fed LOX/hydrogen lander Descent Stage and pressure-fed LOX/methane engines for both the lander Ascent Stage and CEV SM.

According to the ESAS Report, P(LOM) generally followed the same trends as P(LOC). LOR and EOR–LOR 2-launch options exhibited the greatest P(LOM), in a range between 7 and 8% per mission. The EOR–LOR 1.5-launch option using the LOX/hydrogen lander Descent Stage engines scored the lowest P(LOM) among the full-up mission options. This same mission mode and propulsion technology combination scored the lowest P(LOC) as well.

Unfortunately, the details of these calculations are very intricate and in many cases are not supported by direct data. These calculations of P(LOC) and P(LOM) are not available to the public, but one might suspect (on general grounds) that there are significant error bars attached to them. That being the case, it would appear that the minor differences between alternatives might be negligible within that uncertainty.

Estimates of IMLEO were made. IMLEO for the EIRA was 184 mT, while IMLEO for EOR–LOR was 173 mT and for EOR–direct return it was 253 mT.

The life-cycle cost (LCC) was estimated for each option. The complete LCC, including design, development, test, and evaluation (DDT&E), flight units, operations, technology development, robotic precursors, and facilities, were all included in this analysis. Generally, the choice of mission mode had only a small effect on the LCC of the exploration program. Of the options modeled, the 1.5-launch EOR–LOR mission using a LOX/hydrogen lander Descent Stage propulsion system exhibited an LCC that was in the same range as the other options.

The ESAS Report says that of the options studied the 1.5-launch EOR–LOR mode yielded both the lowest P(LOM) and the lowest P(LOC) when flown with a LOX/hydrogen lander Descent Stage and common pressure-fed LOX/methane propulsion system for both the lander Ascent Stage and CEV SM. Cost analysis was less definitive, but also showed this same EOR–LOR 1.5-launch option being among the lowest cost of all the alternatives studied. Based on the convergence of robust technical performance, low P(LOC), low P(LOM), and low life-cycle costs, the 1.5-launch EOR–LOR option using LOX/hydrogen lander Descent Stage propulsion was selected as the mission mode to return crews to the Moon.

It appears to this writer as if the comparison of safety and life-cycle cost for various mission options provides very narrow differences between several architectural options. Considering the great uncertainties remaining in these safety and cost estimates, it would appear that almost any option could be justified. One thing that was missing in the ESAS analysis was the potential benefits of various options from use of ISRU. The direct return option will benefit the most from ISRU.

7.4 THE ESAS LUNAR ARCHITECTURE

7.4.1 Description

The lunar architecture selected by NASA was the so-called 1.5-launch EOR–LOR architecture with the LSAM and EDS pre-deployed in a single launch to LEO with the heavy-lift CaLV. A second launch of a 25 mT class CLV delivers the CEV and crew to orbit, where the two vehicles initially rendezvous and dock. The EDS then performs the TLI burn for both the LSAM and CEV and is then discarded. Upon reaching the Moon, the LSAM performs the LOI for the two mated elements, and the entire crew transfers to the LSAM, undocks from the CEV, and performs descent to the surface. The CEV is left unoccupied in LLO. After a lunar stay of up to 7 days, the LSAM returns the crew to lunar orbit, where the LSAM and CEV dock and the crew transfers back to the CEV. The CEV then returns the crew to Earth with a direct or skip-entry-and-land touchdown, while the LSAM is disposed of via impact on the lunar surface. This is illustrated in Figure 7.9.

7.4.2 Independent mass estimates

NASA has not provided adequate information on mass estimates for Habitats and vehicles that would allow independent analysis of the merits of alternate options. Instead, an individual without access to secret data would have to rely on fragmentary data from several sources.

The following vehicles are defined:

Crew Exploration Vehicle (CEV)—consisting of Service Module (SM) and Crew Module (CM).

Lunar Surface Access Module (LSAM)—consisting of Descent Stage (DS), Habitat (H), and Ascent Stage (AS).

Earth Departure Stage (EDS).

Table 7.6 provides some estimates of vehicle masses.

The Launch Vehicle used to transport the crew and CEV to LEO is specified as capable of delivering 22,900 kg to a 30 × 160 nmi delivery orbit. The capability of the Launch Vehicle to transport the LSAM to the delivery orbit (30 × 160 nmi) is stated as 124,600 kg and the capability of the EDS to deliver mass to trans-lunar injection is 54,600 kg.

Table 7.7 provides an overview of the steps involved in the proposed architecture for lunar human missions.

The estimates of Δv from Table 7.4 provided by the Focused Study are used here. It is assumed that LOX/LH$_2$ propulsion is used for Earth departure, lunar orbit insertion, and descent. LOX/methane is assumed for ascent and return to Earth. As Table 7.4 shows, there are seven cases to be considered. For each case, a spread-

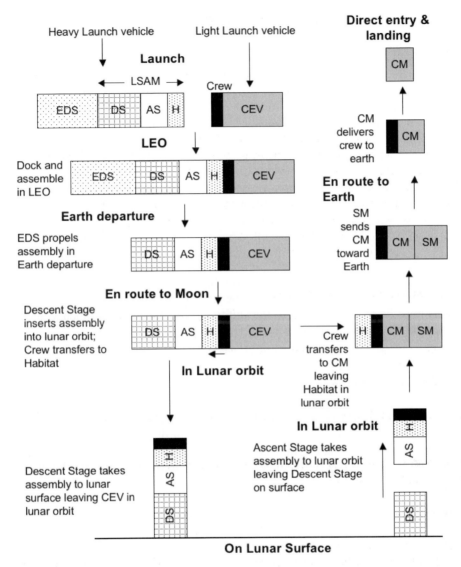

Figure 7.9. Depiction of the 1.5-launch EOR–LOR architecture selected by NASA.

Table 7.6. Vehicle mass estimates (in kg).

	CM	SM	CEV = CM + SM	Ascent Stage	Descent Stage	Habitat	LSAM = AS + DS + H	Earth departure stage
Dry mass	9,640	5,830	15,470	5,640	4,460	2,700	12,800	10,290

Table 7.7. Steps in human lunar mission. If a check appears in a cell, the corresponding vehicle is involved in propulsion for that step. Shading in a column shows vehicles that are joined in any step. Cells without shading represent vehicles that have been discarded.

Step ⇒ Vehicle ⇓	In LEO	Dock in LEO	TLI	LOI	Descent	Ascent	Dock in LLO	TEI	Earth entry
CM									✓
SM								✓	
DS				✓	✓				
AS						✓	✓		
H									
EDS		✓	✓						

sheet was prepared. A typical spreadsheet (for Case 4) is displayed as Table 7.8. A summary of estimates of IMLEO for the seven cases is given in Table 7.9. Serious doubts must be raised in regard to the ability of NASA-planned Launch Vehicles to deliver the various vehicles in their architecture.

7.5   DO WE NEED A CEV?

Supposedly, the overall plan provides a stepwise evolution from lunar sorties to lunar outposts to Mars missions. However, there are major differences between these three types of missions that cause difficulty in providing smooth connectivity between them. The great disparity between the time required to travel to Mars (6 to 9 months) vs. the time required to travel to the Moon (~3 days) results in a huge difference in requirements for transporting the crew in the two cases. These differences include differences in mass of consumables, radiation protection, low-gravity mitigation, and space and facilities.

One major difference lies in life support consumables. As Table 4.2 shows, roughly 36 mT of consumables are needed for a transit to Mars. An optimistic projection is that these can be supplied by an ECLSS of mass ~5 mT (see Table 4.3). However, the actual requirement is likely to be greater than 5 mT. By contrast, the consumables required for a transit to the Moon are far more modest. Not only is there a significant difference in mass requirements, but the volume requirements for consumables for a Mars transit are considerable. Furthermore, long-term reliability testing of ECLSS systems is needed for Mars and this will take many years to validate.

Another difference lies in the need for radiation protection. As Section 4.2 explains, the impact of radiation exposure is very serious for Mars missions but is much smaller for short-duration lunar missions.

Table 7.8. Example spreadsheet for Case 4. Cells with shaded background indicate propulsion activity. Similar spreadsheets were prepared for each of the seven cases.

		Dock in LEO	TLI	LOI	Descent	Ascent	Dock in LLO	TEI	Earth entry
I_{SP}		452	452	452	452	360	360	360	360
Δv (km/s)		0.10	3.07	1.42	1.88	1.87	0.10	1.41	0.10
Rocket equation ratio			1.0228	2.0016	1.3767	1.5290	1.6980	1.0288	1.4913
Propellant used (kg)		4,395	96,318	20,153	10,530	6,452	251	7,601	
STEP									
CM	Mass	9,640	9,640	9,640	9,640	9,240	9,240	9,640	9,640
SM	Dry mass	5,830	5,830	5,830	5,830	5,830	5,830	5,830	
	Propellant	7,601	7,601	7,601	7,601	7,601	7,601	7,601	
	Wet mass	13,431	13,431	13,431	13,431	13,431	13,431	13,431	
DS	Dry mass	4,460	4,460	4,460	4,460				
	Propellant	30,683	30,683	30,683	10,530				
	Wet mass	35,143	35,143	35,143	14,990				
AS	Dry mass	5,640	5,640	5,640	5,640	5,640	5,640		
	Propellant	6,703	6,703	6,703	6,703	6,703	251		
	Wet mass	12,343	12,343	12,343	12,343	12,343	5,891		
H	Mass	2,700	2,700	2,700	3,100	3,100	3,100		
EDS	Dry mass	22,904	22,904						
	Propellant	100,713	96,318						
	Wet mass	123,617	119,222						
Total mass		*196,873*	*192,479*	*73,256*	*53,503*	*38,114*	*31,662*	*23,071*	*9,640*
CEV mass		23,071	23,071	23,071	23,071	22,671	22,671	23,071	9,640
Cargo mass		173,803	169,408	50,186	30,433	15,443	8,991		

Table 7.9. Summary of mass estimates for seven cases from Tables 7.4 and 7.8.

	"Loitering" used to minimize Δv for Earth–Moon–Earth transfers			Plane change requires higher Δv for Earth-Moon-Earth transfers			Anytime departure and return
	Set 1	Set 2	Set 3	Set 4	Set 5	Set 6	Set 7
Total IMLEO	174,244	177,033	183,094	196,873	199,951	206,643	254,641
CEV mass	19,768	19,768	19,768	23,071	23,071	23,071	19,763
Cargo mass	154,476	157,264	163,326	173,803	176,881	183,572	234,878
Launch opportunities per month	3	7	11	3	7	11	28
Lunar orbit departure opportunities per month	1	1	1	3	7	11	28
Total days in lunar vicinity	23	23	30	5	9	13	28

Transit Habitats for Mars should be designed to maximize use of facilities and stored resources for shielding. This is not applicable for lunar missions. The effects of extended low-gravity exposure are discussed in Section 4.3. Mars missions probably require some form of artificial gravity. Lunar missions do not.

Finally, the crew requires a much greater volume as well as many more facilities in the transit to Mars than it does in transit to the Moon. The crew could easily put up with cramped quarters in a small module for transfer to the Moon in 3 days, whereas a complete Habitat is needed for transfer to Mars over 6+ months. Even so, the psychological impacts of confinement are not well explored. In fact, NASA planning for Mars utilizes a full-scale Habitat as a transfer vehicle and the empty CEV merely hitches a ride to Mars and back, and is only utilized for entry to Earth upon return from Mars. This is a serious waste of mass and capability. We must therefore draw the conclusion that the CEV is more capable than is needed to transfer crew to the Moon and far less capable than is needed to transfer crew to Mars, and is therefore sub-optimal for either application. The truth is that we do not need a CEV.

7.6   THE NASA LUNAR ARCHITECTURE PROCESS

The NASA lunar architecture development process is so complex and intricate that it is very difficult to critique the whole enterprise. Only a few comments will be made here. NASA's approach to lunar architecture development has clearly been based on assembling enough assets to land humans on the Moon at the earliest possible date, mainly using methods, facilities, technologies, and systems that are either already in existence or can be relatively quickly developed from existing systems. NASA's plans rely to an absolute minimum on new technology. Selection of the 1.5-launch EOR–LOR architecture is based on estimates of narrow margins of benefit compared with other options, and totally ignores the potential impact of ISRU. This architecture maximizes the number of rendezvous and crew transfers. NASA's system-engineering function is suspect. It is not clear that the Launch Vehicles that NASA proposes to use will do the job needed for lunar missions, and it is even more doubtful that they will suffice for Mars missions. Details on mass figures are missing from the reports, despite the great length of some of the reports (>750 pages). Most of NASA's work is kept secret. It is doubtful that the ESAS Report would have ever been released to the public if a copy had not been leaked to *NASA Watch* and distributed on the Internet. It is not clear whether independent non-advocate boards have reviewed these ESAS plans, and the mass and other resource margins adopted at this early stage seem to be inadequate—although it is not always clear what these margins actually are. Although JSC has spent much time "selling" the program, that time would have been better spent doing the detailed engineering. Good engineering like good science must remain fundamentally skeptical. The word "Mars" appears 287 times in the ESAS Report (as can be verified by a pdf search), but a minimum of data or information about Mars missions is actually contained in the ESAS Report.

There are deep-rooted contradictions in NASA's lunar plans. On the one hand, there is a great deal of emphasis on locating an outpost at the south pole, ostensibly to tap putative lunar water ice resources for ISRU, although more likely to seek a more benign thermal environment. However, there are no plans of any substance to carry out *in situ* prospecting to validate observations from orbit, locate resources, and determine requirements for excavating and extracting water in any quantity. Furthermore, the robotic precursor program, such as it was, now appears to be defunct. Nor do there seem to be any plans to develop nuclear power, which appears to be required to exploit these resources in the dark craters. Instead, we have been presented with schemes to beam solar energy from hilltops into these dark craters. And if extraction of water proves to be impractical would there still be the need for an outpost? Could it then be equatorial? Or would thermal stresses be too great? What purpose does it serve in either case? In addition, NASA has eliminated its initial plan to use LOX/methane for ascent, which would wipe out the small benefit of ISRU if LOX is produced from indigenous resources.

Because of the requirements for global access, and anytime return, the propulsion requirements for sorties are actually greater than they would be for crew delivery to a polar or equatorial outpost. Therefore, the design and sizing of the Launch Vehicle and the Descent and Ascent Vehicles is determined by the sorties, not the outpost. This brings up the question as to whether sorties are needed at all. But, since NASA's goal seems to be to send humans back to the Moon at the earliest possible date, the virtue of sorties within this context can be understood.

The NASA efforts in planning lunar missions seem to be concentrated in CEV design and, to a slightly lesser extent, Launch Vehicle design. It is clear that NASA prefers developing vehicle designs to developing advanced technologies. Fortunately, there seems to be enough capability on hand to go back to the Moon; however, such is not the case for Mars.

8

Why the NASA approach will likely fail to send humans to Mars prior to c. 2080

8.1 MAJOR DIFFERENCES BETWEEN MARS AND LUNAR MISSIONS

The great disparity between the time required in traveling to Mars (6 to 9 months) vs. the time required to travel to the Moon ($\sim$3 days) results in a huge difference in requirements for transporting the crew in the two cases. These differences include differences in mass of consumables, requirements for longevity and durability of ECLSS systems, radiation exposure, low-gravity exposure, and Habitat volumetric space and facilities. When the entire mission is taken into account, including transit to Mars, surface stay, and return, the differences between Mars missions and lunar missions are very significant. There are also significant differences in Δv requirements for various transits, although these can be partly mitigated by aero-assist technologies at Mars.

Some of the requirements for long-stay Mars missions include:

- Develop new technologies and validate at Mars where necessary.
- Provide propulsion systems and propellants and/or aero-assist technologies for all transfers.
- Develop and validate pinpoint landing capability.
- Provide life support consumables and environmental control for:
  - ○ $\sim$200 days traverse to Mars;
  - ○ $\sim$550 days on surface (after 200 days in space);
  - ○ $\sim$200 days return to Earth (after 750 days in space).
- Mitigate adverse effects of low-gravity exposure.
- Protect from excessive exposure to radiation.
- Provide Habitats for transfer to and from Mars, as well as on the surface.

- Provide nuclear power, equipment, and supplies for surface operations.
- Define abort requirements at all stages—to the maximum extent possible.
- Define approach to ISRU—what, how, and when?

Mars missions are fundamentally different from lunar missions for the following reasons:

- Mars missions require ~950 days in space, while lunar missions can be as short as ~20 days.
- No "sorties" are feasible at Mars. Once committed, the crew must spend ~950 days in space or on Mars.
- Most abort options available to lunar missions are not available at Mars. Instead of providing options for abort if something goes wrong (as in lunar missions), Mars missions require an intense long-term program of validation and precursors to provide extremely high reliability.
- Radiation and low-gravity effects are far more serious problems in Mars missions due to the much longer exposure times.
- Life support systems for Mars must function in a fail-safe mode for very long periods compared with ISS or lunar missions. Testing and validation will not only be expensive, but will require many years.
- ISRU on Mars is far more feasible and has much greater mission impact.
- Mars launch opportunities are spaced at 26-month intervals with significant variance in launch and return characteristics at each opportunity. Significant Δv variations from opportunity to opportunity require that a design be made for the worst case.

As a result, excessive use of the lunar paradigm for Mars is inappropriate.

ISRU on Mars is a very different situation than it is on the Moon. The main differences are discussed in Section 5.3.1.

Let us repeat (from p. 173) the major unknowns regarding Mars ISRU: What are the requirements for excavating water-bearing, near-surface regolith and extracting water? In the case of equatorial, water-bearing, near-surface regolith, is the water in the form of ground ice or mineral hydrates? Recall also that neither ESAS nor the Mars Exploration (Science) Program has any specific plans to investigate these questions.

8.2 IS NASA TOO BUSY FOR MARS MISSIONS?

Recall (p. 246) that if one examines the many specific occurrences[218] where the ESAS Report mentions the word "Mars", it is found that essentially all of them are cursory

---

[218] Sections 1–4 of the ESAS Report mention the word "Mars" 232 times, Sections 7–10 mention it 27 times, and Sections 11–17 mention it 30 times (see also Footnote 160).

and lacking in content. It seems quite apparent that NASA has not revisited the realm of Mars human missions in depth. The likely reason for this is that NASA had its hands full (in 2006) attempting to deal with lunar missions, and Mars missions have been put on the back-burner.

The ESAS Report suggests: "The ESAS architecture does not address the Mars phase in detail, but it is recognized that traditional chemical propulsion cannot lead to sustainable Mars exploration with humans. Nuclear Thermal Propulsion (NTP) is a technology that addresses the propulsion gap for the human Mars era."[219] When the ESAS Report uses the word "sustainable" it seems to be a way of saying that we will eventually need it, but perhaps not on the early missions. However, the ESAS Report states its intent to use NTP on early missions,[220] so it is not clear what they mean. Yet there is no indication that anyone understands the requirements (fiscal, logistic, health, and political) to develop, test, validate, and implement NTP, and whether the whole concept is feasible and affordable at all. Clearly, the crucial issue of the minimum altitude for start-up[221] does not seem to have been addressed.

Despite the fact that short-stay Mars missions are clearly impractical, JSC continues to list the short-stay mission as a viable possibility in their public reports and intra-NASA presentations. For example, the ESAS Report says: "As the MTV approaches Earth upon completion of the 1.5- to 2.5-year round-trip mission, . . ." In a 2006 NASA presentation, the "short-stay" mission with 40 days on the Mars surface is still listed as a possibility. Compared with a long-stay mission, the short-stay mission (a) only reduces overall mission length by 27% while requiring almost twice as much duration in zero-g, (b) only provides 40 days on the surface (plant flag and run mode), (c) passes within 0.4 AU of the Sun in the return flight, and (d) requires exotic nuclear propulsion and then some to meet the excessively demanding Δv requirements.

A fundamental element of current plans for Mars missions is to include an "abort-to-orbit" option after ~30 days on the surface, implying that (a) the crew will land despite the fact that there is no certainty that surface systems will function properly, (b) the Earth Return Vehicle is required to provide an extra ~600 days of life support in the event of an abort-to-orbit adding significantly to IMLEO, (c) after abort, the crew, spending ~600 extra days in Mars orbit, will face exposure to severe radiation, zero-gravity, and incredible boredom.

We are repeatedly assured by the ESAS Report that a 125 mT-to-LEO Launch Vehicle is adequate for Mars missions, but it seems likely that such a capability will be inadequate.

NASA documents indicate that the expected IMLEO for Mars human missions is 446 mT, which is low by at least a factor of 2, and possibly more, depending on assumptions that are made.

---

[219] *ESAS Report*, Section 9.3.3, p. 631.
[220] *ESAS Report*, Section 2.4.7, p. 69.
[221] See Section 3.4.3.

8.3   MARS ARCHITECTURE PLANNING

It is too early (in 2007) to schedule mission phasing and reliably estimate vehicle masses for Mars missions. We need to define a campaign for Mars, rather than a mission. The campaign will consist of several sequential missions preceded by a long technology development and validation program (see Figure 8.1). This campaign is likely to require ~20 years of technology development and validation, costing many billions of $ prior to any actual landings on Mars. The most critical technologies are aero-assisted entry, descent, and pinpoint landing. Without efficient aero-assist, IMLEO may increase by a few thousand metric tons. Other vital technologies include life support, radiation protection, low-gravity mitigation, ISRU, nuclear power, and Habitat design. We also need an honest evaluation of the nuclear thermal rocket (NTR)—its benefits, costs, time-line for development, and risks.

Despite previous work, the feasibility of Mars missions remains murky. DRM-3 pointed the way toward Mars missions, but it suffered from a number of anomalies:

- Mission staging and phasing may not satisfy safety requirements.
- No consideration of radiation or low-gravity effects was included.
- The nuclear thermal rocket was assumed to launch from LEO with an optimistic dry mass.
- Optimistic assumptions were made regarding aero-assisted entry, descent, and landing. Section 4.6 suggests that aero-entry systems are far more massive than was assumed by JSC.
- Storage and specific impulse of cryogenic propellants were optimistic.
- Vehicle masses in DRM-3 were "scrubbed" from DRM-1, but the methodology used is not easily traceable to any sources.
- Transport of H_2 to Mars for ISRU utilized mass-less tanks with zero boil-off.
- A surface nuclear reactor was installed and connected by a ~1 km cable.
- Inflatable Habitats were used.
- Very efficient, long-life ECLSS was assumed.

The new Mars mission architecture c. 2006 is illustrated in Figure 8.2 (color section).

The new Mars mission architecture c. 2006 is based somewhat on DRM-3 in that: it retains NTR; it retains optimistic aero-assisted entry, descent, and landing; it is unclear on surface power; and, as before, no consideration was given to radiation or low-gravity effects.

It differs from DRM-3 in that:

- It includes an "abort-to-orbit" option after ~30 days on the surface, requiring the Earth Return Vehicle to provide an extra ~600 days of life support.
- It changes phasing so that the crew lands with the Ascent Vehicle.
- It emulates lunar descent with the Ascent Vehicle atop the Descent Vehicle.
- It does not launch the Ascent Vehicle 26 months prior to crew launch, so it is not feasible to have ISRU fill the ascent propellant tanks prior to crew departure from Earth.

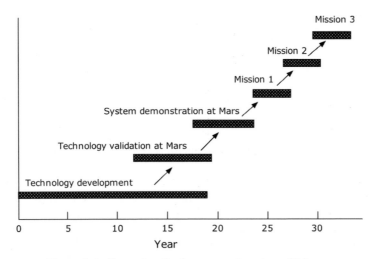

Figure 8.1. Campaign for human exploration of Mars.

- It uses a nuclear thermal rocket for Mars orbit insertion and Earth return (requires transport of many tens of mT of hydrogen to Mars, and storage for ~2 years in Mars orbit).

The elements that are similar include:

- Pre-deployed surface elements 26 months prior to crew arrival.
- Three vehicles: Surface Habitat, Mars Lander, and Mars Transit Vehicle (a.k.a. Earth Return Vehicle).
- General timing of mission: 500–600 day surface stay.
- Launch vehicle with ~100–125 mT capability to LEO.
- Use of nuclear thermal rocket for Earth departure.[222]

The elements that are different include:

- In DRM-3 the Mars Ascent Vehicle and its accompanying ISRU plant land 26 months early and fill the Ascent Vehicle tanks prior to crew arrival. In the ESAS 2006 DRM, the Ascent Vehicle lands at the same time as the crew.
- There is no provision for ISRU. Indeed, ISRU is impractical unless the Ascent Vehicle lands 26 months early. Oddly enough, the NASA exploration technology program calls for development of lunar ISRU (but not Mars ISRU). The ironic thing is that lunar ISRU is far more difficult to implement and has a much lesser mission impact than Mars ISRU.

---

[222] The fact that—in the ESAS Report—rendezvous of the NTR with the MTV takes place at 800–1200 km altitude suggests that the benefits of NTR are small. The NASA technology plan does not seem to call for development of NTR.

- The nuclear thermal rocket and hydrogen storage must persist for 9 months to be used for Mars orbit insertion and for 3 years prior to Earth return.

Elements that are unclear include:

- Surface power system: nuclear or solar? The NASA exploration technology plan does not seem to call for development of a reactor.
- Elliptical or circular orbit for MTV? This has a significant impact on the assignment of propellant requirements for ascent, Mars orbit insertion, and Mars orbit departure.

NASA estimates for IMLEO for vehicles in the Mars mission illustrated in Figure 8.2 (color section) are shown in Figure 8.3 (color section). Neither NASA nor JSC provided any further breakdown of how the masses in Figure 8.3 were derived, and there was an implication that even JSC did not have an available archive where the details were retained. It is of interest to examine the question of what these gross masses imply regarding individual flight systems. To do this, one can reverse-engineer these values into a model that fits the totals given in the figure. However, the result is not unique. There may be several starting points that lead to the same result. The fundamental question is: Can a Mars mission be implemented with 446 mT in LEO? If not, what is a more likely figure?

In August 2006 the informal reverse-engineering study of the mission described in Figure 8.2 was carried out at JPL to determine what assumptions had to be made to obtain the values of IMLEO for the vehicles in Figure 8.3. In general, JPL assumed many of the optimistic assumptions typically made by JSC:

- No boil-off of cryogenic H_2, despite $\sim$180 days to get to Mars plus 550–600 days in orbit about Mars.
- Nuclear thermal rockets ($I_{SP} \sim$900–950 s) allowed to be fired up in LEO.
- Large-scale aerocapture and EDL systems with low entry system mass.
- LOX/methane engines for ascent/descent with a very optimistic $I_{SP} \sim$379 s.[223]
- No additional mass was allocated for closed-loop ECLSS, radiation, and low-g solutions.
- No Δv allocated for rendezvous, TCMs, 2 years of station keeping, etc.

Additional assumptions are summarized in Tables 8.1 to 8.3.

According to Table 8.1, in order to obtain an IMLEO $\sim$152 mT for the CTV, JPL had to:

- Assume that NTR could start up in LEO.
- Use an NTR dry mass $= 36/(49.5 + 20.1 + 16.4 + 3.0) = 40\%$ of (hydrogen) propellant mass.
- Assume that a 15.5 mT CTV suffices for the two-way trip to Mars and back.

[223] Not that JPL believed this; they were using JSC estimates.

Table 8.1. JPL's reverse-engineering of CTV (masses in mT).

Crew Transit Vehicle	In LEO	En route to Mars	In Mars orbit	En route to Earth	At Earth	Comments
CTV	15.5	15.5	15.5	15.5		Houses crew during transfer
Crew	0.6	0.6	0.6	0.6	0.6	
CEV	6.5	6.5	6.5	6.5	6.5	Used for Earth entry
NTR (dry)	36.0	36.0	36.0			Total dry mass = 45% of total propellant mass
NTR fuel tank for Earth departure	3.0					6% of propellants— jettisoned after TMI
NTR propellants— Earth departure	49.5					From LEO, $\Delta v = 3.7\,\mathrm{km/s}$, $I_{SP} = 950\,\mathrm{s}$
NTR propellants— MOI	20.1	20.1				$\Delta v = 2.1\,\mathrm{km/s}, I_{SP} = 950\,\mathrm{s}$
NTR propellants— Mars departure	16.4	16.4	16.4			$\Delta v = 2.3\,\mathrm{km/s}, I_{SP} = 950\,\mathrm{s}$
Consumables	4.4	4.4	4.4			For ascent and rendezvous
Total	*152.0*	*99.5*	*79.4*	*22.6*	*7.1*	

Table 8.2. JPL's reverse-engineering of Surface Habitat (masses in mT).

Surface Habitat element	In LEO	En route to Mars	In Mars orbit	On Mars surface	Comments
Habitat (payload)	30.0	30.0	30.0	30.0	
NTR (dry)	30.0	0.0	0.0	0.0	58% of propellant mass
NTR propellants	51.7	0.0	0.0	0.0	From LEO $\Delta v = 4.45\,\mathrm{km/s}$, $I_{SP} = 900\,\mathrm{s}$
Aeroshell for MOI	8.0	8.0	0.0	0.0	20% of orbit mass
Descent system (dry)	5.7	5.7	5.7	5.7	
Descent system (propellants)	5.1	5.1	5.1		$I_{SP} = 379\,\mathrm{s}$ (methane/oxygen)
Total	*130.5*	*48.8*	*40.8*	*35.7*	

Table 8.3. JPL's reverse-engineering of Descent/Ascent Vehicle (masses in mT).

Descent/ Ascent Vehicle	In LEO	En route to Mars	In Mars orbit	On Mars surface	In Mars orbit	Comments
NTR dry mass	30.0					
NTR propellant— Earth departure	61.0					
Aeroshell	18.1	18.1				
Descent system (dry)	22.7	22.7	22.7			
Descent propellants	8.4	8.4	8.4			$\Delta v = 0.63$ km/s, $I_{SP} = 379$ s
Ascent system (dry)	7.2	7.2	7.2	7.2	7.2	
Ascent propellants	16.1	16.1	16.1	16.1		$\Delta v = 4.0$ km/s, $I_{SP} = 379$ s
Crew				0.6	0.6	
Total	*163.5*	*72.5*	*54.4*	*23.9*	*7.8*	

Table 8.4. Comparison of NTR systems for TMI for three vehicles.

	Propellant mass	K	F (200 km)	F (1,250 km)	F (chem)
CTV	86	0.45	0.45	0.35	0.31
DAV	61	0.49	0.43	0.33	0.31
Habitat	52	0.58	0.40	0.30	0.31

- Provide no allocation for radiation protection or zero-g mitigation.
- Carry an empty, stripped-down 6.5 mT CEV to Mars and back, solely for the purpose of Earth entry.
- Provide a mere 4.4 mT of consumables. Note that according to Table 4.2 the requirements for the round trip to Mars include 4.4 mT of food and waste disposal materials. No allowance was made for the ECLSS to provide water and breathable air during the round trip. This ECLSS mass may actually be somewhere between 5 mT and 50 mT depending on the efficiency.
- Assume no boil-off of LH_2 despite storage for up to ~800 days.
- Assume an unreasonably low Δv for trans-Mars injection.

According to Table 8.2, in order to obtain an IMLEO of ~131 mT for the Surface Habitat, JPL had to:

- Assume that the NTR could start up in LEO.
- Use an NTR dry mass $= 30/(51.7) = 58\%$ of (hydrogen) propellant mass.
- Use an aeroshell for MOI $= 8/(30 + 5.7 + 5.1) = 20\%$ of injected mass (17% of approach mass). However, according to Section 4.6, the mass of the heat shield, backshell, and allocated margin is likely to be about 70% of the mass injected (41% of approach mass).
- Use a descent system from Mars orbit involving a mass $= 10.8/(30) = 36\%$ of payload to surface. However, according to Section 4.6, the mass of the descent system is likely to be about 160% of the payload to the surface. Note that JPL assumed an unreasonably large I_{SP} for LOX/methane $\sim$379 s.[224]

Assume that a 30 mT Habitat not only houses and supports the crew, but provides all life support for up to 600 days, all EVA and mobility equipment, all power and utilities, as well as communication systems and other facilities. No specific mass allocation is made for power systems.

According to Table 8.3, in order to obtain an IMLEO of $\sim$163 mT for the Descent/Ascent Vehicle, JPL had to:

- Assume that the NTR could start up in LEO.
- Use an NTR dry mass $= 30/(61) = 49\%$ of (hydrogen) propellant mass.
- Use an aeroshell for MOI $= 18.1/(22.7 + 8.4 + 7.2 + 16.1) = 33\%$ of injected mass (25% of approach mass). According to Section 4.6, however, the mass of the heat shield, backshell, and allocated margin is likely to be about 70% of the mass injected (41% of approach mass).
- Use a descent system from Mars orbit involving a mass $= 31.1/(23.3) = 133\%$ of payload to surface. However, according to Section 4.6, the mass of the descent system is likely to be about 160% of the payload to the surface. Note, again, that the model assumed an unreasonably large I_{SP} for LOX/methane of $\sim$379 s.
- Use ascent to a circular orbit.

There are many reasons this NASA 2006 conceptual Mars mission will not be feasible. Use of the NTR is a major factor. Table 8.4 presents data relevant to NTR usage (based on Table 3.20). In Table 8.4, F is the fraction of mass initially in LEO sent on its way as payload toward Mars by trans-Mars injection (TMI), the 4th and 5th columns are for TMI using the NTR from either starting altitude, the 6th column is for LOX/LH$_2$ TMI and K is the NTR dry mass/propellant mass. It can be seen that the benefit of using NTR for TMI (compared with chemical propulsion) is quite small if the NTR must be fired up from an altitude of 1,250 km. The assumptions that the NTR can be turned on in LEO and that hydrogen can be stored during the trip to Mars as well as during the $\sim$600 days that the crew is on the surface seems to be very optimistic, especially since no mass has apparently been assigned to the storage system. A strange aspect of this is provided by a comparison with lunar propellants. There does not seem to be any intent by NASA to use hydrogen as an

---

[224] In order to agree with JSC estimates.

ascent propellant or Earth-return propellant from the Moon, presumably because of storage problems, yet NASA proposes to store hydrogen for much longer periods in Mars missions!

Another important factor is the question of how much mass is required to supply consumables (food, waste disposal materials, water, and breathable air) in transit to and from Mars and on the surface of Mars, including allowance for an extra 550–600 days of consumables in the CTV if the crew decides to "abort-to-orbit" instead of remaining on the surface. Even the optimistic JSC estimates of ECLSS systems (Table 4.3) require 5 mT for each one-way transit, 3 mT for ascent or descent, and 22 mT on the surface. The actual required masses are likely to be much higher, particularly when the need for back-up redundancy is taken into account. None of these masses appear to be included in the NASA 2006 Mars mission.

We have independently estimated IMLEO for a human mission to Mars in Section 6.6. The most likely scenario using LOX/LH_2 for Earth departure, methane/LOX propulsion thereafter, full aero-assist for orbit insertion and descent, but no ISRU, leads to an IMLEO of ~1,400 mT. That is about triple the NASA estimate of 446 mT.

8.4   NEED FOR NEW TECHNOLOGY

The critical advanced technologies needed to enable human missions to Mars include aerocapture and aero-assisted descent, long-term cryogenic propellant storage, methane/oxygen propulsion systems, nuclear reactor for surface power, safety—radiation protection, reduced gravity mitigation, life support systems (ECLSS and recycling)—and, more speculatively, ISRU and/or a nuclear thermal rocket. However, the utility and value of some of these technologies depends on factors difficult to pin down at this time. In regard to the NTR, for example, these include the minimum altitude for turn-on, dry mass, hydrogen storage requirements, development cost/time and risk, and *in situ* resource accessibility.

Unfortunately, long, sustained, expensive technology developments are rarely carried through to completion at NASA.

8.4.1   Space Science Enterprise (SSE)

8.4.1.1   *SSE scope of technology*

One could imagine two extreme positions that NASA might take on advanced technology. On the one hand, NASA could take a narrow point of view that it is an organization dedicated to developing spacecraft for exploration of space, and emphasize technologies of importance to space missions, regardless of the world movements and trends in technology. On the other hand, there is a danger of NASA developing the best spacecraft, but filled with obsolete technology. For the world as a whole, the ascendant areas of technology in the 21st century tend to be in the areas of microelectronics, computing, robotics, and biotechnology. Should NASA follow these world movements and develop technology in the most rapidly expanding technologies, or should it take a more constricted view and develop technology for better spacecraft? The record shows that NASA has tended to emphasize "relevant" technologies, although JPL has leaned heavily toward following world trends. In fact, a review of JPL technology discretionary funds in the 1980s and 1990s shows that the majority of funds were allocated to ideas of little relevance to space exploration that ended up going nowhere.[225]

For many years, NASA technology programs were fairly independent of NASA missions. However, by about 1990, it was widely recognized that a gulf existed between the two cultures. In the mission culture, the emphasis was on producing a workable mission with a minimum use of new technology in order to avoid risk, whereas the technology sector, more and more, sought "unprecedented, breakthrough, revolutionary" concepts that rarely if ever panned out, and in any case would require decades for maturation. As a result, a new paradigm became infused in the NASA culture during the 1990s, that technologies would be prioritized in proportion to their putative importance in enabling and enhancing proposed future missions.

However, this did not prove to be as simple as it may seem. One problem was that the list of proposed new missions was always much longer than the number that could possibly be affordable. The technology needs of this excessive list of missions swamped the technology budgets. Another problem was that planners of new missions usually attempted to minimize their needs for new technology, considering NASA's struggle in developing new technology to maturity. Quite often, they downplayed their needs for technology. A third problem was that the people who worked on developing new technology were culturally very different from those who did mission planning, and communication between the two was not usually as clear as would be desired. But, perhaps the greatest problem was the fact that the priorities for missions seemed to change faster than technologists could keep up with them.[226] Nevertheless, the basis of technology prioritization during the 1990s and early 2000s was relevant to future mission needs.

The culmination of this philosophy was the 2003 NASA Technology Blueprint, chartered by Harley Thronson (at that time NASA Technology Director) and Giulio Varsi (NASA HQ), prepared by the author (with help from many NASA personnel).[227] The purpose of this Blueprint was to document the technologies required to implement the future missions of the NASA Space Science Enterprise.

---

[225] The author carried out that study at JPL *sub rosa*, but did not disseminate it.

[226] When the author arrived at JPL in 1979, he met a fine fellow at JPL, Bob Miyake. He asked what he was working on. Miyake said: "Solar Probe, it is planned for mission start in two years." In 2006, he met Bob Miyake walking across the campus at JPL and asked him what he was doing. Miyake said: "I am still working on Solar Probe and we are still hoping for a project start."

[227] This 118-page report is available at *http://www.mars-lunar.net/References/Blueprint.2003.pdf*

It provided the development status of these technologies both within and outside the Agency, and the adequacy of ongoing programs within the Enterprise and elsewhere to develop these technologies to maturation on an appropriate time scale. The Blueprint also identified the areas that were insufficiently funded and those that are inadequately defined. The Blueprint provided a synoptic view of the technology needs and gaps in then-existing programs addressing these needs, providing a good perspective for guiding investment in new and evolving technologies for future missions.

It remains to be seen whether these principles will be applied effectively in developing NASA technology for the NASA Space Science Enterprise (SSE). The advent of the Human Exploration Initiative of 2004, followed by NASA's decision to withhold SSE funds for the Human Exploration Initiative, has cast a pall over SSE technology efforts in 2005–2006. All the plans, analyses, and ideologies in the world will produce nothing if they are not backed up with funding and continuity.

The problem of defining a meaningful advanced technology program for an organization as diverse as NASA is a very difficult one. Over the years, NASA SSE has probably done a decent job of defining many of the needed technology areas and providing some funding to at least some of these areas. If one scans the list of "RTOP" titles, one cannot help but feel that many important topics (though perhaps not all) are included. On the other hand, the NASA budget for advanced technology has been woefully small for many years. NASA's approach to dealing with a multitude of needs and an inadequate budget has been to provide some funding for projects that need considerably more funding to do properly, then getting disillusioned with the lack of progress, and after a few years terminating the effort altogether. Then, the funds are transferred to another area, which is again under-funded, leading to the same result. Thus, while the list of tasks supported by NASA is very impressive, the list of what NASA develops to space flight readiness is rather short.

Technology transfer has not been effective within the present bureaucratic structure, nor will it be in the future in an environment where programmatic priorities change frequently, where in-projects become out-projects at regular intervals (and *vice versa*), where the listed schedule of future projects is so lengthy that it would require trebling the NASA budget to implement it, and where even the main thrust of NASA seems to change dramatically every couple of years. In this environment, a separate establishment with a charge to carry out a program in advanced technology for the benefit of these applications is severely lacking in clear directions. Such a research organization, divorced from the programmatic applications that are elusive and variable, will tend to respond to situational imperatives rather than edicts from above on technology transfer. These situational imperatives include: (1) the research and advanced technology budget of NASA is woefully small by any yardstick; (2) fairness requires that within budgetary constraints many groups be under-funded, rather than concentrate funds in the hands of a few who are most capable; (3) researchers are motivated by published papers and recognition of their peers, not approbation from project managers; (4) research establishments can often win more plaudits by counting papers and reporting conceptual "breakthroughs" for their various small research efforts than they ever could by concentrating the same amount of funds to produce a single, workable piece of hardware.

One area of spacecraft technology that has had some successes (and some failures) is power generation. Most space missions carried out up to now were located in space within a reasonable distance of the Sun. As a result, they utilized solar cell arrays for power combined with batteries for occasional or periodic outages (such as passing into shadow behind a planet). This system has been very successful and the space solar array business has profited from advances by the terrestrial solar array business that is considerably larger in size and funding. Similarly, the terrestrial battery business has made huge advances that have been adopted by NASA for space use. However, the requirements for batteries for space missions differ from terrestrial requirements, particularly in the need for tolerance of very low temperatures. JPL engineers have developed new electrolytes and electrodes to enable commercial batteries to be modified for low-temperature operation, and working with industry have transferred this technology to industry that has manufactured batteries for space missions. This is a bright spot in the dismal picture of NASA technology.[228]

However, some space missions cannot suffice with only solar energy. These include missions to outer planets, missions to polar craters on the Moon, long-term Mars surface missions, and missions that utilize nuclear electric propulsion. All of these require nuclear energy in one form or another. The non-solar energy source used by previous missions is the so-called radioisotope thermoelectric generator (RTG). This device converts about 5% of the heat released by radioactive plutonium isotopes into electrical energy. After some rather abortive ups and downs, NASA has embarked on a program to try to increase the conversion percentage and the specific power. However, for some applications the power requirements are far too high for RTGs to suffice and a nuclear reactor is needed. There are numerous political difficulties in developing a nuclear reactor for space, because the space part is the responsibility of NASA, but the nuclear part is the responsibility of the Department of Energy (DOE). These two organizations have not always been able to work effectively with one another. In addition, development of a space nuclear reactor is a complex technological challenge that involves industrial contractors as well as NASA and DOE. The Department of Defense attempted to develop a space nuclear reactor (SP-100) during the 1980s, but that program got bogged down and finally terminated incomplete. In 2003, with great fanfare, NASA decided to resurrect the SP-100 in a new form to enable a complex mission called the Jupiter Icy Moons Orbiter (JIMO). After expenditure of a considerable amount of effort in preliminary study and analysis, there was a regime change at the top of NASA, and JIMO was canceled and work on nuclear reactors was stifled.

Currently, space nuclear reactor technology is in limbo, despite the fact that NASA seems intent on exploring dark polar craters on the Moon where a nuclear reactor seems almost certain to be needed.

---

[228] This successful program is due to the efforts of Rao Surampudi, Marshall Smart, Ratnakumar Bugga, and others at JPL.

8.4.1.2   Lead Centers

NASA Centers are either designated as research and technology (R&T) centers (LaRC, ARC, and GRC) or applications centers (GSFC, JPL, MSFC, JSC). The NASA approach to advanced technology development is based on the "Lead Center" concept, in which one center (usually an R&T center, but not always) is appointed to carry out the needed research and technology development in any given area of technology for application to missions developed at all the other Centers. For example, LaRC is the Lead Center for structures and structural materials, GRC is the Lead Center for electric propulsion and space communications, and GSFC is the Lead Center for cryogenics. These organizations are chartered to maintain expert staff and facilities in their assigned domains, and carry out relevant R&T to enable and enhance future NASA missions.

The Lead Center concept was created to try to avoid duplication of capabilities in advanced technology between NASA Centers. The theory was that by setting up one Center with the best facilities, and an exclusive budget, this Center would have the capability to meet the challenges and supply the technology needs of all the applications centers for their various future missions. In actual fact, it is widely known, though rarely acknowledged or admitted, that the designated Lead Centers tend to have their own agendas, while mainly paying lip service to the programmatic needs of applications centers. Furthermore, much of the work is on futuristic concepts at early stages of emergence, and rarely (if ever) are concepts developed to a sufficiently mature state for use in space missions. This leaves the applications centers with very few useful new products. As a result, the various applications centers have set up small- to moderate-scoped technology development programs, in order to work on needs for their planned future missions. These efforts at non-Lead Centers are extremely important for many planned projects. Over the past several decades, NASA's handling of these non-Lead Center programs has varied considerably. During the 1980s, NASA discouraged funding of these activities and centralized their technology funding into the Lead Centers. However, under Dan Goldin's "smaller, cheaper, faster" regime in the 1990s, technology development was opened up to competition with the goal that the best team with the best ideas should win, regardless of where it was located.[229] Under Michael Griffin's leadership in 2005–2006, the centralization of technology within Lead Centers has been reinstituted.

The difficulties created by the Lead Center concept may be illustrated by the role of LaRC in materials work on the PSR (precision segmented reflectors) Task to develop reflector systems for sub-millimeter astronomy in space.[230] From the start of the PSR Task, which was conceived and implemented by JPL, a fundamental stumbling block was the need to acknowledge that the Langley Research Center (LaRC) is the officially designated Lead Center for NASA work on structural materials. Therefore, a separate sub-task of PSR was created called "advanced

---

[229] Competitive technology proposals are not without their problems—often the organization that promises the most gets the award, but they may not be the group that can deliver the most.
[230] See *http://www.mars-lunar.net/The PSR Story.pdf* for a detailed review.

materials" and this was assigned to LaRC. A much smaller task, "alternate materials", was given to JPL. Supposedly, JPL was only mandated to work on modifications and improvements of existing materials, whereas LaRC was supposed to look at more far-reaching materials. Little value was added to the Task by the LaRC "advanced materials" subtask. Nevertheless, LaRC received about $2M of precious funds from the PSR Task for materials work, only because of their mandate.

Another example of the effect of the Lead Center culture is the way NASA has funded the excellent JPL cryogenics group on sorption refrigeration development and other related technologies.[231] There are many underlying reasons this cryogenics research was not continued. Part of the problem was that NASA Headquarters had no official "home" for cryogenics technology, and cryogenics was made a sub-element of its sensor development organization. However, the scientists and engineers who develop sensors are solid-state physicists, whereas those who develop cryogenic coolers are usually mechanical engineers. As a result, cryogenics has always been a "step-child" in the sensor organization. A second reason was that the Goddard Space Flight Center (GSFC) was the appointed "Lead Center" for cryogenics, and JPL was considered to be an interloper in this area of technology. At the same time, the high-priority areas of research at JPL in the 1980s were microelectronics, sensors, and supercomputers, and cryogenic cooling of sensors did not receive the support and advocacy from JPL management that was given to these areas of technology. This, in turn, was due to the fact that JPL was designated as a Lead Center in microelectronics. From time to time, claims were made by the Director of the JPL Center for Space Microelectronics that cryogenics technology is not needed because new breakthroughs would supply sensors that did not require cryogenic cooling. This claim violates many basic principles of physics, but was nevertheless proclaimed loudly by JPL's Center for Space Microelectronics. As a result, JPL's hydride refrigerator and solid-hydrogen stage concepts languished in the doldrums from about 1983 until the DOD provided funding for it in late 1990.

NASA's viewpoint that GSFC is *their Center* for sensor-cooling technology seems rather near-sighted. Quality and ascendancy occur where the best people are, not where NASA Headquarters chooses to appoint an official *Center*. Deciding in advance where the work is going to be is not very effective if the people with good ideas and the ability to develop them are at another Center. This does not mean that one can't centralize large facilities at one Center. But, just because major facilities may be located at one Center is no reason not to fund the best people to do research and development wherever they are. At JPL, the sensor-cooling group was world-

---

[231] NASA's start-again, stop-again funding in sensor cooling at JPL has a long history. In 1979, NASA funded an RTOP at JPL, but unfortunately this was decomposed into a number of subtasks, all of which were under-funded. However, the most important accomplishment of this effort was the investigation of sorption refrigeration technology. This RTOP was discontinued in 1984, and JPL's effort in sorption refrigeration would have died had JPL internal funds not backed the effort to develop sorption refrigeration for the HIMS instrument. NASA again began funding in this area in 1988, but then abruptly canceled it after 1990. Then, NASA promised very substantial funding for 1993, but at the last minute cut this to zero.

class and unexcelled in several areas of cryogenics. NASA's on-again, off-again funding of this group during the 1980s and 1990s was detrimental to NASA interests.

Spacecraft power had been an exceptional area where a fairly good accommodation had been secured between JPL and the Lead Center, GRC. For example, JPL worked on arrays while GRC worked on solar cells. This has required the incredible and sustained efforts of Rao Surampudi at JPL to keep this fragile arrangement from breaking.

The Lead Center approach never has worked, it doesn't work now, nor does it have prospects of working in the future. Spacecraft technology needed for NASA space science missions will develop slowly as long as the Lead Center system remains in force.

8.4.1.3 On again/off again

The following is an excerpt from a memo that I wrote at JPL on November 2, 1993:

"There are some things that are seasonal, like the annual migrations of birds and whales, the onset of winter, or the monsoons in India. At JPL, we have come to expect that every spring, when the NASA research and advanced technology budgets are being deliberated, there will be the usual move to zero-out JPL technology efforts which overlap with the nominal sacred territories of NASA Lead Centers. JPL will respond by sending Art Murphy and others back and forth to Washington with pleas, reclamas, petitions, appeals, defenses and explanations of why the zeroed-out work should be reinstated. Sometimes these efforts are partly successful, and sometimes they are not. Whether funding is restored or not, the inevitable effect of these annual exercises is demotivation and disillusionment of JPL technologists, set-backs to progress in advanced technology, creation of an atmosphere of fear, uncertainty and insecurity, and a generally negative outlook for JPL technology work. A few examples of these annual occurrences are:

1. Zeroing-out of the Microprecision Control-Structure Interaction (CSI) Task for FY 1993, and subsequent reclamas and restoration of funds.
2. History of on-again/off-again funding for sensor-cooling work at JPL:
   - RTOP funded 1980–1984
   - Zeroed-out 1985–1987
   - RTOP reinstated 1988–1989
   - Zeroed-out 1989–1992
   - Promise of major funding for new RTOP for FY1993.
   - Zeroed-out 1993–?
3. Zeroing-out of Materials and Structures RTOP in FY 1994.
4. Zeroing-out of the JPL Power Program for FY 1994, and subsequent reclamas and restoration of funds.
5. Zeroing-out of High Performance Photovoltaics Array and Solar Cell/Environmental Interactions RTOP in FY 1994.
6. Potential zeroing-out of JPL Electric Propulsion for FY1995.

I want to make it clear that I do not argue that all RTOPs should be continued *ad infinitum*, nor that NASA should not review the quality and importance of the work to determine whether to continue RTOPs, and at what level of funding. What I do argue against is the arbitrary zeroing-out of very productive technology activities at JPL which have significant potential to provide important needs to planned JPL missions, simply because the subject matter overlaps (at least nominally) with the assigned responsibilities of a NASA Lead Center. I do not argue here against the general concept of Lead Centers as central hubs of activity in selected areas (although the inherent effectiveness of the Lead Center system must remain as a subject for future discussion). I argue here against the concept of totally exclusive Lead Centers with no allowance for benefits of limited specialized efforts at Centers other than the Lead Center. These specialized efforts are justified at JPL when unique JPL talent, capabilities and facilities are joined with a strong motivation to fulfill critically needed technology developments for future JPL missions."

The author has little reason to believe that this memo, written 14 years ago, is not appropriate today in 2007.

8.4.1.4   SSE technology summary

The problems facing SSE technology development include:

- Too many future missions planned.
- Future missions minimize their stated needs for new technology.
- Changing priorities for future missions.
- The mission–technology cultural gap.
- Time and funds needed to develop new technology to maturity.
- Emphasis on incredible, breakthrough, revolutionary technologies, to the detriment of evolutionary technologies.
- Lead Centers vs. competition by the best people with the best concepts.
- Start, stop, and begin again, *ad infinitum*.
- Long-term plans—plan and replan and re-replan.
- Organization, reorganization, and lack of continuity in management.
- Separation of control (management) and knowledge (technologists).

Interestingly enough, it appears that (at least in the past) most technology maturation has been carried out within space projects, rather than within the technology community. The Mars Pathfinder Rover is an excellent example. In the past decade or so, the New Millennium Program has been chartered to carry out ultimate validations of technologies in space in the hope of bridging the gap between the technology and mission cultures. They had a notable success in first demonstrating solar electric propulsion in space. It seems likely that this program will play an important role in future technology maturation.

   The strangest thing about technology maturation is that, even though it is rarely accomplished, almost everyone seems to discuss the process for achieving it as if it were obvious.

8.4.2   Human exploration technology

For a number of years, JSC has carried out research and technology in various areas related to human exploration. Like JPL, JSC has adopted a very broad approach to technology, much of which is only vaguely related to the JSC mission (e.g., an inordinate preoccupation with nano-tubes). JSC maintains a website with pdf versions of technical reports that are downloadable.[232] This is a very nice feature. However, only a few of the reports seem to directly contribute to JSC's mission to develop human spaceflight. Three critical areas relevant to human missions where significant research has been carried out are life support, artificial gravity, and radiation effects. However, the amount of work accomplished does not seem to be in consonance with the importance of these topics, and the reporting of results has been sporadic and sparse.

One critical area of research that has been ongoing for many years at JSC is the Advanced Life Support (ALS) program. A positive factor is that a number of reports are available from JSC. Section 4.1 provides a discussion of these. However, the missing link in all of the ALS reports on life support systems is the lack of connection to experimental data. NASA has embarked on a program to acquire relevant data via the Advanced Life Support Project. This was preceded by the Lunar–Mars Life Support Test Project in the 1990s that consisted of a series of closed-chamber, human tests that demonstrated operation of closed-loop life support systems for increasingly longer durations. The final test in the series, the Phase III test, incorporated the use of biological systems as well as physicochemical (P/C) life support system technologies to continuously recycle air, water, and part of the solid-waste stream generated by a four-person crew for 91 days. These tests appear to have been a good start, but only a small step toward proving that an ECLSS for Mars is feasible. In 2002 this work seems to have evolved into another program called the Advanced Life Support Project. A plan for this project is available on the ALS web site. The plan seems very ambitious, perhaps overly so. However, it is difficult to discern what progress has been achieved since the plan was published 4 years ago.

Another critical area of research at JSC is artificial gravity. This is discussed in Sections 4.3.2 and 4.3.3. So far, the work seems to be mainly conceptual although some experiments with centrifuges appear to be included.[233] However, the author has not been able to find any reports of results.

Radiation effects were discussed in Section 4.2. Over the years, some very good work was done by NASA teams, and most recently the work by Cuccinotta and co-workers has been very revealing. However, NASA mission planners seem to view radiation effects through rose-tinted glasses.[234]

---

[232] *http://ston.jsc.nasa.gov/collections/TRS/*

[233] *Artificial Gravity for Exploration Class Missions?* W. H. Paloski, NASA–JSC, September 28, 2004.

[234] "Estimating the Integrated Radiation Dose for a Conjunction-Class Mars Mission Using Early MARIE Data," John F. Connolly, *Earth & Space 2004, Engineering, Construction, and Operations in Challenging Environments, 9th Biennial Conference of the Aerospace Division.*

8.4.3 Exploration Systems Research and Technology (ESRT)

Soon after the President announced the new Human Exploration Initiative in 2004, the Exploration Systems Research and Technology Director issued two calls for proposals for research and development in human and robotic technology.[235] The first call was for "intramural" proposals within NASA Centers and that was followed by a call for "extramural" proposals led by industry. Quite a large number of grants, totaling several hundred million dollars, were awarded in both categories. Most of these awards were for innovative technology concepts, but their connection to exploration mission needs were at best indirect in most cases. It is noteworthy that these calls for proposals were issued hurriedly, and proposers were given only 1 month to submit proposals. There seemed to be a headlong rush to get the proposals funded as soon as possible, and the value of the selected research projects to the needs of exploration systems remained rather ephemeral. These research efforts had been in effect for about a year when Michael Griffin took the helm at NASA. One of his early actions was to terminate most of those ongoing technology development efforts.

In a memo dated Friday, October 28, 2005, 4:44 PM by Lon Forehand (HQ-NC010):

Subject: NASA Realigns Research and Technology to Accelerate CEV/CLV

"NASA has completed the Exploration Systems Architecture Study (ESAS), ... CEV and CLV development requirements now directly drive the content of ESMD's R&T components ... Focus is shifted from advancing technologies for long-term requirements to directed research and maturing technologies for near-term use. As a result of these R&T requirements, ESMD is undertaking ... to suspend expenditures on specific R&T tasks that will not be continued in FY 2006. FY 2006 funding made available as a result of this transition will be redirected to Project Constellation to enable timely development of the CEV and CLV."

He went on to say that $292M would be moved from R&T activities into Project Constellation for CEV and CLV acceleration during FY 2006 and an additional $493M was identified from the R&T activities for acceleration of CEV and CLV. This yielded a total shift from R&T to Constellation in FY 2006 of $785M, relative to original plans for FY 2006. In order to achieve this level of funding realignment within Exploration Systems, NASA initiated actions immediately with respect to those affected activities. Thus, a total of $785M allocated in 2004–2005 for exploration R&T was canceled in 2006 to support CEV development. How can an organization make huge awards in one year, only to rescind them the next year?

Another concern is whether NASA "threw out the baby with the bathwater." Lon Forehand's memo says that NASA changes as of late 2005 include the

---

[235] Broad Agency Announcement, dated July 28, 2004.

following:

- De-emphasis of radiation-shielding materials development.
- Substantial reduction in artificial gravity research for crew physiological counter-measures to support Mars transits.
- Deferment of investments in long-duration, closed-loop life support technologies until lunar outpost or Mars transit requirements dictate the need for future investments.

These are all giant steps in the wrong direction. The outgrowth of the great debacle in exploration technology was installment of a new plan beginning in FY 2006. The technology work included in this new plan is summarized in Table 8.5. Clearly, this set of tasks is far more closely tied to the immediate engineering needs of the exploration program. However, the "devil is in the details", and examination of the details is not always convincing. For example, consider the set of tasks 10A to 10H dealing with lunar ISRU. These tasks stem from a study conducted in 2005 by the ISRU Capability Team as part of a NASA-wide "roadmap" activity.[236] The ISRU Capability Team Report is discussed in Section 5.2. The optimism exuded in this report is highly questionable. Nevertheless, this report provided a springboard to establishment of the ISRU tasks.[237] But, major problems still exist in regard to lunar ISRU as discussed in Sections 5.2.4 and 5.2.5. These include the facts that: (1) recovery of oxygen by high-temperature processing of regolith appears to be imprac-tical; (2) plans are inadequate to prospect for ground ice in polar craters; (3) there are no credible plans to develop nuclear power for operations in dark polar craters; (4) lunar mission planners have relegated ISRU to a late program add-on with minimum benefits and the benefit/cost ratio for lunar ISRU in any form is far less than 1.0; (5) without LOX as an ascent propellant, there is no short-term ISRU benefit; and (6) if abort-to-orbit is a requirement during descent, there is no short-term ISRU benefit.

The tasks listed in Table 8.5 are predominantly engineering tasks to develop prototype hardware for near-term use. Most of this work is based on general principles and methods that are already well known, and the challenges typically involve new configurations or extensions of previous systems. Development of a full-scale aero-entry, descent, and landing system for >40 mT vehicles on Mars would require an effort far greater than anything in Table 8.5. It remains to be seen whether NASA will carry out such an ambitious program that might require two decades of development and demonstration.

It should be noted that according to Table 8.5 the NASA ESRT is putting 100% of its effort into lunar technology and 0% into Mars technology. While it is true that lunar missions are nearer-term and require the majority of technology funds, total exclusion of Mars from funding seems extreme. In the specific case of ISRU where Mars ISRU appears to be far easier to implement and has much greater mission impact, this seems to be self-defeating.

---

[236] See *http://www.mars-lunar.net/References/ISRU.capability.team.roadmap.pdf*
[237] The ISRU Capability Team Report provided a basis for NASA to fund the current $80M Lunar ISRU Technology Program.

8.5 VISION FOR 2020

The author's vision for 2020 is that prior to ~2020, almost all ESAS funding will go into lunar programs, and Mars will be on the far "back-burner". After 2020, the ESAS Report calls for parallel funding of the lunar outpost(s) and "Mars development".[238] The probability that this plan will be implemented by 2020 is not great, despite NASA pushing their plan forward as fast as possible. However, there has never been a plan of this sort that did not encounter delays, overruns, and other difficulties. Therefore, if NASA predicts that it will be ready to begin work on Mars in 2020, we must expect that it might be closer to 2025 or 2030. The requirements for lunar outposts will likely usurp most Mars funds between 2020 and 2025, and maybe out to 2030. Nevertheless, if we adopt an optimistic posture, we might assume that they are ready to do some Mars development around 2020. However, we must take stock of the likely starting point in 2020.

The expected state of Mars technology in the 2020 era is summarized below:

On the negative side:
- Relatively little actual experimental progress on technology for Mars aero-capture and aero-assisted descent for large (>30 mT) vehicles.
- Little actual progress on nuclear thermal propulsion beyond paper studies.
- No *in situ* experiments on Mars to map near-surface water resources via rover-mounted neutron spectrometers, validate via subsurface sampling, and determine requirements for excavation by prototype excavators.
- Little progress on Mars ISRU technology.
- Uncertain lifetimes and reliability of long-term (200–1,000 day) life support systems.
- Little actual progress on development of a surface nuclear reactor.
- Little progress on construction and testing of artificial gravity hardware.
- Little progress on the radiation threat.

On the positive side:
- Lunar Habitat designs will be available that can be modified for use on Mars.
- Life support systems will be in place that function reliably for several months (efficiency and lifetime remain uncertain).
- Regolith-moving equipment that has been field-tested on the Moon.
- Earth departure systems based on LOX/LH_2 propulsion.
- Rendezvous, assembly, and disassembly capabilities in LEO and lunar orbit.

Unlike the lunar case in 2004–2006, the state of technology in 2020 will be inadequate to allow immediate design of Mars vehicles and missions. It is premature in 2006 to attempt to define vehicle masses and mission architectures for human exploration of Mars. In planning for human exploration of Mars, NASA needs to take a broad view of the end-to-end campaign. A much more thorough system analysis of campaign

---

[238] *ESAS Report*, pp. 58, 59 and 62.

Table 8.5. 2006 Exploration Systems Research and Technology (ESRT) Program.

No.	Field	Specific task
1A	Structures	Lightweight structures—pressure vessel, insulation (vehicle)
2A	Protection	Detachable, human-rated, ablative, environmentally compliant TPS
2C		Lightweight radiation protection for vehicle
2E		Dust and contaminant mitigation
3A	Propulsion	Human-rated, 5–20 K lbf class in space engine and propulsion system
3B		Human-rated, deep throttlable engine 5–20 K lbf engine (lunar descent, pump-fed LOX/LH_2 baseline)
3C		Human-rated, pump-fed LOX/CH_4 5–20 K lbf thrust-class engines for upgraded lunar LSAM ascent engine)
3D		Human-rated, stable, monoprop, 50–100 lbf thrust-class RCS thrusters
3F		Manufacturing and production facilities
3G		Long-term, cryogenic storage and management (for CEV)
3H		Long-term, cryogenic storage, management, and transfer (for LSAM)
3K		Human-rated, non-toxic 900 lbf thrust-class thrust class RCS thrusters
4B	Power	Fuel cells (surface systems)
4E		Space-rated lithium-ion batteries
4F		Surface solar power (high-efficiency arrays, and deployment strategy)
4I		Surface power management and distribution
4J		Launch Vehicle power for thrust vector and engine actuation
5A	Thermal	Human-rated, non-toxic, active thermal control system fluid
5B	control	Surface heat rejection
6A	Avionics	Radiation-hardened/tolerant electronics and processors
6D	and	Integrated system health management
6E	software	Spacecraft autonomy (vehicles and Habitat)
6F		AR&D (cargo mission)
6G		Reliable software/flight control algorithms
6H		Detector and instrument technology (Mars precursor measurements)
6I		Software/digital defined radio
6J		Autonomous precision landing and GN&C (lunar and Mars)
6K		Lunar return entry guidance systems (skip entry capability)
6L		Low-temperature electronics and systems (permanent shadow region operations)
7A	ECLSS	Atmospheric management (CO_2, contaminants, and moisture removal)
7B		Advanced environmental monitoring and control
7C		Advanced air and water recovery systems
8B	Crew	EVA suit (surface including portable life support system)
8E	support and	Crew healthcare systems
8F	accommo-dation	Habitability systems (waste management, hygiene)

No.	Field	Specific task
9C	Mechanisms	Autonomous/teleoperated assembly and construction
9D		Low-temperature mechanisms (lunar permanent shadow region ops)
9E		Human-rated airbag or alternative Earth landing system for CEV
9F		Human-rated chute system with wind accommodation
10A	ISRU	Demonstration of regolith excavation and material handling
10B		Demonstration of oxygen production from regolith
10C		Demonstration of polar volatile collection and separation
10D		Large-scale regolith excavation, manipulation, and transport
10E		Lunar surface oxygen production for human systems or propellant
10F		Extraction of water/hydrogen from lunar polar craters
10H		*In situ* production of electrical power generation
11A	Analysis	Architecture/mission/technology analysis/design, modeling and simulation
11B		Technology investment assessment and systems engineering and integration
12A	Operations	Supportability (commonality, interoperability, maintainability, logistics, ...)
12B		Human–system interaction (including robotics)
12C		Surface handling, transportation, and operations equipment (lunar or Mars)
12E		Surface mobility

architecture alternatives is needed that integrates alternative two-decade development and validation programs and their impact on mission capabilities with mission architecture alternatives and their requirements for new technology. Any viable concept for a human Mars mission requires large-scale, fully aero-assisted entry, descent, and landing. This will require a costly two-decade development and validation program. Failure to recognize this will lead to unrealistic long-term plans. In 2020, NASA will not be able to assemble elements of existing technology to enable a Mars mission. NASA will be faced with the need to develop new technologies that are very costly, and require many years to validate at Mars. NASA will not be able to immediately plan Mars missions and design vehicles—too many unknowns will remain and questions arise about the feasibility of NASA undertaking a ~two-decade precursor program leading up to eventual human missions to Mars.

Since NASA seems to utilize the lunar paradigm in planning for Mars, the 2020 activity may adopt plans for something less than a 550–600 day stay on the Mars surface for the initial missions. One possibility is to merely send a crew to Mars orbit and teleoperate equipment on the surface from orbit for 550–600 days. That would give NASA more time to develop entry, descent, and landing technologies. Another possibility is that NASA might land the crew for a short period (~30 days) and then return them to orbit for another $\sim$500 days prior to Earth return. That would eliminate the need for a massive long-term surface Habitat and life support system, as well as power and mobility capabilities appropriate for a long-term surface mission. The mass delivered to the surface would be far less than for a long-term surface mission. Such a "sortie" has actually been mentioned in NASA plans as a possibility. However, any mission where the crew spends $\sim$200 days in transit to Mars, $\sim$550 days in Mars orbit, and another $\sim$200 days in transit back to Earth would have to contend with (1) exposure to excessive radiation, (2) long-term exposure to zero-g, (3) long-term requirements for life support systems, and (4) incredible boredom and confinement.

8.6   HUMAN EXPLORATION STRATEGY

8.6.1   Background

8.6.1.1   *Destination-driven vs. constituency-driven programs*

In order to assess the relative merit of alternate potential exploration strategies, it is useful to begin by looking back a bit into the history of NASA's activities involving humans in space. Robert Zubrin[239] distinguishes between programs that are "destination-driven" vs. programs that are "constituency-driven". As Zubrin says:

> "In the Apollo Mode, business is (or was) conducted as follows: First, a destina-tion for human spaceflight is chosen. Then a plan is developed to achieve this objective. Following this, technologies and designs are developed to implement that plan. These designs are then built and the missions are flown. The Shuttle Mode operates entirely differently. In this mode, technologies and hardware elements are developed in accord with the wishes of various technical com-munities. These projects are then justified by arguments that they might prove useful at some time in the future when grand flight projects are initiated. Con-trasting these two approaches, we see that the Apollo Mode is destination-driven, while the Shuttle Mode ... is constituency-driven ... In the Apollo Mode, the space agency's efforts are focused and directed. In the Shuttle Mode, NASA's efforts are random and entropic."

Zubrin goes on to say that: "The Shuttle Mode is hopelessly inefficient because it involves the expenditure of large sums of money without a clear strategic purpose." He quotes Sean O'Keefe (the NASA Administrator from 2001 until early 2005) as repeatedly rebutting critics by saying that "NASA should not be destination-driven." As Zubrin points out, advocates of the Shuttle Mode claim that by avoiding the selection of a destination they develop generic technologies that will supposedly allow

---

[239] *Getting Space Exploration Right*, Robert Zubrin, The New Atlantis, 2005; also *http://www.thenewatlantis.com/archive/8/zubrin.htm*

us to go anywhere, anytime. Unfortunately, the Shuttle Mode has not gotten us anywhere, and does not seem likely to ever get us anywhere.

After completion of the Apollo Program, NASA elected to develop a generic access to space in the form of the Space Shuttle. Unfortunately, the Space Shuttle has been used over the years mainly to transport crews and minor experiments back and forth, with very little significant accomplishment. In fact, weighed against the funds expended to fly them many of the payloads were so trivial. As Zubrin points out, "Columbia was lost on a mission that had no significant scientific objectives, certainly none commensurate with the cost of a Shuttle mission, let alone the loss of a multi-billion dollar shuttle and seven crew members ... The Columbia flight program included conducting experiments in mixing paint with urine in zero-gravity, observing ant farms, and other comparable activities—all done at a cost greater than the annual federal budgets for fusion energy research and pancreatic cancer research, combined." In fact, the Shuttle has reached the point where the latest flights seem to have a major mission objective of doing enough inspection and repairs in space to assure a safe landing on return!

When he took the helm at NASA, Michael Griffin tried to phase out the Shuttle. As unproductive as the Shuttle Program is, the ISS produced less. Good sense would dictate that NASA should kill both the Shuttle and the ISS and get on with the Exploration Program, but that seems to be fraught with political difficulties.

8.6.1.2   *The 2004–2005 Human Exploration Initiative*

After several abortive attempts during the 1990s, President Bush (i.e., the 2nd) began pushing NASA back toward the destination-driven mode starting in 2004. Unfortunately (or fortunately for those on the receiving end of funding), the initial thrust of this movement in 2004 was not planned in any cohesive way, and was heavily fragmented and disorganized. In an almost frantic attempt to get generic technology started, large amounts of funding were distributed based on hurriedly prepared and reviewed proposals that were evaluated one at a time, in the absence of any cohesive plan for exploration other than the general notion that we would return to the Moon as a stepping-stone toward Mars. Although this program was nominally destination-driven, in actual fact it was administered in the same old way as support to constituencies independent of any specific mission plan, because there was no mission plan. Somewhat after this, in late 2004, a NASA-wide road-mapping activity under cognizance of the National Research Council (NRC) was instituted to lay out the future mission structure for human exploration of the Moon and Mars. This process was a good illustration of the aphorism that "the road to hell is paved with good intentions". A total of 26 committees were formed, of which 13 were strategic and 13 were concerned with capabilities, and each of these had typically 20–40 members. Communications between committees were rare, and no credible plan for assembly of the pieces into a whole was ever devised. Nor could there be, considering that guidelines, constraints, objectives, and funding profiles were not provided to the committees. At least some of the capability teams operated in constituency mode—that is to say, they included every conceivable technology advocated by a constituency in an over-bloated plan that would break the NASA bank and not work technically or programmatically. As Zubrin says: "Mention humans-to-Mars within the NASA community, and you will be deluged with proposals for space stations and fuel depots in various intermediate locations, fantastical advanced propulsion technologies, and demands that billions upon billions of dollars be spent on an infinite array of activities that define themselves as necessary mission precursors. Representatives of such interests sit on various committees that write multi-decade planning 'roadmaps' and exert every effort to make sure that the 'roads,' as it were, go through their own home towns." The roadmap process became increasingly divergent with time, and as each committee accommodated its constituencies, thus increasing its scope and implicit cost well beyond reality, it became clear that this whole process had become untenable.

At about this point in the spring of 2005, Dr. Michael Griffin was appointed as the new NASA Administrator. To his credit, two of the first things that he did were (1) he canceled out most of the fragmented technology development tasks awarded in 2004, and (2) he declared victory on the road-mapping activity and filed the 26 reports away in an archive, never to be used again. As already stated (see p. 244), following this the NASA Administrator assembled a set of technical and system experts to work together as a cohesive team to plan out sequential missions and technology needs for the Human Exploration Initiative—the "60-day study". The first appearance of the ESAS Report was on *NASA Watch* via a leaked version. While NASA probably intended to keep this document secret, the appearance on *NASA Watch* forced their hand and they soon afterward released it to the public (without a table of contents or appendices).[240]

The following discussion of human exploration strategy will concentrate on "how" rather than "why". Considering the cost compared with robotic missions, human missions, particularly to Mars, seem to offer a relatively small return on investment. There are many discussions in the literature of "why" to send humans into space, none of which are very convincing. For example, Bob Zubrin provides a passionate rationale for humans to Mars based on science return, social significance, and its implications for the future. His scientific rationale is based on a search for life on Mars, which is the backbone of the current robotic Mars Exploration Program. This is bolstered by the widely prevalent belief that formation of life from inorganic matter is a likely process "wherever there is liquid water, a temperate climate, sufficient minerals, and enough time." This belief does not appear to have any basis in fact or theory. Much has been made of the fact that there is evidence that life first appeared on Earth in simplest forms soon after it cooled down from the bombardment phase, about 3.5 billion years ago. As the argument goes, if life appeared that soon, then formation of life must be highly probable. But, this argument is specious. If we have a time span over which an event might (or might not) take place, the only way we can determine the probability of the event occurring vs. time is to observe a large number of parallel systems and build up statistics. If we observe only one system and observe that the event took place relatively early in that time span, it tells us

---

[240] *http://www.nasa.gov/mission_pages/exploration/news/ESAS_report.html*

nothing about the innate probability of the event occurring. In fact, it is precisely because we are here to record the event that we are aware of it. Furthermore, the innate mathematical probability that inanimate matter would rearrange itself spontaneously into living matter is exceedingly low, as Hoyle has shown. In addition, we don't even know if life arose spontaneously on Earth, or if it was "seeded" from without. One of the irrational aspects of all this is the number of Mars scientists who seek to look for liquid water on Mars (a necessity for life) when the probability of finding liquid water at any reachable depth is very low.

The NASA MEPAG (Mars Exploration Program Analysis Group) was asked by NASA to work with the science community to establish consensus priorities for the future scientific exploration of Mars. Those discussions and analyses resulted in a report entitled *Scientific Goals, Objectives, Investigations, and Priorities*, that is informally referred to as the *Goals Document*. In a recent version of this report,[241] there are four over-arching goals, each of which is supported by more focused objectives and suggested investigations to fulfill these objectives and goals. The goals are: (1) determine if life ever arose on Mars, (2) understand the processes and history of climate on Mars, (3) determine the evolution of the surface and interior of Mars, and (4) prepare for human exploration. It is noteworthy that goal #1 (determine if life ever arose on Mars) has the greatest number of subordinate objectives and investigations and appears to be the top priority. However, our concern here is with goal #4 (prepare for human missions).

It should be noted that an alternate basis for our current exploration program that has been suggested by some is to exploit planetary resources and thereby create outposts and establish bases in the solar system using primarily indigenous space resources. While this makes more sense to this writer than the search for life, proponents of this philosophy appear to have grossly underestimated the technical difficulty and cost of pursuing this approach.

There is an even better reason to concentrate on "how" rather than "why". In several of his recent speeches, Michael Griffin has assured us that we are going to the Moon (and by implication, Mars) and that it is a waste of time to argue whether we should or should not. The question then arises: "based on all the work summarized in this report, how should NASA plan its exploration program to the Moon and Mars?"

8.6.2 Current NASA plans (2006–2007)

In the current (2006–2007) NASA Human Space Initative, mission plans for exploration of the Moon have taken a back seat to immediate plans for developing the replacement for the Space Shuttle, the so-called Crew Exploration Vehicle (CEV). What seems to be evolving initially is development of the CEV, and then building the lunar missions around the capabilities of the CEV. In addition, a new Heavy

---

[241] *Mars Scientific Goals, Objectives, Investigations, and Priorities: 2005*, 31 p. white paper posted August, 2005 by the Mars Exploration Program Analysis Group (MEPAG) at *http://mepag.jpl.nasa.gov/reports/index.html*

Lift Launch Vehicle, based primarily on Shuttle technology, will be developed. The capability is claimed to be 125 mT to LEO. The plan for lunar exploration involves an initial phase with short-term sorties to the lunar surface, followed by establishment of a long-term outpost with gradually increasing infrastructure, and crews rotated every few months. However, this plan was very sketchy. One possibility is that this outpost would be placed near the lunar South Pole in the hope of prospecting for water deposits in permanently shaded areas, although it is far from clear how they will obtain power. A more tangible benefit might be a more benign thermal environment. Nevertheless, it remains unclear where the lunar missions will land, what they will do there, and why.

Lurking well in the background are vague ideas about extending exploration to Mars. However, plans for exploration of Mars have received only cursory attention, and it is a source of concern to those who doubt that such a lunar program feeds forward to Mars very smoothly.

8.6.3   Recommended strategy

Based on all the studies described in previous sections, we can evolve an overall strategy that is quite different than NASA's. One might begin by taking at face value the assertion made many times by NASA: that the Moon is mainly a stepping-stone toward Mars. Therefore, it is necessary to elucidate a fairly definite plan for human exploration of Mars at the very beginning, and then use this as a basis for defining missions to the Moon that could serve as test-beds and validations for technologies intended for use in Mars missions. Instead of pursuing lunar missions in detail and hoping to someday extend lunar technologies to Mars exploration, one could totally reverse this emphasis and derive plans for the Moon based on unknowns, challenges, and uncertainties associated with human missions to Mars. While lunar missions may have value in their own right, nevertheless the prime motivation for lunar missions would be to enable human exploration of Mars.

In addition, it does not make sense to go ahead with design and implementation of the CEV for lunar missions at this time. The requirements for a Mars mission that involves transfers that take >6 months are likely to be quite different from requirements for transfer to the Moon that require just a few days. It seems clear that the lunar CEV cannot easily be upgraded for Mars transfers. Indeed, current NASA plans for Mars use the wasteful approach of carrying the lunar CEV all the way to Mars and back merely as a reentry capsule on return to Earth.

The key to a better strategy for human exploration is to begin with an in-depth systems analysis of options for human exploration of Mars, as the driving force for the exploration program. In this first step, NASA should carry out a thorough, detailed analysis of options for human exploration of Mars. Unlike previous design reference missions (DRMs),[242] this should not be an advocacy study pre-ordained to present a sales pitch to convince managers that human exploration of Mars is feasible and affordable. Nor should it relentlessly hold onto propulsion concepts that may

---

[242] See Section 6.2.

have merit on paper, but which are hopelessly impractical in reality. Instead, it should provide a sober, neutral evaluation of options and challenges, utilizing propulsion systems that are very likely to be technically feasible, affordable, and politically acceptable. The Mars mission analysis should make a clear and neutral evaluation of the alternatives to either carry out multiple visits to the same site while adding infrastructure, vs. visits to several dispersed sites to maximize global coverage which do not provide the value and safety associated with repeat visits to one site. If, as seems likely, the strong preference is for repeat visits to one site, but feasibility of water-based ISRU affects this conclusion, a final decision on this issue may have to wait several years until such feasibility can be established by robotic precursor missions. The "short-stay" concept should be discarded. The present state of Mars DRM analysis is inadequate and lacking in connectivity to the Moon. While the MIT Study[243] did some good preliminary work in this regard, NASA has done very little.

Only after completion of this study, which is likely to require the good part of a year by a large team, should an overall plan be produced for development of the wherewithal to explore Mars using lunar tests and demonstrations to the fullest extent. As a part of this initial study, analysis is needed of options in critical technology areas. However, these analyses must not be allowed to cater to the fancies of constituencies bent on working on their pet projects. They must utilize non-advocate reviews to assure balance and feasibility.

As of April, 2007, NASA began a new Mars mission study, but it did not appear to have sufficient breadth or depth, and it seems likely to repeat the same old fallacies: (a) unjustified assumptions regarding NTP, ECLSS, and aero-entry/pinpoint landing; (b) inadequate or omitted mitigation of radiation and zero-g, reliance on solar power; (c) optimistic mass estimates for everything; and (d) treatment of ISRU as an embellishment to be added late in the campaign.

8.6.3.1   Propulsion systems

The predilection of MSFC propulsion technologists for electric propulsion must be countered by sober non-advocate reviews. Use of electric propulsion for Mars missions, whether nuclear or solar-powered, while appealing in some ways on viewgraphs, introduces a number of technical, programmatic, and logistic difficulties that have been glossed over in studies by advocates. Either use of electric propulsion must be shown to have pragmatic feasibility and mission merit, or the idea ought to be scrapped (this is the likely outcome). Amongst the many difficulties are cost to develop and validate, requirements for xenon fuel, feasibility of low mass and low-cost arrays for SEP, feasibility of firing up a reactor in LEO for NEP, extended duration of transfers, and need for a fast Crew Express.

Use of a nuclear thermal rocket (NTR) should be rigorously evaluated on a non-advocate basis with particular emphasis on the cost to develop and validate this

[243] "The Mars-Back Approach: Affordable and Sustainable Exploration of the Moon, Mars, and Beyond Using Common Systems", *International Astronautical Congress, October 17–21, 2005.*

technology, requirements for hydrogen storage, and likely altitude where it can be safely turned on. Unless these can be shown to be within acceptable and useful limits, the NTR should be dropped from consideration. Conversely, if the NTR appears to be practical, its development should be one of the higher priority technology developments for human exploration. Since a surface nuclear reactor is clearly needed for power on long-stay human missions, development of such a reactor should be initiated as soon as possible, based on previous work done on SP-100 and Prometheus.

If NASA embarks on development of the NTR, any overlaps between the technologies involved in the NTR reactor and the surface power reactor should be emphasized in the technology plan.

While propulsion technologists may be enamored of electric propulsion, tethers, solar sails, and even more advanced technologies, the fact is that human exploration over the next several decades is highly likely to rely mainly, if not entirely, on chemical propulsion. Cryogenic propellant storage then becomes a crucially important technology.

8.6.3.2   *Aero-assisted entry, descent, and landing*

Aero-assisted entry, descent, and landing (EDL) is the most important technology (by far) to enable future human missions to Mars. As Table 3.17 shows, the values of IMLEO for Mars missions are too high to contemplate unless aero-assisted EDL is implemented. This technology is discussed in Section 4.6. Pinpoint landing accuracy must be incorporated into the EDL scheme. Due to the need to test systems of gradually increasing scale at Mars and the 26-month interval between launch opportunities, it is likely that the program to develop EDL systems will not only cost many billions of $, but may take two decades to complete.

During the 2005 NASA roadmap exercise, the Aerocapture, Entry, Descent, and Landing (AEDL) Capability Team outlined a program to assure that human-scale vehicles can be landed on Mars.[244] They pointed out the flight dynamics differences between the Moon and Mars. At the Moon, ballistic "entry" is followed by a long (11 min) propulsive descent to surface. The terminal descent burn starts around 18 km at 1.7 km/s. However, this cannot be used at Mars because of the much higher entry velocity at Mars, and the fact that the atmosphere starts high up (>100 km) requiring aero-thermal protection at these speeds. Natural variations (density and winds) in the atmosphere strongly perturb the system (much worse than the gravity variations at the Moon). The system needs to muscle through these uncertainties. In landing humans, there are flight dynamics differences that must be taken into account. In a sense, the problem at Mars is that there is too much atmosphere to land as we do on the Moon, and too little atmosphere to land as we do at Earth (with the Mars atmosphere being ~1% of Earth's, imagine landing the Shuttle at 100,000 ft).

---

[244] *Aerocapture, Entry, Descent and Landing (AEDL) Capability Evolution toward Human-Scale Landing on Mars*, R. Manning (ed.), Report of the Capability Roadmap: Human Planetary Landing Systems, March 29, 2005,

But, we absolutely need the atmosphere so that we can use aero-assist and avoid all-propulsive entry that would force us into an unreasonably large IMLEO. As the AEDL Team pointed out, "so far no feasible human scale AEDL system has been found, but there are promising ideas that need assessment and testing."

The AEDL Team outlined a multi-decade roadmap for development of appropriate entry and descent capabilities that proceeds in stages:

- Validate Mars atmosphere and interaction models: EDL *in situ* measurements and a 3-Mars-year (~6-Earth-year) atmosphere monitor mission.
- Decide what could work: human-scale AEDL architecture systems assessment.
- Decide what to baseline: sub-Scale AEDL component development and architecture evaluation tests (Earth).
- Validate AEDL models: scaled Mars AEDL validation flight(s).
- Develop and qualify the full-scale hardware: Earth-based full-scale development and Earth flight test program.

The cost of this program does not seem to have been estimated, but it is probably very high. In addition, at least one full-scale validation at Mars is likely to be necessary.

Pinpoint landing is a critical need for human exploration because multiple payloads will have to be delivered to the same site, and mobility of heavy cargo will be limited. Pinpoint landing is closely connected to AEDL. A detailed roadmap for pinpoint landing capability needs to be established (see Section 4.6.5).

8.6.3.3 Habitats and capsules

A critical part of the Mars human mission study involves analysis of options for the minimum of four vital Habitats involved in such a mission: (1) the Trans-Habitat element that delivers the crew from Earth orbit to Mars in about 6–8 months, (2) the Earth Return Vehicle (ERV) that transfers the crew back to Earth from Mars with a trip time of perhaps 6–8 months, (3) the Surface Habitat that the crew inhabits ~1.7 years on Mars, and (4) the small capsule used for descent and ascent from the surface to rendezvous with the ERV, if a Mars orbit rendezvous return architecture is used. However, the Descent and Ascent Capsules may be different and distinct. If a direct return from Mars is used, there is no ERV *per se*, but the Trans-Habitat is landed and is used to house the crew during return from Mars all the way back to Earth.

If the Mars orbit rendezvous return architecture is utilized, these Habitats must be designed so as to minimize their mass, and yet provide the needed volume, life support, and other facilities necessary for safety and survival. These Habitat masses are critically important in defining mission IMLEO requirements. The Surface Habitat can likely be based on previously proven lunar Habitats.

8.6.3.4 Life support consumables

Consumables include four major elements: atmosphere, water, waste disposal, and food. As Section 4.1 shows, the requirements for a crew of six for 600 days (corresponding to the surface phase of a Mars mission) include more than 100 mT of water,

3.6 mT of oxygen, and 10.8 mT of atmospheric buffer gas on the surface. These figures are enormous and, if translated at face value back to IMLEO, would make missions unaffordable. All DRMs generated so far have dealt with these consumables rather blandly by optimistically projecting use of recycling technologies based on approaches developed for the Space Station. Unfortunately, not much detail has been provided. A critical factor is the percentage of consumables recycled in each cycle. This determines the size of the back-up cache that is needed. In addition, the mass of the recycling plant must be included. But, most importantly, the reliability of recycling plants needs to be clarified. Such reliability requirements are likely to require considerable redundancy and spares. None of the ECLSS systems developed to date come anywhere close to the longevities required for Mars missions. Because there is a need for multi-year, long-term testing of these systems under simulated mission conditions, validation of ECLSS systems will be expensive and take a long time.

An alternative to recycling on the surface would utilize indigenous water for water needs and, after electrolysis, for oxygen. On Mars, atmospheric buffer gas can in principle be separated from atmospheric CO_2, but recycling might be more desirable. Certainly, the availability of *in situ* water would eliminate almost all recycling and the risks that come along with it. When the benefits for consumables and propellants are added together, water-based ISRU is a very attractive possibility, particularly for Mars. Nevertheless, recycling is required for transit to and from Mars. If NASA persists in requiring an abort-to-orbit option from the surface, the Earth Return Vehicle in orbit would have to include the possibility of providing life support for an extra 500–600 days. This will be clearly impossible without an extremely efficient ECLSS that has a long life.

8.6.3.5   In situ *resource utilization*

Mars ISRU is discussed in Section 5.3. Visions of the future use of ISRU in the NASA community run the gamut from those (like the author) who address only propellant and life support consumable production, with prime emphasis on utilizing water as a feedstock, to visionaries who would like to begin the transfer of the industrial and electronic revolutions to the Moon and Mars by fabricating solar cells and spare parts from regolith in the short term, and go on from there. From the viewpoint of ISRU visionaries, ISRU does not merely enhance mission capabilities and lower mission cost, but, rather, exploitation of indigenous materials becomes the focus and rationale for the whole exploration program. Thus, in their eyes, the exploration program serves ISRU, not *vice versa*.

A campaign is needed to implement ISRU on Mars:

(1) The first step in this campaign, to use a neutron spectrometer (NS) from orbit to locate hydrogen signals in horizontal spatial pixels of dimension ~200 km, has already been accomplished (see Section C3.4). The next step would be to use improved NS instruments from a lower orbit to improve the resolution and

accuracy. The NS instrument planned for the Lunar Reconnaissance Orbiter appears to be a significant improvement over the NS used on Mars Odyssey.

(2) Despite the fact that NASA does not seem to have the slightest intention of doing this, the second step would be to send several long-distance rovers equipped with dynamic active NSs to several of the regions on Mars indicated from orbit as fruitful for near-surface water availability. These would cover several tens of km, in order to determine at the outset: (a) to what extent the hydrogen signals from orbit are substantiated by the more reliable ground measurements, (b) how the hydrogen signal is distributed within each realm to ~1 m pixel size (is the distribution fairly uniform or some kind of checkerboard?), and (c) a much better estimate of the vertical distribution of hydrogen signal to a depth of perhaps 1 to 1.5 m, and in particular the thickness of any desiccated upper layer covering the H_2O-containing layer.

(3) From the results of (2), a decision can be made as to which specific site (or sites) on Mars will be selected for more detailed measurement and verification. NASA would send a short-range rover system to the selected site(s) to (a) map out the site with NS in great detail, (b) take subsurface samples to validate rover-mounted dynamic active NS measurements of water-equivalent content, (c) determine the actual form of subsurface water–ground ice or mineral hydrates, (d) extract water from some samples and determine the water purity and the potential need for purification, and (e) determine the soil strength and requirements for excavation of the site. If this step can be done robotically, why send a crew to do it?

(4) Develop a ~1/10 scale lunar ISRU demonstration system for use at this site, deliver it, get it started, and leave it to operate autonomously.

8.6.3.6 Power systems

It appears to be inevitable that a nuclear reactor will be required to support human exploration of Mars and long-term occupation of the Moon. Detailed power requirements remain unclear. Rough estimates suggest that the power requirement for ISRU will be similar to that needed to support the crew, so that the same power system can be used to produce propellants and life-support consumables prior to crew arrival, and then be used to support the crew directly after the crew arrives. This needs to be verified.

There are numerous political difficulties in developing a nuclear reactor for space, because the space part is the responsibility of NASA but the nuclear part is the responsibility of the Department of Energy (DOE). These two organizations have not always been able to work effectively with one another. In addition, development of a space nuclear reactor is a complex technological challenge. Previous attempts to develop a space nuclear reactor got bogged down and finally terminated incomplete.

8.6.3.7 Site selection

While it is clearly inappropriate at this time to attempt to select specific sites for human exploration of Mars, it is important for NASA to define the factors that enter into such a decision, and to constrain and delimit the range of options for landing sites to the extent possible at this early date. Although near-surface ground ice is clearly more abundant at higher latitudes, and such resources could have a significant impact on mission feasibility, there may be other reasons not to go to higher latitudes. Rather than waste a good deal of time studying options that will eventually be rejected for other, perhaps even non-technical reasons, NASA should strive to restrict the range of landing options as early as possible so that system-engineering studies can concentrate within a framework that is likely to ultimately be acceptable.

8.6.3.8   *Precursor missions at Mars*

It is clear that a series of robotic precursor missions to Mars will be required to ensure that the critical technologies needed for reliable human missions are indeed feasible. The 26-month delay between launches requires that the precursors be planned very carefully to maximize the return and avoid (as much as possible) surprises that might throw off the schedule.

MEPAG (Mars Exploration Program Analysis Group) provided its views on preparing for human exploration of Mars.[245] According to MEPAG, "Robotic missions serve as logical precursors to eventual human exploration of space. In the same way that the Lunar Orbiters, Ranger and Surveyor landers paved the way for the Apollo moon landings, a series of robotic Mars Exploration Program missions is charting the course for future human exploration of Mars."

Goal IV of the MEPAG document addresses science and engineering questions specific to increasing the safety, decreasing the cost, and increasing the productivity of human crews on Mars. To address these issues MEPAG defines Objective A as "obtain knowledge of Mars sufficient to design and implement a human mission with acceptable cost, risk and performance" and Objective B as "Conduct risk and/or cost reduction technology and infrastructure demonstrations in transit to, at, or on the surface of Mars." MEPAG goes on to say: "Specifically, robotic precursor missions will be used in part to acquire and analyze data for the purpose of reducing cost and risk of future human exploration missions, to perform technology and flight system demonstrations for the purpose of reducing cost and risk of future human exploration missions, and to deploy infrastructure to support future human exploration activities."

Within the realm of measurements, MEPAG identified the following areas of need:

- Properties of Martian dust. Martian dust poses a significant hazard to mechanisms and joints, and is a potential hazard to human health if ingested. This is clearly an important issue. However, MEPAG appears to have been recommended an excessive and unaffordable program to characterize Mars dust:

[245] *Mars Scientific Goals, Objectives, Investigations, and Priorities: 2005*, 31 p. white paper posted August, 2005 by the Mars Exploration Program Analysis Group (MEPAG) at *http://mepag.jpl.nasa.gov/reports/index.html*

"A complete analysis, consisting of shape and size distribution, mineralogy, electrical and thermal conductivity, triboelectric and photoemission properties, and chemistry (especially chemistry of relevance to predicting corrosion effects), of samples of regolith from a depth as large as might be affected by human surface operations."

- Atmospheric fluid variations from ground to >90 km that affect entry, descent, and landing (EDL) and take-off and ascent to orbit (TAO) including both ambient conditions and dust storms. As in the case of dust properties, atmospheric properties are also very important, but, once again, the suggested measurements laid down by MEPAG appear to be excessive and unaffordable.
- Determine if each Martian site to be visited by humans is free—to within acceptable risk standards—of biohazards that may have adverse effects on humans and other terrestrial species. The major measurement under this goal is "Determine if extant life is widely present in the martian near-surface regolith, and if the airborne dust is a vector for its transport. If life is present, assess whether it is a biohazard."
- Characterize potential sources of water to support ISRU for eventual human missions. Although MEPAG suggests that this is an important thing to do, they also indicate the following caveats: (a) the range of allowable landing sites is unknown, (b) it requires further system engineering to resolve how vital water-based ISRU is to Mars missions. They suggest a very appropriate and reasonable set of measurements of water availability at various latitudes.

Within the realm of technology demonstrations and validations, MEPAG suggests the following:

- Conduct a series of three aerocapture flight demonstrations.
- Demonstrate an end-to-end system for soft, pinpoint Mars landing with 10 m to 100 m accuracy using systems characteristics that are representative of Mars human exploration systems.
- Conduct a series of three ISRU technology demonstrations.

The aerocapture and pinpoint landing demonstrations are well chosen. However, it would seem likely that (at least in the latter stages of demonstration) both would be demonstrated in the same flight.

The three ISRU technology demonstrations require some discussion. The first suggested demonstration is ISRU atmospheric processing. While acquisition of a dust-free atmosphere and producing pure, compressed CO_2 needs to be demonstrated, this presumably ought to be delayed until dust properties are measured. Yet, the MEPAG Report suggests that this be done "early", which seems to be in conflict with the need to study dust first. Furthermore, in situ production of O_2 from the collected CO_2 has no technical value, because it can be amply demonstrated in the laboratory. The second suggested ISRU demonstration is "ISRU regolith-water processing (early)." This is a good demonstration, but it must await prospecting for water deposits and the 26-month delay between missions will dictate that there is no way that the demonstration can be done "early". The final ISRU demonstration is a "human-scale application dress rehearsal (late)." This requires demonstration of a 1/20 scale end-to-end ISRU system. That is still a system of significant size, and quite costly. Nevertheless, such a demonstration is vitally needed.

Other recommended demonstrations (communications, navigation, material degradation) do not seem quite as pressing as the above-mentioned demonstrations.

While the MEPAG Report does a good job of suggesting what ought to be done at a high level, no timetable or plan for implementing the required measurements is provided. Fulfillment of the MEPAG recommendations will be expensive and time-consuming. The fact that NASA recently canceled robotic precursors for the Moon does not augur well for robotic precursors at Mars.

8.6.3.9   Mitigation of dust effects

As MEPAG has pointed out, Martian dust poses a significant hazard to mechanisms and joints, and is a potential hazard to human health if ingested. Although MEPAG recommends an extensive program in dust characterization, from a pragmatic point of view, we really only need enough data on Mars dust to be able to design systems that either reject dust or are tolerant of it. It is not clear why photoemission and corrosion are included, since photoemission sounds like it is irrelevant, and there is no reason to believe that corrosion is a factor on Mars. Size distribution and composition would seem to be the major factors. What is not clear is how these properties vary from site to site and from season to season.

MEPAG does not seem to have recommended demonstration of dust-resistant and dust-tolerant systems on Mars. Perhaps much of this can be done in chambers on Earth?

8.6.3.10   Use of Moon for testing and demonstrations

Whereas current NASA planning seems to be heading in a direction of first developing a CEV, then determining how to use this for lunar missions, then deciding what to do while on the Moon (and why), and then adapting this experience to Mars missions, it would make more sense to start from a Mars implementation plan, and use the Moon as appropriate to test and demonstrate systems ultimately applicable to Mars exploration. The following sections identify some of the systems amenable to lunar test or demonstration.

Deployment and interconnection of assets

Any human Mars mission or long-term lunar mission requires power from a nuclear reactor. For Mars missions, the reactor would be launched some 26 months prior to departure of the crew from Earth, and it would have to deploy autonomously and provide power to an ISRU system that would fill the Ascent Vehicle propellant tanks prior to crew departure from Earth. Protection from radiation from the reactor requires that the reactor be deployed a good distance away ($\sim$1 km?) or shielded by a regolith berm or piled up regolith, or perhaps a water jacket. An autonomous system is needed for deploying the reactor, and connecting it via cable to the main base. In addition, the ISRU system must be separate from the Ascent Vehicle, but deliver cryogenic propellants to the MAV tanks. All of these systems need to operate autonomously on Mars, but deployment and interconnection can be tested first on Earth, and later under human supervision on the Moon. This will assure that an autonomous system will work on Mars.

ISRU demonstrations

While ISRU on the Moon may not resemble ISRU on Mars very closely, nevertheless three areas of overlap can be demonstrated on the Moon: regolith excavation, water recovery from regolith, and cryogenic fluid storage and transfer. While lunar regolith is known to differ from Martian regolith, and environmental conditions are also different, nevertheless it is likely that the equipment to be used at both sites will be similar. The thermal environment on the Moon is different than on Mars, but, again, the problem of condensing and storing cryogenic fluids should be addressable with similar technologies.

Habitats, capsules, and EVA

From the point of view of structural strength and leakage, it is the difference between interior and exterior pressures (not the ratio) that matters. Therefore, except for the different environment (which is actually more stressful on the Moon), Habitats, capsules, and EVA suits can be tested adequately on the Moon. Certainly, the Habitat that is intended to house the Mars crew for about 600 days should be fully tested on the Moon.

8.6.3.11   Testing in Earth orbit

Testing in Earth orbit should be utilized wherever it is appropriate. One of the principal applications is to AEDL systems. The following subsections suggest some additional applications.

In-space assembly

What we have learned from the high gear ratios (see Section 3.12) for sending vehicles to Mars is that it is unlikely that full-size vehicles can be launched in one fell swoop and delivered to Mars without assembly in LEO, unless a much bigger Launch Vehicle is developed. The gear ratio for delivery to Mars orbit from LEO is roughly 5:1 and for delivery to the surface it is roughly 10:1.[246] For a Launch Vehicle with capability to send, say, 150 mT to LEO, this would indicate that the largest vehicle that can be sent to Mars orbit without assembly is $150/5 \sim 30$ mT, and to the surface it is $150/10 \sim 15$ mT.

---

[246] Note that JSC used much lower gear ratios in their estimates. However, my estimates of EDL systems masses are based on the latest estimates of entry system mass from Georgia Tech (Section 4.6).

It is quite certain that vehicles heavier than this will be needed at Mars, and therefore some in-space assembly is required. The simplest form of assembly is just joining two parts. Other, more complex assemblies can be conjectured. These can be tested in LEO where they would ultimately take place anyway.

Artificial gravity

Providing artificial gravity is closely related to in-space assembly. It has been suggested that the spent upper stage of the Launch Vehicle be coupled to the crew transport vehicle via a truss or a tether, to allow slow rotations of the dumbbell-like structure on its way to Mars. Assembly procedures can be tested in LEO.

8.7 CONCLUSIONS

1   The NASA approach to human exploration seems to be extrapolative in the sense that all the initial focus is on the lunar CEV and Launch Vehicle. Gradually, attention will shift to the Lunar Lander (LSAM). Then, presumably, NASA will determine where to land, what to do there, and why. All of this will primarily utilize existing technology, and most of the exploration technology program is really engineering of improved and advanced versions of systems that either exist or are well understood. Vehicle design (as opposed to mission design) seems to preoccupy NASA. As Lon Forehand's memo[247] said: "Focus is shifted from advancing technologies for long-term requirements to directed research and maturing technologies for near-term use."

2   At some point in the future, predicted by NASA to be 2020 but likely to be 2025–2030, NASA will have to face the problem of planning Mars missions. And if NASA attempts to implement the same kind of process used for the Moon, beginning by designing vehicles, there will be a "tough row to hoe" to enable human missions to Mars. NASA will face the need for a two-decade technology development program, with multiple robotic precursors, costing many billions of $.

3   Based on past performance, will NASA be capable of meeting this challenge?

---

[247] Section 8.4.3.

Appendix A

Solar energy on the Moon

A.1 FIRST APPROXIMATION TO LUNAR ORIENTATION

As a first approximation, we may regard the Moon as if it moves in a circular orbit about the Earth in the plane of the ecliptic, with an axis of rotation perpendicular to the plane of the ecliptic. The Moon always faces the Earth. The Moon rotates in the direction of its motion about the Earth (see Figure A.1).

The Moon makes one full revolution about its axis in 27.32 days. However, because the Earth moves during this period, it takes 29.53 days for the Moon to complete a 360° revolution about the Earth. This is illustrated in Figure A.2. In this figure, the Moon starts off at position (A) and the Earth is at (1). A month later, the Earth is at (2). When the Moon makes a full rotation about the Earth to point (B), it returns to its original orientation relative to the stars, and so it has made one full rotation about its axis. But, from the point of view of an observer on Earth, the Moon will not reach an equivalent point where the angle S–1–A equals the angle S–2–C until point (C). Thus, point (C) defines the "synodic" month (29.53 days) and point (B) defines the "sidereal" month (27.32 days). In a similar manner, the orientation of the solar rays to an observer at (A) will not repeat at point (B), but does repeat at point (C). Therefore, it takes 29.53 days for the apparent position of the Sun to repeat itself (to an observer on the Moon)—the synodic month is the time scale of interest for solar energy on the Moon.

In actuality, the plane of the lunar motion is not quite in the ecliptic (it is tilted 5.15°), and the rotational axis of the Moon is offset by 1.54° from perpendicular to the ecliptic. The orbit of the Moon has a significant eccentricity (0.055). Furthermore, the lunar orbital plane precesses quickly (i.e., its intersection with the ecliptic rotates clockwise), in 6,793.5 days (18.60 years), mostly because of the gravitational perturbation induced by the Sun. These effects are neglected in the first approximation.

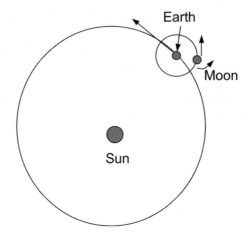

Figure A.1. Simple model of the Moon in the ecliptic plane revolving about the Earth (while the Earth is revolving about the Sun).

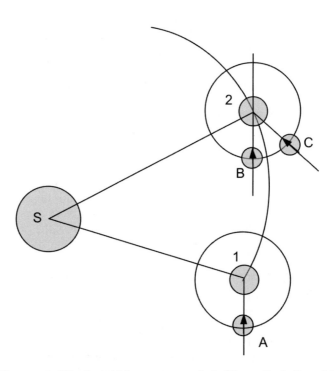

Figure A.2. Movement of Earth and Moon over a period of 1 month. A, B, and C are positions of the Moon, and 1 and 2 are positions of the Earth, separated by one month. Arrows indicate a fixed orientation of the Moon always pointed at the Earth.

A.2   SOLAR INSOLATION ON A HORIZONTAL SURFACE

If we consider a place on the equator of the Moon, as the Moon rotates about its axis, it will proceed through a continuum of variable orientations relative to the Sun. For a period, this place will rotate on the far side of the Moon and no sunlight will fall on it. At some point, the place will emerge from the far side and suddenly be illuminated. This place will continue to rotate in view of the Sun for $29.53/2 = 14.77$ days and then go back behind the Moon again.

For a plane mounted horizontally on the equator, the insolation will be

$$S = So \cos \xi \tag{A.1}$$

where ξ is the angle measured from $\xi = 0$, shown in Figure A.3, that goes from $-90°$ to $+90°$ as the Moon rotates from P1 to P2, and So is the solar constant $1{,}367\,\text{W/m}^2$ at 1 AU. Since it takes 14.77 days to go from $\xi = -\pi/2$ to $+\pi/2$, we can put:

$$S = So \cos(\pi D/14.77) \tag{A.2}$$

where D is a variable (measured in days) that goes from $-14.77/2 = -7.38$ at P1 to $+7.38$ at P2.

A side view of the lunar orbit is shown in Figure A.4.

Because of the curvature of the Moon, any place defined by a latitude angle as shown in Figure A.4 will have a solar insolation on a horizontal surface given by:

$$S = So \cos(\pi D/14.77) \cos(L) \tag{A.3}$$

If the rotation of the Moon is measured in hours, we have $D = H/24$ and:

$$S = So \cos(0.00886H) \cos(L) \tag{A.4}$$

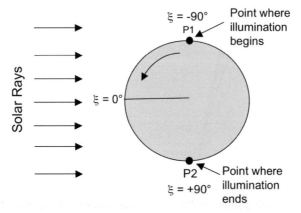

Figure A.3. View looking down on the ecliptic plane showing the lunar orbit and the period when a place on the equator is illuminated (between P1 and P2).

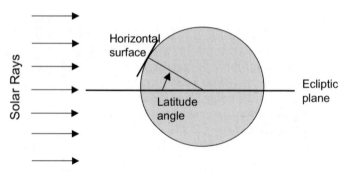

Figure A.4. Side view of horizontal surface looking edge-wise at ecliptic plane.

Over the entire span of 14.77 days of illumination the total solar energy falling on one square meter of surface is:

$$E = So\cos(L)\int_{-177}^{+177}\cos(0.00886H)\,dH \qquad (A.5)$$

$$So\frac{2}{0.00886}\cos(L) = 225.7So\cos(L) \qquad (A.6)$$

Figure A.5 shows the expected variation of solar intensity on a horizontal surface with time for several latitudes. There is a repetitive cycle of on/off periods, each of 14.77-day duration. A maximum solar intensity is reached when the surface is perpendicular to the solar rays, halfway through each cycle. When a surface is tilted up toward the equator at the latitude angle, it has the solar intensity curve of the

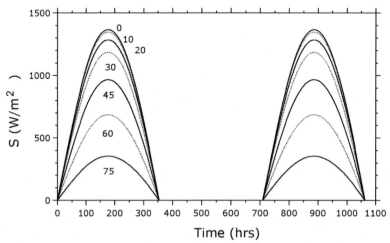

Figure A.5. Variation of solar intensity (W/m^2) on a horizontal surface with time for several latitudes. If the surface is tilted toward the equator at the latitude angle it takes on the values for latitude $= 0°$ regardless of the latitude.

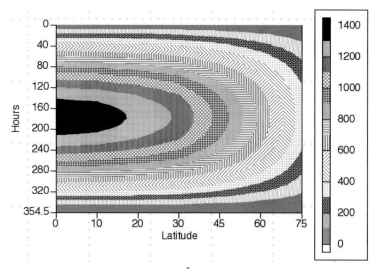

Figure A.6. Variation of solar intensity (W/m^2) on a horizontal surface with time for several latitudes. If the surface is tilted toward the equator at the latitude angle it takes on the values for latitude $= 0°$ regardless of the latitude. Values for a vertical surface are obtained by inverting the latitude axis (90 at left and 0 at right).

equator, regardless of its actual latitude. Figure A.6 shows the solar intensity vs. time and latitude as a contour plot. Figure A.7 shows the solar energy (W-hr) falling on a horizontal surface per 14.77-day cycle as a function of latitude.

A.3 SOLAR INSOLATION ON A VERTICAL SURFACE

The insolation on a vertical surface oriented so that it is perpendicular to solar rays at the start of a 14.77-day cycle, is

$$S = So \cos(0.00886H) \sin(L)$$

See Figure A.8. The dependence of solar intensity on a vertical surface is similar to that on a horizontal surface except that the phase is changed by $90°$. The maxima occur at time 0 and 354.5 hours and the solar intensity goes to zero at 177.3 hours.

A.4 INSOLATION ON A SURFACE TILTED AT LATITUDE ANGLE TOWARD THE EQUATOR

A surface tilted up toward the equator at the latitude angle (Figure A.9) always has the same orientation as a point on the equator at the same longitude. Thus, it has the same solar energy input as a horizontal surface at the equator:

$$S = So \cos(0.00962H)$$

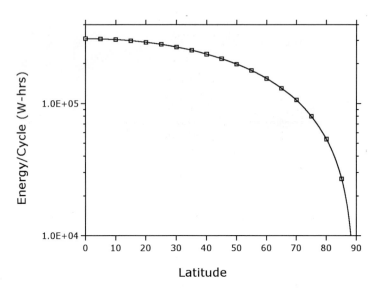

Figure A.7. Total energy (W-hr per square meter) falling on a horizontal surface per 14.77 day cycle.

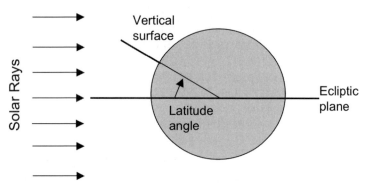

Figure A.8. Orientation of a vertical surface at an arbitrary latitude.

A.5 A SURFACE THAT IS ALWAYS PERPENDICULAR TO THE SOLAR RAYS

Relative to a horizontal plane at any point on the Moon, we can define the orientation of a tilted plane by two angles (as shown in Figure A.10):

> θ = angle between (1) a vertical line, and (2) a line in the tilted plane whose projection is a north–south line in the horizontal plane. (*Note*: $\theta = 90° - T$, where the tilt angle is up from horizontal.)

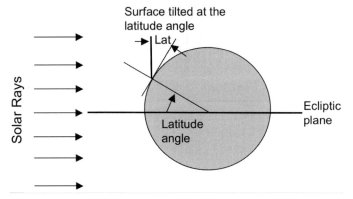

Figure A.9. Orientation of a surface tilted up toward the equator at the latitude angle is same as horizontal surface at the equator.

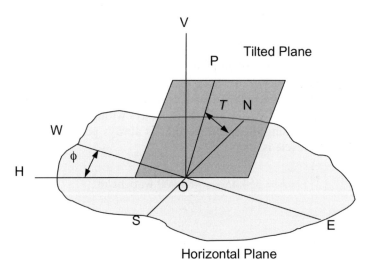

Figure A.10. Tilted plane is located by angles $T = (90° - \theta)$ and azimuthal angle ϕ in the horizontal plane. Lines EW and NS are east–west and north–south lines in the horizontal plane. Line HO is the intersection of the tilted plane with the horizontal plane.

ϕ = azimuthal angle in the west $\Rightarrow$ east direction between the EW line in the horizontal plane and the line of intersection of the tilted plane with the horizontal plane.

A surface with a tilt angle equal to the latitude angle $(T = L)$ always has the orientation of a surface at the equator. Now, referring to Figure A.3, we can draw Figure A.11, where we show plane OA is perpendicular to the Sun. Hence, when angle ϕ in Figure A.10 is set equal to $-(90° - \xi)$ in Figure A.11, the plane tilted at T

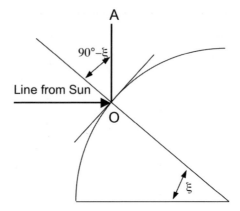

Figure A.11. View looking down on the ecliptic plane from above the lunar north pole showing a plane tilted in the east–west direction by $-(90° - \xi)$ in order to make it perpendicular to the line to the Sun. The change in orientation vs. time is illustrated in Figure A.12.

in the north–south direction will be perpendicular to the Sun. Since ξ rotates at the rate of 180° per 14.77 days, the tilted plane will always be perpendicular to the Sun if it is rotated at the rate of 0.508° per hour (see Figures A.12 and A.13). For practical purposes, this could be accomplished with a stepping motor that steps perhaps 3° every 5.91 hours, assuring that the surface is always within 3° of the solar rays.

The total energy falling on a surface that is always perpendicular to the Sun during the 14.77-day cycle is simply

$$E = 354.4So = 484,000 \text{ watt-hr/m}^2$$

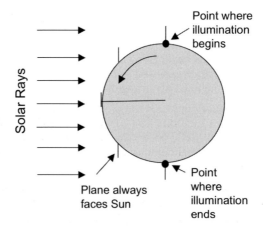

Figure A.12. View looking down on the ecliptic plane from above showing rotation of a plane so that it is always facing the Sun.

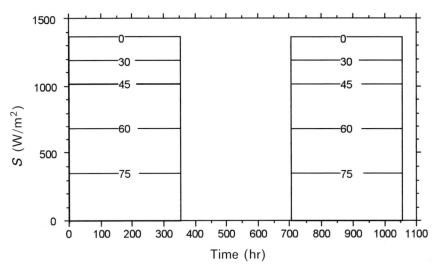

Figure A.13. Solar intensity vs. time for various latitudes for horizontal surfaces that rotate in the east–west direction at 0.508°/hr to always face the Sun. If the surface is tilted toward the equator at the latitude angle it achieves the same intensity as at the equator.

A.6 EFFECT OF NON-IDEAL LUNAR ORBIT

As we pointed out in Section A.1, the plane of the lunar motion is not quite in the ecliptic (it is tilted 5.15°), and the rotational axis of the Moon is offset by 1.54° from perpendicular to the ecliptic. The orbit of the Moon has a significant eccentricity (0.055). Furthermore, the lunar orbital plane precesses quickly (i.e., its intersection with the ecliptic rotates clockwise), in 6,793.5 days (18.60 years), mostly because of the gravitational perturbation induced by the Sun. These effects were neglected in the first approximation considered in previous sections.

Although the Internet was searched extensively, no references could be found to the geometrical layout of these effects and their effect on the orientation of solar rays on the Moon. A figure on the Internet suggests that the plane of the Moon is fixed in space (except for precession), and tilted at 5.15° to the ecliptic in a constant direction. In this case, it would appear that correction for this 5.15° offset could easily be made as a single tilt of 5.15°. Precession would change the orientation of the tilted plane by 360°/6,793.5 days, or 0.053° per day. Over a 14.77-day solar collection cycle, the total variation is 0.78°, which is negligible for non-concentrating solar collectors.

The 1.54° angle of the lunar rotational axis is too small to be a concern for planar solar arrays, although it could pose problems for concentrating solar collectors depending on the concentration ratio.

Concentrating collectors on the Moon will require proper orientation to take these effects into account.

A.7 OPERATING TEMPERATURE OF SOLAR ARRAYS ON THE MOON

Landis and co-workers[248] pointed out that the lunar soil is a good insulator, and therefore a solar array on the Moon will be able to radiate from only one side, whereas a solar array in space can radiate from two sides. Due to the T^4 radiation law, this causes a 19% increase in array temperature compared with an array in space. Arguing that a typical array in GEO operates at around 305 K, Landis (in 1990) suggested that an array on the Moon would operate at $\sim$90°C. He indicated that Apollo 11 and 12 observed temperatures in this range. The solar cells available in 1990 produced decreased output at the rate of roughly 0.3% per °C, so this represents a $\sim$17% reduction in power output compared with an array in space.

However, as the cell efficiency increases, the amount of heat generated in the cell decreases, and the rise in operating temperature is reduced. A recent analysis using modern, three-junction, high-efficiency solar cells predicts an operating temperature of about 69°C.

A.8 SOLAR ENERGY SYSTEMS AT THE EQUATOR

As we have seen previously, an equatorial solar array can be configured in several ways. The apparent position of the Sun will vary from due east at the beginning to directly overhead at the middle, to due west at the end of a 354-hour daylight cycle. If a solar array is fixed and laid out horizontally, the electric power produced will vary as a cosine function over the 354-hour daylight period and will be zero for the subsequent 354-hour period, etc. If the array is continuously rotated about a horizontal north–south line from east to west at the rate of 180° per 354 hours, so it is vertical facing east at the beginning of a cycle, horizontal at the middle of the cycle, and vertical facing west at the end of the 354-hour cycle, the power produced will be constant and a maximum at all times during the 354-hour cycle of daylight.

A.8.1 Short-term systems (<354 hours)

To avoid the need for movable tracking arrays, Landis and co-workers (*loc. cit.*) dealt with the problem of defining a solar array with fixed orientation that has the most nearly constant power output during the 354-hour daylight period. The point of this is that energy storage tends to be massive, and an array configuration that produces as nearly constant an output vs. time as possible will minimize the need for storage during the daylight cycle. This might be valuable for short stays on the Moon (<354 hours) even though it never produces the maximum possible power. The solution involves a "tent" structure with two solar arrays, one tilted toward the east, and the other tilted toward the west, as shown in Figure A.14. The arrays make an angle θ with the horizontal. The interior of the tent is always dark. The angle of the Sun's rays

[248] "Design Considerations for Lunar Base Photovoltaic Power Systems," NASA Technical Memorandum 103642, 1990, and *Acta Astronautica*, Vol. 22, 197–203, 1990.

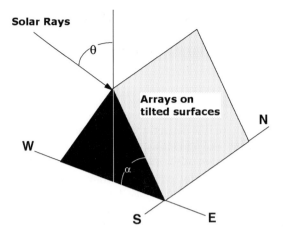

Figure A.14. Tent array structure. N, S, E, and W denote directions. Solar arrays are placed on both tilted surfaces.

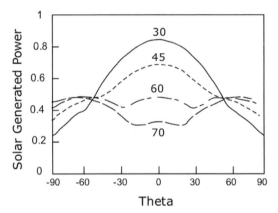

Figure A.15. Dependence of solar power generated by tent array on time for several values of the tent angle (α) as measured by the angle θ relative to a vertical. The angle θ varies from $-90°$ to $+90°$ over the 354-hour daylight cycle. Power generated by a single array of area equal to both sides of the tent that rotates to always face the Sun is 1.0.

is denoted by the angle θ relative to a vertical, and this angle varies from $-90°$ to $+90°$ over the 354-hour daylight cycle. Landis and co-workers worked out the variation of solar power output vs. θ and this is plotted in Figure A.15 for several values of α. It can be seen that the most level power output for the full 354-hour daylight period is for $\alpha = 60°$. The average power for the 354-hour daylight period is $(1 + \cos\alpha)/\pi$ which is 0.48 for $\alpha = 60°$, based on setting the power equal to 1.0 for an array of area equal to both sides of the tent that tracks the Sun at all times during the 354-hour daylight period. While a horizontal array would provide a higher average power (0.637 of a tracking array), the variation across the 354-hour daylight period would be much greater.

The tent with $\alpha = 60°$ produces half as much power as an array of the same area that rotates to always face the Sun, but the tent array is fixed and has no tracking mechanism. Power output of the 60° tent varies by about $\pm 7\%$ during a 14.77-day cycle.

A.8.2   Long-term systems (>354 hours)

For an equatorial site where it is desired to provide power using solar arrays through several cycles of daylight and darkness, it is necessary to have an energy storage system. Landis and co-workers (*loc. cit.*) evaluated the case of a triangular tent solar array as described in the previous section, combined with an energy storage system that is described by:

F = power fraction
 = (average power generated during 354-hour night period)/(average power generated during 354-hour day period);
E = storage efficiency
 = energy provided by storage/energy deposited into storage;
P_D = average power load during the 354-hour *day*;
P_N = average power load during the 354-hour *night*;
P_G = average power that must be *generated* during the 354-hour day to supply day and night average power requirements.

In this case, the solar array must be sized so that the energy stored in each daylight cycle can provide the energy needed in each subsequent night cycle. Thus:

$$P_N = FP_D$$
$$P_G = (1 + F/E)P_D$$

To minimize the storage, Landis *et al.* set the array power at sunrise equal to the daytime load P_D (i.e., immediately at sunrise no power is drawn from the storage system). They show that this provides an equation for the array tilt angle α:

$$\alpha = \cos^{-1}[(k^2 - 4/\pi^2)/(k^2 + 4/\pi^2)]$$

where $k = F/E$.

As an example, suppose night and day power requirements are equal, and the energy storage efficiency is 100%. Then, the sunrise power must be exactly half the average daytime power, and the angle α is 35.3°. The array considered provides 58% of the power per unit area of a tracking array. For a more realistic example, suppose the required night power is half the daytime power and the round-trip storage efficiency is 60%. Then, $F/E = 0.833$, and the array angle $\alpha = 38.4°$. This provides 57% of the power per unit area of a tracking array. As can be seen, the required angle α increases as F/E decreases. This method yields the array tilt angle such that the average power integrated over the lunar day is sufficient for daytime load and nighttime storage requirements. Care must be taken, however, in cases where the

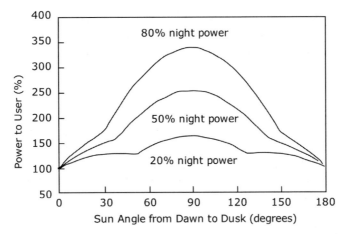

Figure A.16. Effect of high nighttime power fraction (20%, 50%, or 80%). The size of the array is driven by the high night power fraction.

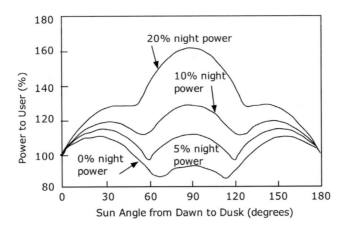

Figure A.17. Effect of low nighttime power fraction (0 to 20%). At low night power fractions, some energy storage is needed during the day when power <100%.

nighttime power requirement is a low percentage of the daytime power (low F). In these cases, the tilt of the arrays from the horizontal is so large that the power variation during the day may drop below the load requirement, requiring use of energy storage during the daytime. This would require additional array area and fuel cell radiators designed to work at the higher daytime temperatures. A tent array with a round-trip storage efficiency of 60% ($E = 0.60$) and a power fraction of only 5% yields a tilt angle of 60.9°. Tent angles above 60° allow the generated power to dip below the load power level. The tilt angle α will equal 60° when $k = (2\sqrt{3})/\pi$. Figures A.16 and A.17 show the variation of power generated during the day required to furnish the energy needed at night, based on the setting of dawn power generation to the power demand during the day, using a tent array.

A.9   EFFECTS OF DUST

Lunar dust is different from dust on Earth: it is almost like fragments of glass or coral—odd shapes that are very sharp and interlocking. On Apollo, in sunlit areas, fine dust levitated above the Apollo astronauts' knees and even above their heads, because individual particles were electrostatically charged by the Sun's ultraviolet light.

There is limited information available for effects of dust on Mars and lunar solar arrays. However, there are significant differences between the two cases, Moon and Mars. On Mars, dust is constantly floating in the atmosphere and is periodically stirred up by regional or global dust storms. On the Moon, there is no atmosphere, so dust raising and settling only occurs as a result of human activity and meteoroid impact, and the dust that is stirred up drops down rapidly compared with Mars, although it can travel a considerable distance. Furthermore, it is known that the properties of lunar dust are rather different than for Mars dust.

In a paper,[249] Katzan and Stidham point out that:

- Dust is stirred up and electrostatically charged by the movement of the "terminator" that demarks the light/dark transition line as the Moon rotates.
- Astronauts walking on the Moon will stir up dust to heights of 4 m and distances as great as 8 m.
- Roving vehicles will stir up even more dust with particles traveling up to 20 m.
- The most significant source of dust above the surface is landing and launch of space vehicles. Apollo 12 coated a Surveyor with dust about 155 m distant. It was estimated that the Surveyor acquired about 1 mg/cm^2 of dust. The mean arrival speed was in the range 40–100 m/s and some particles were as fast as 2 km/s.

Katzan and Stidham carried out tests using a lunar dust simulant and acquired the data shown in Figure A.18. Their data suggest that it takes about 2.7 mg/cm^2 of lunar dust to produce an obscuration of 50%. Katzan and Stidham predict that the effect of an Apollo lunar module touchdown is as shown in Figure A.19. Katzan and Stidham conclude that the threat of lunar dust to solar arrays is significant.

[249] "Lunar Dust Interactions with PV Arrays," Cynthia M. Katzan and Curtis R. Stidham, *22nd IEEE Photovoltaic Specialists Conference, Las Vegas, NV, October 7–11, 1991*, New York, Institute of Electrical and Electronics Engineers, Conference Record Vol. 2 (A92-53126, 22-44), pp. 1548–1553, 1991.

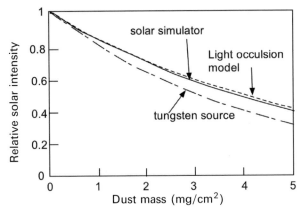

Figure A.18. Relative solar intensity through a dust layer of 20–40 micron particles. Data are for transmittance, short-circuit current, and an optical occlusion model.

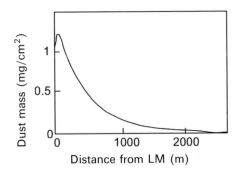

Figure A.19. Predicted deposition of dust on surfaces near an Apollo lunar module touchdown.

A.10   SOLAR ENERGY SYSTEMS IN POLAR AREAS

A.10.1   Polar sites

The axis of rotation of the Moon is tilted at 1.54 degrees relative to the ecliptic plane and is fixed in space. As the Moon travels with the Earth around the Sun, there is a seasonal change in orientation of this tilt, so that in local summer a pole tilts toward the Sun and six months later (in local winter) it tilts away from the Sun. In between these extrema, at 3-month intervals, the lunar tilt axis is perpendicular to the line to the Sun, corresponding to vernal and fall equinoxes.

An observer standing at a lunar pole during the 6-month period when that pole tilts toward the Sun (local summer) would see the Sun apparently circle the horizon once a month. This is precisely the same as the "midnight sun" effect on Earth. However, because of the low angle of inclination of the Moon, the altitude of the Sun would never be higher than 1.54 degrees (three solar diameters) and would only reach

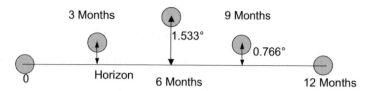

Figure A.20. Apparent elevation of the Sun from a point on the lunar pole. During local summer, the Sun, with an apparent diameter of ~0.5°, is three diameters above the horizon at summer solstice (June 21).

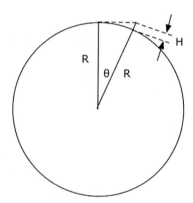

Figure A.21. The height (*H*) that a polar site needs to be to see over the Moon in winter.

this at summer solstice in the middle of the 6-month period (see Figure A.20). At the beginning and end of this 6-month period, the Sun would circle with its center on the nominal horizon. During the "winter" the Sun would drop no more than 1.54 degrees below the horizon. Thus, for a very flat area near a pole, solar illumination would be limited to about half the year, and the elevation of the Sun would vary from 0 to 1.54° during that half-year period. The Sun would drop below the horizon for the other half-year. Obviously, this situation would be improved if the observer was situated atop a peak with a 360° view.

A rotating solar collector at a lunar pole with an altitude of 622 meters (2,042 feet) higher than the terrain out to a flat horizon (only 45 km away) would always see at least half of the solar disk, summer and winter (see Figure A.21 and the discussion in the accompanying text). If it were higher than 1,244 meters, it would always see the full Sun, but at winter solstice the bottom of the Sun would be tangent to the horizon. This presumes that the observer is on top of a tower with a 360° view of flat terrain:

$$(R + H) \cos \theta = R$$

$$H = (R \sec \theta - 1)$$

$$\theta = 1.533°, \quad \text{and}$$

$$R = 1{,}738 \text{ km}$$

so

$$H = 622 \text{ meters}$$

As the observational site is moved away from the actual pole, the Moon tends to block the solar rays in local winter unless the site is at the top of a peak of sufficient height with steep side slopes so that it can "peer over the top of the pole in local winter." In preliminary papers Bussey et al.[250] discuss the available data. The 1999 paper says: "An analysis of the lunar south polar lighting using Clementine image data revealed some interesting illumination conditions" (Figure A.22, in color section).

> "No place appears to be permanently illuminated. However several regions exist which are illuminated for greater than 70% of a lunar day in [mid-]winter. Two of these regions, which are only 10 km apart, are collectively illuminated for more than 98% of the time [summed over a whole year]."

It has been estimated that the temperature in a region with constant grazing illumination is approximately 220 K ± 10 K, a benign thermal environment relative to the rest of the Moon. This borderline between heavily illuminated and heavily shadowed regions has been suggested as a good place for an outpost.

Bussey et al. estimate that at least 7,500 km² and 6,500 km² at the north and south poles, respectively, are in permanent shadow. Additionally, permanent shadow can exist in craters more than 10° latitude away from a pole (Figure A.23, in color section). Therefore, there is potential for more cold traps and volatile deposits.

In their most recent (2005) paper, these authors have further refined their analysis of the north polar area. The lunar north pole is in a highland region, between three large impact craters—Peary (88.6°N, 33.0°E; diameter, 73 km), Hermite (86.0°N, 89.9°W; diameter, 104 km), and Rozhdestvensky (85.2°N, 155.4°W; diameter, 177 km). Because the pole lies just outside the crater rims, it is likely to be at a relatively high elevation, increasing the likelihood that some areas could be permanently illuminated. By contrast, the lunar south pole is just inside the rim of the South Pole–Aitken impact basin (2,500 km diameter), and there is no area at this pole that is constantly illuminated during the southern winter, as measured at the scale of the Clementine UVVIS data (500 m per pixel). They identified 53 images taken by the craft's UVVIS camera that cover the north pole with a spatial resolution of roughly 500 m per pixel. Each image encompasses an area of about 190 × 140 km. The images show which areas are illuminated as a function of solar azimuth during a lunar day in summer. Their quantitative illumination map for the north polar region shows the percentage of time that a point on the surface is illuminated during a lunar summer day (Figure A.24, in color section). There are several regions, all on the rim of Peary crater, that are illuminated for the entire day (white areas). With the information

[250] "Illumination conditions at the lunar south pole," Bussey et al., Geophys. Res. Lett., Vol. 26, No. 9, 1999; and "Permanent shadow in simple craters near the lunar poles," Bussey et al., Geophys. Res. Lett., Vol. 30, No. 6, 2003.

available, it is not possible to state definitively that these areas are permanently sunlit, because the data correspond to a summer rather than a winter day. But, the authors are certain that they are the most illuminated regions around the north pole and that they are also the areas on the Moon most likely to be permanently sunlit, given that there are no constantly illuminated areas in the south polar region. Their quantitative illumination map also identifies permanently shadowed areas. These are associated with small impact craters (3 km or less in diameter) on the floor of Peary crater, with two larger craters (14 and 17 km in diameter) on the rim of Peary, and with the area just outside Peary's rim, in the highland region.

Geoffrey Landis (GRC) has studied lunar solar energy over a period of years starting in the 1980s. He has assessed energy storage media for survival over lunar night periods and found that the mass requirements would be excessive. At the equator, with its 354-hour periods of darkness, energy storage would be very difficult. Near the poles with a potential 6-month period of darkness, energy storage would be even more difficult. Regenerative fuel cells appeared to be the best approach for storage although the technology is still not very mature. As Landis points out, even during mid-summer, when the Sun appears to be a mere 1.5° above the horizon, "inky black shadows will cover most of the surface making exploration (and even walking) difficult."

A.10.2   GRC solar polar study

In 2004 a team at GRC prepared a design of a solar power plant for near the south pole of the Moon. Power requirements to the Habitat interface were 50 kW continuous during daytime (about $\frac{1}{2}$ year in continuous sunlight and $\frac{1}{2}$ year in a period with "sunset events". Sunsets were claimed to last up to 199 hours. It is not clear how this was derived.

A south polar landing site was assumed at Malapert Mountain (86°S lat., 0° long., 5 km height). This site is 60–100 km distant from potential water ice locations. No allocation was made for solar cell obscuration by settling lunar dust.

An elevated solar array was employed in a vertical orientation, with perfect Sun tracking (single-axis, azimuthal, Sun-tracking gimbal) and negligible solar-pointing loss. Solar cell meteoroid damage was assumed to be negligible. Solar cell radiation damage loss for current and voltage was based on the "JPL-91" solar flare model; 90% confidence level. A single solar array was deployed vertically from the Habitat with multi-junction crystalline cells having 35% conversion efficiency (AM0, 1-Sun, 28°C). (*Note, however, that such cells are not available although predictions have been made that they should be commercially available within a decade.*) They assumed a Ge substrate, glass cover, AEC-Able "Aurora" flexible panel, and flat copper interconnects. The areal mass was estimated to be 1.34 kg/m^2 and the deployment and support structure mass was estimated to be 1.51 kg/m^2. A planar configuration with a 6:1 aspect ratio was used with a solar array operating at 160 V.

Energy storage was set at 240 kW-hr based on regenerative proton exchange membrane (PEM) fuel cells (RFCs) with fuel cell conversion efficiency 55%, and electrolyzer conversion efficiency 90%. Gaseous reactants were stored in composite tanks. The storage system mass was estimated based on a performance level of 412 W-hr/kg including tanks, reactants, ancillaries, electronics, and thermal control. A summary of masses is:

- Solar array   505 kg
- Gimbal and electronics   40 kg
- RFC system mass   582 kg

The solar array was 5.4 m wide by 32.6 m tall with 177 m^2 of area.

The end-of-mission solar array power was 56.7 kW based on a 5.6% array power loss from solar flare proton irradiation, and the solar array panel operating temperature was estimated to be 69°C.

The baseline 240 kW-hr energy storage system could support a ~1 kW continuous load for the estimated maximum night period at the top of Malapert Mountain. This was deemed insufficient power to support a crew, and probably even insufficient for base keep-alive power. To support a 35 kW continuous load (estimate based on Mars DRM) for 199 hours requires a 7,000 kW-hr energy storage system. At an optimistic RFC system mass of 627 W-hr/kg (*Note: estimates of RFC performance by power technologists are notoriously optimistic.*) with a 30-day recharge period, the RFC energy storage system mass would be >11,103 kg. To support a 35 kW continuous load at a lunar equatorial site (328-hour night) would require a 11,500 kW-hr energy storage system. With an optimistic RFC system mass of 631 W-hr/kg and a 354-hour (day time) recharge period, the RFC energy storage system mass would be >19,635 kg.

By contrast, a 50 kW nuclear reactor mass is estimated at 6,000 kg (we might double this optimistic estimate, but the message is still the same: long-term solar energy on the Moon does not look good, and a nuclear reactor makes more sense).

The impacts of terrain masking must be considered for any selected landing site. There are two types: (1) local terrain and/or features, and (2) far-field terrain. Local terrain masking casts transient shadows onto arrays. The consequent array power loss is typically much greater than the shadowed area fraction, depending on the array string layout. The array power loss is reduced by elevating the array (e.g., on top of the Habitat). Far-field terrain masking blocks a portion of the solar disk and reduces the overall insolation intensity. The consequent array power loss would be proportional to the insolation loss.

The GRC report concludes that a solar power system is feasible for a lunar south polar mission during the continuous sunlight season (~180 days). The solar power components (excluding PMAD) would have a mass of ~1 metric ton (mT) for a 50 kW system. The report concludes that a solar power system would not be feasible during the Sun eclipse season, because excessive energy storage mass would be required. Similarly, long-term (>14 days) equatorial lunar missions would require excessive energy storage.

Appendix B

Solar energy on Mars

B.1 SOLAR INTENSITIES IN CURRENT MARS ORBIT

B.1.1 Introduction

The Mars axis of rotation is tilted on its axis by about 25.2° relative to the plane of its motion, similar to the tilt of the Earth on its axis. The tilt of Mars is illustrated in Figure B.1.

On the surface of Mars, the solar elevation angle is the angle between the line to the Sun and the horizontal plane. The sum of the zenith angle and the elevation angle is 90°. The fixed orientation in space of the tilt of Mars causes the Sun to be $2 \times 25.2° = 50.4°$ higher in the sky at noon in mid-summer than it is at noon in mid-winter at moderate latitudes. At the equator, the swing from winter to summer is from −25.2° to +25.2°, so the Sun is always within a moderate range of being overhead at noon. Since the cosine of 25.2° is 0.90, the variation in solar intensity on a horizontal surface with season tends to be modest at the equator. As one moves away from the equator, the effects of differences of the elevation of the Sun between summer and winter become accentuated.

The position of Mars in its orbit around the Sun is characterized by a parameter sometimes called the heliocentric longitude, or longitude of the Sun, and is represented by the symbol L_S. L_S is defined to be 0° at the vernal equinox, and increases to 90° at summer solstice, 180° at the autumn equinox, and 270° at the winter solstice. Values of L_S equal to 90°, 180°, and 270° correspond to northern winter, northern autumn, and northern winter, respectively. Seasons are reversed in the southern hemisphere. The actual dates at which L_S occurred in 2006–2007 were: January 22, 2006, $L_S = 0$; August 8, 2006, $L_S = 90$; February 8, 2007, $L_S = 180$; July 4, 2007, $L_S = 270$.

Mars has a much more elliptical orbit about the Sun than Earth does, with the solar intensity being 45% greater at the nearest point to the Sun than it is at the

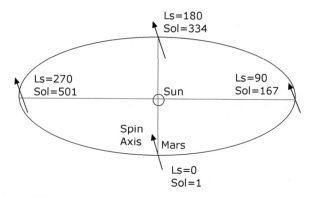

Figure B.1. Mars axis of rotation as Mars moves about the Sun.

furthest point. Mars is closer to the Sun when the southern hemisphere is undergoing summer, producing high solar intensities in summer and comparatively low solar intensities in southern winter. In the northern hemisphere the Sun is closer to Mars in winter, thus partly mitigating the differences in solar intensity between winter and summer. One diurnal cycle of day/night (i.e., one rotation about its axis) on Mars is called a "sol" and one sol $= 24.62$ Earth hours. It takes Mars 668.60 sols (686.98 Earth days) to complete its orbit around the Sun.

The distance of Mars from the Sun is a minimum at $L_S \sim 250°$, and is a maximum at $L_S \sim 70°$. The distance is given by:

$$\frac{r}{r_{av}} = \frac{1 + 0.0934 \cos(L_S - 250°)}{0.9913}$$

where r_{av} is the Mars mean distance from the Sun (1.52 AU), and r is the distance of Mars from the Sun on any sol. Since the extraterrestrial solar intensity at 1 AU is 1,367 W/m^2, the direct solar intensity impinging on Mars is:

$$I_{ext} = 592 \left(\frac{r}{r_{av}}\right)^2 = 592 \left(\frac{1 + 0.0934 \cos(L_S - 250°)}{0.9913}\right)^2$$

in units of W/m^2. This extraterrestrial solar intensity is on an element of area perpendicular to the line to the Sun.

The zenith angle (z) is the angle that a line to the Sun makes with the vertical as measured from any place on Mars. This angle is given by:

$$\cos z = \sin d \sin L + \cos d \cos L \cos\frac{2\pi t}{24.6}$$

where d is the declination (in degrees), L is the latitude (in degrees) at the point of observation, t is the time of day (in hours with solar noon taken as zero), and 24.6 hours is the length of a diurnal day on Mars. The declination is an angle that starts at $0°$ when $L_S = 0$, and varies with L_S according to the formula:

$$\sin d = \sin 25.2° \sin L_S = 0.4258 \sin L_S$$

The time t is in hours, and is measured from local solar noon, so that t ranges from -12.3 to $+12.3$ during one day. At $t = 0$ corresponding to local solar noon, the zenith angle is

$$z = |L - d|.$$

In local mid-summer, when d is near $+25.2°$, the zenith angle at noon is a minimum. In mid-winter, when d is near $-25.2°$, the zenith angle at noon is a maximum for the year. If the latitude exceeds $64.8°$, there will be a period during mid-winter when the Sun never rises above the horizon.

B.1.2 Irradiance in a clear atmosphere

The solar intensity (I_h) impinging on a horizontal surface (assuming no atmosphere) at any latitude for any day of the year and time of day may now be calculated. We have

$$I_h = I_{ext} \cos z$$

The daily total irradiance on a horizontal surface assuming no atmosphere can be calculated from

$$S = \int_{-12.3}^{+12.3} I_h \, dt$$

Tables of peak solar noon solar intensity and daily total irradiance on a horizontal surface can now be prepared assuming there is no atmosphere. These calculations provide an extreme upper limit to the irradiance on Mars. The results are given in Tables B.1 and B.2. These tables show that:

- The highest peak noon intensities are reached at southern latitudes of $15°$ to $30°$ when Mars is closest to the Sun.
- The variation of solar intensity with season is a minimum at northern latitudes of around $15°$.
- As one moves toward the poles, there are long seasonal periods when the Sun never sets, alternating with long seasonal periods when the Sun never rises. Reasonably high irradiances are possible in polar regions for limited periods of time.
- Because of the long solar day, daily totals can be very high at high latitudes in local summer.

B.1.3 Effect of atmosphere

For a perfectly clear atmosphere, the difference in solar elevation between local summer and winter manifests itself merely as a cosine effect on the irradiation of a horizontal surface. However, the turbidity of the Mars atmosphere assures that the effect of differences in solar elevation will be accentuated by the absorption and scattering that takes place along the longer path length when the solar elevation is lower in winter. The processes of scattering and absorption are discussed in the sections that follow.

Table B.1. Peak irradiance (W/m^2) impinging on a horizontal surface on Mars at local solar noon assuming no atmosphere.

	Latitude (°)										
L_S	−75	−60	−45	−30	−15	0	15	30	45	60	75
0.0	145.6	281.3	397.9	487.3	543.5	562.7	543.5	487.3	397.9	281.3	145.6
22.5	51.6	185.6	307.0	407.4	480.0	520.0	524.5	493.3	428.4	334.4	217.5
45.0	0.0	108.7	232.1	339.7	424.2	479.7	502.6	491.2	446.3	371.0	270.5
67.5	0.0	58.7	183.6	295.9	388.1	453.8	488.6	490.1	458.2	395.0	305.0
90.0	0.0	41.8	169.2	285.1	381.6	452.1	491.7	497.9	470.1	410.3	322.5
112.5	0.0	61.9	193.6	312.1	409.3	478.6	515.3	516.9	483.2	416.6	321.7
135.0	0.0	119.7	255.6	374.1	467.1	528.2	553.4	540.9	491.5	408.6	297.8
157.5	58.3	209.7	346.9	460.3	542.4	587.6	592.7	557.4	484.1	377.8	245.8
180.0	165.5	319.7	452.1	553.7	617.6	639.4	617.6	553.7	452.1	319.7	165.5
202.5	280.1	430.5	551.6	635.1	675.3	669.5	618.1	524.5	395.2	239.0	66.5
225.0	379.8	521.0	626.7	689.7	705.7	673.6	595.6	477.0	325.9	152.6	0.0
247.5	443.5	574.4	666.2	712.6	710.4	659.8	564.3	430.3	266.9	85.4	0.0
270.0	458.6	583.4	668.4	707.9	699.2	642.8	542.6	405.4	240.6	59.4	0.0
292.5	423.9	549.0	636.7	681.1	679.0	630.6	539.3	411.2	255.1	81.6	0.0
315.0	348.8	478.5	575.6	633.4	648.1	618.6	547.0	438.1	299.3	140.2	0.0
337.5	249.8	384.0	492.0	566.5	602.4	597.2	551.3	467.9	352.5	213.2	59.3
360.0	145.6	281.3	397.9	487.3	543.5	562.7	543.5	487.3	397.9	281.3	145.6

Table B.2. Daily total solar irradiance (W-hr/m^2) impinging on a horizontal surface on Mars assuming no atmosphere.

	Latitude (°)										
L_S	−75	−60	−45	−30	−15	0	15	30	45	60	75
0.0	1,141	2,202	3,114	3,815	4,256	4,406	4,256	3,815	3,114	2,202	1,141
22.5	241	1,205	2,172	3,014	3,663	4,071	4,209	4,071	3,665	3,036	2,280
45.0	0	554	1,474	2,376	3,159	3,756	4,123	4,239	4,108	3,781	3,496
67.5	0	224	1,058	1,978	2,836	3,552	4,074	4,367	4,435	4,359	4,207
90.0	0	133	935	1,872	2,770	3,540	4,123	4,487	4,637	4,667	4,529
112.5	0	236	1,114	2,086	2,991	3,747	4,295	4,605	4,676	4,598	4,438
135.0	0	610	1,621	2,615	3,478	4,135	4,541	4,667	4,521	4,162	3,850
157.5	273	1,360	2,453	3,405	4,140	4,600	4,758	4,600	4,143	3,429	2,578
180.0	1,296	2,504	3,540	4,337	4,836	5,006	4,836	4,337	3,540	2,504	1,296
202.5	2,937	3,906	4,718	5,240	5,422	5,242	4,716	3,882	2,797	1,552	310
225.0	4,910	5,309	5,766	5,951	5,791	5,274	4,435	3,336	2066	777	0
247.5	6,116	6,339	6,448	6,347	5,921	5,166	4,125	2,876	1,538	325	0
270.0	6,440	6,635	6,593	6,381	5,865	5,033	3,938	2,662	1,331	189	0
292.5	5,845	6,059	6,162	6,066	5,660	4,937	3,943	2,748	1,469	310	0
315.0	4,509	4,876	5,296	5,466	5,319	4,844	4,074	3,063	1,899	713	0
337.5	2,620	3,486	4,209	4,674	4,836	4,676	4,207	3,461	2,494	1,385	276
360.0	1,141	2,202	3,114	3,815	4,256	4,406	4,256	3,815	3,114	2,202	1,141

B.1.3.1   The direct beam

The direct beam solar intensity is the flux of solar irradiance that proceeds to the ground level unscattered and unabsorbed by the atmosphere. The diminution of the direct beam is easy to model. If one considers a thin slab of atmosphere of thickness dx, as shown in Figure B.2, the path length in a vertical element dx is $du = \csc z\, dx$.

If the incident flux of irradiance impinging on the slab of thickness dx is I, and a fraction ($a\,du$) of the incident flux is removed by absorption and scattering in the slab, the amount removed in du is

$$dI = Ia\,du$$

Integrating over the entire thickness of the atmosphere, one obtains the direct normal intensity at ground level

$$I_{\text{ground}} = I_{\text{extraterrestrial}}\, \exp(-a\lambda)$$

where λ is the path length through the atmosphere and a is the loss coefficient per unit length.

For vertical rays, the dust optical depth (D) is defined as:

$$D = a\lambda_0$$

where λ_0 is the vertical path length through the atmosphere. (Note that most published papers use the symbol τ for optical depth.) The optical depth is the sum of cross-sectional areas of all the particles in a vertical column of unit cross-sectional area of atmosphere. Thus, if all the particles in a vertical column were to be placed on a single plane of unit area, the fraction occluded would be the optical depth (see Figure B.3).

Optical depths greater than 1 imply that, if all the particles in a vertical column were to be placed on a single plane of unit area, it would require more than one layer of particles.

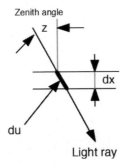

Figure B.2. A vertical element dx gives rise to an element of path length $du = \csc z\, dx$.

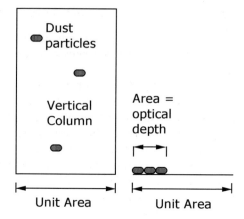

Figure B.3. Optical depth is the sum of areas of particles in a vertical column of unit area.

For non-vertical rays, the path length through the atmosphere is $\lambda = \lambda_0/\cos z$. Thus, in general, the direct normal beam intensity at ground level is

$$I_{\text{ground}} = I_{\text{extraterrestrial}} \exp(-D/\cos z)$$

$$= \text{direct component of irradiance at ground level}$$

or

$$Q(D, z) = \exp(-D/\cos z)$$

where $Q(D, z)$ is the transmission coefficient for the direct component. This is usually known as "Beer's law".

If absorption were the only process that occurred, Beer's law would describe the fraction of extraterrestrial irradiance that reaches the Mars surface as a function of D and $\cos z$ (i.e., the fraction not absorbed). However, it turns out that most of the absorption of sunlight by the Mars atmosphere occurs in the UV, and, overall, absorption is only a secondary factor in the passage of the full spectrum of sunlight through the Mars atmosphere. Scattering by dust is much more important. Multiple scattering events can send "diffuse" light to the surface in addition to the "direct" component of sunlight that passes unscathed through the atmosphere in a straight line. Each scattering event spreads the scattered radiation, which then undergoes a tortuous path as meandering rays undergo multiple scattering events. Ultimately, some of this scattered radiation finds its way to the surface and adds to the direct beam as the "diffuse" component of irradiance. The effects of scattering will be discussed in the following sections.

The direct beam irradiance on a horizontal surface is the product of the direct normal beam intensity times $\cos z$. The total irradiance on a horizontal surface is the sum of the direct normal beam irradiance and the diffuse component scattered from all over the sky.

B.1.3.2   Simple "two-flux" model of scattering and absorption of sunlight in the Mars atmosphere

In this section we describe a relatively simple model of scattering and absorption of sunlight in the Mars atmosphere.

First, consider a vertical column of atmosphere with unit cross-sectional area. The number of dust particles in such a column is estimated in Section B.5.4. At an optical depth of 0.5 (corresponding to moderately clear weather on Mars), there are roughly 2.5×10^6 particles in the column. The total optical cross section of these particles is half the area of the column—that is, the definition of optical depth. Thus, the passage of light through such a column involves the sequential interaction of the light with millions of small dust particles on a statistical basis.

If the average density of particles in the atmosphere is n particles/unit volume, the total number of particles (N) in a column of unit area is determined from:

$$D = (Nq) = (nTq)$$

where q is the cross section of a particle for optical interaction, and T is the thickness of the atmosphere. The optical depth is D. The thickness of the atmosphere, T, times unit area is the volume of the column, and, when multiplied by n, leads to

$$N = nT.$$

When light interacts with a single particle, some light can be scattered into the forward cone, some light can be scattered into the backward cone, and some light can be absorbed. We define the single-scattering albedo as the ratio:

$$w = \frac{\text{amount of light scattered}}{\text{total amount of light scattered and absorbed}}$$

Of the scattered light, the fraction (f) is scattered in the forward direction, and the fraction $(b) = (1 - f)$ is scattered in the backward direction.

Note that in any interaction of light with a particle:

$$wf = \frac{\text{amount of light forward-scattered}}{\text{total amount of light scattered and absorbed}}$$

$$(1 - wf) = \frac{\text{amount of light absorbed and back-scattered}}{\text{total amount of light scattered and absorbed}}$$

$$wb = \frac{\text{amount of light back-scattered}}{\text{total amount of light scattered and absorbed}}$$

The reality of the situation is that scattered light interacts with additional dust particles in a complex three-dimensional system. Pollack *et al.* (1990)[251] and Crisp

---

[251] "Simulations of the General Circulation of the Martian Atmosphere, I. Polar Processes," J. B. Pollack, R. M. Haberle, J. Schaffer, and H. Lee, *J. Geophys. Res.*, Vol. 95, 1447–1473, 1990. Also see "Atmospheric Effects on the Utility of Electric Power on Mars," R. M. Haberle *et al.*, *Resources of Near-Earth Space*, University of Arizona Press, 1993.

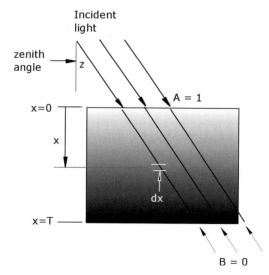

Figure B.4. Simple model of the Mars atmosphere. The vertical distance is measured by $x = 0$ at the top of the atmosphere, and $x = T$ at the ground. The zenith angle of the Sun is z.

et al. (2001)[252] have performed detailed analyses of this three-dimensional radiative transfer problem. However, it is known that forward scattering dominates over backward scattering, and further that much of the forward scattering tends to be projected into a fairly narrow cone of angles. The fact that some light originally incident toward some point "A" on Mars ends up hitting point "B" somewhat removed from "A" is of no great consequence. We are primarily interested in what fraction of incident irradiance on the atmosphere hits the ground, regardless of where it emanated from above the atmosphere. Therefore, it is appropriate to consider a "two-flux" model in which we neglect the three-dimensionality of the problem, and make the approximation that, at each dust interaction, light is either absorbed, scattered backward along the original direction, or scattered forward along the original direction. To the extent that scattered light tends to be grouped around the original direction in forward or backward cones, this should be a good approximation.

Consider a simple model of the Mars atmosphere as shown in Figure B.4. The incident radiant flux is represented by $A = 1$ above the atmosphere. This is the flux moving downward along the original path of the rays.

Within the atmosphere, the existence of back-scattering results in fluxes of light at any point x in the forward and backward directions given by $A(x)$ and $B(x)$, respectively. At ground level, there can be no backward flux, so another boundary condition is that $B(T) = 0$. For an oblique entry of light with zenith angle z, the path length along the rays changes from dx to $(\sec z \, dx) = du$.

[252] "The Performance of Gallium Arsenide/Germanium Cells at the Martian Surface," D. Crisp, A. V. Pathare, and R. C. Ewell, *Acta Astronautica*, Vol. 54, No. 2, 83–101, 2004.

In any small element of path du, located at $u = x \sec z$, the differential equations that govern the changes in A and B are determined by the scattering and absorption that take place in the element du. In any element du:

- the loss in downward solar flux due to scattering and absorption of A
  $= (1 - wf)A$;
- the gain in downward flux due to scattering of B
  $= wbB$;
- the loss in upward solar flux due to scattering and absorption of B
  $= (1 - wf)B$;
- the gain in upward flux due to scattering of A
  $= wbA$.

Thus, we have the coupled differential equations:

$$\frac{dA}{(D/T)\, du} = (-(1 - wf)A + wbB)$$

$$\frac{dB}{(D/T)\, du} = (-(1 - wf)B + wbA)$$

The solutions to these coupled equations were given by Chu and Churchill.[253] At the ground level, $u = T/\cos z$, and the solution for A at this point is:

$$A = \frac{e^{-p/\cos z}(1 - G^2)}{1 - G^2\, e^{-2p/\cos z}}$$

where

$$G = \frac{D(1 - wf + wb) - p}{D(1 - wf + wb) + p}$$

$$p = D((1 - wf)^2 - w^2 b^2)^{1/2}$$

The value of A at ground level is the predicted transmission coefficient since the incident flux was taken as unity.

If there is no absorption, $w = 1$, $p \to 0$, $G \to 1$ and the expression for A becomes indeterminate $(0/0)$. In this case, the differential equations simplify to

$$\frac{dA}{(D/T)\, du} = (-(1 - f)(A - B))$$

$$\frac{dB}{(D/T)\, du} = (-(1 - f)(B - A))$$

If we now transform from variables A and B to $(A + B)$ and $(A - B)$, the differential

[253] "Multiple Scattering of Electromagnetic Radiation," C. M. Chu and S. W. Churchill, *J. Phys. Chem.*, Vol. 59, 855, 1955.

equations are simple:

$$\frac{d(A+B)}{(D/T)\,du} = 0$$

$$\frac{d(A-B)}{(D/T)\,du} = (-(2)(A-B))$$

The solutions are simple:

$$A+B = 1$$

$$A-B = \exp\frac{-2D(1-f)}{\cos z}$$

Adding these two equations, we find that the value of A at ground level is

$$A = 0.5\left(1 + \exp\frac{-2D(1-f)}{\cos z}\right)$$

Interestingly enough, this two-flux model says that, in the absence of absorption, the transmission coefficient cannot drop below 50%. The reason for this physically is that, even at high values of optical depth, light is reflected forward and backward many times until half the light is moving upward and half is moving downward.

For Mars dust, the best approximations are:

$$w = 0.93 \qquad b = 0.17 \qquad f = 0.83$$

A plot of the two-flux model with these parameters is given in Figure B.5. Both the full expression for the transmission coefficient including absorption and the simple expression neglecting absorption are plotted for several optical depths. Neglect of absorption leads to serious errors as the optical depth and zenith angle increase.

B.1.3.3   *Sophisticated model of scattering and absorption of sunlight in the Mars atmosphere*

A fairly rigorous treatment of the process of radiant transmission through the atmosphere was carried out by Pollack *et al.* (1990) in which a detailed analysis of scattering was made for spherical dust particles. Their results are available in tabular form giving the net downward flux onto a horizontal surface as a fractional transmission coefficient for any dust optical depth and solar zenith angle. Because this transmission coefficient is for net irradiance, it represents the difference between the downward flux and the upward flux leaving the surface. To obtain the net downward flux, one should divide it by $(1-\text{albedo})$ to obtain the downward flux $T(D, z)$. A table of $T(D, z)$ is given in Table B.3.

A plot of these transmission coefficients is provided in Figure B.6. These $T(D, z)$ data show a very slow and gradual fall-off of transmission with increasing solar zenith angle and increasing optical depth. This is due to the fact that the scattering of light

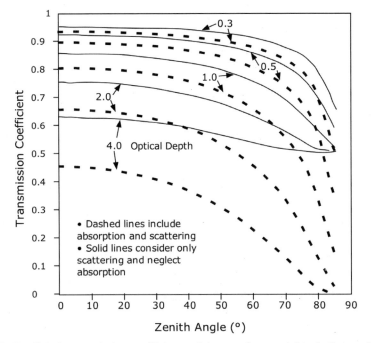

Figure B.5. Predicted transmission coefficients of the two-flux model including and excluding absorption.

by dust particles is primarily in the forward direction, giving rise to relatively large diffuse components even when the direct component is greatly reduced by high D and/or z. In "clear" weather on Mars with $D \sim 0.5$ when the Sun is overhead ($z \sim 0$), the direct component of irradiance is determined by the transmission coefficient

$$\exp(-D/\cos z) = \exp(-0.5) \sim 0.6$$

But the Pollack transmission coefficient (including direct and diffuse light) is estimated from Table B.3 to be 0.9. We may conclude that when the Sun is directly overhead on Mars at an optical depth of 0.5, 60% of the extraterrestrial solar intensity (ET) reaches the ground as the direct beam, and another 30% of extraterrestrial reaches the ground as diffuse light. By contrast, if $D = 3$ and $z = 50°$, the direct flux is determined by $\exp(-3/0.64) = 0.009$ of ET, while the downward flux is 0.367 of ET, so the downward flux is mainly composed of diffuse irradiance and the direct beam is negligible. This demonstrates the importance of scattering models for diffuse irradiance on Mars.

The results of the Pollack model are compared with predictions of the simple two-flux model in Figure B.7. It can be seen that at lower optical depths and lower zenith angles, the two-flux model is amazingly good.

Table B.3. Transmission coefficient $T(D, z)$ **for down**ward flux as a function of zenith angle and optical depth (OD). These values were obtained by dividing Pollack's (1990) net downward flux by $(1 - \text{albedo})$.

OD	0	10	20	30	40	50	60	70	80	85
0.1	0.983	0.981	0.980	0.978	0.973	0.967	0.952	0.922	0.839	0.706
0.2	0.962	0.961	0.956	0.953	**0.946**	0.929	0.903	0.842	0.711	0.522
0.3	0.941	0.940	0.934	0.929	0.918	0.896	0.860	0.787	0.624	0.458
0.4	0.920	0.919	0.912	0.906	0.891	0.864	0.822	0.741	0.558	0.414
0.5	0.900	0.900	0.891	0.884	0.864	0.836	0.787	0.698	0.502	0.380
0.6	0.881	0.879	0.872	0.861	0.839	0.806	0.752	0.659	0.460	0.353
0.7	0.862	0.859	0.851	0.839	0.814	0.778	0.718	0.617	0.426	0.331
0.8	0.844	0.840	0.833	0.818	0.789	0.750	0.684	0.578	0.400	0.311
0.9	0.828	0.822	0.814	0.797	0.767	0.722	0.652	0.541	0.373	0.293
1.0	0.813	0.806	0.797	0.778	0.744	0.698	0.622	0.506	0.352	0.280
1.1	0.792	0.788	0.778	0.758	0.723	0.671	0.599	0.481	0.333	0.266
1.2	0.774	0.769	0.759	0.736	0.702	0.650	0.576	0.459	0.320	0.256
1.3	0.758	0.752	0.741	0.718	0.681	0.630	0.553	0.438	0.303	0.244
1.4	0.740	0.734	0.722	0.699	0.662	0.607	0.531	0.421	0.291	0.233
1.5	0.723	0.718	0.703	0.680	0.644	0.589	0.511	0.402	0.279	0.224
1.6	0.708	0.700	0.687	0.663	0.626	0.569	0.490	0.387	0.267	0.217
1.7	0.691	0.683	0.668	0.646	0.607	0.549	0.471	0.369	0.258	0.209
1.8	0.677	0.667	0.651	0.631	0.590	0.533	0.453	0.353	0.249	0.201
1.9	0.662	0.652	0.634	0.612	0.571	0.516	0.437	0.338	0.241	0.196
2.0	0.647	0.637	0.620	0.597	0.556	0.498	0.420	0.326	0.231	0.189
2.25	0.613	0.602	0.580	0.557	0.513	0.456	0.381	0.294	0.211	0.173
2.50	0.576	0.566	0.547	0.521	0.478	0.420	0.351	0.269	0.193	0.161
2.75	0.540	0.531	0.513	0.489	0.446	0.392	0.326	0.249	0.176	0.151
3.00	0.511	0.500	0.482	0.460	0.418	0.367	0.303	0.229	0.167	0.142
3.25	0.482	0.471	0.456	0.433	0.393	0.342	0.282	0.214	0.156	0.133
3.50	0.457	0.444	0.430	0.408	0.370	0.322	0.267	0.200	0.147	0.122
4.00	0.411	0.400	0.386	0.367	0.329	0.287	0.236	0.178	0.131	0.111
5.00	0.327	0.318	0.306	0.287	0.256	0.226	0.184	0.144	0.104	0.089
6.00	0.253	0.248	0.239	0.222	0.198	0.170	0.144	0.114	0.089	0.076

B.2   SOLAR INTENSITIES ON HORIZONTAL AND TILTED SURFACES

Previous work on Mars solar irradiance was carried out by NASA-GRC.[254] These papers provide the basis for much of the discussion in this section.

[254] "Solar Radiation on Mars," J. Appelbaum and D. J. Flood, *Solar Energy*, Vol. 45, No. 6, 353–363, 1990.

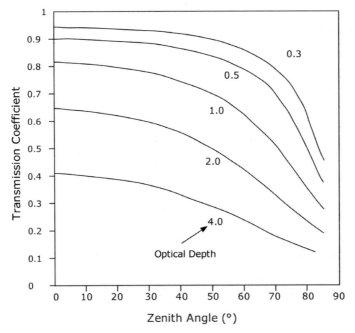

Figure B.6. Dependence of Pollack's $T(D, z)$ on zenith angle for various optical depths.

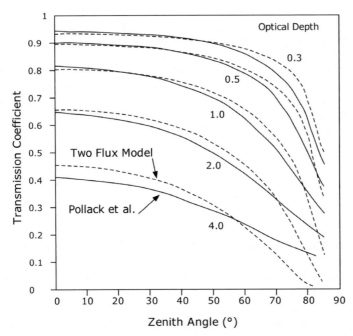

Figure B.7. Comparison of the two-flux model with the three-dimensional model of Pollack *et al.* (1990).

B.2.1   Nomenclature

$\lambda =$ actual path length

$\lambda_0 =$ vertical path length

$a =$ absorption coefficient in Beer's law

$A =$ downward solar flux in two-flux model

$\alpha =$ albedo

$b =$ fraction of light that is scattered into the backward direction

$B =$ upward solar flux in two-flux model

$d =$ declination of Mars

$D =$ optical depth

$f =$ fraction of light that is scattered into the forward direction

$G =$ parameter in two-flux model

$GB =$ beam solar intensity at the ground normal to the Sun after passing through the atmosphere

$GBH =$ beam solar intensity on a ground horizontal surface after passing through the atmosphere $= GB \cos z$

$GBT =$ beam intensity on the tilted surface at the ground after passing through the atmosphere

$GDH =$ diffuse solar intensity at Mars on a horizontal surface from sources other than the direct beam $= GH - GBH$

$GDT =$ diffuse solar intensity at Mars on a tilted surface from sources other than the direct beam

$GDTI =$ estimate of GDT assuming diffuse irradiance is isotropic across the sky

$GDTS =$ estimate of GDT assuming diffuse irradiance emanates from the position of the Sun

$GH =$ total solar intensity at Mars on a horizontal surface after passing through the atmosphere including beam and diffuse components

$GRT =$ contribution to solar intensity on a tilted surface by reflection from the ground

$GT =$ total solar intensity at Mars on a tilted surface after passing through the atmosphere including beam and diffuse components

$H =$ hour of the day measured from 1 to 25 with hour 13 spanning solar noon symmetrically (note actual length of day is 24.7 hours)

$I =$ solar intensity

$I_{ext} =$ extraterrestrial solar intensity

$I_h =$ solar intensity on a horizontal surface

$K_1 =$ coefficient for $GDTI$

$K_2 =$ coefficient for $GDTS$

$L =$ latitude

$L_S =$ heliocentric longitude (degrees) based on $L_S = 0$ at the vernal equinox (L_S varies from $0°$ to $360°$ along a Martian year)

$n =$ average density of particles in column of atmosphere (particles/unit volume)

$N =$ total of particles in a vertical column of atmosphere of unit area

$p =$ parameter in two-flux model

$q =$ cross-sectional area of a particle for interaction with light

$Q =$ transmission coefficient for direct beam from Beer's law

$r =$ current distance of Mars from the Sun

$r_{\mathrm{av}} =$ average distance of Mars from the Sun

$SH =$ solar intensity at Mars on a horizontal surface if there were no atmo-sphere

$\sin d = \sin 25.2°$

$\sin L_S = 0.4258 \sin L_S$

$SN =$ solar intensity at Mars on a surface normal to the Sun if there were no atmosphere

$ST =$ solar intensity at Mars on a tilted surface (angle $= TT$) if there were no atmosphere

$t =$ hour of the day from -12 to $+12$ with $t = 0$ at solar noon

$T =$ thickness of the atmosphere

$T(D, z) =$ transmission coefficient for total irradiance on a horizontal surface as a function of optical depth and solar zenith angle (as tabulated by Pollack *et al.*)

$TT =$ tilt angle of tilted collector surface

$u =$ oblique element of path $= x \csc z$

$w =$ single-scattering albedo

$\quad =$ (amount of light scattered)/(amount of light scattered and absorbed)

$x =$ vertical element of path

$z =$ solar zenith angle (angle between vertical and line to Sun)

$z_T =$ the solar zenith angle relative to a tilted plane

B.2.2 Solar intensity on a horizontal surface

It is desired to estimate the solar intensity on a horizontal surface as a function of latitude, season, and optical depth. It was previously shown that:

$$\sin d = \sin 25.2° \sin L_S = 0.4258 \sin L_S$$

$$\cos z = \sin d \sin L + \cos d \cos L \cos[2\pi t/(D)]$$

$$GH = SN \cos z T(D, z)$$

$$GB = SN \exp(-D/\cos z)$$

$$GBH = GB \cos z$$

$$GDH = GH - GBH$$

B.3   SOLAR INTENSITIES ON A FIXED TILTED SURFACE

By tilting the collector plane, it is possible to change the angles at which solar rays impinge on an array plane (see Figure B.8). In typical installations of flat-plate solar collectors, one tilts the plane of the array upward by the latitude angle, toward the south in the northern hemisphere and toward the north in the southern hemisphere. In this case, the angles at which the Sun strikes the collector are identical to those encountered at the equator, although the path lengths of the solar rays through the atmosphere are much longer than for the equator.

In a clear environment, it is possible to get significant enhancement of the solar intensity in winter on a plane at higher latitudes by tilting the plane upward toward the Sun. This is well known in solar applications on Earth. However, such tilting reduces the solar intensity in summer. And, in a dusty environment at high latitudes, most of the irradiance is diffuse, so tilting may be counter-productive because a tilted surface "sees" less of the sky than a horizontal surface.

When a plane is tilted upward toward the Sun at tilt angle TT, the solar intensity on the plane is affected in three ways:

(1) The direct beam intensity is changed because the solar zenith angles are changed compared with a horizontal plane. The solar zenith angle relative to a tilted plane is labeled z_T. This can be estimated by replacing the latitude angle by the latitude angle minus the tilt angle (replace L by $L - TT$):

$$\cos z_T = \sin d \sin(L - TT) + \cos d \cos(L - TT) \cos(2\pi t/24.6)$$

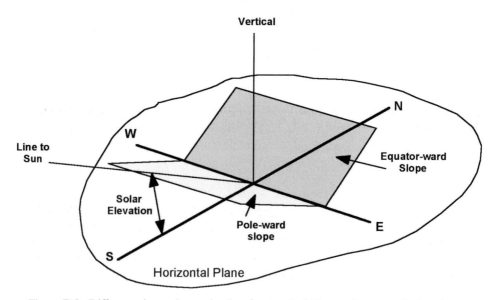

Figure B.8. Difference in angles to the Sun from pole-facing and equator-facing slopes.

If $TT = L$, then the formula for $\cos z_T$ becomes very simple:

$$\cos z_T = \cos d \cos[2\pi t/24.6]$$

The Sun appears to rotate from east to west at $14.6°$ per hour on Mars. The beam intensity on the tilted surface is

$$GBT = GB \cos z_T$$

(2) The diffuse component is affected because one obtains a projection of the diffuse component onto a tilted plane. This effect is difficult to estimate accurately. This will be discussed more fully in later sections.

(3) There is an additional input from the reflection of solar rays by the ground in front of the tilted plane onto the tilted plane, proportional to the albedo of the ground.

B.3.1   The diffuse component on a tilted surface

The diffuse component on a tilted surface can easily be estimated in two extremes. In one extreme, it is assumed that the diffuse irradiance emanates isotropically from the whole sky ($GDTI$). This extreme would be approached for high optical depths and high solar zenith angles (the Sun is low in the sky and the atmosphere is thick). The other extreme is to assume that the diffuse irradiance emanates from near the Sun ($GDTS$). This extreme might be approached at very low values of optical depth and low solar zenith angles (the Sun is overhead and the atmosphere is thin). Photographs taken on Mars indicate that the actual case is somewhere between these extremes (see Figures B.9 and B.10, both in color section).

Our procedure is to estimate the two extreme possible values of the diffuse component on a tilted surface and take a linear combination of the two estimates:

$$GDT \sim K_1(GDTI) + K_2(GDTS)$$

where

$$K_1 + K_2 = 1$$

There is no absolute way to estimate the coefficients. We use a heuristic method based on the belief that, as $D/\cos z$ becomes large, $K_1 \Rightarrow 1$, and, as $D/\cos z$ becomes small, $K_2 \Rightarrow 1$. A handy function that seems to make some sense is:

$$K_2 = \frac{1}{1 + D/(\cos z)}$$

A plot of this function is given in Figure B.11. In the extreme case that the diffuse component on a horizontal surface is isotropic (i.e., the diffuse light comes equally from all parts of the sky) the tilted collector only "sees" a portion of the sky that varies from 100% when it is horizontal, to 50% when it is vertical. It follows that the diffuse component on a surface at tilt angle TT is given by

$$GDTI = GDH \cos^2(TT/2)$$

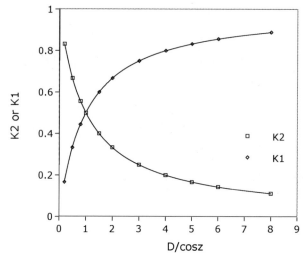

Figure B.11. Variation of parameters with $D/\cos z$.

since $\cos^2(TT/2)$ is the fraction of space subtended by the tilted collector (see Figure B.12). This reduces the diffuse component on a tilted surface compared with that on a horizontal surface.

However, it has been observed that diffuse light on Mars does not emanate uniformly from the whole sky, but, rather, has a source function that is most intense near the Sun and falls off with angle away from the Sun. That being the case, the part of the sky that the tilted collector does not "see" is the part of the sky away from the Sun, and therefore the diffuse component on a tilted surface (GDT) is undoubtedly greater than the estimate based on the isotropic assumption.

In the other extreme, we assume that scattering of the Sun's rays by the Mars atmosphere is all small-angle scattering, and the diffuse rays arrive in a bundle that emanates from near the Sun. In this case, we may regard the entire solar input as if it were all coming from the direction of the Sun. We then estimate $GDTS$ as:

$$GDTS = GDH \frac{\cos z_T}{\cos z}$$

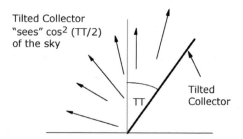

Figure B.12. Fraction of sky "seen" by tilted collector.

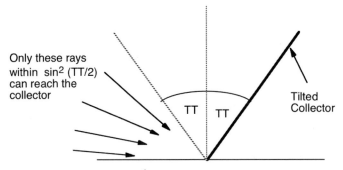

Figure B.13. Only rays within $\sin^2(TT/2)$ can reflect onto the tilted collector.

As mentioned previously, our final estimate for the diffuse component on a fixed tilted surface is

$$GDT \sim K_1(GDTI) + K_2(GDTS)$$

B.3.2   Reflection from ground in front of tilted collector

In addition to the direct beam and diffuse components, there is a contribution to the solar intensity on a tilted collector from reflection from the ground in front of the collector. The albedo of the ground is designated as α. Therefore, the contribution from reflection is

$$GRT = \alpha(GH) \sin^2(TT/2)$$

See Figure B.13.

B.3.3   Total intensity on a tilted surface

The beam intensity on the tilted surface is:

$$GBT = GB \cos z_T$$

The total intensity on a tilted surface is the sum of the direct beam input, the diffuse input, and an input reflected from the ground in front of the tilted collector.

The total intensity on a tilted surface is then

$$GT = GBT + GDT + GRT$$

$$GT = GB \cos z_T + K_1(GDTI) + K_2(GDTS) + \alpha(GH) \sin^2(TT/2)$$

B.3.4   Rotating tilted surfaces

The treatment of a rotating tilted surface is similar to that for fixed tilted surfaces except that the solar collector plane always "faces" the Sun. For rotating solar collectors, the time factor t in the equation

$$\cos z_T = \cos d \cos[2\pi t / D]$$

is always zero (effective value at solar noon) and therefore

$$\cos z_T = \cos d$$

Other than this, the remainder of the calculation is the same.

B.4 NUMERICAL ESTIMATES OF SOLAR INTENSITIES ON MARS

B.4.1 Solar energy on horizontal surfaces

B.4.1.1 Daily total insolation

With the present Mars orbit (eccentricity = 0.093, obliquity = 25.2°, L_S at closest approach to Sun = 250°), solar energy falling on horizontal surfaces was calculated using the Pollack transmission coefficients (Table B.3). Contour plots of daily total insolation as a function of L_S and latitude are shown in Figures B.14A through B.14E for optical depths of 0.3, 0.5, 1.0, 2.0, and 4.0, respectively. An optical depth of 0.3 corresponds to extremely clear weather, 0.5 corresponds to normal clear weather, 1.0 and 2.0 correspond to dust storms, and 4.0 corresponds to extreme peak optical depth at the height of a global dust storm at equatorial latitudes.

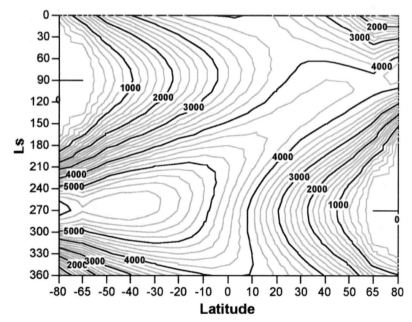

Figure B.14A. Daily total insolation on horizontal surfaces for an optical depth = 0.3 (watt-hr/m^2).

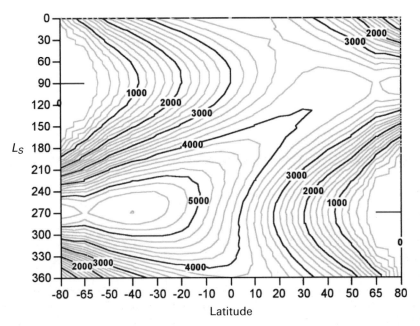

Figure B.14B. Daily total insolation on horizontal surfaces for an optical depth $= 0.5$ (watt-hr/m^2).

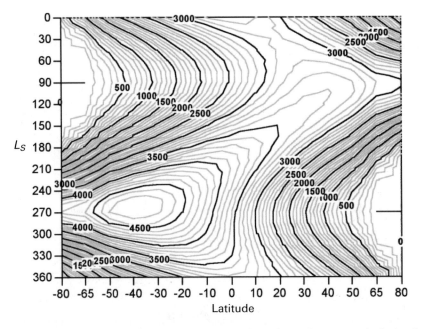

Figure B.14C. Daily total insolation on horizontal surfaces for an optical depth $= 1.0$ (watt-hr/m^2).

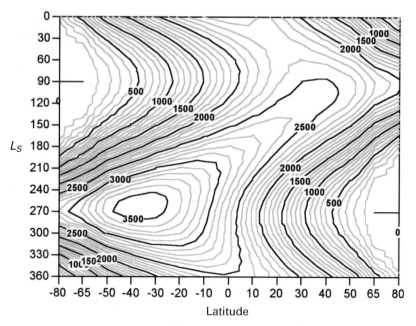

Figure B.14D. Daily total insolation on horizontal surfaces for an optical depth = 2.0 (watt-hr/m^2).

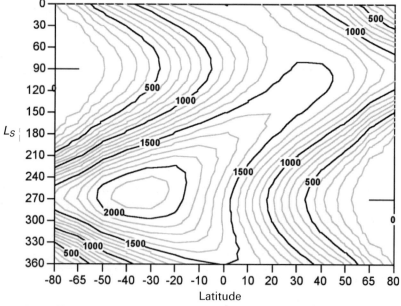

Figure B.14E. Daily total insolation on horizontal surfaces for an optical depth = 4.0 corresponding to the peak of a global dust storm (watt-hr/m^2). Insolation at the height of a global dust storm (optical depth ∼4) is about 30% of that in a clear atmosphere.

There are three major factors that determine the levels of insolation. One factor is the latitude, which controls the range of solar zenith angles for passage of rays through the atmosphere. Another is the fact that Mars is closest to the Sun at $L_S = 250°$, and is farthest from the Sun at $L_S = 70°$. This skews the contour plots in these figures toward higher values of insolation in the lower left quadrant of Figures B.14. The third factor is the seasonal variation in the declination which varies from $+25.2°$ at summer solstice to $-25.2°$ at winter solstice. Summer solstice occurs at $L_S = 90°$ in the northern hemisphere and at $L_S = 270°$ in the southern hemisphere, resulting in a skewing of Figures B.14 toward higher solar intensities in the upper right and lower left quadrants.

The highest daily total solar intensities are recorded in the southern hemisphere in local summer when Mars is closest to the Sun. The peak daily total insolation is reached at a fairly high latitude when the days are longest. As the optical depth increases, insolation decreases globally, but it decreases more at high latitudes due to the longer path lengths through the atmosphere. Therefore, at higher optical depths, insolation at high latitudes is impacted more than at equatorial latitudes.

B.4.1.2 Hourly insolation patterns on a horizontal surface

Hourly insolation patterns on horizontal surfaces are presented in Figures B.15 for several latitudes. At higher latitudes—as, for example, in Figure B.15A—the days are long in summer and there is no Sun at all near winter solstice. The variation in hourly insolation over the course of a year is quite large. At 15°N, the variation in hourly patterns during the course of a year is a minimum because Mars lies closer to the Sun in northern winter and this counterbalances the fact that the solar elevation is lower in winter (cf. Figure B.15C). At intermediate latitudes such as 45°, there is always some insolation at all times of the year, but the summer/winter variation is significant. The summer/winter variation is always greater in the southern hemisphere at all latitudes because Mars is closer to the Sun in southern summer and farther from the Sun in southern winter.

B.4.1.3 Total insolation on a horizontal surface over a Martian year

The yearly total solar intensity (Martian year = 668 sols) for a horizontal surface is shown in Figure B.16. Note the asymmetry between the hemispheres. The southern hemisphere receives more annual solar input because solar intensities are higher during southern summer.

B.4.2 Solar intensities on sloped surfaces

B.4.2.1 Fixed slope surfaces

Solar intensities on pole-facing and equator-facing slopes of 20° were calculated and plotted for an optical depth = 0.5:

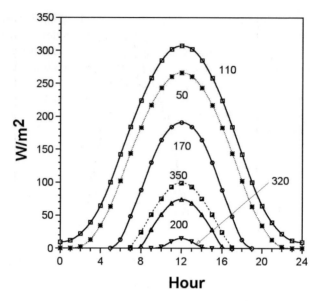

Figure B.15A. Hourly solar intensities on a horizontal surface for an optical depth $= 0.5$ at a latitude of 70°N for various L_S (W/m^2). The curves for $L_S = 230$, 260, and 290 are all on the x-axis (there is no solar intensity in mid-winter at this latitude).

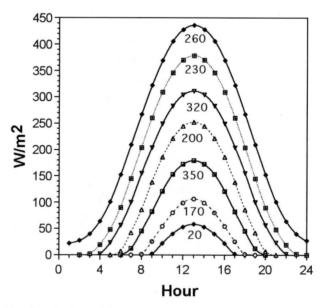

Figure B.15B. Hourly solar intensities on a horizontal surface for an optical depth $= 0.5$ at a latitude of 70°S for various L_S (W/m^2).

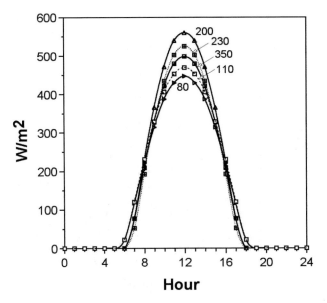

Figure B.15C. Hourly solar intensities on a horizontal surface for an optical depth $= 0.5$ at a latitude of 15°N for various L_S (W/m^2).

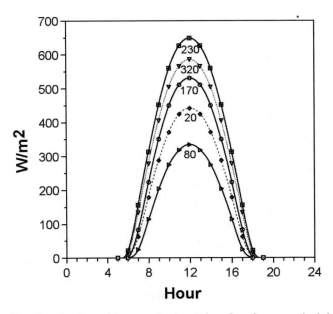

Figure B.15D. Hourly solar intensities on a horizontal surface for an optical depth $= 0.5$ at a latitude of 15°S for various L_S (W/m^2).

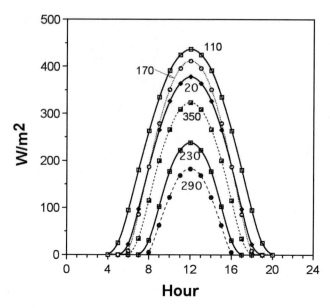

Figure B.15E. Hourly solar intensities on a horizontal surface for an optical depth = 0.5 at a latitude of 45°N for various L_S (W/m^2).

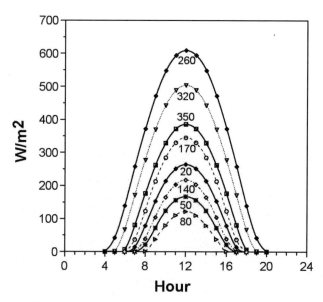

Figure B.15F. Hourly solar intensities on a horizontal surface for an optical depth = 0.5 at a latitude of 45°S for various L_S (W/m^2).

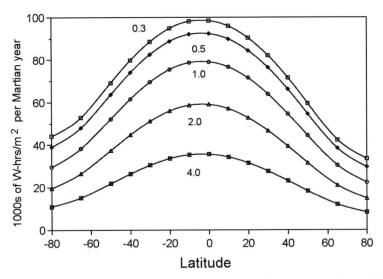

Figure B.16. Total insolation on a horizontal surface for various optical depths (shown by different data points) during a Martian year (1,000s of watt-hr/m^2).

- Figure B.17 shows the solar intensity for equator-facing slopes. The solar intensity on a 20° slope at latitude L is roughly equivalent to the solar intensity on a horizontal surface at $L + 10°$.
- Figure B.18 shows the solar intensity for pole-facing slopes. The solar intensity on a 20° slope at latitude L is roughly equivalent to the solar intensity on a horizontal surface at $L - 10°$.
- Figure B.19 shows the difference between solar intensities on pole-facing and equator-facing slopes. The difference between daily total solar intensity on an equator-facing slope of 20° and a pole-facing slope of 20° can be higher than 1,000 watt-hr/m^2 per Martian sol.

B.4.2.2 Rotating Sun-facing tilted planes

At high latitudes, in clear weather, considerable gains in acquisition of solar energy can be achieved by tilting the solar array up from the horizontal at the latitude angle, and rotating the array at 14.6°/hr so it always faces the Sun. These gains are diminished as the optical depth increases, which increases the percentage of diffuse irradiance. Tilting the collector plane only improves the direct irradiance and, furthermore, tilting decreases the diffuse irradiance because a tilted solar array "sees" only part of the sky. Figures B.20 and B.21 show daily total solar irradiance for rotating tilted arrays with horizontal arrays at 80°S and 65°N for optical depths of 0.3, 0.5, and 1.0. At optical depths of 0.3 to 0.5, the improvement with rotating tilted planes is considerable; at an optical depth of 1.0 it is minimal.

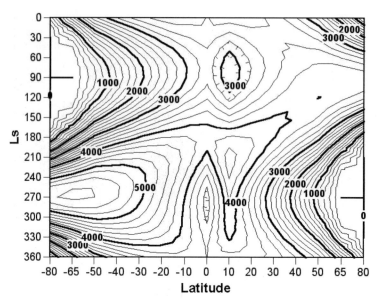

Figure B.17. Daily total solar intensity (watt-hr/m^2) on an equator-facing slope $= 20°$ at an optical depth $= 0.5$.

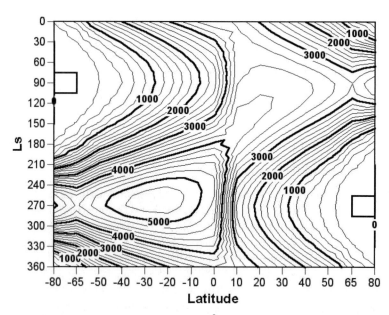

Figure B.18. Daily total solar intensity (watt-hr/m^2) on a pole-facing slope $= 20°$ at an optical depth $= 0.5$.

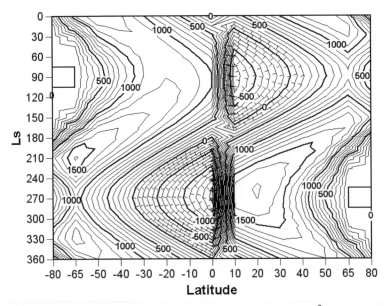

Figure B.19. Differential between daily total solar intensities (watt-hr/m^2) on an equator-facing slope $= 20°$ at an optical depth $= 0.5$ and a pole-facing slope $= 20°$ at an optical depth $= 0.5$.

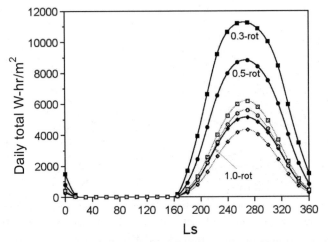

Figure B.20. Daily total solar intensity on a surface at latitude 80°S. The surface is tilted up from the horizontal by 80° and rotates at 14.6°/hr to always face the Sun. The various curves are for different optical depths. Rotating tilted surfaces denoted by RT; horizontal surfaces by H.

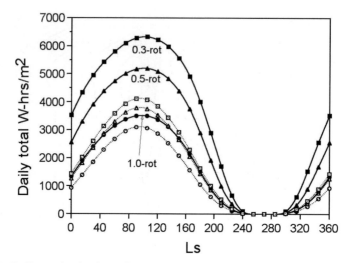

Figure B.21. Daily total solar intensity on a surface at latitude 65°N. The surface is tilted up from the horizontal by 65° and rotates at 14.6°/hr to always face the Sun. The various curves are for different optical depths. Rotating tilted surfaces denoted by *RT*; horizontal surfaces by *H*.

B.4.3   Solar energy on Mars over the last million years

B.4.3.1   *Variations in the Mars orbit*

The orbit of Mars goes through quasi-periodic variations with a period of the order of roughly 100,000 years. Figure B.22 shows the variation in obliquity and eccentricity over one million years. In addition, the historical variation of precession of the longitude of perihelion introduces an asymmetry in peak summer insolation between the poles exceeding 50%, with the maximum cycling between poles every 25,500 years.

Since the extraterrestrial solar intensity at 1 AU is 1,367 W/m^2, the solar intensity impinging on Mars is:

$$I_{ext} = 592 \left(\frac{r}{r_{av}}\right)^2 = 592 \left(\frac{1 + \varepsilon \cos(L_S - L_{min})}{(1 - \varepsilon^2)}\right)^2$$

The solar zenith angle (z) is given by:

$$\cos z = \sin d \sin L + \cos d \cos L \cos(2\pi t/24.6)$$

where d is the declination (in degrees), L is the latitude (in degrees) at the point of observation, t is the time of day (in hours with solar noon taken as zero), 24.6 hours is the length of a diurnal day on Mars. The value of L_S when Mars is closest to the Sun is L_{min}. The declination is an angle that starts at 0° when $L_S = 0$, and varies with L_S according to the formula:

$$\sin d = \sin \varphi \sin L_S$$

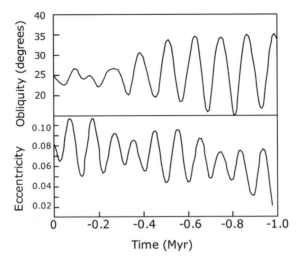

Figure B.22. Historical variation of the obliquity and eccentricity of Mars over the past million years. [Based on data from "Orbital forcing of the martian polar layered deposits," Jacques Laskar, Benjamin Levrard, and John F. Mustard, *Nature*, Vol. 419, 375–376, September 26, 2002.]

where φ is the obliquity angle (currently 25.2°) that has varied widely in the past. The precession of the equinoxes was estimated as a constantly varying quantity with a period of 51,000 years.

Thus, the value of L_{min} was approximated by

$$\cos(L_{min}) = \cos[250° - 360°(Y/51{,}000)]$$

where Y is the number of years prior to the present. L_{min} is presently 250° and varies uniformly with time with a period of 51,000 years.

Numerical fits were made to the graphs for obliquity (φ) and eccentricity (ε) as a function of years in the past, and, when combined with the approximation for precession of the equinoxes, this allows calculation of solar intensities at any time in the past million years.

B.4.3.2   *Insolation on horizontal surfaces over a million years*

The insolation on horizontal surfaces for various northern and southern latitudes is shown in Figures B.23 and B.24, respectively. Periods of high obliquity increase solar intensities at high latitudes and decrease solar intensities at equatorial latitudes. The percentage effect is much greater at high latitudes. At 80° latitude the variation in insolation on horizontal surfaces spans almost a factor of 3 over the obliquity/ eccentricity/precession cycle. The variation in equatorial insolation is only about 12%. A comparison of the insolation on horizontal surfaces for high northern latitudes, high southern latitudes, and the equator is provided in Figure B.25. The

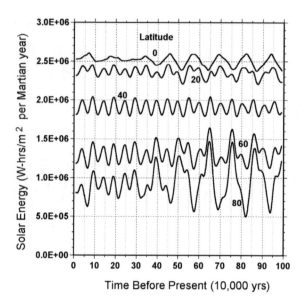

Figure B.23. Insolation on a horizontal surface summed over a Martian year for several latitudes in the northern hemisphere. Peaks at higher latitudes are aligned with troughs at equatorial latitudes. Variation in insolation at 80°N is from 5×10^5 to 1.4×10^6 W-hr/m^2 per Martian year—nearly a factor of 3.

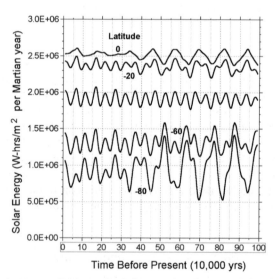

Figure B.24. Insolation on a horizontal surface summed over a Martian year for several latitudes in the southern hemisphere. Peaks at higher latitudes are aligned with troughs at equatorial latitudes. Variation in insolation at 80°S is from 6×10^5 to 1.4×10^6 W-hr/m^2 per Martian year—nearly a factor of 3.

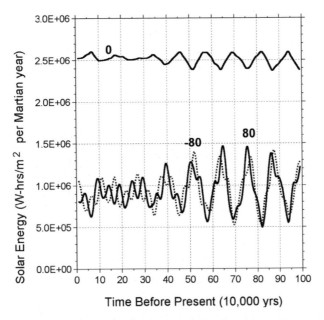

Figure B.25. Comparison of insolation on a horizontal surface summed over a Martian year at the equator with a high southern latitude and a high northern latitude. Peaks at higher latitudes are aligned with troughs at equatorial latitudes. The alignment of peaks and troughs at high northern and southern latitudes depends upon the precession of the equinoxes. When Mars is closer to the Sun in northern summer, the northern peak is higher, and when Mars is closer to the Sun in southern summer, the southern peak is higher.

difference between the northern and southern peaks and valleys is due to the precession of the equinoxes, which continuously changes L_{min} (the value of L_S at which Mars is closest to the Sun). When L_{min} is closer to $90°$, the northern peak will be higher, and when L_{min} is closer to $270°$, the southern peak will be higher. For example, if the peak near 750,000 years ago in Figure B.25 is examined in greater detail, we obtain Table B.4. At 750,000 years before present (YBP) L_{min} is $74°$ (close to $90°$), and therefore the insolation at $80°N$ is quite a bit higher than it is at $80°S$. By contrast, at 780,000 YBP, L_{min} is $285°$ (close to $270°$), and therefore the insolation at $80°S$ is quite a bit higher than it is at $80°N$. This continual reversal of which pole is closer to the Sun continues throughout the history of Mars with a 51,000-year period.

B.4.3.3 Insolation on tilted surfaces over a million years

Insolation on pole-facing and equator-facing $30°$-tilted surfaces over the past million years are presented in Figures B.26 to B.29.

A comparison of solar insolation on a surface tilted at $30°$ toward the equator or the pole with solar insolation on a horizontal surface in the northern hemisphere is

Table B.4. Variation of insolation and orbit properties of Mars near 750,000 years ago.

Time before present	Obliquity	Eccentricity	L_{min}	Insolation at 80°N (W-hr/m^2) per Martian	Insolation at 80°S (W-hr/m^2) per Martian
(10,000 yr)	(°)			year	year
70	17.3	0.065	81	666,552	548707
71	20.0	0.058	151	741,965	680,754
72	22.5	0.062	222	761,971	866,552
73	25.5	0.066	292	851,592	1,031,847
74	29.0	0.070	3	1,097,855	1,085,439
75	33.0	0.075	74	1,409,831	1,121,310
76	34.2	0.081	144	1,416,605	1,217,938
77	32.6	0.076	215	1,162,053	1,333,291
78	29.0	0.072	285	974,361	1,214,174
79	24.6	0.067	356	895,897	909,488
80	20.7	0.062	66	806,128	676,107
81	17.1	0.058	137	634,086	562,618

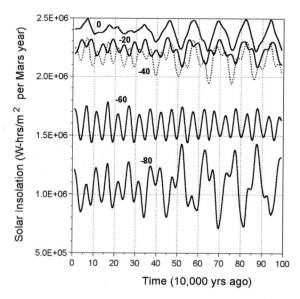

Figure B.26. Insolation on a surface tilted at 30° toward the equator in the southern hemisphere over the past million years. Latitudes are given in the figure.

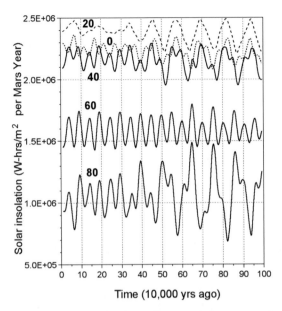

Figure B.27. Insolation on a surface tilted at 30° toward the equator in the northern hemisphere over the past million years.

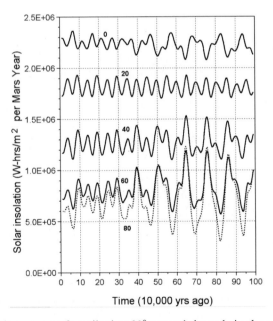

Figure B.28. Insolation on a surface tilted at 30° toward the pole in the northern hemisphere over the past million years.

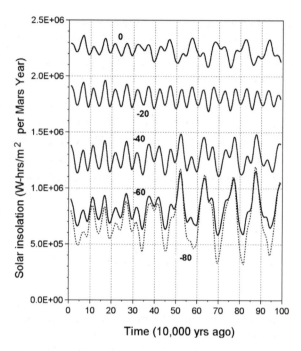

Figure B.29. Insolation on a surface tilted at 30° toward the pole in the southern hemisphere over the past million years.

provided in Table B.5. The positive effects of tilting toward the equator are partially mitigated at high latitudes where much of the solar intensity is diffuse, and tilting the collector plane reduces the fraction of the sky "seen" by the collector plane. A similar table for the southern hemisphere is Table B.6.

Thus, we see that, at latitudes greater than about 40°, pole-facing slopes receive about 30–40% less insolation than a horizontal surface and an equator-facing slope receives about 15–30% more insolation than a horizontal surface.

Table B.5. Comparison of solar energy on surfaces tilted at 30° toward the equator, and at 30° toward the pole, with solar energy on horizontal surfaces in the northern hemisphere.

Latitude	Effect of tilting at 30° toward equator compared with horizontal	Effect of tilting at 30° toward pole compared with horizontal
0	~5% reduction in yearly solar input	~10% reduction in yearly solar input
20°N	Small reduction in yearly solar input	30–35% reduction in yearly solar input
40°N	~15% increase in yearly solar input	30–35% reduction in yearly solar input
60°N	~25% increase in yearly solar input	35–40% reduction in yearly solar input
80°N	~30% increase in yearly solar input	~30% reduction in yearly solar input

Table B.6. Comparison of solar energy on surfaces tilted at 30° toward the equator, and at 30° toward the pole, with solar energy on horizontal surfaces in the southern hemisphere.

Latitude	Effect of tilting at 30° toward equator compared with horizontal	Effect of tilting at 30° toward pole compared with horizontal
0	~5% reduction in yearly solar input	~10% reduction in yearly solar input
20°S	~5% reduction in yearly solar input	25% reduction in yearly solar input
40°S	~15% increase in yearly solar input	~35% reduction in yearly solar input
60°S	~20–25% increase in yearly solar input	~40% reduction in yearly solar input
80°S	~20% increase in yearly solar input	~30% reduction in yearly solar input

B.5   EFFECT OF DUST ON ARRAY SURFACES—SIMPLE MODELS

B.5.1   Introduction

In previous work, Landis[255] estimated the rate at which dust settles out of the Mars atmosphere onto a horizontal solar array. The present section follows his method, with updated parameters, and a few embellishments. The basic approach taken by Landis involves the following steps:

- Assume an optical depth.
- Adopt a particle size distribution for Mars dust from other studies.
- Estimate the cross-sectional area of an average dust particle.
- Estimate the number of particles per cm^2 in a vertical column on Mars at the chosen optical depth.
- Estimate the settling time for dust to settle from dust storm decay rates.
- Estimate the number of particles that settle onto a horizontal surface per sol.
- Estimate the reduction in solar irradiance produced by this settled dust.

Since Landis published his paper, new data suggest a different average particle size.[256] This leads to small changes in the calculation. In addition, an analysis has been added that describes how settled dust becomes distributed into segments one particle thick, two particles thick, etc. The net result is a model that predicts the obscuration produced by settled dust if no dust is removed. It is not possible at this time to estimate rates of dust removal due to lack of data.

[255] "Dust Obscuration of Mars Solar Arrays," G. A. Landis, *Acta Astronautica*, Vol. 38, 885–891, 1996.
[256] "Properties of Dust in the Martian Atmosphere from the Imager on Mars Pathfinder," M. G. Tomasko, L. R. Doose, M. Lemmon *et al.*, *J. Geophys. Res.-Planet*, Vol. 104, 8987–9007, 1999.

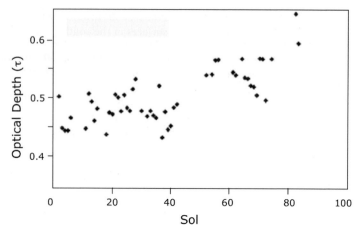

Figure B.30. Measured optical depths on the Pathfinder mission.[257] [Based on data from Crisp *et al.*]

B.5.2   Optical depth

The optical depth is defined as the cross-sectional area for scattering and absorption presented by all the particles in a vertical column of atmosphere of unit area. If the cross-sectional areas for scattering and absorption of each of the particles in the vertical column are summed, the result is the optical depth. It was well established by Viking and Pathfinder data that the optical depth during repetitive "clear" weather is around 0.5 (see Figure B.30 showing Pathfinder data). More recently, MER data indicate that optical depths can drop to as low as ∼0.3 (see Figure B.31 showing MER data). The optical depth can rise to about 4–5 during the peak of a major global dust storm (see Figure B.32). An optical depth of 0.5 implies conceptually that if all the particles in a vertical column are arrayed in a horizontal plane as a monolayer, half of the area will scatter and absorb incident light.

B.5.3   Particle size distribution

Landis (1996) pointed out that the estimated distribution of particle sizes for Mars dust has a narrow peak and a rather long tail extending to much larger sizes. If the most probable particle radius is r_m, then the long tail out to larger sizes assures that the average radius will be about $2.7 r_m$, and the average cross section of a particle will be $\pi(6.9 r_m)^2$. The average mass of a particle will be $\frac{4}{3}\pi(9.8 r_m)^3$. At the time Landis did his calculations, the evidence pointed toward $r_m \sim 0.4$ micron. However, more recent data appear to indicate a smaller value of about 0.25 micron.

[257] "The Performance of Gallium Arsenide/Germanium Cells at the Martian Surface," D. Crisp, A. V. Pathare, and R. C. Ewell, *Acta Astronautica*, Vol. 54, No. 2, 83–101, 2004.

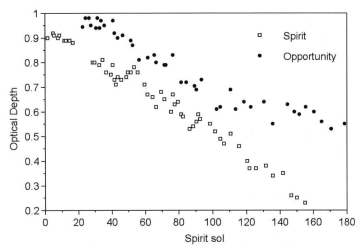

Figure B.31. Measured optical depths on the MER mission. [Based on data from Crisp *et al.*]

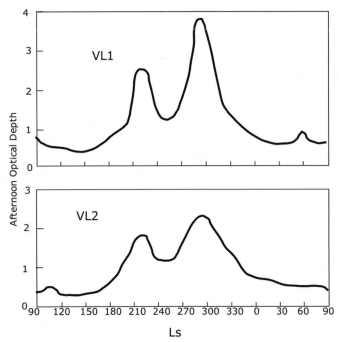

Figure B.32. Optical depth measured by Viking Landers during a major global dust storm.[258] [Based on data from Kahn *et al.*]

[258] "The Martian Dust Cycle," R. Kahn, T. Z. Martin, R. W. Zurek, and S. W. Lee. In: *Mars*, H. H. Kieffer *et al.* (eds.), University of Arizona Press, pp. 1017–1053, 1992.

We may now contrast the average particle sizes to be used in this section with those used by Landis:

	Present calculation	Landis (1996)
Most probable radius (microns)	0.25	0.4
Average radius (microns)	0.7	1.1
Area-weighted average radius (microns)	1.6	2.75
Mass-weighted average radius (microns)	2.5	3.9

B.5.4   Number of dust particles in vertical column

We now follow Landis' method to calculate the number of dust particles in a vertical column of $1\,m^2$ area. We assume an optical depth $= 0.5$ throughout.

The definition of optical depth is

$$D = \text{(number of particles in vertical column)}$$

$$\times \frac{\text{(average area for scattering and absorption per particle)}}{1\,cm^2}$$

Thus

$$N = \frac{D}{A_s}$$

where A_s is the effective area of a particle for light scattering and absorption. A_s is related to the geometrical area of a particle, A_p, by the relation:

$$A_s = A_p Q_{ext}$$

where Q_{ext} is the so-called extinction efficiency or scattering efficiency. It turns out that, because of diffraction effects, Q_{ext} is >1. Pollack's 1990 model (used by Landis) had $Q_{ext} = 2.74$, but a more recent model by Tomasko (1999) based on PF data sets $Q_{ext} = 2.6$. Thus, we have

$$A_s = (\pi(1.6 \times 10^{-4})^2)(2.6) = 2 \times 10^{-7}\,cm^2$$

$$N = 2.6 \times 10^6 \text{ particles in the column}$$

B.5.5   Rate of settling of dust particles

The rate of settling of dust on a horizontal surface is calculated next.

The rate at which dust settles onto a horizontal surface was estimated by Landis as the number of dust particles in the column divided by the settling time (average time it takes for the dust to settle out of a vertical column). Landis estimated the settling time in two ways. One used Stokes' law and the other relied on the observed rough rate of decay of a dust storm. An example of the decay rate of a dust storm is provided in Figure B.32. Based on this, it appears that the settling time is about 80 ± 40 sols. This implies that the settling rate is about 30,000 particles per sol for a $1\,cm^2$ column.

B.5.6   Initial rate of obscuration

It seems likely that, initially, almost all the dust lands as discrete particles without agglomeration. The effective obscuration of a dust particle is estimated as follows. The geometric area is multiplied by Q_{ext} to obtain the scattering area. This, in turn, is multiplied by the fraction of light that is either absorbed or scattered backwards ($\sim$0.23). Thus, the effective blocking area of a particle on the array is

$$0.23 \times 2 \times 10^{-7} \text{ cm}^2 \sim 0.5 \times 10^{-7} \text{ cm}^2.$$

For 30,000 particles, the fractional obscuration per sol is 0.0015, or 0.15%. It is interesting to note that, using the larger particle size adopted by Landis, one obtains nearly the same result. With a larger particle size, one obtains fewer particles in the column and the bottom line is that the fractional obscuration per sol is 0.0012, or 0.12%. Landis used an optical depth of 1.0 in his original calculations that gave about double this rate.

B.5.7   Longer term buildup of dust

As dust continues to fall on the surface, some of the dust starts to fall on other dust particles as well as on the surface of the cell. Gradually, the surface can be divided into segments:

1. Part is free of dust.
2. Part contains a dust monolayer (area $= A$).
3. Part contains a bilayer of dust (area $= B$).
4. Part contains a trilayer of dust (area $= C$).
5. Part contains a quadrilayer of dust (area $= D$), etc.

The rate of buildup of these layers is controlled by the differential equations:

$$\frac{dA}{dt} = 0.0065 * (1 - A - B - C - D \cdots - A)$$

$$\frac{dB}{dt} = 0.0065 * (A - B)$$

$$\frac{dC}{dt} = 0.0065 * (B - C)$$

$$\frac{dD}{dt} = 0.0065 * (C - D) \quad \text{etc.}$$

The factor 0.0065 is 0.0015/0.23 to convert to geometrical (not optical) area. The result of the integration is shown in Figure B.33. According to this model, about 30% of the array becomes covered by a monolayer and about 7% by a bilayer after 100 sols.

The optical obscuration due to these layers can only be crudely estimated. For that part of the surface covered by a monolayer, we assume that the obscuration is

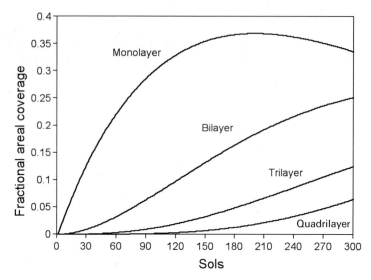

Figure B.33. Predicted distribution of dust particles by layers as a function of sols at OD = 0.5 assuming that each dust particle falls randomly on the surface.

0.23 times the area of the monolayer. For that part of the surface covered by a bilayer, we assume the transmission is 0.77 ∗ 0.77, so the obscuration is 0.41 times the area of the bilayer. Similarly for multiple layers, we take higher powers of 0.77. With this crude assumption we obtain the curve of obscuration vs. sol shown in Figure B.34 assuming no dust is removed.

The mass of dust on the array corresponding to each state can be estimated by multiplying the number of dust particles by the mass of a dust particle. The number of dust particles is

$$\frac{A + 2B + 3C + 4D + \cdots}{0.8 \times 10^{-7} \text{ cm}^2}$$

To calculate the mass of an average dust particle we must use the mass-weighted average radius, 2.5 microns. Then, the average mass of a dust particle is roughly:

$$\tfrac{4}{3}\pi(2.5 \times 10^{-4})3\delta = 5.8 \times 10^{-11}\delta \text{ grams}$$

where δ is the density of the dust particle. The mass of dust on the array for any state is

$$(A + 2B + 3C + 4D + \cdots)(\delta)(7.3 \times 10^{-4}) \text{ g/cm}^2$$

The resulting curves of obscuration vs. dust loading for a few possible densities of the dust are shown in Figure B.35.

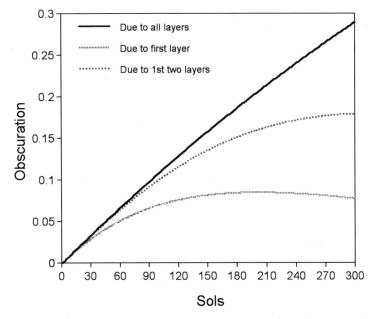

Figure B.34. Obscuration vs. sol for multiple layers if no dust is removed. The upper curve is for all layers. The lower curves are for one and two layers only (OD = 0.5).

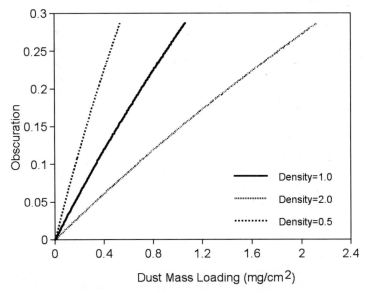

Figure B.35. Stimated dependence of obscuration on dust loading for several possible dust densities, assuming no dust is removed.

B.6   PATHFINDER AND MER DATA ON DUST OBSCURATION

The Mars Pathfinder rover provided several sources of data on the effect of dust accumulating on photovoltaics. Ewell and Burger[259] analyzed the main photovoltaic power system that produced about 175 watts at solar noon and found approximate agreement between the profile of power generated during a typical day and the predictions of power generated from solar and cell models. This is an absolute comparison of the performance of the solar cells. Other tests that do not depend upon cell performance but which provide relative efficiencies from sol to sol are more valuable. One of these was the Materials Adherence Experiment (MAE).[260] The purpose of the MAE instrument was to make a measurement by which degradation of the electrical output of a cell due to dust coverage could be reliably separated from degradation due to other causes or changes in output due to variations in the solar intensity at the surface. With this instrument, the first quantitative measurement was made of the amount of dust deposited and the effect of settled dust on solar cell performance. The MAE solar cell experiment used a GaAs solar cell with a removable cover glass to measure optical obscuration caused by settling dust. During the course of the mission, the cover glass on the cell is occasionally rotated away from its normal position in front of the solar cell, and the short-circuit current (I_{sc}) is measured. Comparing the cell current with and without the cover glass in place measures the optical obscuration of the glass surface by dust on the cover, plus the reflectance of the cover glass itself. The known reflectance of the cover glass is then subtracted out, to give the amount of obscuration due to dust. Measurements were made at noon and at 14:00 local solar time (LST). Unfortunately, the tilt of the rover relative to the Sun is dependent on the local terrain, and therefore the measurement has slight variations due to non-normal angle of incidence. For measurements taken at solar noon, the angle of incidence was typically within reasonable tolerances, but there was considerably more scatter in measurements taken later in the day. The experiment requires that the rotating arm fully remove the cover glass from in front of the solar cell. The rover energizes the actuator, waits a predetermined time, and then measures the solar cell. The measurement is then repeated after the cover is closed. The cover glass reflects 6% of the incident light. Therefore, a verification of whether the cover glass has been removed is obtained by measuring at least a 6% increase in I_{sc} when the cell is uncovered. After 36 sols of operation on Mars, the moving cover glass jammed, and further data on deposition rate from the MAE was lost. Most of the results are confined to the first 20 sols.

---

[259] "Solar Array Model Corrections from Mars Pathfinder Lander Data," R. C. Ewell and D. R. Burger, *Photovoltaic Specialists Conference, 1997*, Conference Record of the 26th IEEE, pp. 1019–1022, 1997.

[260] "Measurement of the Settling Rate of Atmospheric Dust on Mars by the MAE Instrument on Mars Pathfinder," G. Landis and P. Jenkins, *J. Geophysical Research*, Vol. 105, No. E1, 1855–1857, January 25, 2000. Presented at the *AGU Fall Meeting, San Francisco CA, December 6–10, 1998*.

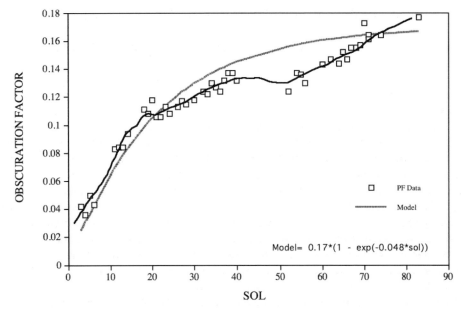

Figure B.36. Obscuration estimated for Pathfinder array by Crisp *et al.* (2001)[261] compared with the simple model: obscuration $= 0.17(1 - \exp(-0.048 \text{ sol}))$.

The results of the MAE experiment appeared to lead to the conclusion that initial dust obscuration increased by about 0.3% per day. Some Mars planners were concerned that at this rate obscuration might reach 27% after 90 sols, and 54% after 180 sols. As we shall see, this seems unlikely.

A single solar cell was included on Pathfinder that was continuously exposed to the environment. Its short-circuit current could be measured typically between 13:00 and 14:00 local time on many sols. This avoided morning ice clouds and shadowing by the mast. By using a model of solar intensity vs. sol, one can predict the mid-day solar intensity on Mars at the Pathfinder latitude for the sols involved in the mission. If one arbitrarily scales this solar intensity to the measured short-circuit current at one of the early sols, the relative effect of dust for the remainder of the mission can be determined by comparing the observed I_{sc} with the scaled solar intensity. In doing this, the daily optical depth should be used for each sol. This calculation was carried out independently by Ewell and Burger (1997) and by Crisp (2001) using slightly different solar models. Depending on which sol is used to scale the solar intensity to the I_{sc}, the curves may differ. The results of Crisp are shown in Figure B.36.

An overly-simple model for this is to assume that the rate of deposition of dust is constant and the rate of removal of dust is proportional to how much dust is on the

[261] D. Crisp, A. V. Pathare, and R. C. Ewell, "The Performance of Gallium Arsenide/ Germanium Cells at the Martian Surface," *Acta Astronautica*, Vol. 54, No. 2, 83–101, 2004.

array. In this case, simple integration of the resultant differential equation leads to the conclusion that the obscuration as a function of time has the form:

$$A(1 - \exp(-B * \text{sol}))$$

The best fit to the data in Figure B.36 is with $A = 0.17$ and $B = 0.048$. This curve is compared with PF data in Figure B.36. It is not at all clear whether the obscuration factor is "saturating" after about 90 sols.

Data were also taken by the MER rovers on the current produced by a standard cell, as well as by the main solar array. Data are available at solar noon from the standard cell and the solar array. Daily total data are available for the solar array. These data can be compared with solar models for the expected variation of current produced, and the difference attributed to obscuration due to dust on the arrays.

Figures B.37 and B.38 show the predicted variation of solar intensity at local noon with day of the mission for several fixed values of optical depth. Figures B.39 to B.42 show the predicted variation of daily total solar intensity and solar intensity at local noon with day of the mission using the daily measured values of optical depth along with the actual measured solar power. The gap between the curves is presumably due to obscuration by dust on the arrays.

Over a longer period of time, it was found that the power generated by the solar arrays on Opportunity recovered considerably and approached levels appropriate to a fairly clean surface (see Figure B.43). It should be emphasized that this was not due to a change in season or a change in optical depth. It represents a true cleaning of the solar array surface. It is not clear what caused the cleaning of the array. An insight

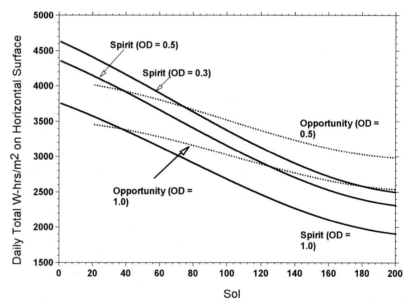

Figure B.37. Predicted variation of daily total solar irradiance at both MER sites for two assumed constant values of optical depth.

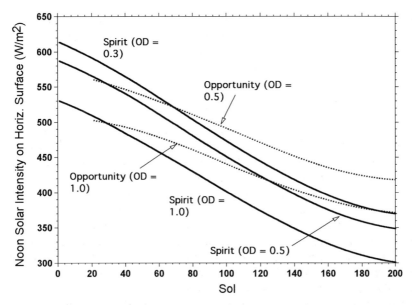

Figure B.38. Predicted variation of solar noon irradiance at both MER sites for two assumed constant values of optical depth.

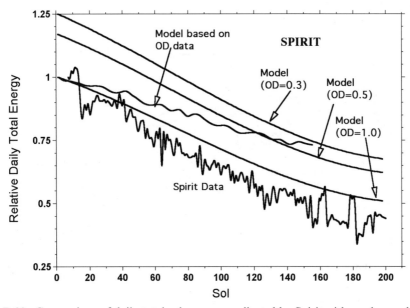

Figure B.39. Comparison of daily total solar energy collected by Spirit with a solar model for three fixed optical depths. The predicted variation of daily total solar intensity and solar intensity at local noon is compared with the daily measured values of optical depth along with the actual measured solar power. The gap between the curves is presumably due to obscuration by dust on the arrays.

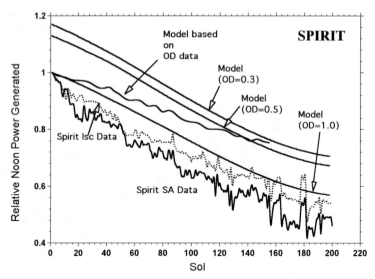

Figure B.40. Comparison of solar noon energy collected by Spirit with a solar model for three fixed optical depths. The predicted variation of daily total solar intensity and solar intensity at local noon is compared with the daily measured values of optical depth along with the actual measured solar power. The gap between the curves is presumably due to obscuration by dust on the arrays.

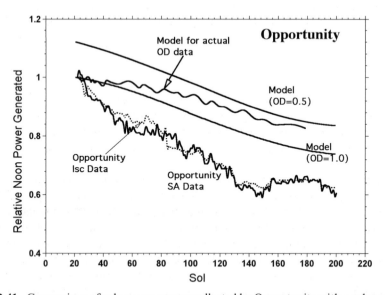

Figure B.41. Comparison of solar noon energy collected by Opportunity with a solar model for three fixed optical depths. The predicted variation of daily total solar intensity and solar intensity at local noon is compared with the daily measured values of optical depth along with the actual measured solar power. The gap between the curves is presumably due to obscuration by dust on the arrays.

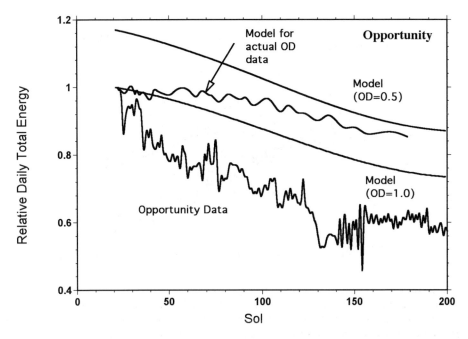

Figure B.42. Comparison of daily total solar energy collected by Opportunity with a solar model for three fixed optical depths. The predicted variation of daily total solar intensity and solar intensity at local noon is compared with the daily measured values of optical depth along with the actual measured solar power. The gap between the curves is presumably due to obscuration by dust on the arrays.

can be obtained, however, by noting that pictures taken from the mast of Opportunity before and after the cleaning took place show that not only was the top deck of Opportunity cleaned off but the ground beneath it had a notable reduction in dust (Figure B.44, in color section). This suggests that the cleaning may have been due to winds.

For Spirit, the ratio of power obtained to power expected if there were no dust on solar arrays appeared to bottom out at around 72% as shown in Figure B.45. For Opportunity, the initial dust loss was similar to Spirit with a 20% dust-induced array current loss over the first ~120 sols. However, after continuing to show increased loss due to dust, at Sol 190 Opportunity suddenly showed an improvement of about 3%. This was followed by another sudden improvement of 5% at Sol 219 and another 4% at Sol 270, resulting in a calculated loss in power due to dust of only 11% at Sol 303 (see Figure B.43). This suggests that dust was removed from Opportunity in a series of distinct actions. Speculation is that possible wind removal, enhanced by the tilted configuration of the array, was instrumental in this. Whatever the reason, this is evidence that the adherence of dust to the solar arrays is weak when it reaches heavy loading.

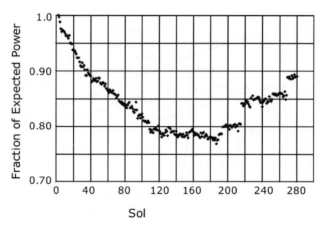

Figure B.43. Long-term power output of Opportunity. [Based on data in "Design and Performance of the MER (Mars Exploration Rovers) Solar Arrays," Paul M. Stella, Richard C. Ewell, and Julie J. Hoskin, *31st Photovoltaic Specialists Conference, January 3–7, 2005.*]

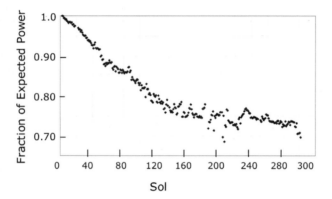

Figure B.45. Estimate of ratio of power obtained to power expected if there were no dust on solar arrays for Spirit. [Based on data in "Design and Performance of the MER (Mars Exploration Rovers) Solar Arrays," Paul M. Stella, Richard C. Ewell, and Julie J. Hoskin, *31st Photovoltaic Specialists Conference, January 3–7, 2005.*]

B.7   AEOLIAN REMOVAL OF DUST FROM SURFACES

Scientists at GRC carried out some experiments on aeolian deposition and removal of dust from surfaces around 1990.

One might guess that the most important variables in such experiments are:

- the size distribution of the dust;
- the amount of dust on the surface;
- the wind velocity;

- the angle of attack;
- the nature of the dust simulant.

In one study, samples were dusted with various Mars dust simulants, and then subjected to winds in a wind tunnel to determine the conditions under which dust could be removed by winds. Samples were held at various angles of attack to the wind for various wind speeds. Three dust simulants were used: an aluminum oxide optical polishing grit, a basalt, and iron oxide. The size distributions of these three dust materials were 7 to 25 microns, 5 to 20 microns, and 0.5 to 2.6 microns, respectively. It appears that the optical polishing grit and the basalt were too large to properly represent Mars dust. However, the iron oxide was in the right size range.

The amount of dust on the surface is likely to be important in aeolian removal processes for small particles, because it is likely that the first layer will adhere more tightly to the surface than subsequent layers. It might be a good deal easier to reduce the obscuration by wind forces from, say, 90% to 50% than to reduce it from 30% to 10%. The experiments on dust removal utilized obscurations [obscuration = (initial transmittance − dusted transmittance)/(initial transmittance)] ranging from 0.2 to 0.8 for the optical polishing grit, 0.3 to 0.4 for basalt, and 0.15 to 0.25 for iron oxide. It is unfortunate that the experiments on the fine iron oxide were limited to such small loading.

The majority of the experiments were conducted on the optical polishing grit (7 to 25 micron size) which is large compared with Mars dust. These experiments showed that:

- It took very high wind speeds (~100 m/s) to remove particles from a horizontal surface.
- Almost no dust was removed by a 10 m/s wind at any angle of attack.
- For wind speeds of >35 m/s, significant dust removal occurred for angles of attack $\gtrsim 20°$.

Experiments on the basalt (5 to 20 microns) yielded results similar to those for optical grit, from which it might be inferred that size (not chemical constitution) is most important in dust stickiness.

The experiments on the iron oxide (0.5 to 2.6 microns) are probably most relevant for Mars because of the small particle sizes. The results were reported only briefly. It was stated that there was little dust removal for wind velocities below about 85 m/s. Even at velocities >85 m/s, the angle of attack had to be $\gtrsim 20°$ to effect dust removal. However, considering the fact that the surfaces were laden with iron oxide at obscuration levels of 25% or less, the surfaces were undoubtedly dominated by one-layer and two-layer regions of dust. At heavier dusting levels, wind removal would likely be easier.

We may conclude from this study that dust particles in the size range 0.5 to 2.6 microns are very difficult to remove by wind for light dustings of surfaces. It was noted that surfaces laden with the smallest dust particles (iron oxide) could be turned upside-down without loss of any dust.

A later study took a somewhat different approach. It was concerned with deposition of wind-blown dust onto clear surfaces, removal of dust from dusty surfaces by wind-blown dust, and abrasion of surfaces by wind-blown dust. The later study repeated some experiments on dust removal by clear winds for larger particles. It confirmed that wind speeds $\gtrsim 30\,\text{m/s}$ for moderate angles of attack can easily remove large particles (>8 microns). In addition to the Mars-simulant materials used in the 1990 paper, they also introduced an artificial glass with size range 5 to 100 microns, well outside the preferred range for Mars simulants. Experiments were also conducted with wind-blown dust and initially clean surfaces. Obscurations were found ranging from 0–10% at low angles of attack, to 20–30% at high angles of attack. Wind-blown dust deposition did not seem to be as effective in depositing dust as merely letting it settle on surfaces. Oddly enough, obscuration was less with lower velocity wind. Abrasion was not significant until wind velocities exceeded 85 m/s, and even then did not contribute much to obscuration.

The actual fate of surfaces in the Mars environment remains somewhat of a mystery. It seems likely that Mars surface winds will remove the larger particles from array surfaces, particularly if they are tilted by 20° or more. Smaller particles, in small amounts, are unlikely to be removed by winds. As multiple layers of small particles build up, or if particles agglomerate, they are likely to be held less tightly and might be more susceptible to wind removal.

B.8   OBSCURATION PRODUCED BY DUST ON SOLAR ARRAYS

B.8.1   JPL experiments—2001

In 2001, preliminary experiments were conducted at JPL to determine the relation between dust loading and obscuration. The spectrum of light transmitted by a dusty glass cover was compared with the spectrum of light impinging on the glass. This showed that when sufficient dust is applied to a glass cover such that transmission is reduced by about 45%, the transmitted spectrum is hardly affected at all. This implies that a solar cell designed to operate in the Mars spectrum will continue to function as planned even though dust accumulates on the surface of the cover glass. The reason that the spectrum is changed by dust in the atmosphere, but not by dust on the array surface, has to do with the cone of forward scattering by dust. While short wavelengths are scattered into a wider cone than long wavelengths in both the atmosphere and on the array surface, it makes little difference on the array surface where essentially all forward scattering is transmitted. In the atmosphere, multiple scattering events will tend to deplete short wavelengths from reaching the ground.

Experimental methods were developed to simulate Mars dust materials, measure dust particle distributions, and apply dust to surfaces. Particle sizes comparable with those on Mars were verified. The amount of dust on the surface at this point, where the obscuration (percent reduction in short-circuit current of cell) was 20%, was estimated to be about $0.3\,\text{mg/cm}^2$. A comparison with Figure B.35 shows this result to be in excellent agreement with the simple model for a density of 0.5. Pictures of

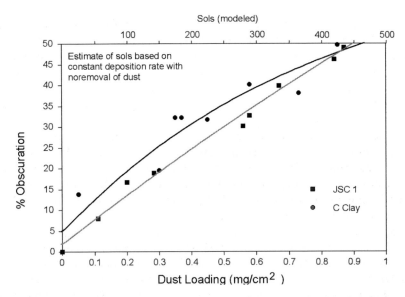

Figure B.47. Measured obscuration of solar cells vs. dust loading for two different Mars dust simulants (JSC 1 and Carbondale Clay).

some of the test cells are shown in Figures B.46A and B.46B (both in color section). It is interesting to note that, visually, a glass surface may appear to the eye to be quite dusty, and still have relatively high transmission characteristics.

The amount of dust on each cell was measured by weighing with and without the dust coating. The resultant plot of obscuration vs. dust loading is shown in Figure B.47. The horizontal scale at the top of this figure represents the number of sols needed to deposit these loadings at a constant rate of 0.0017 mg/sol with no dust removal.

These results provide insights into expectations for dust effects on horizontal Mars arrays. If dust is deposited at a constant rate at optical depth 0.5, and no dust is removed, it seems likely that obscuration might reach about 25–30% after about 200 sols. Laboratory handling of cells with 20–30% obscuration indicates that much of the dust can be blown off easily, and some of it can be removed by vibration. Therefore, it seems unlikely that the actual obscuration will rise to these levels in that amount of time.

B.8.2 Summary and conclusions on dust obscuration

The following conclusions can be drawn:

1   The solar intensity on a horizontal surface on Mars at any latitude in any season for any optical depth can be estimated reasonably well. The solar intensity on

tilted surfaces can be estimated if an approximation is made for the diffuse component. Data have been tabulated and plotted.

2   The main unknown in estimating electrical power that can be generated on Mars is the uncertain effect of dust accumulation on solar arrays.

3   A simple model (originally developed by Landis, 1996) can be used to estimate the initial rate of obscuration of solar arrays on Mars if all the dust that falls on an array accumulates. The result is that the predicted initial rate of obscuration is about 0.15% per sol when the optical depth is 0.5.

4   As dust accumulates on a cell, it gradually becomes distributed into regions of monolayer, bilayer, trilayer, etc. If a simple extension of Landis' simple model is applied to all layers, the dependence of obscuration on sol can be estimated, assuming that no dust is ever removed. The calculation predicts that after about 100 sols, the surface is about 30% covered by a monolayer, about 7% covered by a bilayer, and about 1% covered by a trilayer of dust, with a total obscuration of about 11%. If the density of the dust is assumed to be $0.5 \, g/cm^3$, this corresponds to a dust loading of about $0.18 \, mg/cm^2$. This model predicts that if no dust is removed, obscuration might reach roughly 20% after 200 sols.

5   Since the model described in Sections B.3 and B.4 above assumes that no dust is removed, the figures given represent predicted upper limits to obscuration for an optical depth of 0.5. At higher optical depths, deposition will be greater.

6   Experiments conducted at JPL indicate that dust obscuration might reach about 12–15% after 100 sols, and about 25–30% after 200 sols if no dust is removed. However, it is possible that, when dust obscuration levels reach greater than 20%, at least some of the dust is potentially removable by blowing or vibration.

7   A single cell on Pathfinder provided daily data on short-circuit current. Based on predictions of solar intensity for the measured optical depths on each sol of the Pathfinder mission, the effective obscuration was calculated. The obscuration curve appeared to approach about 15% after 80 sols.

8   Data taken by the MER rovers indicate that obscuration after 200 sols amounts to roughly 25–30%. The optical depth started out at $\sim$1 and decreased with time at both sites.

9   Comparing our models for dust deposition rates with the measured extinction curves, we predict that, if no dust is removed, dust will build up on arrays on Mars in clear weather such that obscuration will reach $\sim$15% after 100 sols, and $\sim$25% after 200 sols. Although dust will continue to deposit after 200 sols, dust removal by aeolian forces and jostling of a rover will probably establish an ultimate equilibrium where obscuration will likely remain in the 20–30% range for extended durations. While dust deposition processes can be modeled to some extent, dust removal processes are much more difficult to comprehend.

10 These results suggest that long-life solar power on Mars is possible. Without overt dust mitigation, obscuration of arrays on a rover is likely to plateau out somewhere between 25% and 30%. With overt mitigation (either vibration or electrostatic), it is likely that obscuration can be maintained in the 5–10% range for long durations.

Appendix C

Water on Mars

C.1   INTRODUCTION

A conflux of theoretical models and experimental data provide a very strong indication that near-surface subsurface H_2O is widespread on Mars at higher latitudes, and reaches down to equatorial latitudes in some regions. This has major implications for *in situ* resource utilization (ISRU) for human exploration of Mars. The H_2O poleward of about 55–60° latitude is undoubtedly ice or snow, while the H_2O at equatorial latitudes may be mineral water of hydration or possibly ground ice.

We are directly aware of H_2O on Mars by observing the polar caps and by measuring concentrations of water vapor in the atmosphere. The water vapor interacts with the porous surface and may (depending on temperatures and water vapor concentrations) act as a source to deposit H_2O in the porous subsurface, or act as a sink to withdraw H_2O from the subsurface. This process has been extensively modeled by a number of scientists.

The morphology of craters on Mars suggests that there may be huge ice reservoirs in the subsurface. If this proves to be correct, such reservoirs would act as a source for formation of subsurface ice at all depths above the source. Only a few models have been developed for such an occurrence.

In order to understand and predict the present-day stability of near-surface subsurface H_2O in the porous interstices of Mars' near-surface regolith, several fundamental scientific principles and properties of Mars must be understood. These include:

- The phase diagram of water and how H_2O behaves under various pressure–temperature regimes.
- Surface temperatures on Mars—mean annual average and seasonal variations as a function of latitude.

- Water vapor concentration on Mars and how it varies with season and latitude.
- Thermal properties of Mars' regolith as a function of porosity and the effect of ice-filled pores.
- Dependence of subsurface temperature profiles on surface temperature seasonal variations and properties of regolith.
- Rates of diffusion of the atmosphere penetrating through the porous regolith medium at subsurface temperatures.

Subsurface ice may have formed in great abundance in temperate zones during previous epochs when the tilt of Mars' axis was greater. Some of this subsurface ice may possibly remain today in localities where, even though ice is no longer stable at equilibrium, receding of the ice by sublimation may be inhibited by various factors. Because of this, it is also necessary to understand:

- Solar energy distribution on Mars, the history of variations in Mars' orbit parameters, and the effect of orbit variations on solar energy at various latitudes.

In addition, it is important to study the morphology of craters on Mars and interpret what they imply regarding distribution of subsurface H_2O as a function of depth and altitude.

These issues have been investigated and discussed in the scientific literature to a greater or lesser degree. A brief summary of the present situation is as follows:

- The level of water vapor concentration in the atmosphere of Mars has been measured in some cases, but data will vary from year to year. It is latitude-dependent and season-dependent, and we have rough estimates of concentrations that are reasonable averages for rough analysis of subsurface ice formation from the atmosphere.
- Measurements and analysis of synoptic temperatures on Mars provide us with mean annual temperatures at the surface as a function of latitude. However, local variations are not well known.
- Models indicate that the rate of diffusion and heat transfer through the pores and channels in the regolith is slow enough that the asymptotic subsurface temperature a few meters down reaches the mean annual surface temperature. The temperature profile through the top few meters of subsurface transitions from the seasonally changing thermal wave of surface temperatures to the constant asymptotic mean annual temperature at a few meters depth.
- The phase diagram for H_2O is well understood. No liquid water can exist below 273.2 K.
- At any location on Mars, if the asymptotic subsurface temperature is low enough that the vapor pressure (over ice) at that temperature is lower than the water vapor partial pressure of the atmosphere, water will tend to diffuse through the regolith and de-sublimate out as ice in the subsurface. This is a fairly slow process, but is rapid enough to allow significant transfers of ground ice to and from the atmosphere over obliquity cycles (tens of thousands of years).

- Given that in some locations the asymptotic subsurface temperature is low enough and the atmospheric partial pressure of water vapor is high enough that subsurface ice will form, the minimum depth at which subsurface ice is stable depends upon the temperature profile as the temperature changes from the surface temperature downward to the asymptotic temperature at deeper depths. A number of authors have made detailed calculations of the temperature profiles (which depend on thermal inertia of the subsurface), and estimated minimum depths for stable ice formation ("ice table") for various soil properties and latitudes.

- Although details vary from author to author, the general outlines of models are similar. These can be summarized as follows:

  (a) The mean annual temperature in the equatorial region ($-30°$ to $+30°$ latitude) is too warm and the water vapor concentrations in the equatorial atmosphere are too low to allow subsurface ground ice to be thermodynamically stable under present conditions. Subsurface ice in this region will gradually sublime away.

  (b) Near the poles, the year-round temperatures are low enough and the average water vapor concentration is high enough to support stable ground ice at the surface and below the surface. The surface ice cap grows and retreats with season. During local summer in the northern hemisphere, a good deal of ice is sublimed, raising the water vapor concentration in the northern atmosphere. A lesser effect occurs in the southern hemisphere.

  (c) As one moves away from the poles, a latitude is reached (perhaps in the range $50°$ to $60°$ depending on soil properties, local temperatures, local water vapor concentrations, slopes, etc.) where the depth of the equilibrium subsurface ice table increases sharply.

  (d) If the regolith is as porous as suspected, and if the measurements of water vapor and temperature on Mars are correct, significant amounts of subsurface ice must form in the pores of the regolith at higher latitudes by the laws of physical chemistry, and the demarcation line where subsurface ice is no longer stable varies with terrain, soil properties, and local weather, but is probably in the range 50–$60°$ latitude. The rate of formation depends on unknown subsurface properties.

  (e) These results for current subsurface ice stability are remarkably different from past epochs when Mars' orbit tilt was much greater, thus enhancing polar heating by the Sun and reducing solar heating in temperate zones. During these periods, near-surface ice may have been stable over much of Mars, and considerable subsurface ice could have been deposited in temperate zones. In addition, the effect of periodic precession of the equinoxes will move water back and forth between the poles, depositing ground ice at intermediate latitudes along the way.

  (f) It must be understood that there is a difference between the models for subsurface ice formation, which *must* occur as a natural consequence of spontaneous physical processes driven by known environmental conditions, as opposed to hypotheses about other forms of water that might be stable under some conditions, but for which no direct cause–effect deterministic process has been identified to assure that such forms of water exist.

(g) The crater record suggests that large reservoirs of subsurface ice exist at depth on Mars and a complete model for the distribution of subsurface H_2O should take this into account.

- The Odyssey Gamma Ray/Neutron Spectrometer has made measurements of H_2O content all over Mars to depths of ~ 1 m, in area elements $5° \times 5°$ (300 km $\times$ 300 km). These measurements support the models that predict near-surface subsurface ice will be prevalent and widespread at latitudes greater than about 55–60°. The data also indicate pockets of locally relatively high H_2O concentration (8–10%) in the equatorial region where albedo is high and thermal inertia is low, suggestive of the possibility that some remnant ice, slowly receding from previous ice ages, might still remain near the surface.
- The equatorial regions with relatively high water content (8–10%) present an enigma. On the one hand, thermodynamic models predict that subsurface ice is unstable near the surface in the broad equatorial region. On the other hand, the Odyssey data are suggestive of subsurface ice. It is possible that this is metastable subsurface ice left over from a previous epoch with higher obliquity. Alternatively, it could be soil heavily endowed with salts containing water of crystallization. The fact that these areas coincide almost perfectly with regions of high albedo and low thermal inertia suggest that it may indeed be subsurface ice. Furthermore, the pixel size of Odyssey data is large, and the 8–10% water average figure for a large pixel might be due to smaller local pockets of higher water concentration (where surface properties and slopes are supportive) scattered within an arid background. Over the past million years, the obliquity, eccentricity, and precession of the equinoxes of Mars has caused the relative solar input to high and low latitudes to vary considerably. Certainly, ground ice was transferred from polar areas to temperate areas during some of these epochs. It is possible that some of this ground ice remains today even though it is thermo-dynamically unstable in temperate areas. In order for remnant subsurface ice from past epochs to be a proper explanation, the process of ice deposition must be faster than the process of ice sublimation in the temperate areas over time periods of tens to hundreds of thousands of years.

C.2   BACKGROUND INFORMATION

C.2.1   Temperatures on Mars

The surface temperature on Mars has been mapped by a number of instruments. The mean annual temperature varies primarily with latitude, but also depends secondarily on elevation and some geological and atmospheric factors. There is a strong latitudinal component with secondary longitudinal variations that reflect variations in surface thermal inertia and albedo. The annual mean temperature ranges from a low of $\sim$160 K in the polar regions to a high of about 220 K near the equator with a global mean of about 200 K. A global plot of average temperature is provided in Figure C.9.

In the absence of ice, the near-surface temperatures depend primarily on the albedo (reflectivity for sunlight) and a quantity called the "thermal inertia". The thermal inertia is defined as [units in square brackets]:

$$I = (k\rho c)^{1/2} [\mathrm{J\ m}^{-2}\ \mathrm{K}^{-1}\ \mathrm{s}^{-1/2}]$$

where $k =$ thermal conductivity [W m^{-1} K^{-1}]
$\rho =$ density [kg m^{-3}]
$c =$ heat capacity [J kg^{-1} K^{-1}]

As the name implies, thermal inertia represents the ability of a material to conduct and store heat, and in the context of planetary science it is a measure of the surface's ability to store heat uniformly. The higher the thermal inertia, the more uniform the temperature distribution and the less temperature rise will occur for any given solar heat input at the surface. A measured map of thermal inertia on Mars is provided in Figure C.1 (color section).[262]

Areas of low thermal inertia exhibit low annual mean temperatures because peak seasonal temperatures are higher and enhance the radiant heat loss [*via* T^4 *radiation law*] to space relative to areas of higher thermal inertia. Daily temperature profiles can reach as high as about 290 K during midday at temperate latitudes. However, overnight temperatures typically dip down to about 200 K even near the equator.

Heat transfer from the surface of Mars downward into the subsurface is a relatively slow process compared with diurnal and seasonal variations in surface temperature. As the surface temperature varies diurnally, and even from season to season, the soil a few meters below the surface does not have time to react to the changing boundary condition at the surface. Therefore, the temperature distribution at depths below a few meters is determined by the average boundary condition: the annual average temperature. Thus, the instantaneous subsurface temperature profile on Mars varies from the highly variable temperature at the surface to a constant asymptotic temperature at sufficient depth (typically a few meters, depending on porosity and thermal inertia).

A number of authors have modeled subsurface heat transfer and obtained temperature distributions vs. depth based on various assumptions about the properties of the subsurface. The subsurface temperature distribution for a particular case (55°S latitude, assumed values of soil and surface properties, no ice included in pores) is shown in Figure C.2.

Figure C.2 shows wide variations in surface temperature with season, but there is a common asymptotic deep temperature that is characteristic of the mean annual temperature. For this particular example, the subsurface is treated as a uniform continuum, and the temperature reaches the asymptotic value (180 K) below about

[262] "Global thermal inertia and surface properties of Mars from the MGS mapping mission," N. E. Putzig, M. T. Mellon, K. A. Kretke, and R. E. Arvidson, *Icarus*, Vol. 173, 325–341, 2005.

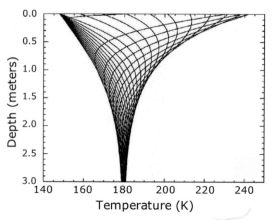

Figure C.2. Calculated subsurface temperature profiles for a homogeneous subsurface at 25-day intervals through a Martian year (55°S latitude, thermal inertia = 250 J/[m² K s^{1/2}]). Despite the fact that surface temperatures swing from 150 to 240 K during the course of a year, the asymptotic deep temperature stays constant at 180 K. [Based on data from Mellon *et al.*[263]]

3 m depth. In a second example, the calculation was repeated except that it was assumed that the porous regolith was filled with ground ice below a depth of 0.5 meter. The high thermal conductivity of regolith filled with ground ice leads to the graph shown in Figure C.3.

Superimposed on the near-surface temperature variations described in Figures C.2 and C.3, there is a slow but inevitable temperature rise (geothermal gradient) with depth due to the outward heat flow from the hot interior of Mars. The geothermal gradient on Earth is the rate of variation of temperature with depth across the lithosphere (upper mantle and crust). The geothermal gradient is closely related to the geothermal heat flow by the simple conduction equation:

$$\text{geothermal heat flow (W/m}^2) = \text{thermal conductivity (W/°C-m)}$$

$$\times \text{ geothermal gradient(°C/m)}$$

Various references give various estimates for these quantities. On Earth, the mean geothermal gradient is 61.5 mW/m² for both oceanic and continental crust, and typical geotherms are around 30–35°C/km. Neither the geothermal gradient nor the geothermal heat flow on Mars is known accurately. Various references have made rough estimates of the geothermal heat flow that range from 15–30 mW/m². However, finite element mantle convection simulations suggest that there can be lateral variations of about 50% relative to the mean value. The thermal conductivity of icy soil is around 2 W/(m-K) but for dry fragmented regolith it could be a factor of 10 (or more) lower. At a thermal conductivity of 2 W/(m-K), a geothermal heat flow of 30 mW/m² translates to a temperature gradient of 15 K per km, so that a 273 K

[263] "The presence and stability of ground ice in the southern hemisphere of Mars," J. T. Mellon, W. C. Feldman, and T. H. Prettyman, *Icarus*, Vol. 169, 324–340, 2003.

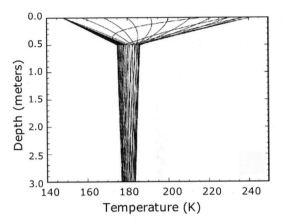

Figure C.3. Subsurface temperature profiles for a two-layer subsurface at 25-day intervals through a Martian year (55°S latitude, thermal inertia $= 250 \, J/[m^2 \, K \, s^{1/2}]$ in upper layer, $2{,}290 \, J/[m^2 \, K \, s^{1/2}]$ in the lower layer). [Based on data from Mellon *et al.*]

isotherm would be reached at a depth of about 5 km if the average surface temperature were 200 K. At a thermal conductivity of 0.2 W/(m-K), this temperature would be reached at a depth of about 500 m.

To illustrate the dependence on material properties, Mellon and Phillips superimposed the geothermal gradient for constant values of thermal conductivity and density consistent with an ice-cemented soil, an ice-free sandstone, and an ice-free soil on the phase diagram of water. A constant mean annual surface temperature of 180 K was assumed. This is shown herein as Figure C.4. The depth at which the melting point of ice is reached is strongly dependent on the thermal conductivity, and the thermal conductivity can vary by orders of magnitude between geologic materials; uncertainties in other parameters such as heat flow are secondary. For high thermal conductivity ice-cemented soil or sandstone, the depth to the melting point is kilometers below the surface, while for the low thermal conductivity uncemented dry soil the depth to the melting point is estimated to be between 100 and 200 m. Clearly, there is a wide variation in behavior, primarily depending upon the thermal conductivity.

C.2.2 Pressures on Mars

The atmosphere of Mars contains about 95.5% carbon dioxide, with argon and nitrogen making up most of the remainder. Viking made measurements of the pressure at two locations on the surface of Mars, one of which was at 48°N and the other at 22°N. These data show:

- There is a significant repeatability of the variation of pressure with season from one year to the next.

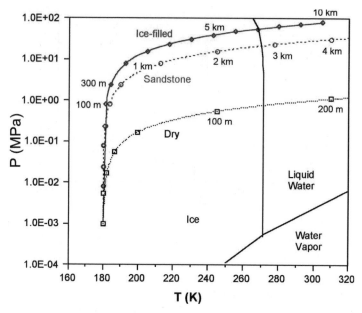

Figure C.4. Superposition of geothermal gradient for ice-filled regolith, dry sandstone, and dry regolith superimposed on the phase diagram of water assuming a near-surface temperature of 180 K. The vertical scale is lithostatic pressure. [Based on data from "Recent gullies on Mars and the source of liquid water," Michael T. Mellon and Roger J. Phillips, *Journal of Geophysical Research*, Vol. 106, 2001.]

- Atmospheric pressure varies significantly with season, the peak occurring in southern summer (northern winter), and the trough occurring in late southern winter (northern summer) with a subordinate peak in northern spring.
- Atmospheric pressure varies significantly with latitude, the pressure being higher at higher latitudes.

The range of pressures is from about 7 to 10 mbar (5 to 8 mm Hg). The seasonal condensation of a significant part of the atmosphere in the polar caps occurs because of the very low polar night temperatures, which allow condensation of carbon dioxide, the principal atmospheric constituent. This phenomenon is responsible for the large-amplitude, low-frequency fluctuations of the Viking pressure measurements as shown in Figure C.5. The first deep minimum of pressure, near Sol 100, occurs during southern winter when a great part of the atmosphere is trapped in the south polar cap . . . . The secondary minimum near Sol 430 corresponds to the northern winter, which is much shorter and less cold than the southern winter because of the high eccentricity of the Martian orbit. The Viking mission has shown that the Martian surface pressure varies by about 25–30% during the course of the year.

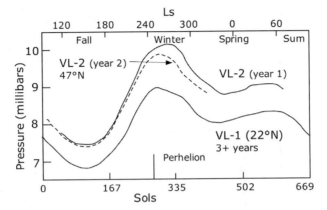

Figure C.5. Viking measurements of pressure on Mars. Horizontal axis is measured in sols. [Based on data from the website *http://www-k12.atmos.washington.edu/k12/resources/mars_data-information/pressure_overview.html*.]

C.2.3 Water vapor concentrations on Mars

Water vapor concentrations are very low on Mars due to the low prevailing temperatures, and they vary significantly with latitude and season. Mars scientists typically present water vapor on Mars in units of *precipitable microns* (pr μm). This is the amount of water (measured as the height of a vertical column of liquid) that would result from condensing all the water vapor in a vertical column of atmosphere of the same area. This unit comes from orbital observations that look through a path length of atmosphere and integrate the water vapor in a column. It turns out that 1 pr μm $= 10^{-3}$ kg/m^2 because the density of water is 1,000 kg/m^3. Using models for vertical distribution of water vapor, it is possible to develop a conversion factor to estimate the partial pressure of water vapor at the surface from the precipitable microns in a column. It is assumed that water vapor is distributed vertically in the atmosphere with a scale height $H = 10,800$ m. Thus, the mass of water (per pr μm) in a column 1 m × 1 m in area is given by

$$M = 0.001H \ (kg)$$

The density is approximately:

$$\rho = \frac{0.001H}{10,800} = 9.3 \times 10^{-8} H \ (kg/m^3) = 9.3 \times 10^{-11} H \ (g/cm^3)$$

Then, using the ideal gas law for an assumed temperature of, say, 220 K, the pressure in mm Hg is:

$$P = \rho RT$$

$$= (9.3 \times 10^{-11}/18) \times 82.1 \ cm^3\text{-atm/mol-K} \times 760 \ mm \ Hg/atm \times 220 \ K$$

$$= 0.00007 \ (mm \ Hg) \ per \ pr \ \mu m$$

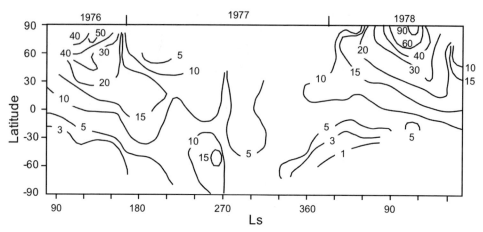

Figure C.5A. Global water vapor as seen by Viking. Contour values in pr μm; shaded areas are regions of no data. Arrows mark dust storms. Shaded rectangle at lower right is size of data bin. Dark lines indicate edge of polar night.[264] As a rough approximation, values of pr μm can be converted to surface partial pressure of water vapor (in mm Hg) by multiplying by 0.00007. [Based on data from Jakosky, 1985).]

This allows us to approximately convert any water vapor partial pressure in precipitable microns to surface partial pressure in mm Hg.

Viking measurements of global water vapor are shown in Figure C.5A. The high water vapor concentrations near the north pole in northern summer indicate that the water vapor arises from sublimation from the north pole and, furthermore, the surface temperature must be ~205 K to support such concentrations. Indeed, we have shown how to convert column densities to surface pressures, and when polar column densities reach ~90 pr μm this corresponds to surface pressures slightly under 0.006 mm Hg, which implies the surface temperature is ~205–210 K (from the phase diagram for water). At this temperature the ice cap must be pure water with no carbon dioxide.

In more recent work,[265] spectra taken by the Mars Global Surveyor Thermal Emission Spectrometer (TES) were used to monitor the latitude, longitude, and seasonal dependence of water from March 1999 to March 2001 (see Figure C.5B, in color section). A maximum in water vapor abundance was observed at high latitudes during midsummer in both hemispheres, reaching a maximum value of 100 pr μm in the north and 50 pr μm in the south. Low water vapor abundances were observed at middle and high latitudes in the fall and winter of both hemispheres.

---

[264] "The seasonal cycle of water on Mars," B. M. Jakosky, *Space Sci. Rev.*, Vol. 41, 131–200, 1985.
[265] "The annual cycle of water vapor on Mars as observed by the Thermal Emission Spectrometer," Michael D. Smith, *Journal of Geophysical Research*, Vol. 107, No. E11, 5115, doi:10.1029/2001JE001522, 2002.

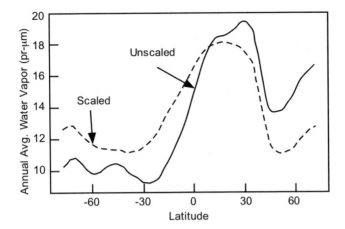

Figure C.5D. The latitude dependence of the annually averaged water vapor abundance. Scaled values are divided by $p_{surf}/6.1$. [Based on data from Jakosky, 1985.]

Smith also presented yearly averaged values of water vapor column abundance as shown in Figure C.5C (color section). The upper plot shows the raw data, while the lower plot is corrected for variable surface pressure due to variable topography. These results clearly show the existence of two regions between $0°$ and $30°N$ where the yearly average water vapor concentration is unusually high. Furthermore, the entire band of latitude from $0°$ to $30°N$ has higher average water vapor abundance than other latitudes. This is shown in Figure C.5D.

C.3   EQUILIBRIUM MODELS FOR SUBSURFACE ICE

C.3.1   Introduction

Over the period from 1966 to the present, a number of studies have analyzed the flow of H_2O between the subsurface and the atmosphere on Mars. This requires understanding the relationship between subsurface temperatures, water vapor pressure over ice at subsurface temperatures, and average local water vapor partial pressures, as a function of latitude. At any location where the average surface water vapor partial pressure exceeds the vapor pressure of ice at the prevailing temperature at subsurface depths below the seasonal thermal wave in near-surface ice will be stable at this depth and below. By "stable" it is meant that (a) ice located in the pores and interstices of the regolith at these depths will not sublime and be lost to the atmosphere, and (b) given enough time water vapor will diffuse from the atmosphere to fill any open pores and interstices with ice at these depths. This will allow preparation of a map that shows the minimum depth where subsurface ice is stable vs. latitude. Some investigators have also studied the rate of H_2O transfer between the subsurface and the atmosphere.

Initially, models were based on the assumption of a uniform subsurface, but, eventually, this was replaced by a two-layer subsurface model with the upper layer desiccated, and the volume below this layer containing ice-filled pores and interstices in the regolith. The demarcation between the layers is called the top of the "ice table". This two-layer model appears to be more physically realistic, and is compatible with observations from orbit. If the upper layer is dry, it has lower thermal inertia, and therefore variations in the surface temperature do not propagate far into the sub-surface. The lower realm reaches a temperature that is an average of the surface temperature over the course of time. The surface temperature is relatively colder (on a yearly average) for high-albedo/low-thermal-inertia surface materials. In those areas where the average surface temperature is too high (and/or the humidity is too low) to allow subsurface ice formation, the ice table will be (to all intents and purposes) infinitely deep.

When average values for soil properties and humidity are used, the models tend to indicate that the ice table will occur near the surface in polar areas, and will slowly descend with decreasing latitude, reaching a depth in the range of perhaps ~1 m at a latitude of roughly 55–60°. Subsurface ice becomes unstable as the latitude is reduced below 55–60°, because the annual average temperature exceeds the frost point of the water vapor in the atmosphere. When soil properties are not taken as averages, but rather as extremes of very low thermal inertia and high albedo, the dividing line between near-surface ice and ice instability moves downward in latitude to perhaps ~50°, depending on the model and the parameters chosen. With these soil properties, the entire calculation becomes sensitive to the water vapor content in the atmosphere, and if a high value (not supported by data) is arbitrarily selected, subsurface ice can be stable down to much lower latitudes. Heterogeneous surfaces with pole-facing slopes and extreme surface properties could support subsurface ice stability over small areas at lower latitudes than would otherwise be expected based on average values. However, the effect of sloping surfaces diminishes as the equator is approached. It does not appear likely that any surface has currently stable subsurface ice below about 40° latitude.

In addition to models that deal with current stability of subsurface ice, there are also several papers that have studied the historical evolution of subsurface ice over long periods. These papers have emphasized that:

(1) The process for subsurface ice formation and sublimation is very slow and requires diffusion through small pores.
(2) Over the past million years (as well as long before that) the orbit of Mars has undergone large quasi-periodic variations in tilt with respect to orbit plane (obliquity), eccentricity of orbit, and positioning of the tilt along the orbit (precession of the perihelion longitude). These variations have had a significant effect on the distribution of solar energy inputs to different regions of Mars, with a consequent variation in distribution of surface temperatures, and redistribution of water assets across the planet.
(3) It is predicted that during periods of high obliquity, when the poles were relatively warmer and equatorial regions were relatively cooler, large amounts

of water were evaporated from the polar caps, and significant deposition of subsurface ice occurred all over the equatorial and mid-latitude regions of Mars. Furthermore, due to periodic precession of the equinoxes, the north and south poles will alternate in phasing relative to distance of closest approach to the Sun of Mars' elliptical orbit, with a period of 51,000 years. This will cause periodic transfer of ice between the poles, potentially depositing considerable ground ice at intermediate latitudes. As Mars entered the post-glaciated period of the past few hundred thousand years, a good deal of this ice has sublimed from equatorial and mid-latitudes. However, in some local areas, where dust inhibited sublimation and temperatures were relatively cooler due to extreme soil properties, subsurface ice might possibly remain from the previous glacial period. This provides a possible hypothesis for the 8–10% water content observed by Odyssey in some local equatorial regions.

Although the flow of H_2O between the subsurface and the atmosphere has been the main concern of theoretical models, there may be deep reservoirs of H_2O in the subsurface of Mars (ice at intermediate depths, and possibly liquid water at deeper depths) and this might be the major source of all H_2O on Mars. In that case, one must couple the analysis of surface–atmosphere interactions to the flow of H_2O between the deep subsurface and the near surface. Although a few investigators have made progress on this, none of the results is entirely satisfying.

C.3.2   Models for stability of subsurface ground ice on Mars—current conditions

Leighton and Murray,[266] in their pioneering 1966 paper, pointed out that based on an assumed water vapor concentration in the atmosphere of about 10^{-3} g/cm^2 (equivalent to about 10 pr μm) the frost point where the water vapor pressure above ice is equivalent to this density is around 190 K. Therefore, for porous soil at temperatures below 190 K, water vapor will condense out as ice in the pores of the regolith. They suggested that such ground temperatures would prevail in regions poleward of about 40–50° latitude and that H_2O would tend to migrate to such regions and condense as permafrost. Leighton and Murray estimated the depth of the ice table as shown in Figure C.6.
They said:

"It would be most surprising if the saturated trapping layer did not extend to depths of at least several tens of meters, so it seems quite possible that several hundred grams of water per square centimeter could be present in the pores of the soil. While we cannot specify the depth to which the permafrost layer extends, we are able to estimate at what depth below the surface its top is situated. At a given latitude, this level will be that at which the vapor pressure of H_2O, averaged

[266] "Behavior of carbon dioxide and other volatiles on Mars," R. B. Leighton and B. C. Murray, *Science*, Vol. 153, 135–144, 1966.

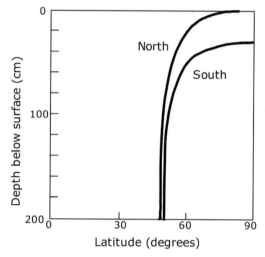

Figure C.6. Leighton and Murray's estimate of latitude dependence of depth of permafrost layer. [Based on data from Leighton and Murray, 1966).]

throughout the year, is equal to the atmospheric average vapor pressure, for then the net annual exchange of water with the atmosphere will be zero."

In 1979, Farmer and Doms[267] utilized Viking data on Mars water abundance and prepared charts of water vapor abundance vs. latitude and season. They said:

"In examining the equilibrium that may exist between the atmosphere and the regolith, we must define the region within the regolith where, at a given season, the temperature never exceeds the local atmospheric frost point temperature. The location of this equilibrium boundary thus identifies the depth, as a function of latitude and season, below which the regolith can retain water in the solid phase. That part of the regolith that is below the frost point boundary throughout the Martian year is the permafrost region; this region can, in principle, act as a reservoir for water on time scales longer than a year."

In addition to permafrost regions, they also identified "tempofrost" regions in the mid-latitudes where ground temperatures remain below the frost point only for the duration of the cold seasons.

The critical factor in determining subsurface distribution of ground ice is the relation between the yearly average water vapor column abundance that determines the frost point temperature at any location. This must then be compared with estimated ground temperatures. Farmer and Doms assumed a global frost point temperature of 198 K based on an assumed equatorial average water vapor concen-

---

[267] "Global seasonal variations of water vapor on Mars and the implications for permafrost," C. B. Farmer and P. E. Doms, *J. Geophys. Res.*, Vol. 84, 2881–2888, 1979.

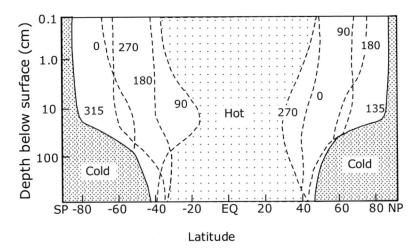

Latitude

Figure C.7. Results of the Farmer and Doms model. Numbers attached to curves are L_S values (90° corresponds to northern summer solstice and 270° corresponds to northern winter solstice). [Based on data from Farmer and Doms, 1979.]

tration of 12 pr μm. In polar regions with higher water vapor concentrations, the actual depth of the ice table will be shallower than their conservative estimate.

Farmer and Doms' results are reproduced in Figure C.7.

The cross-hatched regions (marked "cold") to the left of the solid line in the south (and to its right in the north) never rise above the frost point temperature (198 K in this model) at any time during the year. The solid line thus marks the boundary of the permafrost region that can act as a permanent reservoir of ice. The shaded area marked "hot" is a region where subsurface temperatures rise above 198 K at some time during every day of the year and no permafrost can exist. The white areas represent regions in which the subsurface temperature drops below 198 K for part of the year, allowing some ice formation, but rises above 198 K for the remainder of the year, permitting sublimation of ice. These are the tempofrost regions that can act as sinks for the vapor at mid-latitudes during the cold seasons.

These results indicate that subsurface ice at various depths should be stable at latitudes >45°, but the depth of the ice table drops sharply below 1 m at latitudes lower than about 52°.

During the seasonal progression from winter to summer in each hemisphere, the tempofrost boundary moves toward the equator (as shown by the dashed lines of constant L_S) reaching its lowest latitude point at the winter solstices.

In 1985, Jakosky[268] wrote an extensive review on the "Seasonal cycle of water on Mars" that dealt with many aspects of water on Mars. In regard to subsurface ice, he quotes the results of Farmer and Doms, but cautions that their model is an

---

[268] "The seasonal cycle of water on Mars," B. M. Jakosky, *Space Sci. Rev.*, Vol. 41, 131–200, 1985.

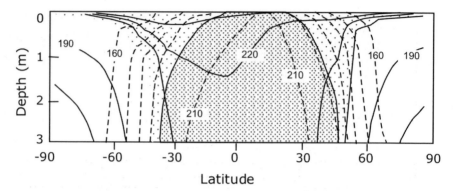

Figure C.8. Paige's results for Mars' average inertia soil with high inertia ice. Regions of permanent ice stability (white areas), transient ice stability for part of a year (light-shaded areas), and no ice stability (dark-shaded areas). Solid lines are maximum temperatures and dashed lines are minimum temperatures, spaced at 10°C intervals. [Based on data from "The thermal stability of near-surface ground ice on Mars," D. A. Paige, *Nature*, Vol. 356, 43–45, 1992.]

over-simplification because the frost point is both seasonally and latitude-dependent whereas Farmer and Doms assumed a global average.

In 1992, Paige carried out a study of subsurface ice on Mars using sophisticated thermal models that included allowance for regolith with high, average, and low values of thermal inertia. Orbital observations of Mars have indicated wide variations in thermal inertia, with regions of low thermal inertia presumably covered by blankets of thermally insulating fine-grained dust. These regions of low thermal inertia will experience higher maximum surface temperatures in local summer, and lower minimum surface temperatures in winter because the amplitudes of their daily temperature variations are larger. In the summer, the higher surface temperatures will allow them to radiate more heat via the T^4 radiation law. Therefore, these regions will develop lower subsurface temperatures below the seasonal thermal wave, and support stable ice at lesser depths. In addition to allowing for variable thermal inertia, he also introduced the notion of the two-level subsurface in which the upper desiccated layer is characterized by a low thermal inertia, and the lower ice-filled layer is characterized by a much higher thermal inertia. His result for average desiccated regolith as an upper layer above an ice-filled regolith lower layer (with high thermal inertia) is shown in Figure C.8.

Mellon and Jakosky (1993) published an important paper on ground ice on Mars. They included the geographical distribution of albedo and thermal inertia, as well as the latitudinal variation of solar intensity in the range from −60° to +60° latitude. Time-dependent models were developed. Thermally driven diffusion of atmospheric water vapor was found to be capable of supplying the top few meters of the regolith with ice in regions where the annual mean surface temperature was below the atmospheric frost point. Over thousands of years, ice will deposit and fill up the pore space, amounting to 30 to 40% of subsurface volume. Their estimate of

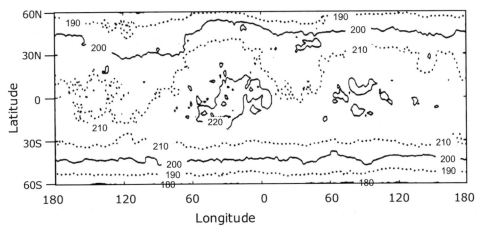

Figure C.9. Annual mean surface temperature—Mellon and Jakosky.

annual mean surface temperature is given in Figure C.9. Figure C.10 shows the latitude dependence of depth of the ice table.

Mellon and Jakosky (1993) modeled the thermal and diffusive stability of ground ice. They concluded that significant geographic variations can occur, in addition to the effects of latitude, and that the boundary poleward of which ground ice is stable can vary by 20° to 30° of latitude due to variations in thermal inertia and albedo. At longitude 90° to 120°, lower temperatures reach down to lower latitudes allowing ground ice to be stable at lower latitudes.

Mellon and Phillips (2001)[269] were primarily interested in explaining the source of gullies (see Section C.7.3.1) in the latitude range 30°S to 70°S. In the process of doing this, they estimated the depth of the ice table in this latitude range as a function of the ground slope. They assumed average values for soil properties and atmospheric water content. Based on this, they found that under current Mars conditions the slope of the surface has a strong effect on the ice table depth in this latitude range. At 30°S, they found that there is no stable ice table for equator-facing slopes, or for pole-facing slopes less than 20°. Pole-facing slopes >20° have a stable ice table about two meters down. At 50°S, the ice table depth ranges from ~60 cm for an extreme pole-facing slope of 40–60° to an ice table depth of about 2 meters for an equator-facing slope of 20°. At 70°S, the ice table depth ranges from ~30 cm for an extreme pole-facing slope of 40–60° to an ice table depth of about 80 cm for an equator-facing slope of 40°.

In addition to these calculations for current conditions, they also explored the dependence on the obliquity of the Mars orbit to infer what might have happened in past epochs. They found that when obliquity exceeds about 31° there is a rather abrupt transition to a stable ice table at 30°S for all slope angles from −60° to +60° at

---

[269] "Recent gullies on Mars and the source of liquid water," Michael T. Mellon and Roger J. Phillips, *Journal of Geophysical Research*, Vol. 106, 2001.

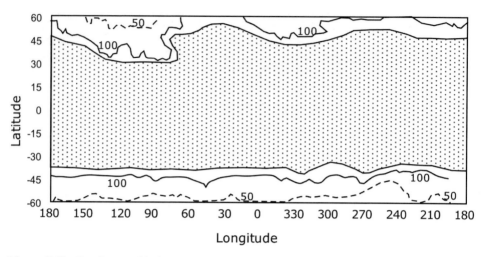

Figure C.10. Depth to stable ice according to Mellon and Jakosky. Contours indicate depth below the surface at 25, 50, and 100 cm. The dark region in the center is where ice is not stable at any depth. The mean atmospheric water abundance is assumed to be 10 pr μm. [Based on data from "Geographic variations in the thermal and diffusive stability of ground ice on Mars," M. T. Mellon and B. M. Jakosky, *J. Geophys. Res.*, Vol. 98, 3345–3364, 1993.]

a depth of about 2 meters. At higher obliquities the ice table is shallower and exists at all latitudes. These results echo other results that lead to the conclusion that, at sufficiently high obliquity, ground ice becomes stable at equatorial latitudes.

Mellon *et al.* (2003) revisited the issue of ground ice on Mars.[270] They made new estimates of ground ice stability and the depth distribution of the ice table and compared these theoretical estimates of the distribution of ground ice with the observed distribution of leakage neutrons measured by the Neutron Spectrometer instrument of the Mars Odyssey spacecraft's Gamma Ray Spectrometer instrument suite. Their calculated ground ice distributions were based on claims of improvements over previous work in that (1) they included the effects of the high thermal con-ductivity of ice-cemented soil at and below the ice table (although Paige had done this previously), (2) they included the surface elevation dependence of the near-surface atmospheric humidity, and (3) they utilized new high-resolution maps of thermal inertia, albedo, and elevation from Mars Global Surveyor observations. All of their results scale with the fundamental (but still uncertain) parameter: the global annual average precipitable microns of water vapor. Comparison of their results with neutron spectrometer results from Odyssey suggests that this parameter is between 10 and 20 pr μm with outer bounds of 5 and 30 pr μm. For the case of 10 pr μm, their results are shown in Figure C.11 (color section). These results indicate that the ice table reaches ∼1 m depth at latitudes northward of $50 \pm 5°$N, and southward of

---

[270] "The presence and stability of ground ice in the southern hemisphere of Mars," J. T. Mellon, W. C. Feldman, and T. H. Prettyman, *Icarus*, Vol. 169, 324–340, 2003.

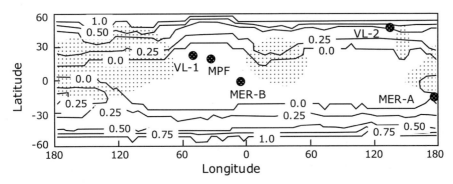

Figure C.12. Fraction of a year for which the frost point is higher than subsurface temperature. Background shading shows regions of high thermal inertia. Mars landing sites are shown as circles. [Based on data from Schorghofer and Aharonson, *loc. cit.*]

$55 \pm 5°$S. If the global annual average water vapor concentration were as high as 20 pr μm, the latitudes where the ice table is at 1 m depth would shift about $4°$ toward the equator.

Schorghofer and Aharonson[271] carried out an analysis of subsurface ice stability. Their work included two parts: (1) depth of the ice table vs. latitude at equilibrium, and (2) rate of change of subsurface due to changing surface conditions. Model predictions were made for ground ice in thermodynamic equilibrium with the water vapor in the present-day atmosphere. Temperatures were obtained with a one-dimensional thermal model of the subsurface, using a thermal inertia map, an albedo map, orbital elements, and partial surface pressures obtained from the Thermal Emission Spectrometer over a Martian year. They found that the depth of the ice table ranged from about $20 \, \text{g/cm}^2$ at $85°$ latitude to $50 \, \text{g/cm}^2$ at $70°$ latitude, to about $100 \, \text{g/cm}^2$ at $60°$ latitude, and then the depth of the ice table plunged at lower latitudes. These depths in g/cm^2 can be converted to linear distance if a density is assumed. For example, if the density is $1.5 \, \text{g/cm}^3$, the depth in cm is the depth in g/cm^2 divided by 1.5. These results for average climatological conditions are in line with previous calculations. They estimated that, due to the rapid exchange of water vapor between the atmosphere and the subsurface, small amounts of subsurface frost will accumulate during the cold season down to latitudes of $25°$, in a layer below the penetration depth of diurnal temperature variations and above the penetration depth of seasonal variations. Figure C.12 shows that, in the regions where subsurface ice is not stable year round, it may be stable for part of a year, particularly where the thermal inertia is low. It can be seen that all of the Mars landers so far, except VL-2, have landed in regions of high thermal inertia where subsurface ice is never stable.

[271] "Stability and Exchange of Subsurface Ice on Mars," N. Schorghofer and O. Aharonson, *Lunar and Planetary Science*, Vol. XXXV (2004), paper 1463; "Stability and Exchange of Subsurface Ice on Mars," N. Schorghofer and O. Aharonson, *Journal of Geophysical Research*, Vol. 110, E05003, doi:10.1029/2004JE002350, 2005 (Preprint).

C.3.3   Long-term evolution of water on Mars

Several papers in the literature deal with the question of how a presumed initial endowment of ground ice on Mars may have evolved over several billion years of exposure to the atmsophere through porous regolith. However, all of these models suffer from our lack of knowledge regarding the Mars environment over billions of years.

In 1967, Smoluchowski used a model that assumed that a layer of ice 10 m thick, protected by an upper layer of thickness L meters, existed 10^9 years ago, and he proceeded to estimate the rate of sublimation as a function of L, porosity, and grain size. From these results, he concluded that, for sufficiently low porosity and small grain size, subsurface ice could persist for 10^9 years when protected by layers of thickness of several meters or more.

In 1983, Clifford and Hillel[272] performed an extensive analysis of the long-term (3.5 billion years) evolution of an initial 200 m thick ice layer buried beneath a 100 m ice-free regolith layer in the equatorial region from $-30°$ to $+30°$ latitude. In the process, they developed detailed models for diffusion through porous regolith. They concluded that it is unlikely that such ice deposits in equatorial regions would have survived the passage of time, depending on soil structure, the geothermal gradient, and historical temperature profiles. All of this ground ice would have had sufficient time to sublime and make its way to the atmosphere.

In 1986, Fanale et al.[273] studied the long-term evolution of subsurface ice on Mars. They concluded that the regolith at latitudes less than $\sim$30–40° is depleted of subsurface ice, whereas the regolith at higher latitudes contains permanent ice.

Mellon et al. (1997)[274] revisited the problem of long-term evolution of ice on Mars, but took into account re-condensation that occurs when water vapor, streaming upward from below, encounters subsurface temperatures cold enough to freeze out the vapors. They assumed that the regolith was initially saturated with ice to a depth of 200 m. They found that, for latitudes in the range $-50°$ to $+50°$, ground ice is initially lost increasingly from the upper levels of the ice deposit. Gradually, the layers near the surface become depleted, and this desiccated layer becomes thicker with time so that an "ice table" is formed that increases in depth with time. [Note: This is very different from the ice tables modeled based on the equilibrium between the subsurface and the atmosphere. This "ice table" is transient, and disappears after sufficient time.] After about 7 Myr, the ice table recedes to a stable depth that depends on latitude and regolith properties. However, after that, further water loss occurs from deeper levels, and re-condensation occurs to maintain the ice table at around this depth. A quasi-steady state is reached in which the upper layer down to the ice table remains desiccated, the depth of the ice table discontinuity remains unchanged, and ice is

---

[272] "The stability of ground ice in the equatorial region of Mars," S. M. Clifford and D. Hillel, J. Geophys. Res., Vol. 88, 2456–2474, 1983.

[273] "Global distribution and migration of subsurface ice on Mars," F. P. Fanale, J. R. Salvail, A. P. Zent, and S. E. Postawko, Icarus, Vol. 67, 1–18, 1986.

[274] "The persistence of equatorial ground ice on Mars," M. T. Mellon, B. M. Jakosky, and S. E. Postawko, J. Geophys. Res., Vol. 102, 19357–19369, 1997.

gradually lost from below the ice table demarcation with the progress of time. As this quasi-steady state progresses, more and more ice is lost to the atmosphere from the lower levels of the ground ice until, after about 19 Myr, all the ground ice is removed. However, the authors say: "If a deeper source provides water vapor at a rate at least equal to the rate of loss to the atmosphere, the steady state distribution can be maintained indefinitely."

These results indicate that, on a geological time scale, diffusion of H_2O in the subsurface is extremely rapid, and a 200 m thick layer of ground ice will disappear in a mere ~19 Myr. Although the paper is not very clear on this, the variation in water content vs. depth with time appears to be almost a step function with zero ice content above the demarcation line. This seems strange. It would seem more natural that there would be a gradient of ice concentration across the upper layer from the fully saturated level at the bottom to zero at the surface. However, the paper says that above the demarcation line, "ice cannot persist because the loss of water to the atmosphere is more efficient than recondensation of vapor from deep within the regolith. Below this depth, ice can persist because the recondensation of water vapor is more efficient (due to higher regolith temperatures) than the loss of vapor to the atmosphere." This description is not clear to this writer. Recondensation cannot be more efficient at higher temperatures. For depths greater than a few meters below the seasonal thermal wave at the surface the temperature increases very slowly with depth. That being the case, there is a driving force to send water vapor from lower depths upward to colder regions where the vapor pressure of colder ground ice is lower. If the surface were sealed, no flow of water vapor would take place if the regolith were already saturated with ice. However, because the surface is open to the atmosphere, water vapor will be lost to the atmosphere if the average surface temperature exceeds the frost point. Water vapor will then diffuse upward to replace the lost water vapor at the surface. Instead of a step function with zero ice content above the demarcation line, it seems likely that there should be a gradient of ice concentration in the upper layer from the fully saturated level at the bottom to zero at the surface.

C.3.4   Effect of Mars orbit variations during the past ~1 million years

As Section B.4.3 shows, the orbit of Mars has undergone rather large variations during the past million years. The most important factor is the variation in obliquity, but variations in eccentricity and periodic precession of the equinoxes are also relevant. Such variations would produce major changes in the distribution of solar energy input to Mars vs. latitude, potentially resulting in redistribution of H_2O resources over this period.

Mellon and Jakosky (1995)[275] extended their previous work[276] by including a consideration of orbital oscillations and found that moderate changes in Martian

---

[275] "The distribution and behavior of Martian ground ice during past and present epochs," M. T. Mellon and B. M. Jakosky, *J. Geophys. Res.*, Vol. 100, 11781–11799, 1995.

[276] "Geographic variations in the thermal and diffusive stability of ground ice on Mars," M. T. Mellon and B. M. Jakosky, *J. Geophys. Res.*, Vol. 98, 3345–3364, 1993.

obliquity can shift the geographic boundary of stable ground ice from the equator (global stability) up to about $70°$ latitude, and that the diffusion of water vapor is rapid enough to cause similarly dramatic changes in the presence of ground ice at these locations on time scales of thousands of years or less. It is estimated that the ice content of the upper 1 to 2 m of the soil can vary widely due to the exchange of atmospheric water at rates faster than the rate of change of Mars obliquity. They provide an analysis of the behavior of near-surface ground ice on Mars through many epochs of varying obliquity during the past ~1 million years. They point out that Mars' climate undergoes two major responses to changing obliquity: (1) temperature change due to redistribution of insolation vs. latitude, and (2) increased summertime water sublimation from polar caps during higher obliquity, thus increasing the atmospheric water abundance and affecting the rate and direction of diffusive transport of water vapor in exchange with regolith at various latitudes. It turns out that the increase in atmospheric water abundance is more important than temperature changes in regard to deposition of ground ice at equatorial and mid-latitudes. A comprehensive thermal/ diffusion model allowed mapping out ground ice formation and depletion as a function of depth, latitude, and Mars' orbital history. Their model describes regions and time periods where ice is stable as well as regions and time periods where ice is not stable, but previously deposited ice remains residual because insufficient time has passed to allow it to sublime. They assumed 40% regolith porosity allowing a maximum of $0.37\,\mathrm{g/cm^3}$ of ice to accumulate.

Even though it may take a few years for water vapor sublimated from polar areas to travel to equatorial regions, the process is fast compared with the rate of variation of obliquity (many thousands of years). Therefore, it was assumed that an increase in polar sublimation rate is matched by an equivalent magnitude increase in mean atmospheric water vapor abundance. Since the current estimated mean atmospheric water vapor abundance is about 10 pr μm for an obliquity of $25.2°$, they scaled the mean atmospheric water vapor abundance to a wide range of obliquities. The result is shown in Figure C.13.

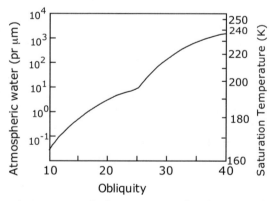

Figure C.13. Estimated mean atmospheric water vapor abundance vs. obliquity. The right side shows the frost point corresponding to each water vapor pressure. [Based on data from Mellon and Jakosky, *loc. cit.*]

According to this model, the mean atmospheric water vapor abundance would be about 35 times greater at an obliquity of 32° than it is today at 25.2°. This raises the frost point from ~195 K to ~218 K, allowing stable deposition of ice in the regolith at all latitudes. Because of the non-linear dependence of water vapor pressure on temperature, and the direct dependence of diffusion on the water vapor pressure at the surface, when Mars enters into a period of increasing obliquity there is a relatively rapid spread of ground ice into lower latitudes, culminating in planet-wide stable ground ice at sufficiently high obliquities. Subsequently, as the obliquity diminishes with time, the near-surface ground ice is gradually depleted due to sublimation. Therefore, during the past million years, it was concluded that there were periods of widespread ice stability alternating with periods where ice was stable only at high latitudes. Because oscillations of the obliquity of Mars have been relatively small during the past 300,000 years, this period has been marked by unusual stability.

It is noteworthy that this study found that "ice accumulates more rapidly during high obliquity than can be lost during low obliquity." Therefore, their curves of depth to the ice table tend to have a characteristic sharp reduction during the early stages of high obliquity, with a longer "tail" extending out as the obliquity diminishes. This appears to be due to the low subsurface temperatures that reduce the vapor pressure and rate of diffusion during the warming period as the obliquity diminishes. The study provides a great amount of data, and it is difficult to summarize all of their findings succinctly. They determined the regions of ground ice stability and the depths of the ice table as a function of obliquity. Their results are summarized in Table C.1.

The periodic transfer of water from the polar caps to temperate zones and back that is likely to take place over the years as the obliquity varies may be tied to the observation of layered deposits near the north pole. It was estimated that the global average of water transferred is about 40 g/cm^2 in the temperate zones, amounting to about 6×10^{16} kg of water.

This study deals mainly with ground ice that is transported from the polar caps via water vapor through the atmosphere. A permanent layer of such ground ice may be built up below the upper desiccated layer by deposition from the atmosphere to the depth of the seasonal thermal wave where the ground temperature is lower than the air temperature, causing condensation of water vapor in the pores of the regolith. This thermal wave may penetrate perhaps up to several meters. Below that level, the geothermal gradient takes over and the temperature slowly increases with depth, removing the driving force for deposition from the atmosphere. The study concludes that, if ice fills the pores of the regolith below this level, it must be ice that was emplaced there a long time ago and is not part of the periodic exchange process between polar caps and near-surface regolith. At sufficient depths, the temperature will exceed 273 K, and it is possible that liquid water may exist at such depths. However, it would appear that in such a case water vapor rising from these depths will condense out in the sub-freezing regolith below ~10 m depth, filling the entire subsurface with ground ice down to the point where $T > 273$ K. It is suggested that in the pattern of variable obliquity over the past million years, during a period of high obliquity, "diffusion is sufficiently rapid to fill the pore space in the near-surface

Table C.1. Regions of ground ice stability within latitude range −60° to +60°.

Obliquity (°)	Northern latitudes where ground ice is stable	Southern latitudes where ground ice is stable	Typical range of ice table depths (cm)	Comments
19.6	≳58°	≳55°	>100	Very little ice stability between −60° and +60°
22.0	≳55°	≳52°	100 and up	North shows beginnings of longitudinal variation
24.6	≳40°- 50°	≳50°	75–100	North shows more longitudinal variation
25.2	≳38°- 48°	≳48°	75–100	Present situation
27.1	Varies from 20° to 40°	≳42°	50–100	Pronounced longitudinal variation in north. Incursions to low latitude at longitudes +30° and −100°
29.3	Varies from 0° to 30°	≳30°	10–50	Ice is stable at all latitudes near −100° longitude, and stable from 0° northward at +30° longitude
31.1	Almost all locations	Almost all locations	5–10	Just a few small pockets near the equator where ice is not permanently stable
32.4	All locations	All locations	5–10	Ice is stable everywhere
33.0	All locations	All locations	2–10	Ice is stable everywhere

regolith completely with ice in just a few thousand years." Conversely, as the obliquity falls, they find that "sublimation can be seen to remove ice down to about a meter before the obliquity completes a cycle and again begins to rise." At a latitude of around 50°, they find a steady build-up of subsurface ice in which each period of high obliquity deposits more ground ice than each period of low obliquity removes ice. It is also noted that it is possible that, during periods when large amounts of water are alternately ingested and released by the regolith as ice, a cyclic inflation and deflation of the surface might cause the small-scale surface features observed on Mars from orbit.

The following conclusions can be drawn:

- While the specific details of the calculations may not be precise due to simplifying assumptions made in the model, the general trends appears to be valid.

- When the obliquity is less than about 22°, ground ice is unstable over most of the region between −60° and +60° latitude.
- When the obliquity is greater than about 30°, ground ice is stable over most of the region between −60° and +60° latitude, and the depth to the ice table tends to be a few tens of cm.
- As the obliquity varies from about 26.5° to 29.5°, the region of ground ice stability in the temperate zone expands greatly and the depth of the ice table drops from >100 cm to a few tens of cm. This is a very sensitive region where small changes in obliquity produce large changes in water ice distribution.
- The present obliquity of 25.2° lies just below this region of high sensitivity, and if the obliquity increases during the next hundred thousand years it may cause very significant changes in the water distribution on Mars.
- The alternating cycles of obliquity tend to deposit deeper ice below about a meter in depth at intermediate latitudes (45–55°). This may produce a long-term build-up of ground ice to the present day, even though ground ice is not thermo-dynamically stable presently. This build-up does not occur at lower latitudes.

Chamberlain and Boynton (2004)[277] investigated the conditions under which ground ice could be stable on Mars based on the past history of changing obliquity of Mars' orbit. They used a thermal model and a water vapor diffusion model. The thermal model finds the temperatures at different depths in the subsurface at different times of the Martian year. Temperatures are functions of latitude, albedo, and thermal inertia. The thermal model can determine the depth to stable ice. Ice is stable if the top of the "ice table" has the same average vapor density as the average water vapor density in the atmosphere. H_2O is allowed to move by the vapor diffusion model. The temperatures from the thermal model are used to partition water between three phases: vapor, adsorbed, and ice. Vapor is the only mobile phase and the diffusion of vapor is buffered by adsorbed water. Vapor diffusion models can have ice-poor or ice-rich starting conditions. Vapor diffusion models are run for long periods to check the long-term evolution of depth to stable ice. As ice distribution in the ground changes, the thermal properties of the ground change too. Thermal conductivity increases as ice fills the pore spaces. Vapor diffusion models here are run iteratively with thermal models to update the temperature profiles as ice is re-distributed. In one set of results, Chamberlain and Boynton present data on the stability of ground ice vs. latitude for various Mars' obliquities. The obliquity of Mars has varied considerably in the past. Two sets of ground properties were utilized:

(a) bright, dusty ground: (albedo = 0.30 and thermal inertia = 100 SI units)
(b) dark, rocky ground (albedo = 0.18 and thermal inertia = 235 SI units).

Their results are shown in Figure C.14.

[277] "Modeling Depth to Ground Ice on Mars," M. A. Chamberlain and W. V. Boynton, *Lunar and Planetary Science*, Vol. XXXV, 2004.

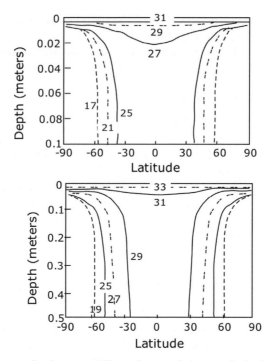

Figure C.14. Minimum depth to stability of ground ice vs. latitude for various Mars' obliquities: (a) upper figure for bright dusty ground, and (b) lower figure for dark rocky ground. [Based on data from Chamberlain and Boynton, *loc. cit.*]

These results indicate that ground ice is never stable at equatorial latitudes at low obliquity. However, as the obliquity is increased, a point is reached (depending on the soil properties) where a discontinuous transition occurs from instability to stability of ground ice. According to this model, this transition occurs between 25° and 27° for bright dusty ground, and between 29° and 31° for dark rocky ground. With the present obliquity at 25.2°, Mars is at the ragged edge of the realm where ground ice could be stable in some locations at equatorial latitudes. According to Figure B.22, there were several periods in the past million years when the obliquity reached 35°. Even in the past ~400,000 years, the obliquity has reached 30°, and was as high as 27° only ~80,000 years ago. During those periods, as Figures B.23 to B.29 imply, solar energy input to equatorial regions was significantly reduced in winter and solar energy input to high latitudes was significantly increased in summer. Although the details have not been worked out, it seems likely based on Figure C.14 that there must have been a major transfer of near-surface ice from high latitudes to temperate latitudes during these epochs. The obliquity has been ≤25.2° over the past ~50,000 years, implying that the subsurface ice deposited in earlier epochs has been subliming, receding, and transferring to polar areas. However, these processes may be slow and could be severely inhibited by dust in some localities. Therefore, it is possible that, in some very bright, low-thermal-inertia regions in the equatorial belt, some of this vestigial subsurface ice from former epochs may remain even today, particularly on surfaces tilted toward the poles. This could possibly explain the areas of higher water content in the equatorial belt observed by Odyssey.

C.3.5 Evolution of south polar cap

The key to ground ice deposition from the atmosphere is the relation between the vapor pressure of ground ice at subsurface temperatures and the water vapor partial pressure in the atmosphere. The possible role of the south polar cap as a source of water vapor pressure in the atmosphere was discussed by Jakosky et al.[278] They discuss the possibility that water vapor pressures were higher in the recent past, depositing ground ice at moderate latitudes, some of which still remains because there has not been sufficient time available for sublimation. The following material is based on the paper by Jakosky et al.

The polar caps represent the major reservoirs for supply of water vapor to and loss from the atmosphere. The peak northern hemisphere water abundance of about 100 pr μm results from summertime sublimation from the exposed north polar water ice cap. Much lower southern hemisphere water content results because the south cap has not been observed to lose its CO_2 ice cover and thereby expose underlying water ice. As a result, the cap never warms to temperatures that would result in sublimation of significant quantities of water. This difference in cap behavior is reflected in the existence of a strong gradient in average atmospheric water content from north to south. Water ice is expected to be present beneath the south polar CO_2 ice because it is less volatile than CO_2 ice. The covering of CO_2 frost acts as a cold trap to remove water vapor from the atmosphere. If the south cap were to entirely lose its CO_2 ice covering at the present epoch, it would expose this underlying water ice. With CO_2 frost no longer present, the south cap would heat up to higher temperatures during summer than does the north cap, as the planet is presently closest to the Sun during southern summer, and significant quantities of water would sublime into the atmosphere. The peak southern water content could be as great as several hundred pr μm. The low-latitude atmospheric water content reflects the control by the polar regions, and lies between the annual average polar amounts. The annual average atmospheric water content over the north polar cap is presently about 30 pr μm; the annual average over an exposed south polar is likely to be closer to 150 pr μm. Thus, it is possible that the annual average atmospheric water content could be 100 pr μm, five or more times greater than the present observed values. This would result in substantial deposition of water ice into the low-latitude subsurface. Can the south cap lose its CO_2 covering at the present time? Theoretical models of the seasonal CO_2 cycle suggest that the south residual cap has two distinctly different stable states. In one state, CO_2 can be present year-round, and the wintertime condensation and

---

[278] "Mars low-latitude neutron distribution: possible remnant near-surface water ice and a mechanism for its recent emplacement," Bruce M. Jakosky, Michael T. Mellon, E. Stacy Varnes, William C. Feldman, William V. Boynton, and Robert M. Haberle, *Icarus*, Vol. 175, 58–67, 2005; Erratum, *Icarus*, Vol. 178, 291–293, 2005.

summertime sublimation exactly or nearly balance. In the second state, the CO_2 ice can disappear by about mid-summer, exposing the underlying material. This material then will heat up substantially during the remainder of summer. One can envision processes by which the cap might jump back and forth between these two stable states. There is some evidence that the cap might occupy each of these stable states at times. Atmospheric water measurements made from Earth in 1969 showed much greater southern hemisphere abundance than has been seen subsequently. The water vapor abundance was high enough to be interpreted at that time as indicating the presence of a residual water ice cap in the south. More recently, observations from the Mars Global Surveyor and the Mars Express spacecraft have suggested that there are portions of the south cap that have exposed summertime water ice. This exposure, if it varies from year to year, might explain the factor-of-10 inter-annual variability in southern summer atmospheric water abundance. If the south polar region contained an exposed water ice cap during summer, the resulting enhancement of atmospheric water content might make water ice stable over a large fraction of the planet. With ice stable, water vapor can diffuse into the subsurface and condense out as ice in a relatively short time. The time for a significant amount of water ice to condense within the regolith could be as short as 10^3–10^4 years; this is short compared with the time over which the orbital elements evolve. If the cap only recently became covered again with CO_2 frost, sufficient time might not have elapsed for the low-latitude water ice, now unstable with respect to sublimation, to diffuse back into the atmosphere. Thus, it is possible that a south-polar-cap configuration in which it is covered with CO_2 year-round might not be representative of the most recent epochs, despite this being the state observed from spacecraft. Instead, the more representative state could be with the south cap losing its CO_2 ice covering during summer. This could have been the case as recently as a few decades ago, a hundred years ago, or a thousand years ago, at which time water ice might have been stable quasi-globally even under the present orbital conditions. Ice deposited 100 or 1,000 years ago might still be present as a transient phase even though it would not be stable today.

C.4   EXPERIMENTAL DETECTION OF WATER IN MARS' SUBSURFACE BY NEUTRON SPECTROSCOPY FROM ORBIT

C.4.1   Introduction

"When cosmic rays strike the atmosphere and surface of Mars, they generate neutrons from other nuclei by various nuclear reactions. The neutrons then lose energy by collision with surrounding nuclei, and in the process they excite other nuclei, which then de-excite by emission of gamma rays. After the neutrons approach thermal energies, they can be captured by nuclei, which then also de-excite by emission of gamma rays. Some of the neutrons escape the planet's surface and can be detected in orbit. The flux of these leakage neutrons is indicative of the amount of moderation and capturing of the neutrons. These processes are a function of the composition of the surface and atmosphere because different elements have different cross sections for capture and have different abilities to moderate neutrons. Hydrogen is especially effective at moderating neutrons because its mass is nearly the same as that of the neutron .... Neutrons are conventionally divided into three different energy bands: fast, epithermal, and thermal."[279]

These energy bands are thermal (energies less than 0.4 eV), epithermal ($0.4\,eV < E < 0.7\,MeV$), and fast ($0.7\,MeV < E < 1.6\,MeV$) neutrons.

By measuring the flux of neutrons in each energy band, it is possible to estimate the abundance of hydrogen in the upper $\sim$1 meter of Mars, thus inferring the presence of water. However, the hydrogen will be obscured if there is a surface layer of carbon dioxide ice.

The neutron spectrometer (NS) returns several neutron spectra and a gamma spectrum about every 20 s, which is the equivalent of one degree of motion or 59 km over the surface. The data are then binned over regions of interest to improve statistics. For much of the data reduction, the data were binned in 5° latitude bands to improve the signal-to-noise ratio. The data record only the abundance of hydrogen regardless of its molecular associations; however, the results are reported in terms of water-equivalent hydrogen (WEH) mass fraction (or just plain H_2O mass fraction).

Relating these data to the actual distribution of H in the surface is not straightforward. If the flux of neutrons is constant and hydrogen is uniformly distributed with depth, then the concentration of hydrogen is directly proportional to the gamma signal strength. Since the hydrogen concentration can vary with depth, the relation between concentration and gamma signal is complex. Similarly, if there is a single layer with constant water content, the neutron fluxes can be estimated as a function of water content. However, if—as is almost certainly the case—vertical distribution of the water content is variable, the dependence of gamma and neutron signals on the water distribution function is also complex. The raw data (counts) can only be converted to water content for a specific model of vertical distribution. For all models, it is assumed that the concentration of elements other than H was that of the soil measured by the Mars Pathfinder Alpha Proton X-Ray Spectrometer. In the original work, the results were normalized to unity for a soil with the equivalent of 1% H_2O by weight at the location of Viking I. In subsequent work, normalization was accomplished by using counts from a polar area covered by a CO_2 cap (that acts as a shield) as a "zero" base. This made a small change in the normalization (basically, the 1% minimum rises to 2%).

The model used to process most of the data is a two-layer model in which a desiccated upper layer containing 2% water by weight of thickness D covers an infinite slab containing $X\%$ water by weight. The neutron count rates can be estimated by detailed modeling for any values of X and D. There are two parameters involved, and—in the more recent treatment—use is made of neutrons with different energy to attempt to resolve X and D.

[279] "Distribution of Hydrogen in the Near Surface of Mars: Evidence for Subsurface Ice Deposits," W. V. Boynton, W. C. Feldman, S. W. Squyres, T. H. Prettyman, J. Bruckner, L. G. Evans, R. C. Reedy, R. Starr, J. R. Arnold, D. M. Drake *et al.*, *Science*, Vol. 297, 81–85, July 2002.

C.4.2   Original data reduction

Data reduction from the NS is complex. If it is assumed that the regolith is uniform with water content independent of depth, a comparison of the measured thermal and epithermal fluxes within the model leads to an estimate of the % water.

However, such a uniform regolith is at odds with models of the subsurface that predict a desiccated upper layer with a water-bearing layer below it. Assuming a simple two-layer model, with the upper layer containing ∼1% water, the fluxes of thermal and epithermal neutrons were modeled as a function of the thickness of the overburden of desiccated regolith (in g/cm^2) for various settings of the water content (% by weight) in the lower layer.[280] A similar set of curves can be generated for the upper level containing 2% water. It was found that the 2% curves fit data better at higher latitudes and the 1% curves fit better at lower latitudes. For high latitudes, the data indicated a lower-layer water concentration of perhaps 35%. The thickness of the upper layer appeared to vary typically from about 20–50 g/cm^2 at higher latitudes (60–80°) to over 100 g/cm^2 at 40° latitude.

C.4.3   Modified data reduction

Further analysis provided a revised procedure for data reduction.[281] It was necessary to remove data that were contaminated by seasonal deposits of CO_2 frost by separating the total data set into three parts. The data measured before the autumnal equinox in the south (corresponding to an areocentric longitude of $L_S \sim 0$) were used to generate the portion of the data set poleward of about 60°S latitude. A similar procedure for the northern boundary used data measured after $L_S = 100$. Data within the middle band of latitudes (∼−60° to +60°) were averaged over the entire data set. Correction of all counting rates for variations in cosmic-ray flux, global variations in atmospheric thickness, and effects of variations in the NS high voltage were made. Corrections were also made for variations in atmospheric thickness due to topography at mid-latitudes.

The initial analysis normalized the data against the Viking I site data that indicated ∼1% water content in the upper layer. In that work it was found that data fits to models were best when the upper layer had about 1% water content at mid-latitudes and about 2% water content at higher latitudes. In the new analysis, calibration was made against areas covered with CO_2 ice that should have no hydrogen signal, and this led to adoption of a uniform 2% water content for the upper layer at all locations.

[280] "Distribution of Hydrogen in the Near Surface of Mars: Evidence for Subsurface Ice Deposits," W. V. Boynton, W. C. Feldman, S. W. Squyres, T. H. Prettyman, J. Bruckner, L. G. Evans, R. C. Reedy, R. Starr, J. R. Arnold, D. M. Drake *et al.*, *Science*, Vol. 297, 81–85, July 2002.
[281] "Global distribution of near-surface hydrogen on Mars," W. C. Feldman, T. H. Prettyman, S. Maurice, J. J. Plaut, D. L. Bish, D. T. Vaniman, M. T. Mellon, A. E. Metzger, S. W. Squyres, S. Karunatillake *et al.*, *J. Geophys. Res.*, Vol. 109, E09006, doi:10.1029/2003JE002160, 2004.

As before, both a single-layer and a two-layer model were utilized. The two-layer model utilized an upper layer of depth D (g/cm^2) with 2% water content by weight, and a lower layer of infinite thickness containing X% water by weight. Since neutron data are not sensitive to depths deeper than $\sim$1–2 m, any assumptions made regarding water content below $\sim$1–2 m are irrelevant. Unlike the initial analysis that used thermal and epithermal neutron data, the revised data analysis combined the data on epithermal and high-energy neutrons. Although a preliminary analysis showed that thermal neutrons should be more sensitive to burial depth than fast neutrons, the authors purposely avoided them at this point because a unique interpretation of thermal neutrons requires knowledge of the composition of surface soils (specifically, the abundances of the strongest neutron absorbers: Fe, Ti, Cl, Gd, and Sm). These abundances are not presently available from Mars Odyssey gamma-ray observations.

C.4.4   Water content based on uniform regolith model

Based on the simple model of a uniform regolith (no layers) containing an unknown mass fraction of water, and utilizing only epithermal neutron data, a global estimate of water content was made assuming that the lowest water concentration is $\sim$2%, as shown in Figure C.15 (color section).[282] The effect of a two-layer model with 2% water in the upper layer on Figure C.15 will lead to increased water concentrations in the lower layer compared with those estimated for a single layer. For middle latitudes ($-45°$ to $+45°$) it was estimated that the correction to the data in Figure C.15 is given by:

Water fraction (lower of two layers) $= 0.998$ (water fraction in figure)

$$+ 1.784 \text{ (water fraction in figure)}^2$$

$$+ 0.000435$$

For an equatorial region with, say, 10% water according to the figure (water fraction $= 0.10$), the estimated water content of the lower layer in a two-layer model is

$$0.0998 + 0.0178 + 0.0004 = 0.118 \text{ or } 11.8\%.$$

This estimate is based on a $10\,\text{g/cm}^2$ upper layer. If the upper layer is thicker, the correction will be greater.

C.4.5   Water content based on a two-layer regolith model at equatorial and mid-latitudes

A first-order approximation to the depth of burial of the lower H_2O-containing layer can be made through a combined study of epithermal- and fast-neutron counting rates. Processing data from epithermal neutrons with an assumed uniform regolith model will under-estimate the water content in the lower layer of a two-layer model.

---

[282] Feldman *et al.*, *loc. cit.*

The fast-neutron indication of water content will be lower than the epithermal indication of water content, because the fast signature of hydrogen drops off more rapidly with depth, and generally there is more water at depth than there is near the surface. As the thickness of the uppermost desiccated layer increases to a value greater than about $100 \, g/cm^2$, both the epithermal and fast-neutron curves flatten out to a constant prediction of 2% H_2O. In other words, a water-rich soil layer buried beneath a relatively desiccated layer can no longer be detected from orbit through measurements of escaping neutrons if its physical thickness is larger than about one meter.

Models were developed to define the relationship between the apparent mass fraction of H_2O from measured fast-neutron currents to that determined from measured epithermal as a function of water content in the lower layer and thickness of the upper layer. As Figure C.15 shows, there are three places in the equatorial zone where the simple uniform regolith model indicates water mass fractions as high as $\sim$10%. The two-layer model was used to elucidate more insight into these and other equatorial and mid-latitude sites. For this purpose, the measured counting rates equator-ward of $\pm45°$ were binned into 5° latitude quasi-equal area spatial elements. Data analysis indicates that most of the water-equivalent hydrogen at near-equatorial latitudes is buried below a desiccated layer. The preponderance of data seems to suggest an upper-layer thickness of typically 10–20 g/cm^2 with occasional points going up to 40 or even 60 g/cm^2. Water content at a few places exceeds 10%.

C.4.6   Interpretation of depth from neutron measurements

There is direct correspondence between the energy of a registered neutron and depth where it was produced. The production rate of gamma rays from fast neutrons has a maximum at depths less than tens of centimeters, while the epithermal neutrons originate from a layer 1–3 m below the surface. Combining measurements in the epithermal energy range with measurements above 1 MeV, one may reconstruct the water abundance distribution at different depths starting from the thin subsurface layer and going down to a meter or two in depth. This allows checking of simple models describing the layered structure of the regolith.

To extract information on regolith structure from neutron data Mitrofanov et al.[283] implemented the two typical types of regolith models. One used a homogeneous distribution of water with depth. The second utilized two layers with the relatively dry ($\sim$2% of water) upper soil layer covering the lower water-rich layer. In the first model there was only one free parameter: water content. In the second model there were two free parameters: thickness of the upper layer and water content of the bottom layer. The calculations were restricted to selected high-latitude provinces of Mars. Some wet equatorial regions inside Arabia Terra were also investigated to find the regions of

[283] "Vertical Distribution of Shallow Water in Mars Subsurface from HEND/Odyssey Data," I. G. Mitrofanov, M. L. Litvak, A. S. Kozyrev, A. B. Sanin, V. Tretyakov, W. V. Boynton, D. K. Hamara, C. Shinohara, R. S. Saunders, and D. Drake, *Microsymposium 38*, MS069, 2003.

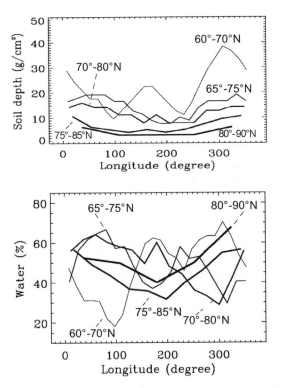

Figure C.16. The top graph shows the depth of the lower layer, while the bottom graph shows the percent water in the bottom layer for the *northern* polar region. [Based on data from Mitrofanov *et al.*, *loc. cit.*]

highest water content at equatorial latitudes, although no specific graphical data were presented for this case. The footprint size was typically 600 km × 600 km. The claim is made that only the two-layer model fits the data but this was not demonstrated.

Figures C.16 and C.17 provide the depth of the ice-filled layer and the water content of the ice-filled layer. The depth of the upper desiccated layer is given in g/cm^2. The actual depth depends upon the density. If the density is, say, $2\,g/cm^3$, then the soil depths in cm are obtained by dividing the ordinates of the upper graphs by 2. Soil depths in the north polar region are small ($0-15\,g/cm^2$ above 70°N) and increase as the latitude is decreased. Variations with longitude are minor above about 70°N, but are highly variable near 60°N. Soil depths are somewhat greater in the southern polar area. The soil depth decreases at higher latitudes. The percentage of water in the lower layer is high enough that it implies that the subsurface could be dirty ice rather than regolith with ice-filled pores, although the two may become indistinguishable at a certain point.

In the south, soil depths are a bit larger ($15-20\,g/cm^2$ pole-ward of 70°S), but water content is similar. However, water content drops more sharply than in the north near 60° latitude. Mitrofanov *et al.* mention briefly that calculations for Arabia

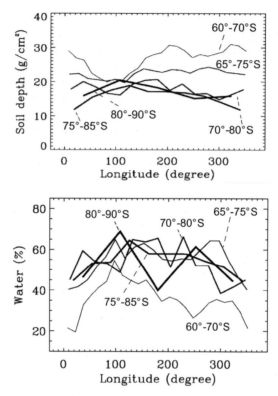

Figure C.17. The top graph shows the depth of the lower layer, while the bottom graph shows the percent water in the bottom layer for the *southern* polar region. [Based on data from Mitrofanov *et al.*, *loc. cit.*]

Terra wet regions show that the water-rich layer lies ~30–40 cm beneath the surface and consists of 9–10% of water. It was claimed that the wettest spot at equatorial latitudes (30°E, 10°N), has about 16% of water placed at a depth below 30 cm.

C.5 COMPARISON OF NEUTRON DATA WITH PHYSICAL PROPERTIES OF MARS

C.5.1 Surface and atmospheric properties

An examination was made[284] of the possibility that there might be statistical correlations between the observation of higher water content in the equatorial area

[284] "Mars low-latitude neutron distribution: Possible remnant near-surface water ice and a mechanism for its recent emplacement," Bruce M. Jakosky, Michael T. Mellon, E. Stacy Varnes, William C. Feldman, William V. Boynton, and Robert M. Haberle, *Icarus*, Vol. 175, 58–67, 2005; Erratum: *Icarus*, Vol. 178, 291–293, 2005.

(±45° latitude) and various parameters that characterize local variations in the surface and climate on Mars. Accordingly, Jakosky et al. prepared the plots shown in Figure C.18 (color section). Jakosky et al. explored the quantitative connections between the regolith water abundance and each of the physical properties that might be controlling the abundance. They did this by calculating the degree of correlation between the epithermal neutron abundance (from which the water abundance is derived) and each of the physical parameters in Figure C.18, after smoothing the latter data sets in order to match the spatial resolution of the neutron measurements.

Data were compared between latitudes of ±45° latitude only. They found a notable lack of statistical correlation between water abundance and any of the parameters, either singly or in groups. They conclude: "Clearly, neither a comparison by inspection of Figure C.18 nor a quantitative comparison shows a compelling relationship, and no single parameter is able by itself to explain a significant fraction of the water distribution."

This statistical analysis showed that there is no cause–effect relationship between the water abundance and the various parameters. Mathematically, we can say that there is no "sufficiency" relation. That is, there is no range of parameters sufficient to assure that water abundance is high. The reason for this is that—if, for example, one considers albedo—there are many areas of high albedo with low water content and a few with high water content.

On the other hand, if we ignore the "sufficiency" of the dependence of water abundance on parameters, and look instead for "necessary" relations, we might conjecture that, whereas there are many areas with high albedo that have low water abundance, nevertheless those areas with high water abundance might tend to have higher albedos. In such a case, high albedo would be necessary but not sufficient for high water abundance. In actual fact, this is the case. In Figure C.18 contours are drawn around the areas where water abundance is highest; in Figure C.18(a) these contours are superimposed on the other plots. These superpositions are shown in Figure C.18.

Starting with Figure C.18(b) and working downward we find:

- Figure C.18(b)   There is a moderate association of higher water abundance with regions where higher peak water vapor concentration occurs in the atmosphere.
- Figure C.18(c)   There is a strong overlap between high water abundance and lower surface temperature.
- Figure C.18(d)   There is a strong correlation of high water abundance and location on the northward side of the pole-facing slope.
- Figure C.18(e)   There is a moderate correlation between regions where higher mean water vapor concentration (corrected for topography) occurs in the atmosphere.
- Figure C.18(f)   There is a good correlation between high water abundance and low thermal inertia.

- Figure C.18(g)   There is a weak correlation between regions where higher mean water vapor concentration (uncorrected for topography) occurs in the atmosphere.
- Figure C.18(h)   There is a fair correlation between high water abundance and high albedo.

We can therefore derive a conclusion (not sanctioned by Jakosky *et al.*) that the equatorial regions of high water abundance are associated with higher values of peak water vapor concentration, lower surface temperatures, location below the tops of north pole facing slopes, and generally somewhat lower values of thermal inertia and higher values of albedo. It must be emphasized that— as Jakosky *et al.* found—if one takes a random site with higher values of peak water vapor concentration, lower surface temperatures, location below the tops of north pole facing slopes, and generally somewhat lower values of thermal inertia and higher values of albedo, the probability of finding high water abundance is almost random. It is only when one works backwards, and takes a site that *does* have high water abundance, that the probability is high that it will be associated with higher values of peak water vapor concentration, lower surface temperatures, location below the tops of north pole facing slopes, and generally somewhat lower values of thermal inertia and higher values of albedo. All of these factors point to the possibility that this water might possibly be ground ice driven by atmospheric deposition. In their erratum, Jakosky *et al.* said:

> "Although such high abundances could be present as adsorbed water in clays or water of hydration of magnesium salts, other measurements suggest that this is not likely. The spatial pattern of where the water is located is not consistent with a dependence on composition, topography, present-day atmospheric water abundance, latitude, or thermophysical properties. The zonal distribution of water shows two maxima and two minima, which is very reminiscent of a distribution that is related to an atmospheric phenomenon. We suggest that the high water abundances could be due to transient ground ice that is present in the top meter of the surface. Ice would be stable at tens-of-centimeters depth at these latitudes if the atmospheric water abundance were more than about several times the present value, much as ice is stable poleward of about $\pm 60°$ latitude for current water abundances. Higher atmospheric water abundances could have resulted relatively recently, even with the present orbital elements, if the south polar cap had lost its annual covering of CO_2 ice; this would have exposed an underlying water ice cap that could supply water to the atmosphere during southern summer. If this hypothesis is correct, then (i) the low-latitude water ice is unstable today and is in the process of sublimating and diffusing back into the atmosphere, and (ii) the current configuration of perennial CO_2 ice being present on the south cap but not on the north cap might not be representative of the present epoch over the last, say, 10,000 years."

Smith provided the distribution of water vapor on Mars in Figure C.5B. Smith also provided contours for albedo and thermal inertia, as shown in Figure C.19.

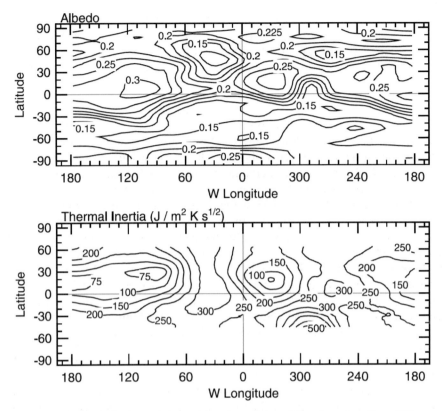

Figure C.19. Maps of albedo and thermal inertia on Mars. [Based on data from "The annual cycle of water vapor on Mars as observed by the Thermal Emission Spectrometer," Michael D. Smith, *Journal of Geophysical Research*, Vol. 107, No. E11, 5115, 2002.]

The correlation between water vapor concentration, albedo, thermal inertia, and the presence of near-surface water in regolith is striking.

C.5.2 Water deposits vs. topography at low and mid-latitudes

A comparison was made between the Mars surface topography and the distribution of near-surface water-equivalent hydrogen (WEH) for six north–south lines on the Mars surface, as illustrated in Figure C.20 (color section).[285]

For each of the vertical white lines in Figure C.20, the comparison of WEH with topography is shown in Figure C.21. The data plotted in Figure C.21 are based on

[285] "Topographic Control of Hydrogen Deposits at Mid- to Low Latitudes of Mars," W. C. Feldman, T. H. Prettyman, S. Maurice, R. Elphic, H. O. Funsten, O. Gasnault, D. J. Lawrence, J. R. Murphy, S. Nelli, R. L. Tokar, and D. T. Vaniman, *Lunar and Planetary Science*, Vol. XXXVI, Paper 1328, 2005.

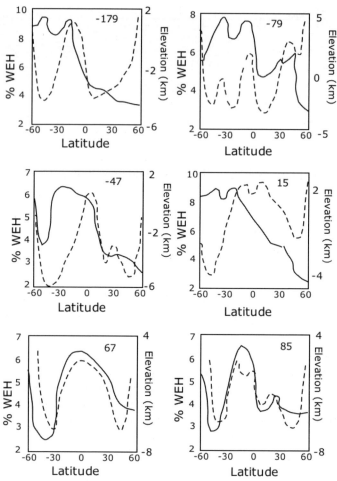

Figure C.21. Comparison of percent WEH vs. elevation for six slices through Mars at constant longitude (shown as white vertical lines in Figure C.20). Dashed lines are percent water and solid lines are elevation in km. [Based on data from "Topographic Control of Hydrogen Deposits at Mid- to Low Latitudes of Mars," W. C. Feldman, T. H. Prettyman, S. Maurice, R. Elphic, H. O. Funsten, O. Gasnault, D. J. Lawrence, J. R. Murphy, S. Nelli, R. L. Tokar, and D. T. Vaniman, *Journal of Geophysical Research*, Vol. 110, E11009, doi:10.1029/2005JE002452, 2005.]

pixels that are typically $2° \times 2° = 160\,km \times 160\,km$. Thus, the data represent averages over such areal segments. Results from the figures are summarized in Table C.2.

Note that in all cases the general trend of elevation is from high to low elevation as the latitude is increased from the southern hemisphere toward the north pole. The magnitude of this drop in elevation can be as large as $\sim 6\,km$ (20,000 ft). Also, note that there is a fairly good correlation between peak water content and peak elevation, and also between peak water content and elevation at which peak water content is found.

Table C.2. Peak water content vs. elevation along six traverses in Figure C.20.

Vertical line	1	2	3	4	5	6
Peak water content (%)	9.6	5.8	5.4	9.5	6	5.8
Peak elevation (km)	10	7.2	6.4	8.7	3.7	3.6
Elevation at peak water (km)	7	5.7	5.2	8 & 6.5	3.7	2
Latitude at peak water (°)	−5	−8	+5	−10 & +12	+5	−20

Some of the interesting highlights from Feldman *et al.* are:

(1) In general, equatorial water concentration is likely to be higher than water concentration at 20–40° latitude.
(2) Over any long, sloping decline from perhaps −15° latitude northward, the peak water concentration is typically found slightly northward of the latitude where the peak elevation is reached.
(3) The higher the peak elevation of the north–south line on Mars, and the higher the elevation at which peak water is found, the higher is the peak water concentration.

It is possible that during northern summer, the flow of "air" southward from the pole might carry water vapor pressures that are high enough even at equatorial latitudes to be higher than the vapor pressure corresponding to subsurface temperatures at high elevations, thus depositing ground ice on the upper reaches of pole-facing slopes. The analogy is not perfect, but it is perhaps a little like Pacific moisture-laden air depositing more precipitation on the flanks of the Angeles Crest mountains surrounding LA than downtown LA.

It is also noteworthy from Figure C.21 that the water content rises rapidly near 60°N in all cases. Northward of 60°, the water surely exists as ground ice.

Whether these equatorial water deposits are mineral hydration or ground ice is difficult to determine. But, this water does appear to be deposited by weather emanating from the north as suggested by Feldman *et al.* It exists at the higher elevations, in regions of low thermal inertia and high albedo, indicating that these regions of high water content have relatively colder subsurface temperatures than other equatorial regions. The temperature profiles show a significant decrease in average temperature with increasing elevation,[286] suggesting greater potential for deposition of ground ice.

[286] "Temperature inversions, thermal tides, and water ice clouds in the Martian tropics," D. P. Hinson and R. J. Wilson, *Journal of Geophysical Research*, Vol. 109, E01002, doi:10.1029/2003JE002129, 2004; "Temperature inversions, thermal tides, and water ice clouds in the Martian tropics," D. P. Hinson and R. J. Wilson, *Journal of Geophysical Research*, Vol. 109, E01002, doi:10.1029/2003JE002129, 2004.

C.5.3   Seasonal distribution of equatorial near-surface water

Kuzmin et al.[287] present data on seasonal variation of equatorial water deposits vs. depth and latitude. The main results are shown in Figure C.22 (color section). The uppermost figure represents the slowest neutrons and therefore the greatest depth (up to 2 m), while the lowermost picture corresponds to the fastest neutrons and therefore the shallowest depth ($\sim$10 cm). Figure C.22 shows that the slower the neutrons, the more horizontal are the lines of constant neutron flux, showing that the deeper water is fairly impervious to seasonal fluctuations, whereas the shallower water varies with season. There seems to be an ebb and flow of near-surface water deposits with "fingers" of higher concentration near $L_S = 160$ and 310. As Kuzmin et al. point out, these seem to be related to the ebbing of the north and south polar caps (at $L_S = 310$ and 160, respectively).

C.6   THE POLAR CAPS

The north and south poles of Mars display polar caps that grow in local winter and shrink in local summer. The growth and shrinkage varies from year to year.

The following material is based on Jakosky's landmark paper.[288] At each pole there are residual caps and seasonal caps. The residual caps remain year-round, while the seasonal caps disappear in summer and re-form in winter.

The north residual cap appears to be water ice due to its surface temperature being elevated above that expected for CO_2 ice and due to the appearance of a relatively large amount of water vapor over the cap. Surface temperatures of the white cap material are about 205–210 K, consistent with an ice albedo of about 0.45. At these temperatures, the vapor pressure of CO_2 ice is >2 bars. Hence, if CO_2 ice were still present, rapid sublimation would use up the incoming solar energy, thereby keeping the temperature near 150 K (the temperature for solid CO_2 in equilibrium with CO_2 gas at a pressure of 6 mbar). Thus, a lack of CO_2 ice is strongly indicated, and the residual white cap material is presumably water ice. The relatively low residual cap albedo, compared with a value much nearer to 1 for a pure water ice cap, suggests that a large amount of contaminating dust has been incorporated into the cap along with the water ice.

The south residual cap is more complicated. The residual covering of CO_2 never disappears from the cap. Certainly, due to the cold nature of the cap, water will be incorporated as ice within or underlying this residual cap. There is some evidence that the CO_2 ice covering may disappear in some years to reveal the underlying water ice cap (see Section C.3.5).

---

[287] "Seasonal Redistribution of Water in the Surficial Martian Regolith: Results of the HEND Data Analysis," R. O. Kuzmin, E. V. Zabalueva, I. G. Mitrofanov, M. L. Litvak, A. V. Parshukov, V. Yu. Grin'kov, W. Boynton, and R. S. Saunders, Lunar and Planetary Science, Vol. XXXVI, Paper 1634, 2005.
[288] "The seasonal cycle of water on Mars," B. M. Jakosky, Space Sci. Rev., Vol. 41, 131–200, 1985.

Ever since the original discussions of the polar cap seasonal cycle by Leighton and Murray,[289] it has been generally recognized that CO_2 ice forms the bulk of the seasonal caps.

It seems surprising that the two residual caps are composed of different frosts, and that the south cap has the colder-condensing ice even though the southern hemisphere has the hotter summer. Among the differences between the two caps are: elevation differences; differences in thermal emission; the seasonal timing of global dust storms; and the lengths of the summer and winter seasons.

The different lengths of winter and summer at the two poles, caused by the elliptical orbit of Mars, will result in different amounts of condensed CO_2. The south cap is in the dark during winter for 372 Mars days, while the north cap is in the dark for 297 days out of 669 Mars days in a Mars year. Because CO_2 condensation is inhibited whenever the cap is sunlit, there will be more CO_2 condensed onto the south cap than onto the north. Global dust storms occur during the southern summer season, when the south seasonal cap is nearly gone and the north seasonal cap is near maximum extent—the dust may play a role in the behavior of the caps.

The following paragraph is based on a private communication from S. Byrne:

"The seasonal cycling of CO_2 into and out of the solid phase at the polar caps amounts to about a 30% variation of the atmospheric pressure. During local winter, frozen CO_2 gas forms the large seasonal caps that are about a meter thick and typically extend down from the pole to the upper mid-latitudes in the winter hemisphere. During the summer, these seasonal caps retreat and eventually disappear. Their disappearance uncovers the much smaller residual ice caps at each pole. The north polar residual ice cap is composed of H_2O and the south polar residual ice is composed mainly of CO_2 ice. These residual caps persist throughout the entire summer until they eventually become covered again by seasonal CO_2 ice the following winter. The large pressure variations in the atmosphere are caused by the sublimation and condensation of the seasonal caps."

The amount of CO_2 in the south polar residual cap (which does not participate in the annual sublimation/condensation cycle) is not well constrained. Leighton and Murray suggested that there was a large reservoir of solid CO_2 in the residual cap that did not take part in the annual cycle, but which could be available if the orbital parameters of the planet changed to provide more insolation to the polar regions. They envisaged this additional reservoir as controlling the long-term mean atmospheric pressure, while the seasonal frost controlled the annual variations. The south polar residual ice was thought to be this additional reservoir. However, Byrne and Ingersoll[290] estimated the mass of this additional reservoir by analyzing the visual thickness and extent of features characterizing the residual cap as seen by high-

[289] "Behavior of carbon dioxide and other volatiles on Mars," R. B. Leighton and B. C. Murray, *Science*, Vol. 153, 135–144, 1966.
[290] "A Sublimation Model for Martian South Polar Ice Features," S. Byrne and A. P. Ingersoll, *Science*, Vol. 299, 1051–1053, 2003.

resolution imaging with the Mars Orbiter Camera. The inevitable conclusion was that this residual cap holds quite a small quantity of CO_2 and would barely be noticeable even if it completely sublimated. The conclusion they reached is that practically all of the observable Martian CO_2 is in the atmosphere–seasonal frost system.

The exchange of carbon dioxide between the atmosphere and the polar caps on Mars creates a seasonal cycle of growth and retreat of the polar caps. CO_2, the major component of the Martian atmosphere, condenses in the polar regions of the planet during the winter seasons, precipitating as CO_2 frost. It then sublimes during the spring and summer seasons in response to solar radiation. Nearly 30% of the atmosphere takes part in this seasonal process. While the northern seasonal CO_2 frost appears to dissipate completely, the south pole has a thin, permanent cover of dry ice over the residual cap.

The underlying residual caps are believed to contain large quantities of water ice.[291] The measured water percentages in the polar areas are so high that it does not seem to be compatible with deposition by vapor into porous regolith. The required porosity to account for the measured ice content is too high to account for the high ice content seen in the polar regions. Another mechanism is therefore needed to emplace ice with a high ice/dust ratio. One mechanism that could operate under different conditions in the past is to deposit ice in the form of snow or frost directly onto the surface of the regolith in the polar regions. The current Mars epoch is not conducive to this mechanism, but sometime in the past it may have been.[292] In a very recent report[293] ground-penetrating radar was used to infer that "the amount of water trapped in frozen layers over Mars' south polar region is equivalent to a liquid layer about 11 meters deep covering the planet."

C.7   LIQUID WATER ON MARS

C.7.1   Regions where surface temperature excursions exceed 273.2 K

According to the phase diagram for water, pure liquid water cannot exist below 273.2 K, although brines can persist as liquids well below this temperature.

---

[291] "Preliminary Thickness Measurements of the Seasonal Polar Carbon Dioxide Frost on Mars," N. J. Kelly, W. V. Boynton, K. Kerry, D. Hamara, D. Janes, I. Mikheeva, T. Prettyman, W. C. Feldman, and the GRS team, *Sixth International Conference on Mars*, Paper no. 3244, 2003; "Abundance and Distribution of Ice in the Polar Regions of Mars: More Evidence for Wet Periods in the Recent Past," W. V. Boynton, M. Chamberlain, W. C. Feldman, T. Prettyman, D. Hamara, D. Janes, K. Kerry, and the GRS team, *Sixth International Conference on Mars*, Paper no. 3259, 2003.
[292] "The State and Future of Mars Polar Science and Exploration," Stephen M. Clifford *et al.*, *Icarus*, Vol. 144, 210–242, 2000.
[293] "Subsurface Radar Sounding of the South Polar Layered Deposits of Mars," J. Plaut, G. Picardi, A. Safaeinili, A. Ivanov, S. Milkovich, A. Cicchetti, W. Kofman, J. Mouginot, W. Farrell, R. Phillips *et al.*, *Science*, March 15, 2007.

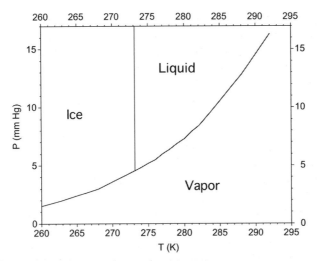

Figure C.23. Phase diagram for water.

Deep within the interior of Mars the temperature certainly rises above 273.2 K due to the geothermal gradient. It is therefore possible for water to exist at such multi-kilometer depths. Brines can exist (at least in principle) as liquids at shallower depths than water.

For near-surface locations, there are two requirements for liquid water to exist in equilibrium. One is $T > 273.2$ K. If the atmospheric pressure is $\sim$6 mm Hg, and the temperature rises above 277.2 K, liquid water will boil and rapidly go into the vapor phase. In the range from 273.2 K to 277.2 K at 6 mm Hg (see Figure C.23), the vapor pressure of liquid water is in the range 4.6 to 6 mm Hg and water will not boil. However, even though water will not boil in this range it will evaporate. The differential between the vapor pressure and the partial pressure provides a large driving force for evaporation. The slightest wind will enhance evaporation of liquid water on the surface of Mars by carrying away water vapor. The same considerations apply to brines. However, they are far less likely to boil on Mars than pure water.[294]

Haberle *et al.* carried out an extensive analysis to identify areas of Mars where the surface temperature can intermittently rise above 273.2. Maximum ground temperatures in the northern hemisphere never exceed 273 K poleward of $\sim$30°N. Thus, the ground temperatures never get warm enough to support liquid water poleward of 30° latitude. Figure C.24 shows the regions where the ground temperature and surface pressure reach above the triple point but below the boiling point. These regions are denoted by the total length of time in a Mars year these conditions are satisfied. According to Figure C.24, there are three broad regions in the northern hemisphere between 0° and 30° latitude where the minimum conditions are met for

[294] "On the possibility of liquid water on present-day Mars," Robert M. Haberle, Christopher P. McKay, James Schaeffer, Nathalie A. Cabrol, Edmon A. Grin, Aaron P. Zent, and Richard Quinn, *Journal of Geophysical Research*, Vol. 106, No. E10, 23317–23326, October 25, 2001.

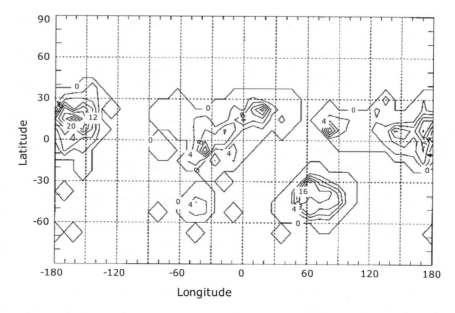

Figure C.24. The length of time during a Mars year (sols) and location where the ground temperature and surface pressure are above the triple point of water but below the boiling point. Contour intervals are 4 sols. [Based on data from "On the possibility of liquid water on present-day Mars," Robert M. Haberle, Christopher P. McKay, James Schaeffer, Nathalie A. Cabrol, Edmon A. Grin, Aaron P. Zent, and Richard Quinn, *Journal of Geophysical Research*, Vol. 106, No. E10, 23317–23326, October 25, 2001.]

liquid water not to boil. The combined area of these regions represents 29% of the planet's surface area.

Haberle *et al.* say: "The existence of these favorable regions does not mean liquid water actually forms in them. To form liquid water, ice must be present, and the energy source (solar heating in this case) must be able to overcome the evaporative heat loss. The first constraint, the presence of ice, is a major issue for the three tropical regions we have identified since ground ice is not stable at these latitudes at the present epoch. It is also an issue for the Hellas and Argyre impact basins, though less so. The second constraint is also a major issue since evaporation rates are likely to be quite high in the Martian environment. However, present estimates of ice evaporation in a low-pressure CO_2 atmosphere are based on theoretical arguments, which need to be validated with carefully conceived laboratory experiments. It is possible, even likely, that solar-heated liquid water never forms on present-day Mars."

It should be noted that, even though Figure C.24 provides multiple sols at some locations where $T > 273.2\,K$, these are not contiguous. There is no place on Mars where the night temperature ever stays above 190 K on any sol of the year. If you placed a bowl of water on the surface of Mars in one of these favored regions, water would gradually evaporate during the day, and the remainder would freeze overnight. If you poured a glass of liquid water onto the regolith it would percolate down through the regolith, until it reached a sub-freezing temperature where it would freeze in the subsurface.

Haberle *et al.* also examined the effect of dissolved salts on the potential for melting, arguing that pure water is unlikely on Mars since salts are believed to be a significant component of the Martian soil. The presence of salts will lower the melting point and reduce the equilibrium vapor pressure of the solution. The figure analogous to Figure C.24 for a 251 K eutectic salt solution has up to 90 sols/year above 251 K. Clearly, the presence of salts greatly expands the regions where melting could occur and increases the total time such conditions might exist. For a NaCl eutectic, virtually the entire planet (except the polar regions) experiences conditions favorable for melting at some point during the year, including the Tharsis plateau. However, once again, there is no place on Mars where the average annual temperature exceeds ~220 K, so any salty water poured on the surface of Mars will percolate down and freeze at a sufficient depth.

There is no place on the surface of Mars where temperatures exceed 273.2 K for more than a few hours a day. It is unlikely that liquid water can exist for more than short transient periods anywhere on the surface of Mars, although high salt concentrations and dust coverings to inhibit evaporation could possibly provide a transient environment for liquid water to persist for a while in limited locations. Liquid water will never be in equilibrium with the atmosphere, and therefore, given enough time, the water will evaporate. In the daily cycle at such locations whereby the surface is heated by the Sun to temperatures >273.2 K for brief periods in the afternoon, the temperature remains below freezing for most of the 24.7-hour sol. The dynamics of the surface layer (down to a skin depth of a few cm) in which temperatures exceed the freezing point for brief periods during the day in local summer may involve the non-equilibrium processes of melting, evaporation, and refreezing.

According to Figure C.23, if surface temperatures rise above 273.2 K, and the atmospheric pressure exceeds 4.5 mm Hg—as it likely will—and there is ice present within the skin depth, some of the ice could melt to form liquid water. This liquid water, in contact with the atmosphere with its low absolute humidity, would tend to evaporate, and whatever amount remained on the surface would freeze overnight. However, the regolith is very porous and therefore it would probably quickly seep down into the cold subsurface where it would freeze in the pores of the regolith. Therefore, liquid water will not last very long (if at all) on the surface of Mars.

Because of the emphasis by the Mars Exploration program on the search for life on Mars, and the requirement of liquid water to support life as we know it, there has been great interest in finding liquid water on Mars—even when it is not there.

Hecht[295] carried out a detailed heat transfer analysis of special-case situations where a surface at the back of a cavity is oriented to face the Sun at some time during the day at a season when Mars is closest to the Sun. He concluded that the rate of solar heat input to the surface can exceed the rate of heat loss from the surface to its surroundings, resulting in temporary melting of ice on or just below the surface. The

[295] "Metastability of liquid water on Mars," M. H. Hecht, *Icarus*, Vol. 156, 373, 2002.

methodology depends upon a detailed heat transfer analysis of the surface of Mars, including solar heat input and various cooling mechanisms that prevail for the special circumstances that he selected. While these arguments are generically sound, the choice of parameters was very extreme, and such melting of surface ice will occur only rarely. In the following sections, several aspects of his methodology are reviewed.

Hecht carried out a thermal analysis of metastable formation of liquid water on Mars by transient melting of surface ice due to solar heat input. However, his estimation of solar heat inputs seems overly optimistic, and more realistic estimates would have the effect of reducing the range of circumstances in which liquid water could form.

The clarity of the atmosphere is determined by the optical depth, commonly represented by the symbol τ. The optical depth during 83 sols of Mars Pathfinder never dropped below 0.45 or rose above 0.65. Over 450 sols, Viking optical depth at both sites never dropped below about 0.5, and reached as high as 5.0 at VL-1 for brief periods during the height of a global dust storm. The optical depth measured by the MER Rovers started out near 0.9 at the start of the mission due to the tail end decay of a dust storm, and these gradually decreased over the subsequent 200 sols to about 0.5 at one MER site and about 0.3 at the other MER site. Hecht used a value of 0.1 in all calculations. This is an extreme value, which might possibly occur under as-yet unobserved conditions, but based on our experience to date it seems very unlikely. It would appear that an optical depth of 0.3 is the absolute minimum that should be considered for these calculations, and that value is used in the following analysis.

For the case considered by Hecht, with $L_S \sim 270°$ corresponding to winter solstice in the north and Mars close to its nearest approach to the Sun, the expected direct normal solar intensity is

$$N = 700 \exp(-0.3/\cos Z) \text{ W/m}^2$$

where Z is the solar zenith angle (measured from the vertical). For the case considered by Hecht in his table 1 (right column) and his figure 2, $Z = 80°$. In this case,

$$N = 700 \exp(-0.3/\cos 80°) = 700 \exp(-0.3/0.174) = 125 \text{ W/m}^2$$

This is considerably smaller than the insolation given in his figure 2 ($>500 \text{ W/m}^2$).

There is another problem with this illustration that has to do with the geometry of the Sun and the planet. Hecht considered a pole-facing slope at a high northern latitude. Consider Figure C.25. At $L_S = 270°$, the entire northern region from latitude 64.8° northward is shrouded in darkness with no direct insolation, although some diffuse insolation could reach this region. The north pole lies below the top of Mars by 270 km (since $R = 3,380$ km). For the case considered at latitude 75°N, this point still lies 53 km below the top of Mars at $L_S = 270°$ and receives no direct insolation. This location is in darkness at $L_S = 270°$.

Whether ice melts locally, and how much melts, is determined by the difference between solar heating and various cooling mechanisms. Hecht estimated heat loss due to radiation, evaporation, convection, and conduction.

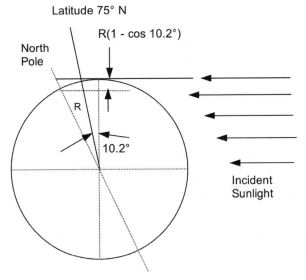

Figure C.25. Illustration of Mars tilted on its axis at 25.2° from the Sun at $L_S = 270°$.

Hecht reached the following conclusions:

"Three factors constrain the melting of ice on Mars. First, the radiative balance must be optimized by a clear, sunny, windless day; a low albedo surface; a slope that faces the Sun for part of the day; and a geometry that shields the surface from part of the sky. Second, conductivity to the bed must be small, a condition met by Martian soil but not by thick ice. Third, the surface temperature must be suppressed a few degrees, perhaps by formation of an ice crust or by incorporation of salts in the water. In addition, the factor that probably determines the *location* of runoff features is the accumulation of ice itself. It is suggested that this accumulation may simply result from a concentration of winter frost from surrounding areas."

These conclusions are sound as generic statements; but detailed analyses are needed to determine whether the rate of heating is sufficient to overcome cooling mechanisms. There appear to be anomalies in his model.

C.7.2 Liquid water below the surface

Hoffman indicated that Mars appears to be much less thermally active than the Earth and has significantly lower heat flux and geothermal gradients ($\frac{1}{4}$ to $\frac{1}{3}$ that for Earth).[296] He points out that there are two principal factors that determine the

---

[296] "Modern geothermal gradients on Mars and implications for subsurface liquids," N. Hoffman, *Conference on the Geophysical Detection of Subsurface Water on Mars*, Paper 7044, 2001.

modern heat flow on Mars: the amount of radiogenic heat and the efficiency of heat loss processes over geologic time. Mars has a mean density of 3.9 compared with Earth's 5.5. Hoffman made the assumption that on balance the radiogenic heat on Mars is 75% less (per unit mass) than that of Earth. This leads to a heat flux per unit area that is 28% of Earth's after correcting for dilution by light elements. On Earth, the mean geothermal gradient is $61.5 \, \text{mW/m}^2$ for both oceanic and continental crust. Therefore, on Mars, Hoffman expected ~28% of this, or $17.5 \, \text{mW/m}^2$. This is near the low end of a range of other estimates from 15 to $30 \, \text{mW/m}^2$.

Heldmann and Mellon[297] used an estimated value for the geothermal heat flux of $30 \, \text{mW/m}^2$. The temperature gradient established by this flux depends on thermal conductivity. The thermal conductivity of icy soil is around $2 \, \text{W/(m-K)}$ but for dry fragmented regolith it could be a factor of 10 (or more) lower. At a thermal conductivity of $2 \, \text{W/(m-K)}$, this flux translates to a temperature gradient of 15 K per km, so that the 273 K isotherm would be reached at a depth of about 5 km if the average surface temperature was 200 K. At a thermal conductivity of $0.2 \, \text{W/(m-K)}$, this temperature would be reached at a depth of about 500 m (geothermal gradient of 150 K per km).

McKenzie and Nimmo said: "The surface temperature gradient [on Mars] depends on the integrated crustal radioactivity as well as the mantle heat flux, and may vary from 6 K/km (where the crustal contribution is not important) to more than 20 K/km (where it dominates). If the surface temperature is 200 K, the base of the permafrost is at a depth of between 11 and 3 km."[298]

Kiefer said: "Estimates of the present-day mean surface heat flux on Mars are in the range 15 to $30 \, \text{mW/m}^2$. Finite element mantle convection simulations suggest that there can be lateral variations of about 50% relative to the mean value. The thermal conductivity for an intact basaltic crust is in the range $2–3 \, \text{W/m-K}$. For a granular regolith, the thermal conductivity would be significantly reduced."[299]

It should be noted that, given a geothermal gradient of $15–30 \, \text{mW/m}^2$, the depth to a temperature level of 273 K from a surface at, say, 200 K would be:

$$\text{Depth to 273 K} = 73 \times 1{,}000 \times \frac{\text{thermal conductivity (W/}^\circ\text{C} - \text{m)}}{(15 \text{ to } 30) \, \text{mW/m}^2}$$

If there is a liquid water reservoir at the 273 K depth, it will exert a water vapor pressure and water vapor will rise and percolate through the porous regolith, filling the pores and interstices with ground ice everywhere above it until a region near the surface may desiccate the near-surface region due to sublimation. Therefore, if there is a deep liquid-water reservoir, the thermal conductivity of the subsurface will be that

[297] "Observations of Martian gullies and constraints on potential formation mechanisms," Jennifer L. Heldmann and Michael T. Mellon, *Icarus*, Vol. 168, 285–304, 2004.
[298] "The generation of Martian floods by the melting of ground ice above dykes," Dan McKenzie and Francis Nimmo, *Nature*, Vol. 397, January 21, 1999.
[299] "Water or Ice: Heat Flux Measurements as a Contribution to the Search for Water on Mars," Walter S. Kiefer, *Conference on the Geophysical Detection of Subsurface Water on Mars*, Paper 7003, 2001.

characteristic of ice-filled regolith, probably in the range 1 to 3 W/m-K. Therefore, the depth to a water reservoir is probably in the range 3 to 15 km. On the other hand, if there is no deep liquid reservoir and the subsurface consists of dry porous regolith, its thermal conductivity might be perhaps 0.05 W/m-K. In this case, the depth to 273 K is a few hundred meters, but the subsurface is dry.

Mellon and Phillips[300] described a model of the temperature profile in the deeper Martian subsurface (of the order of a few hundred meters) and evaluated the potential for liquid water at this depth due to nominal geothermal heating. The thermal conductivity of the Martian regolith at depths of a few hundred meters is unknown. Thermal conductivity can vary over extremely wide ranges, depending on the porosity and compaction of the soil, and particularly whether ice fills the pores. The thermal conductivity of dry particulate soil where gas conduction dominates the heat transfer is approximately 0.05 W/m-K. This is probably a good value to use for the desiccated surface layer, but the thickness is not known. Densely ice-cemented soil would have a thermal conductivity more like 2.5 W/m-K, depending on the ice-to-soil ratio and the temperature. Intermediate values of thermal conductivity between these extremes are likely due to the many processes that can act to raise the thermal conductivity: ice cementing, densification, lithostatic compression, and induration.

To illustrate the dependence on material properties, Mellon and Phillips superimposed the geothermal gradient for constant values of thermal conductivity and density consistent with an ice-cemented soil, an ice-free sandstone, and an ice-free soil on the phase diagram of water. A constant mean annual surface temperature of 180 K was assumed. This was shown as Figure C.4. For high-thermal-conductivity ice-cemented soil or sandstone, the depth to the 273 K isotherm is 3–7 km below the surface, while for low-thermal-conductivity uncemented dry soil, the depth to the 273 K isotherm is estimated to be between 100 and 200 m.

C.7.3   Imaging indications of recent surface water flows

C.7.3.1   *Gullies*

The advent of spectacular photography of Mars—first by Mariner 9, followed by Viking and then Mars Global Surveyor and Odyssey, and now Mars Express and MRO—has provided a wealth of evidence that water once flowed on the surface of Mars. These pictures have provided grist to the mill of planetary scientists who have made great efforts to analyze the photos in order to interpret the past history of the action of surface water flows. Examples include "gullies", cold-based tropical glaciers, paleolakes, and youthful near-surface ice. Amongst these various observations, the occurrence of gullies has received perhaps inordinate attention.

The observation of gullies on Mars suggests the presence of liquid water near the surface in recent times, which is difficult to reconcile with the current cold climate.

---

[300] "Recent gullies on Mars and the source of liquid water," Michael T. Mellon and Roger J. Phillips, *Journal of Geophysical Research*, Vol. 106, 2001.

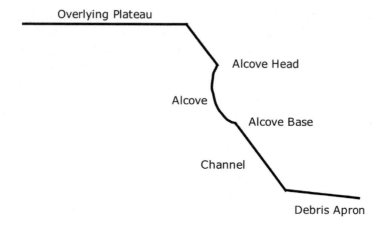

Figure C.26. Schematic illustration of gully profile. The depth from the alcove head to the alcove base varies widely from less than 100 m to over 1 km. [Based on Heldmann and Mellon, *loc. cit.*]

A schematic illustration of a gully is shown in Figure C.26.[301] These features exhibit a characteristic morphology indicative of fluid-type erosion of the surficial material and liquid water has been suggested as a likely fluid. However, the relatively young geologic age of these gullies is somewhat of a paradox for the occurrence or production of liquid water under present-day cold Martian conditions. Thus, their formation mechanism remains controversial. Several models for a source of water have been proposed, along with other potential erosional agents than water. Small young gullies commonly occur in clusters on slopes between 30° and 70° latitude in both hemispheres of Mars, although much more frequently in the southern hemisphere. It is notable that there is a dearth of gullies in the 30°S to 30°N equatorial zone. Gullies consist of alcoves several hundred meters wide, and channels up to several kilometers long and several tens of meters deep. Gullies typically originate within several hundred meters of the slope crest, and can occur on crater walls that are raised above the surrounding terrain or near the summit of isolated knobs. Gullies most probably result from the flow of liquid water. Christensen suggests that gullies can form by the melting of water-rich snow that has been transported from the poles to mid-latitudes during periods of high obliquity within the past 10^5 to 10^6 years. He suggests that melting within this snow can generate sufficient water to erode gullies in about 5,000 years.[302]

Heldmann and Mellon carried out a detailed review and analysis on 106 geologically young small-scale features resembling terrestrial water-carved gullies

[301] "Observations of Martian gullies and constraints on potential formation mechanisms," Jennifer L. Heldmann and Michael T. Mellon, *Icarus*, Vol. 168, 285–304, 2004.
[302] "Formation of recent Martian gullies through melting of extensive water-rich snow deposits," Philip R. Christensen, *Nature*, Vol. 422, 45–48, March 6, 2003; doi:10.1038/nature01436.

in the region 30°S to 72°S. This is the region in which gullies predominantly occur. They compared various proposed models for gully formation with numerous observations of the physical and dimensional properties of gully features. Martian gullies occur in a wide variety of terrain types. Within the latitude range studied (30°S to 72°S) the highest number of gullies was found between ~33°S and ~40°S. There were also significant numbers of gullies in the highest latitude bin (69°S to 72°S). The dimensions of gullies were tabulated extensively, but it is not clear that any conclusions can be drawn from dimensions.

Unfortunately, since a total of only 106 gullies were catalogued, the number in each statistical bin was small, and the statistics are therefore approximate. It was noted that there was a "conspicuous lack of gullies between 60°S and 63°S." Heldmann and Mellon made a point of the lack of gullies in this narrow latitude range, and used it to help justify their conclusion that liquid aquifers are the most probable basis for gullies. This was based on the belief that the latitude range from about 60°S to 63°S is an intermediate range that is least affected by changes in obliquity; it is neither polar nor equatorial. The lack of gullies in this range was attributed to a lack of alternation in the thermal environment during obliquity cycles. However, it seems likely that this could be a statistical quirk due to sparse data.

Heldmann and Mellon provided an extensive discussion and appraisal of each of several postulated theories (liquid CO_2 reservoir, shallow liquid-water aquifer, melting ground ice, dry landslide, melting snow, and deep liquid aquifer) in an effort to find the most convincing mechanism for gully formation. The problem for explaining the formation of gullies revolves about finding some means for liquid water to occur at shallow depths as low as ~100 m. This, in turn, requires a zone where $T > 273$ K at such depths. If the subsurface has a relatively high thermal conductivity characteristic of ice-filled regolith, the 273 K isotherm will not be reached until a depth of many km. However, if the subsurface consists of porous, fragmented, loosely congregated regolith, the 273 K isotherm could be as shallow as a few hundred meters. But, the confounding aspect of all this is that such a porous regolith would fill with ice if it was exposed to water vapor from a pool of water at lower depths. Hence, a possible configuration to allow the 273 K isotherm to be at shallow depths and make liquid water stable at such depths is to encase a wet soil aquifer as shown in Figure C.27.

This model, favored by Heldmann and Mellon, seems to be heavily contrived. They reach the conclusion that ". . . it is very possible that all the gully alcove bases lie within the stability field of liquid water. Such a finding implies that liquid water could exist in an aquifer at these depths." This conclusion is based on the assumption of an extremely low thermal conductivity for the overburden with an ice-filled plug to seal the aquifer from the alcove. The best argument for liquid aquifers is that it explains why the gullies appear where they do on the upper part of the slopes. But, none of the arguments is completely convincing. There is no evidence that such structures exist, and the probability that they would occur with great frequency would intuitively seem to be small.

The orientation of gullies is interesting. Gullies in the 30°S to 44°S latitude range tend to face the pole. This might indicate that, in this latitude range, only relatively

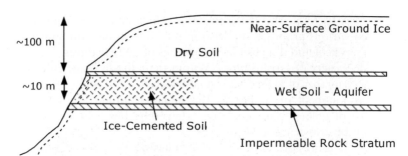

Figure C.27. Model for underground aquifer.[303] [Based on Heldmann and Mellon, *loc. cit.*]

cold areas can store up enough ground ice or snow during periods of high obliquity. Unfortunately, Heldmann and Mellon do not provide thermal inertia and albedo data for each latitude group; they only provide this data for a larger group including all latitudes. Gullies in the 44°S to 58°S latitude range tend to face the equator, and this might indicate that only these have sufficient solar input to melt ground ice during periods of high obliquity. Gullies in the 58° to 72° latitude range primarily face the pole. This is difficult to relate to melting of ground ice. However, according to Heldmann and Mellon: "An inherent bias may exist in the orientation results due to a potential preference for MOC targeting of pole-facing slopes (particularly, in images targeted shortly after the initial discovery of gullies) and the common saturation of more brightly lit equator-facing slopes in MOC images which makes gully identification difficult."

The discussion by Heldmann and Mellon on melting of ground ice is difficult to follow. It begins with a reference to a paper[304] that advocates that pole-facing slopes are the warmest at high latitudes—a proposition that is patently incorrect.

It is strange that Heldmann and Mellon do not seem to have explained why gullies are mainly restricted to the southern hemisphere.

C.7.3.2 *Surface streaks*

Dark slope streaks were seen from orbit, concentrated around Olympus Mons volcano, in the region from 90°W to 180°W and 30°S to 30°N. The dark streaks always appear on slopes, mostly inside craters and valleys, but also on small hills. They are almost always located below Martian sea level (zero elevation). The dark streaks occur in clusters of parallel streaks, wherein the upslope ends of the streaks are aligned with a common rock layer (Figures C.28 and C.29).

---

[303] "A Model for Near-Surface Groundwater on Mars," Kelly J. Kolb, Herbert V. Frey, and Susan E. H. Sakimoto, *http://academy.gsfc.nasa.gov/2003/ra/kolb/poster_details.pdf*; "Recent gullies on Mars and the source of liquid water," Michael T. Mellon and Roger J. Phillips, *Journal of Geophysical Research*, Vol. 106, 2001.
[304] "Formation of Recent Martian Debris Flows by Melting of Near-Surface Ground Ice at High Obliquity," F. Costard, F. Forget, N. Mangold, and J. P. Peulvast, *Science*, Vol. 295, 110–113, 2002.

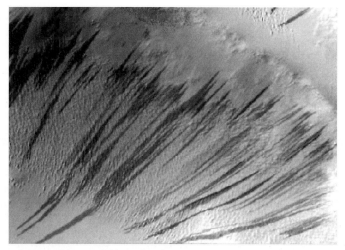

Figure C.28. Dark streaks inside crater. Top right of image is crater rim, bottom left is crater floor. MOC Image E03-02458.[305]

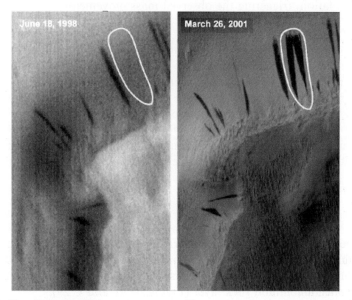

Figure C.29. Two images of the same area, with new streaks present in the later photo. MOC Images SP2-37303 and E02-02379.

[305] "Currently Flowing Water on Mars," T. Motazedian, *Lunar and Planetary Science*, Vol. XXXIV, Paper 1840, 2003. This was a precocious piece of work by an undergraduate at the University of Oregon.

Motazedian suggests that liquid flow is the most promising process for explaining these features, possibly based on geothermal activity surrounding Olympus Mons that causes ice to melt or otherwise driving liquid water from aquifers. This reference suggests that the liquid dissolves salts in the aquifer to form a brine. The salts in the solution lower the freezing point, allowing water to flow at the Martian surface. As the brine flows down slopes it leaves a trail of rock varnish from dark minerals that precipitate from solution.

Schorghofer *et al.* also examined slope streaks on Mars. They systematically analyzed over 23,000 high-resolution images and reached the following conclusions:

"Slope streaks form exclusively in regions of low thermal inertia, steep slopes, and only where peak temperatures exceed 275 K. The northernmost streaks, which form in the coldest environment, form preferentially on warmer south-facing slopes. Repeat images of sites with slope streaks show changes only if the time interval between the two images includes the warm season. Surprisingly (in light of the theoretically short residence time of water close to the surface), the data support the possibility that small amounts of water are transiently present in low-latitude near-surface regions of Mars and undergo phase transitions at times of high insolation, triggering the observed mass movements."

This model presumes that solar heating can melt near-surface ice and, even though liquid water is not stable at the surface of Mars, it can persist for short periods to produce these flows. It is mentioned that temperatures above melting can occur only in the upper $\sim$0.5 cm of the regolith—the depth of the diurnal thermal wave. They note that "streaks do not appear to penetrate deeply, since pre-existing surface textures are often preserved beneath the feature and no accumulated debris is visible at their termination. The small penetration depth of melting temperatures is consistent with the interpretation that the mass flow is restricted to a thin layer."

C.8 EVIDENCE FROM CRATERS

C.8.1 Introduction

Squyres *et al.*[306] provide a review as of 1990 of craters as indicators of subsurface H_2O on Mars. Based on rough guesses for the geothermal gradient, they developed a simple representation of the subsurface as shown in Figure C.30.

[306] "Ice in the Martian Regolith," S. W. Squyres, S. M. Clifford, R. O. Kuzmin, and J. R. Zimbelman, in H. H. Kieffer, B. M. Jakosky, C. W. Snyder, and M. S. Matthews (eds.), *Mars*, University of Arizona Press, 1992.

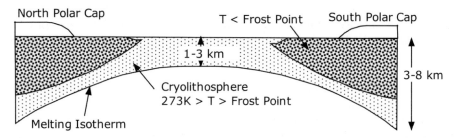

Figure C.30. In the black areas, the temperature is always below the frost point and ground ice is permanently stable relative to sublimation to the atmosphere. In the "cryolithosphere" the temperature is always below 273 K, so ground ice can form but it may be above the frost point for part of the year. Below the "melting isotherm", $T > 273\,\mathrm{K}$ and liquid water can exist. [Based on "Ice in the Martian Regolith," S. W. Squyres, S. M. Clifford, R. O. Kuzmin, and J. R. Zimbelman, in H. H. Kieffer, B. M. Jakosky, C. W. Snyder, and M. S. Matthews (eds.), *Mars*, University of Arizona Press, 1992.]

At high latitudes, the subsurface always remains below the frost point from the surface down to a significant depth. At sufficient depth, the geothermal gradient raises the temperature to $>273\,\mathrm{K}$ and liquid water may exist. The depth to this melting isotherm increases toward the poles. Between these two regions is the so-called "cryolithosphere", where the temperature is always below the freezing point of water, but may only dip below the frost point on a seasonal basis.

They suggest that the variation of subsurface porosity varies from about 20% at the surface to 10% at 2 km depth, to 5% at 4 km depth, to 2.5% at 8 km depth, etc. Squyres *et al.* did not discuss this, but considering the fact that porosity persists to depths of several km it must be concluded that, if liquid water exists below the melting isotherm, it will exert its vapor pressure upward, causing condensation of ice at higher levels. Hence, if liquid water exists at depth, significant amounts of ice must also exist above it. In fact, the upper regions will be filled with ice in the interstices, for the upper regions will act as a cold trap for vapors rising from below.

Large Martian craters typically have an ejecta sheet, and some have a pronounced low ridge or escarpment at their outer edge. Craters of this type are referred to as "rampart" craters. It has been observed that, in general, rampart craters account for a significant fraction of fresh craters on Mars. The majority of craters on Mars are degraded to the point of no longer displaying an ejecta morphology. But, among those that do show an ejecta blanket, layered ejecta morphologies, including those which show ramparts, dominate. In any local area, it has been found that rampart craters do not form at crater diameters below a critical onset diameter, D_0, where this onset diameter varies with location. Thus, for example, in an area where D_0 happens to be, say, 4 km, all craters of diameter <4 km in that area will lack the marginal outflow and have the appearance of lunar craters. Craters with diameters >4 km in that area will mainly be rampart craters. It is widely believed that rampart craters are produced by impacts into ice-laden or water-laden regoliths, although an alternative explanation for the morphology of rampart craters is based on the interaction of dry ejecta with the atmosphere.[307]

Earlier work by Kuzmin and Costard characterized the morphology and location of over 10,000 rampart craters based on Viking data. They found that, near the equator, onset diameters are typically 4 to 7 km, whereas at latitudes of 50° to 60°, onset diameters decrease to as low as 1 to 2 km. Since 1990 a number of other studies have been conducted on Martian craters.

C.8.2   The work of Nadine Barlow (and friends)

The cornerstone of understanding of Mars rampart craters and the relationship of crater morphologies to subsurface H_2O was laid down by Barlow and Bradley[308] with a broad study of Martian craters over a wide range of latitudes and morphologies. Previous studies had led to inconsistent results, probably due to limited areal extent and limited photographic resolution. Barlow and Bradley undertook "a new study of Martian ejecta and interior morphology variations using Viking images from across the entire Martian surface in an attempt to resolve some of the outstanding controversies regarding how and where these features form."

As Barlow and Bradley pointed out, "if a uniform distribution of a morphology across the planet is found, this would suggest that either the target properties are uniform on a global scale (unlikely) or that impact velocity dictates formation of the feature. Alternatively, if craters with a particular morphology are concentrated in certain regions of the planet, target properties are then the likely cause." With the goal of distinguishing between these two theories, they utilized data from over 3,800 craters of diameter ≥ 8 km across the Martian surface (mainly between 60°N and 50°S) "to obtain statistically valid results on the distribution of ejecta and interior morphologies associated with Martian impact craters." Most images used in this analysis had resolutions between 200 and 250 m/pixel, regardless of latitude. Clouds or haze can obscure the details of crater morphology. Due to cloud cover, no studies were made poleward of $-50°$ latitude.

Seven different types of crater morphology have been defined:

(1) single-layer ejecta (SLE);
(2) double-layer ejecta (DLE);
(3) multiple-layer ejecta (MLE);
(4) radial delineated (Rd);
(5) diverse (Di);
(6) pancake (Pn);
(7) amorphous (Am).

---

[307] "Assessing Lithology from Ejecta Emplacement Styles on Mars: The Role of Atmospheric Interactions," H. Schultz, *Workshop on the Role of Volatiles and Atmospheres on Martian Impact Craters, July 11–14, 2005*, LPI Contribution 1273.
[308] "Martian impact craters: Dependence of ejecta and interior morphologies on diameter, latitude, and terrain," N. G. Barlow and T. L. Bradley, *Icarus*, Vol. 87, 156–179, 1990.

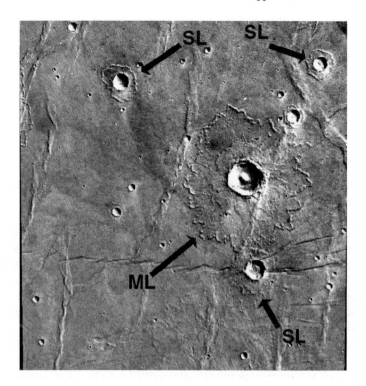

Figure C.31. Examples of the SLE and MLE ejecta morphologies. Image is centered at 19°S, 70°W. The MLE crater is 24 km in diameter. [Modified from "Impact Craters as Indicators of Subsurface Volatile Reservoirs," N. G. Barlow, *Conference on the Geophysical Detection of Subsurface Water on Mars*, Paper 7008, 2001.]

SLE craters have a single layer of layered ejecta deposits surrounding the crater, while DLE craters are surrounded by two complete layers, one superposed on the other. Multiple-layer craters consist of three or more partial to complete layers. The SLE, DLE, and MLE craters may exhibit ramparts. Examples of SLE and MLE craters are shown in Figure C.31.[309] The Rd morphology consists of linear streaks of ejecta radiating outward from the crater and have some similarities to the ballistically emplaced ejecta blankets surrounding lunar and Mercurian craters. The Di morphology consists of radial ejecta superposed on a layered morphology, and the Pn morphology involves the crater and ejecta located on a raised pedestal above the surrounding terrain, likely the result of erosion. Am morphology involves oddities that cannot be placed into any of the other categories. These seven classes have been the basis of most subsequent ejecta classifications utilized in studies of how the ejecta morphologies may reveal information about the distribution of subsurface volatiles.

[309] "Impact Craters as Indicators of Subsurface Volatile Reservoirs," N. G. Barlow, *Conference on the Geophysical Detection of Subsurface Water on Mars*, Paper 7008, 2001.

Table C.3. Excavated depths of craters.

Morphology	Diameter (km)	Depth (km)	Occurrence
SLE	8–20	0.75–1.63	All latitudes but diminishing in the north
DLE	8–50	0.75–3.55	Mainly from 40°N to 65°N
MLE	16–45	1.35–3.25	All latitudes with a peak from 0–30°N
Pn	<15	<1.27	
Rd	>64	>4.34	All latitudes
Di	45–128	3.25–7.89	All latitudes with preference for 30–60°N
Am	>50	>3.55	

Barlow and Bradley found that 2,648 (or 69%) of all ejecta craters in their set displayed one of the three types of layered ejecta morphologies (i.e., single, double, or multiple layer), supporting the general observation that most Martian impact crater ejecta blankets exhibit a layered structure. Crater size–frequency distribution analyses indicated that craters date from the end of heavy bombardment (Lower Hesperian Epoch) or the post-heavy bombardment period (Upper Hesperian and Upper, Middle, and Lower Amazonian Epochs).

Single-layer craters (SLEs) are the most abundant ejecta morphology on Mars. They show a strong correlation with crater diameter and occur mainly at diameters less than 30 km. A summary of diameters and depths for the various morphologies is given in Table C.3.

The interpretations made by Barlow and Bradley are summarized in Figure C.32. It was hypothesized that there is an upper desiccated layer (not shown in Figure C.32), below which is an ice layer, and below that a brine layer in some latitudes, with a volatile-poor layer beneath all these. Very small impact craters never penetrate to the ice layer and produce primarily pancake morphologies. Small craters of sufficient size to reach the ice layer produce primarily SLE morphologies. Larger craters may excavate both ice and brine layers, producing MLE morphologies. Very large craters primarily excavate the desiccated region resulting in radial morphologies.

Further studies extended the analysis of Barlow and Bradley. Barlow[310] performed a global study of how specific ejecta morphologies depend on crater diameter, latitude, and terrain. She found that SLE morphologies dominated at all latitudes and on all terrains, but the diameter range was dependent on latitude. In the equatorial region, the SLE morphology dominated among craters in the 5 to 25 km diameter range, but at higher latitudes the range extended up to 60 km in diameter. In the equatorial region, the MLE morphology was found around craters in the 25 to 50 km diameter range. Craters displaying the MLE morphology were rare

[310] "Subsurface Volatile Reservoirs: Clues from Martian Impact Crater Morphologies," N. G. Barlow, *Fifth International Conference on Mars, July 18–23, 1999, Pasadena, CA*; "Impact Craters as Indicators of Subsurface Volatile Reservoirs," N. G. Barlow, *Conference on the Geophysical Detection of Subsurface Water on Mars*, Paper 7008, 2001.

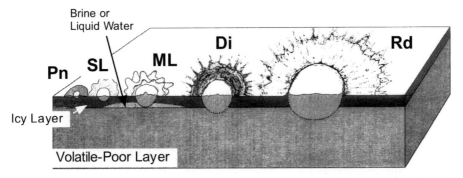

Figure C.32. Diagram showing the relationship proposed by Barlow and Bradley between distribution of volatiles and ejecta morphology. Excavation into ice-rich substrate results in formation of pancake (Pn) and single-layer (SLE) ejecta morphologies. Excavation in liquid water or brines is responsible for multiple-layer (MLE) craters, and radial (Rd) ejecta morphologies result from ejecta consisting primarily of volatile-poor material. Diverse (Di) ejecta morphologies form when an impact event excavates to a depth near the volatile-rich/volatile-poor boundary. [Modified from "Martian impact craters: Dependence of ejecta and interior morphologies on diameter, latitude, and terrain," N. G. Barlow and T. L. Bradley, *Icarus*, Vol. 87, 156–179, 1990.]

at higher latitudes. These results appear to confirm earlier ideas that MLE morphologies correspond to excavation of both ice and brines and that MLE morphologies are restricted to mainly lower latitudes.

Barlow and Perez[311] improved upon previous crater studies by utilizing the best Viking and MGS images in a global study of SLE, DLE, and MLE craters with the $-60°$ to $+60°$ latitude range divided into $5° \times 5°$ latitude–longitude segments to enable identification of local area effects. They used the information in the *Barlow Catalog of Large Martian Impact Craters* to conduct this study. This catalog "contains information on 42,283 impact craters (>5-km-diameter) distributed across the entire planet. Catalog entries contain information about the crater location, size, stratigraphic unit, ejecta morphology (if any), interior morphology (if any), and preservational state. The data were originally obtained from the Viking Orbiter missions but are currently being revised using Mars Global Surveyor and Mars Odyssey data. The resolutions were typically around 40 m/pixel."

Figure C.33 (color section) shows the percentage of craters displaying any type of ejecta morphology attributable to subsurface volatiles. It was found that the percentage of ejecta craters was less than about 30% for the oldest (Noachian-aged) units. Hesperian-aged units had ejecta crater percentages in the range 30–70% and the youngest (Amazonian-aged) units showed either high percentages (>70%) or very

[311] "Martian impact crater ejecta morphologies as indicators of the distribution of subsurface volatiles," Nadine G. Barlow and Carola B. Perez, *Journal of Geophysical Research*, Vol. 108, No. E8, 5085, doi:10.1029/2002JE002036, 2003.

few total craters of any type. It seems clear from Figure C.33 that there is a significant preponderance of ejecta craters in the northern hemisphere compared with the southern hemisphere, but there are also significant longitudinal variations as well as local areas where high probabilities occur in the southern hemisphere. Similar plots are given separately for SLE, DLE, and MLE craters. The map for SLE craters shows that SLE craters are widespread across Mars. However, the map for DLE (see Figure C.34, in color section) shows that DLE craters tend to be concentrated in the north between about 35° and 60°N.

Barlow upgraded previous studies[312] with new photographic evidence from the Mars Global Surveyor (MGS), Mars Odyssey, and Mars Express (MEx) "to investigate the characteristics of Martian impact craters in more detail than has previously been possible in an attempt to better understand the role of target volatiles and the atmosphere on the formation of these features." Analysis of data in this revised catalog provides new details on the distribution and morphologic details of 6,795 impact craters in the northern hemisphere of Mars. Some of the findings are:

- Craters displaying the SLE morphology dominate among the layered ejecta morphologies. Of 3,279 craters with ejecta morphology, 1,797 (55%) display an SLE morphology. The only regions where SLE craters do not dominate are in the areas where double-layer ejecta (DLE) morphology craters are concentrated (40° to 60°N).
- DLE craters dominate within the 40° to 60°N range and display diameters between 5 km (and smaller) and 115 km. They comprise 28% of the ejecta morphology (920 craters). Studies of the ejecta blankets suggest that the material in the outer blanket was much more fluid than that in the inner blanket at the time of impact.
- Multiple-layer craters account for 506 of the 3,279 craters with ejecta morphologies in this study (15%). They range in diameter from 5.6 to 90.7 km, but are most common at diameters greater than 10 km.

Barlow's paper discusses two points of relevance to the interpretation of volatiles as a prime factor in the morphology of rampart craters. One is the depth from which ejecta material derives and the other is the relationship of crater morphology distributions to the age of an area. There is some uncertainty as to whether the full depth of excavation contributes to the ejecta blanket or only the top portion. The low variability observed in circularity and size of crater ejecta for ancient and young sites "suggests that the ice content within the upper 3 km does not vary considerably throughout the northern plains nor has it varied considerably over the time scales recorded by these craters."

[312] "Impact craters in the northern hemisphere of Mars: Layered ejecta and central pit characteristics," N. D. Barlow, *Meteorics and Planetary Science*, Vol. 41, 1425–1436, 2006.

C.8.3   Other crater studies

Several investigators carried out localized studies of craters over small areas on Mars at higher resolution than global studies.

Reiss et al.[313] used crater size frequencies to estimate the age of two equatorial areas on Mars. In both areas the ages were typically 3.5 to 4 billion years ago (BYA). At both sites, a number of small rampart craters with an onset diameter of 1 km were found. The depth/diameter ratios were typically about 0.12 to 0.15, whereas this ratio is expected to be closer to 0.20 for fresh pristine craters.

Hirohide et al.[314] studied a 200 km × 200 km area of Mars in the region around 30°N, where all features had been resurfaced and all craters were formed after the most recent resurfacing event. Their results indicate the onset of ramparts occurs around 0.3 km diameter, and the greatest number of rampart craters occurred for diameters of 0.5 to 0.7 km. These values are considerably smaller than the results from global studies.

Boyce et al.[315] developed a map of rampart crater onset size measured in 5 × 5 degree grid cells. Their results are shown in Figure C.35 (color section). It seems evident from Figure C.35 that:

(a) There is a broad region from 30°N to 60°S where intermediate values (5–11 km) of the onset diameter tend to prevail, except for scattered localities from the equator to −45° where the onset diameters are smaller (1–4 km).
(b) Onset diameters are large (12–20 km) poleward of 30°N.

In the northern hemisphere the onset diameters of rampart craters increase continuously toward the pole. This was interpreted as implying that rampart craters are due to liquid water that occurs at increasing depth at higher latitudes. The onset diameters of undifferentiated fluidized ejecta craters show a progressive decrease in size poleward from about 30°N, consistent with the stability of ground ice. Boyce et al. suggest that "rampart ejecta craters are produced by water-rich target materials while the other major types of fluidized ejecta craters are mainly produced by ice in the target." But, in the southern polar regions these trends are not observed. It does not seem likely that liquid water occurs at shallow depths in the southern hemisphere, so the situation remains unresolved.

[313] "Small rampart craters in an equatorial region on Mars: Implications for near-surface water or ice," D. Reiss, E. Hauber, G. Michael, R. Jaumann, G. Neukum, and the HRSC Co-Investigator Team, Geophysical Research Letters, Vol. 32, L10202, doi:10.1029/2005GL022758, 2005; "Rampart Craters in Thaumasia Planum, Mars," Lunar and Planetary Science, Vol. XXXVII, Paper 1754, 2006.
[314] "A shallow volatile layer at Chryse Planitia, Mars," Hirohide Demura and Kei Kurita, Earth Planets Space, Vol. 50, 423–429, 1998.
[315] "Global Distribution of On-Set Diameters of Rampart Ejecta Craters on Mars: Their Implication to the History of Martian Water," Joseph M. Boyce, David J. Roddy, Lawrence A. Soderblom, and Trent Hare, Lunar and Planetary Science, Vol. XXXI, Paper 1167, 2000.

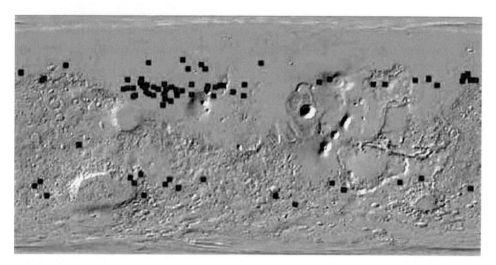

Figure C.36. Locations of DLE craters. (Presumably, the vertical scale is latitude from 90°N to 90°S and the horizontal axis is longitude). [Modified from "The Unique Attributes of Martian Double Layered Ejecta Craters," Peter J. Mouginis-Mark and Joseph M. Boyce, *Lunar and Planetary Science*, Vol. XXXVI, Paper 1111, 2005.]

Mouginis-Mark and Boyce[316] used THEMIS data to improve upon earlier Viking-based observations of DLE craters. Their map of DLE craters is shown in Figure C.36.

C.8.4   Commentary

The various papers by Barlow and co-workers generally take a *phenomenological point of view* in analysis. By this is meant that craters are characterized and categorized by morphology, latitude and longitude dependence, size, and in some cases, terrain characteristics. From this, patterns emerge. Based on these patterns, interpretations are made regarding the nature of the terrain impacted and the distribution of subsurface H_2O. For example, SLE craters occur widely across Mars, but tend to occur in smaller sizes in equatorial areas, but they can occur in larger diameters at higher latitudes. MLE craters tend to occur in larger sizes. This would be consistent with the notions that (1) SLE craters are associated with subsurface ice with the subsurface ice reaching down to greater depths at higher latitudes, and (2) MLE rampart craters are due to liquid water because they are only observed at equatorial latitudes for very large craters that penetrate deeply (3–6 km depth) to where (presumably) a liquid layer may exist, and they are not observed at higher latitudes where the liquid layer would be much deeper. It is also consistent with the notion that SLE rampart craters are due to ground ice because (a) they account for

[316] "The Unique Attributes of Martian Double Layered Ejecta Craters," Peter J. Mouginis-Mark and Joseph M. Boyce, *Lunar and Planetary Science*, Vol. XXXVI, Paper 1111, 2005.

essentially all rampart craters at high latitudes, and (b) they account for smaller rampart craters at equatorial latitudes where the depth of the presumed liquid layer has not been reached. However, Barlow[317] notes that with higher resolution THEMIS and MOC data, they have recently been finding MLE craters down to ~10 km in diameter. Barlow indicates that the multiple layer (and probably also the outer layer of the double-layer morphology) may "indicate some interaction of the volatile-rich ejecta plume with the Martian atmosphere."

Since all the data indicate that some minimum onset diameter for rampart craters occurs in any locality, the implication is strong that there is a comparatively desiccated upper layer that must be penetrated to greater depths before reaching an H_2O-rich layer that can generate a rampart pattern.

As mentioned previously, a somewhat different interpretation of phenomenological data was given by Boyce et al.[318] They suggest that "rampart ejecta craters are produced by water-rich target materials while the other major types of fluidized ejecta craters are mainly produced by ice in the target."

It would be desirable to supplement the phenomenological interpretations with detailed models of the entire interaction of an impactor with the surface and sub-surface of Mars showing temperature distributions, ejecta patterns, and formation of the resultant crater. This should include interactions of the ejecta with the Mars atmosphere. This is a complex topic and no attempt is made here to review the various publications that exist in the field. Only a few very brief comments will be made.

O'Keefe et al.[319] show that very high temperatures are produced in the impacts of extraterrestrial bodies with planetary surfaces and ground ice will be vaporized. Analyses were made of impacts on rock, ice, and rock–ice mixtures. For rock–ice mixtures, energy is preferentially deposited into the more compressible volatile component and hydrothermal fracturing takes place. The results seem to indicate that the composition of the target (ice content) and physical state play an important role in shaping the ejecta pattern.

Pierazzo et al.[320] present "preliminary results of three-dimensional simulations of impacts on Mars aimed at constraining the initial conditions for modeling the onset and evolution of a hydrothermal system on the red planet. The simulations of

[317] "Impact Craters in the Northern Hemisphere of Mars: Layered Ejecta and Central Pit Characteristics," N. G. Barlow, *Meteoritics and Planetary Science*, Special Issue on the Role of Volatiles and Atmospheres on Martian Impact Craters, Vol. 41, 1425–1436, 2006.

[318] "Global Distribution of On-Set Diameters of Rampart Ejecta Craters on Mars: Their Implication to the History of Martian Water," Joseph M. Boyce, David J. Roddy, Lawrence A. Soderblom, and Trent Hare, *Lunar and Planetary Science*, Vol. XXXI, paper 1167, 2000.

[319] "Damage and RockVolatile Mixture Effects on Impact Crater Formation," John D. O'Keefe, Sarah T. Stewart, Michael E. Lainhart, and Thomas J. Ahrens, *Int J. Impact Engng.*, Vol. 26, 543–553, 2001.

[320] "Starting Conditions for Hydrothermal Systems underneath Martian Craters: 3D Hydrocode Modeling," E. Pierazzo, N. A. Artemieva, and B. A. Ivanov, in T. Kenkmann, F. Horz, and A. Deutsch (eds.), *Large Meteorite Impacts III*, Geological Society of America, Special Paper 384, pp. 443–457, 2005.

the early stages of impact cratering allow us to determine the amount of shock melting and the pressure–temperature distribution in the target caused by various impacts on the Martian surface. The late stages of crater collapse are then necessary to determine the final thermal state of the target, including crater uplift, and the final distribution of the melt pool, heated target material and hot ejecta around the crater." They indicate that it is "necessary to follow the entire crater-forming event, from impact to the final crater" to generate "a complete picture of the thermal field underneath an impact crater." They "found that for the smallest craters the 373 K isotherm, corresponding to the boiling point of water, is buried only $\sim$1 km below the crater floor. For craters in the 20–40 km range, the 373 K isotherm reaches a depth of about 5 km. For even larger craters the extent of a post-impact hydrothermal system seems to be controlled by the permeability of the rocks under lithostatic pressure."

Schultz[321] provides a recent summary of analysis that suggests that soil–atmosphere interactions play a significant role in shaping Mars' crater ejecta, and may even account for most if not all of the morphology observed, without necessarily invoking subsurface volatiles. This topic is difficult to understand.

Finally, the relationship between rampart crater onset data, neutron spectrometer data, and theoretical models of ground-ice stability seems to be ephemeral. The global crater data indicate minimum onset diameters of several km, while studies of specific areas at higher resolution indicate some onset diameters as low as 300 m. These correspond to depths of many hundreds of meters in the global case, on down to perhaps 100 meters for a few specific localities. But, the neutron spectrometer data indicate the existence of some H_2O in the top $\sim$1 m of subsurface, and theoretical models indicate that between about $-50°$ and $+50°$ latitude, ground ice is generally not stable. The H_2O found from rampart crater onsets is quite a bit deeper than that observed by the neutron spectrometer. Furthermore, theoretical models suggest that ground ice is not presently stable at any depth in equatorial areas, although that might have been quite different in the past. There does not seem to be any direct connection between crater data and current neutron spectrometer observations.

The theoretical models for subsurface ice formation mainly pertain to the interaction of cold regolith with an atmosphere containing water vapor, although a few attempts were made to deal with the long-term effects of large subsurface water resources (see Section C.3.3). In most models, the atmosphere is the source of H_2O and the regolith is the sink. The various models estimate whether ground ice is stable in any locality by modeling the thermal and diffusive interaction of ground ice with the local atmosphere containing water vapor. At higher latitudes, near-surface (top few meters) ground ice is stable and will form spontaneously when exposed to the local atmosphere. At equatorial latitudes, the vapor pressure of the ground ice is too high and the ground ice will gradually sublime, adding water vapor to the atmosphere that eventually makes its way to the poles. Over tens to hundreds of thousands of years, as Mars' orbit changes, the balance between solar heat input to higher and

---

[321] "Assessing Lithology from Ejecta Emplacement Styles on Mars: The Role of Atmospheric Interactions," P. H. Schultz, *Workshop on the Role of Volatiles and Atmospheres on Martian Impact Craters, July 11–14, 2005*, LPI Contribution 1273.

lower latitudes changes, as does the roles of the south and north poles relative to the elliptical orbit of Mars. This causes significant variations in the stability of ground ice at different locations over such time spans (see Section C.3.2).

But, there is a problem with all of these models. The models deal with the local interaction of ground ice with the atmosphere over a few meters. However, the rampart crater record suggests that there is a global, thick, heavily loaded, ice-filled regolith layer typically at a depth of several hundred meters, but which might in some localities reach to within less than 100 m of the surface even in temperate zones. More speculatively, but still implicit in the crater record, the implication is that liquid-water reservoirs might occur at greater depths (1.5 km to 5 km), particularly in temperate zones. The problem with all of these models is that they only deal with the regolith–atmosphere connection, but they do not adequately connect to the boundary conditions at depth. Suppose there is a liquid-water reservoir at depths of several km, presumably bottomed out by an impervious layer of rock. If the regolith above this liquid water is extensively fractured (as is widely believed), then water vapor will diffuse upward toward the surface. Water vapor will freeze in the pores and interstices of the regolith, but this ice will continue to exert a vapor pressure. Therefore, water vapor will continue to diffuse further upward, although the temperature decreases as the depth is reduced, resulting in exponential decreases in vapor pressure as one moves upward toward the surface. In this picture, given enough time, the regolith will be filled with ground ice in the deep subsurface at all latitudes. A hypothetical attempt to join the crater record to near-surface models leads to the diagram shown in Figure C.37 (color section).

At sufficiently high latitudes, where the subsurface temperature (at depths greater than the "skin depth" over which diurnal temperatures control the subsurface temperature) always remains below the atmospheric frost point, the regolith is filled with ground ice right up to this shallow depth. By contrast, at equatorial latitudes, the subsurface temperature is always above the atmospheric frost point, and therefore there will be a net outflow of water vapor from the subsurface to the atmosphere. Thus, there will be a gradient in H_2O content in the subsurface that is essentially zero below the small depth over which diurnal temperatures control the subsurface temperature, and slowly rising with depth until at some depth the pores and interstices are probably filled with ice. This is likely to be several hundred meters, but it might be as shallow as tens of meters in a few locations. This description is purely hypothetical, but it does correlate with the observed occurrences of SLE and MLE rampart craters. A few studies attempted to deal with this;[322] however, the results are not entirely satisfactory.

---

[322] "The stability of ground ice in the equatorial region of Mars," S. M. Clifford and D. Hillel, *J. Geophys. Res.*, Vol. 88, 2456–2474, 1983; "Global distribution and migration of subsurface ice on Mars," F. P. Fanale, J. R. Salvail, A. P. Zent, and S. E. Postawko, *Icarus*, Vol. 67, 1–18, 1986; "The persistence of equatorial ground ice on Mars," M. T. Mellon, B. M. Jakosky, and S. E. Postawko, *J. Geophys. Res.*, Vol. 102, 19357–19369, 1997.

C.9   SUMMARY

Based on the material reviewed in previous sections, the following conclusions are drawn:

1   The conditions under which near-surface subsurface ice may exist in equilibrium with the atmosphere on Mars have been modeled by a number of prominent Mars scientists for 40 years, and similar results have been obtained by all. The prediction is that subsurface ice is stable in the pores and interstices of Martian regolith at sufficiently high latitudes. Obviously, subsurface ice is stable at, and just below, the surface in polar regions. At lower latitudes, an "ice table" forms in which a desiccated regolith covers an ice-filled layer, with the depth of the ice table increasing with decreasing latitude. At some latitude near 55–60° (or perhaps as low as 45° depending on soil properties and slope), the ice table may be 1–3 meters down. At lower latitudes the depth of the ice table increases sharply, and at latitudes less than typically ∼55° subsurface ice is not thermodynamically stable relative to sublimation to the atmosphere. These are equilibrium models and they do not preclude the possibility of non-equilibrium ice from previous epochs that is very slowly disappearing in regions where ice is not thermodynamically stable.

2   The Mars Odyssey neutron spectrometer has been used to scan the upper ∼1 m layer of the Mars surface in elements 5° × 5° latitude × longitude. These data support the predictions of models for latitudes ≳ 55°. High water concentrations are detected with apparent shallow ice tables approaching the surface toward the poles.

3   In the region of latitude from −45° to +45°, it is found that there is a residual water content that never drops below ∼1–2%, probably representing chemically bound water in the minerals of the soil. In various localized areas within this region, the measured water content in the top 1 m can be as high as 8 to 10%. Comparison of fast neutron data with epithermal neutron data suggests that there is an upper layer that is desiccated, with a higher water content layer below it. The thickness of the desiccated layer is suggested to be >20–30 cm.

4   The localized equatorial regions with relatively high water content (8–10%) present an enigma. On the one hand, thermodynamic models predict that subsurface ice is not stable near the surface in the broad equatorial region. On the other hand, some aspects of the Odyssey data are suggestive of subsurface ice. It is possible that this is metastable subsurface ice left over from a previous epoch with higher obliquity. Alternatively, it could be soil heavily endowed with salts containing water of crystallization. The fact that these areas overlap somewhat with regions of high-albedo and low-thermal inertia suggest that it is indeed subsurface ice. Furthermore, the pixel size of Odyssey data is large, and the 8–10% water figure might represent small local pockets of higher water concentration (where surface properties and slopes are supportive) scattered within an arid background. Over the past million years, the obliquity, eccentricity, and precession of the equinoxes of Mars has caused a variable solar input to the planet in which the relative solar input to high and low latitudes has varied considerably. Certainly, ground ice was transferred from polar areas to temperate areas during some of these epochs. It is possible that some of this ground ice remains today even though it is thermodynamically unstable in temperate areas. In order for remnant subsurface ice from past epochs to be a proper explanation, the process of ice deposition must be faster than the process of ice sublimation in the temperate areas over time periods of tens or hundreds of thousands of years.

5 We have one data point at the poles. Phoenix should provide another data point at a high latitude (65–75°N). We need ground measurements of ice content down to a few meters at latitudes in the 45–65° range to confirm the predicted ice table.

6 We need exploration in the 8–10% water equatorial regions to determine the state of the water in these areas. The first step could be improved spatial resolution of orbital observations. Eventually, a landed mission is needed for ground truth.

7 Deep within the interior, the temperature will rise to the point where liquid water could exist. Presently, there is no convincing evidence that it does. The MARSIS and SHARAD instruments may provide some data in this regard. If liquid water exists deep within Mars (down several km), then the water vapor rising from this liquid water will pass through porous regolith at sub-freezing temperatures. Hence, you cannot have liquid water at depth unless there is a huge thick layer of ice-filled regolith above it.

8 The crater record suggests that the interior of Mars down to several km is mainly filled with H_2O. The connection of this reservoir to near-surface H_2O has not yet been adequately investigated.

Index

(figures are **bold**, tables are *italic*, color figures are **extra-bold**)

Printing: Mercedes-Druck, Berlin
Binding: Stein+Lehmann, Berlin